Advancements in Cancer Theranostics

Rajendra Prasad • Murali Mohan Yallapu
Editors

Advancements in Cancer Theranostics

Integrating Imaging and Intervention

Springer

Editors
Rajendra Prasad
School of Biochemical Engineering
Indian Institute of Technology (BHU)
Varanasi
Varanasi, Uttar Pradesh, India

Murali Mohan Yallapu
School of Medicine
University of Texas
Rio Grande Valley, TX, USA

ISBN 978-981-95-0937-9 ISBN 978-981-95-0938-6 (eBook)
https://doi.org/10.1007/978-981-95-0938-6

This Springer imprint is published by the registered company Springer Nature Singapore Pte Ltd.
The registered company address is: 152 Beach Road, #21-01/04 Gateway East, Singapore 189721, Singapore

We dedicate this book to our parents and mentors, especially Sanjiv Sam Gambhir who was a great leader in molecular imaging and theranostics research.

Foreword

Nanoparticles-based medicines have gained tremendous attention for disease diagnosis and therapeutics either individually or in combination, the combined forms are named as theranostics. Among several thousand nanoparticles, about 200 have been considered as safe for pre-clinical and clinical applications, especially in the case of diseases like cancer. Moreover, about 20 have been evaluated by FDA and EMA, and considered further for clinical studies and clearance. However, there are several major issues that are being questioned about nanoparticles-based systems during bioimaging and therapeutics applications. For example, poor tumor accumulation, low retention time, nonspecific biodistribution, rapid clearance, etc., are known hurdles that are being explored by Warren Chan from University of Toronto and Rajendra Prasad from Indian Institute of Technology (BHU) Varanasi. However, several other researchers are also trying to validate these issues prior to translate nanoparticles-based medicine. On the other hand, understanding imaging and therapeutics response from disease site at the molecular level is also a major task. So far, stimuli-responsive therapeutics such as light-mediated photothermal therapy, photodynamic therapy, photo-immunotherapy, combined photothermal-photodynamic therapy, etc., have been applied to ablate cancer site completely. However, the clinical success is quite slow and the clear mechanism is not yet well explored. To address the blood circulation and specific distribution, various surface conjugation methodologies and targeting agents have been applied for nanoparticle's surface engineering. But it should be noted that scalability, reproducibility, stability, toxicity, etc., are major critical issues observed while engineering nanoparticle systems for biomedical imaging and therapeutics applications. This book focuses on various functional nanohybrid particles for imaging and therapeutic or theranostics applications along with different conventional and modern approaches. The chapters describe the in-depth knowledge of nanoparticles design and related medical applications. Various national and international experts in the field have described the fundamental and applied concepts of nanomaterials with the translational approaches in theranostics "bioimaging and therapeutics by a single system". This book could be considered as the first comprehensive book on various aspects of nanoparticles with the advancements in cancer theranostics. Therefore, this book will benefit biomedical researchers, students, and scientists of all fields in both academia and biotechnology industries.

Jonathan Lovell
Professor of Biomedical Engineering
University at Buffalo
NY, USA

Preface

Nanoparticles face major issues in engineering them and testing them for medical applications like imaging, therapy, or theranostics, though they play an important role in the multimodality of imaging and therapeutics of diseases like cancer. A few nanoparticles have been considered as safe by the regulatory bodies, may be due to their lack of physicochemical characteristics and reliability. Moreover, there are concerns of batch-to-batch variability, scalability, reproducibility, etc. It should be noted that nanoparticles-based imaging or therapeutics or theranostics systems, all show insignificant tumor accumulation (less than 1%) and low tumor retention. Remarkably, a majority of nanoparticles-based systems failed in clinical trials and show massive side effects. However, there is no clear understanding yet on low tumor uptake and side effects though there are thousands of papers published so far. It should also be noted that the dynamic nature of nanoparticles-based imaging and therapeutics depends on the surrounding medium. To the best of the authors' knowledge, the surface chemistry, charge, and overall morphology alter the colloidal stability when nanoparticle medicines are in biological liquid medium. It is essential to mention that the theranostics concept could reduce the dose burden and involved overall cost of imaging and therapeutics. In today's time, imaging-integrated therapeutic agents known as theranostics have been proposed as smart platforms for localized imaging and therapeutics with better reliability and safety. This could be the first book covering a wide range of nanoparticle-based imaging and therapeutics platforms with theranostics advancement. We have gathered various national and international experts in the field to provide their in-depth understanding on nanoparticles-based theranostics systems for cancer diagnosis and therapy in combination, which could reduce the overall cost of planning during imaging and therapy. Moreover, various major issues related to nanotheranostics design and their translational advantages have also been covered in this book. The basic components required to design theranostics systems at nanoscale have been discussed in Chaps. 1–3. Chapter 4 explains the next generation of multimode imaging and synergistic therapeutics for cancer. The importance of molecular imaging and therapeutic approaches along with the choice of nanoparticle-based theranostics systems have been discussed in Chaps. 5 and 6. One of the clinical experts has described about the clinical validations of localized cancer nanomedicine in Chap. 7. Among various therapeutics, NIR light-mediated photothermal therapy has shown promising outcomes for tumor ablation, which has been discussed in detail in the Chap. 8. Various leading scientists have explained nanoparticles-based bioimaging and therapeutics, and theranostics approaches for cancer with nanoparticle's multifunctionality in Chaps. 9–13. Associated biosafety and reliability of nanoparticles-based theranostics systems have been discussed in the Chap. 14. Clinicians have described cancer imaging and theranostics intervention with current status and future perspective in the Chap. 15. This book would surely help various researchers and investigators in the area of translational research on nanoparticles-based imaging, therapy, and theranostics who are working in the area of early-stage diagnosis and treatment. We also believe that this knowledge will surely guide in the designing of safe cancer nanomedicines.

Varanasi, Uttar Pradesh, India Rajendra Prasad
Rio Grande Valley, TX, USA Murali Mohan Yallapu

Acknowledgment

We thank our reviewers for reviewing chapters and contributing authors from national and international top tier institutes. We also acknowledge the support of the School of Biochemical Engineering, Indian Institute of Technology (BHU), Varanasi, India.

Contents

Editors and Contributors

About the Editors

Rajendra Prasad, Ph.D., is an Assistant Professor at the School of Biochemical Engineering, Indian Institute of Technology (BHU) Varanasi, India. Rajendra's main focus is to diagnose cancer at an early stage utilizing ultrabright nanoimaging agents. His lab has been working majorly in the area of molecular imaging and theranostics. Biomimetic theranostics is a recent finding from his Bright NanoGhost group. Rajendra is more toward understanding the site-selective targeting of solid tumors with entry-exit mechanism of administrated nanosized medicines/or theranostics without affecting healthy tissues. Rajendra has gained postdoctoral trainings from Tufts University, Medford, Boston, MA, USA; NOVA Medical School, Faculdade de Ciências Médicas Campo Mártires da Pátria, Lisboa, Portugal; Technion-Israel Institute of Technology, Haifa, Israel; and Indian Institute of Technology, Bombay, India, in the area of cancer nanomedicine and targetable cancer theranostics. Rajendra has contributed for more than (a) 40 research articles published in top-tier journals such as *ACS Nano*, *Advanced Functional Materials*, *Small*, *Nano Letters*, *Cell Reports Physical Science*, *Nanoscale*, *Bioconjugate Chemistry*, *Communications Biology*, *Journal of Controlled Release*, etc.; (b) 17 granted/or published patents nationally and internationally; (c) 7 trademarks and 8 journal cover art images; and (d) 4 book chapters and 2 edited books. Currently, he is also a part of the editorial board at *Nano Letters*, *npj Imaging*, *Chemical & Biomedical Imaging*, and *Nanotheranostics*. Rajendra holds memberships for World Molecular Imaging Society and Royal Society of Chemistry.

Murali Mohan Yallapu, Ph.D., is a tenured Associate Professor in the Department of Immunology and Microbiology and Medical Oncology-ISU, and a core member of the South Texas Center of Excellence in Cancer Research (ST-CECR) at the School of Medicine, University of Texas Rio Grande Valley (UTRGV), McAllen, Texas, USA. Dr. Yallapu is a Director of NIH-funded U54-Research Core Capacity and Director of CPRIT-funded Drug Delivery and Nanomedicine Core. Dr. Yallapu's academic journey is distinguished by his receipt of the *Prof. A. Kameswara Rao Gold Medal (1999)*, awarded for academic excellence. Dr. Yallapu has completed extensive postdoctoral training at internationally recognized research institutions, including the Cleveland Clinic, University of Nebraska Medical Center, Sanford Research, and the Gwangju Institute of Science and Technology, focusing on materials science, drug delivery systems, nanomedicine, and cancer biology. Before his appointment at UTRGV, he served as an Assistant Professor in the Department of Pharmaceutical Sciences at The University of Tennessee Health Science Center. A recognized authority in nanotechnology-driven therapeutics, Dr. Yallapu serves on the editorial boards of numerous high-impact journals in the fields of nanomedicine, drug delivery, and oncology. He also actively contributes to the scientific community as an ad hoc reviewer for the National Institutes of Health (NIH) and various international funding agencies. His scholarly contributions include over 205 peer-reviewed publications and book chapters, along with more than 165 abstracts presented at national and international scientific meetings. His work has garnered significant recognition, with over 18,536 citations, an H-index of 69, and an i10-index of 169. Dr. Yallapu's current research is centered on the design and development of multifunctional nanomaterials to enhance the therapeutic index of clinical drugs, as well as on engineering novel nanoconstructs for diverse biomedical applications. In collaboration with the ST-CECR at UTRGV, his team is also advancing innovative nanoformulations tailored for cancer immunotherapy, with an emphasis on translational safety and efficacy.

Contributors

Pamela Vanessa Meixueiro Aguilar Centro de Nanociencias y Nanotecnología, Universidad Nacional Autónoma de México, Tijuana-Ensenada, Baja California, Mexico

Ankush UNF Sustainable Solutions Laboratory, University of North Florida, Jacksonville, FL, USA

Anushka Balakrishnan School of Chemistry, Indian Institute of Science Education and Research Thiruvananthapuram, Thiruvananthapuram, India

Anjali Bedse Department of Pharmaceutics, K. K. Wagh College of Pharmacy, Nashik, Maharashtra, India

Berivan Cecen Department of Biomedical Engineering, Newark College of Engineering, New Jersey Institute of Technology, Newark, NJ, USA

Stem Cell and Tissue Engineering Application and Research Center (ISUKOK), Istinye University, Istanbul, Turkey

Hitesh Harsukhbhai Chandpa Immuno Engineering and Therapeutics Laboratory, Department of Pharmaceutical Engineering and Technology, Indian Institute of Technology (Banaras Hindu University) Varanasi, Varanasi, India

Kanchan Chauhan Centro de Nanociencias y Nanotecnología, Universidad Nacional Autónoma de México, Tijuana-Ensenada, Baja California, Mexico

Vivek P. Chavda Department of Pharmaceutics and Pharmaceutical Technology, L.M. College of Pharmacy, Ahmedabad, Gujarat, India

Alev K. Erdi Department of Biomedical Engineering, Newark College of Engineering, New Jersey Institute of Technology, Newark, NJ, USA

Amol Gholap Department of Pharmaceutics, St. John Institute of Pharmacy and Research, Palghar, Maharashtra, India

Prabhanjan Giram Department of Pharmaceutical Sciences, University at Buffalo, The State University of New York, Buffalo, NY, USA

Deeksha Hattila Intestinal Biology and Liver Disease Lab (IBLD), Department of Biotechnology, All India Institute of Medical Sciences (AIIMS), New Delhi, India

Hemant Institute of Nano Science and Technology, Mohali, Punjab, India

Academy of Scientific and Innovative Research (AcSIR), Ghaziabad, India

Gustavo A. Hirata Centro de Nanociencias y Nanotecnología, Universidad Nacional Autónoma de México, Tijuana-Ensenada, Baja California, Mexico

Krishna Jadhav Pharmaceutical Nanotechnology Lab, Institute of Nano Science and Technology, Mohali, Punjab, India

Bela Jain Department of Nuclear Medicine, All India Institute of Medical Sciences, New Delhi, India

S. Karthik School of Chemistry, Indian Institute of Science Education and Research Thiruvananthapuram, Thiruvananthapuram, India

Navneet Kaur Department of Chemical and Biomedical Engineering, FAMU-FSU College of Engineering, Tallahassee, FL, USA

National High Magnetic Field Laboratory, FSU, Tallahassee, FL, USA

Eknath Kole Department of Pharmaceutical Technology, University Institute of Chemical Technology, KBC North Maharashtra University, Jalgaon, Maharashtra, India

Nagavendra Kommineni Thermo Fisher Scientific Inc., Cincinnati, OH, USA

Aleena Sabu Koshi School of Chemistry, Indian Institute of Science Education and Research Thiruvananthapuram, Thiruvananthapuram, India

Surbhi Koul Department of Pharmaceutics and Pharmaceutical Technology, L.M. College of Pharmacy, Ahmedabad, Gujarat, India

Siddharth Krishnan Department of Bioengineering, University of Pittsburgh, Pittsburgh, PA, USA

Deepak Kulkarni Department of Pharmaceutics, Srinath College of Pharmacy, Bajajnagar, Chhatrapati Sambhajinagar, India

Vikas Kumar Department of Pharmaceutical Engineering and Technology, Indian Institute of Technology (IIT) BHU, Varanasi, Uttar Pradesh, India

Sathyaprakash Kumarasamy Department of Biosciences and Bioengineering, Indian Institute of Technology Jammu, Jagti, Jammu, India

Prakash Ashok Kumbhar Immuno Engineering and Therapeutics Laboratory, Department of Pharmaceutical Engineering and Technology, Indian Institute of Technology (Banaras Hindu University) Varanasi, Varanasi, India

Rajendra Kurapati School of Chemistry, Indian Institute of Science Education and Research Thiruvananthapuram, Thiruvananthapuram, India

Pavithra Kurungottu School of Chemistry, Indian Institute of Science Education and Research Thiruvananthapuram, Thiruvananthapuram, India

A. S. Lakshmi School of Chemistry, Indian Institute of Science Education and Research Thiruvananthapuram, Thiruvananthapuram, India

Shaheen Mahira Department of Obstetrics, Gynecology and Reproductive Sciences, University of Pittsburgh, Pittsburgh, PA, USA

Ankit Kumar Malik Department of Pharmaceutical Engineering and Technology, Indian Institute of Technology (IIT) BHU, Varanasi, Uttar Pradesh, India

Avtar Singh Meena Intestinal Biology and Liver Disease Lab (IBLD), Department of Biotechnology, All India Institute of Medical Sciences (AIIMS), New Delhi, India

Jairam Meena Immuno Engineering and Therapeutics Laboratory, Department of Pharmaceutical Engineering and Technology, Indian Institute of Technology (Banaras Hindu University)Varanasi, Varanasi, India

Shweta Meena Department of Electronics and Communication Engineering, National Institute of Technology Kurukshetra, Kurukshetra, Haryana, India

Gayatri More Department of Pharmaceutics, Srinath College of Pharmacy, Bajajnagar, Chhatrapati Sambhajinagar, India

Kashish Mulani Department of Pharmaceutics, Srinath College of Pharmacy, Bajajnagar, Chhatrapati Sambhajinagar, India

Madaswamy S. Muthu Department of Pharmaceutical Engineering and Technology, Indian Institute of Technology (IIT) BHU, Varanasi, Uttar Pradesh, India

Jitendra Naik Department of Pharmaceutical Technology, University Institute of Chemical Technology, KBC North Maharashtra University, Jalgaon, Maharashtra, India

Dibyani Nanda Intestinal Biology and Liver Disease Lab (IBLD), Department of Biotechnology, All India Institute of Medical Sciences (AIIMS), New Delhi, India

Shovan Naskar Immuno Engineering and Therapeutics Laboratory, Department of Pharmaceutical Engineering and Technology, Indian Institute of Technology (Banaras Hindu University) Varanasi, Varanasi, India

Prabhakar Panzade Department of Pharmaceutics, Srinath College of Pharmacy, Bajajnagar, Chhatrapati Sambhajinagar, India

Rohankumar Patel Analytical Development, Amneal Pharmaceuticals, Bridgewater, NJ, USA

Paloma Patra Department of Biomedical Engineering, Indian Institute of Technology, Hyderabad, India

Rajendra Prasad School of Biochemical Engineering, Indian Institute of Technology (BHU) Varanasi, Varanasi, Uttar Pradesh, India

Sasvat Sayee Ram Ramesh Department of Biomedical Engineering, Indian Institute of Technology, Hyderabad, India

Nandini Vinodrao Randhave Department of Pharmaceutical Engineering and Technology, Indian Institute of Technology (IIT) BHU, Varanasi, Uttar Prasdesh, India

Komal Rani Department of Pharmaceutical Engineering and Technology, Indian Institute of Technology (IIT) BHU, Varanasi, Uttar Pradesh, India

Satyawati Rawat Department of Botany, Kurukshetra University, Kurukshetra, Haryana, India

Aravind Kumar Rengan Department of Biomedical Engineering, Indian Institute of Technology, Hyderabad, India

Satish Rojekar Department of Pharmacological Sciences, Icahn School of Medicine at Mount Sinai, New York, NY, USA

Agnes Sebastian School of Chemistry, Indian Institute of Science Education and Research Thiruvananthapuram, Thiruvananthapuram, India

Prakhar Sengar Centro de Nanociencias y Nanotecnología, Universidad Nacional Autónoma de México, Tijuana-Ensenada, Baja California, Mexico

Aseem Setia Department of Pharmaceutical Engineering and Technology, Indian Institute of Technology (IIT) BHU, Varanasi, Uttar Pradesh, India

Rahaman Shaik Department of Pharmacology, Jamia Hamdard, New Delhi, India

Ashwin T. Shaji School of Chemistry, Indian Institute of Science Education and Research Thiruvananthapuram, Thiruvananthapuram, India

Lipy Sharma Department of Zoology, Central University of Punjab, Bathinda, Punjab, India

Nancy Sharma Institut Parisien de Chimie Moléculaire (IPCM), CNRS, Sorbonne Université, Paris, France

Vinay Sharma Department of Biosciences and Bioengineering, Indian Institute of Technology Jammu, Jagti, Jammu, India

Santosh Shelke Department of Pharmaceutics, Srinath College of Pharmacy, Bajajnagar, Chhatrapati Sambhajinagar, India

Zain Siddiqui Department of Biomedical Engineering, Newark College of Engineering, New Jersey Institute of Technology, Newark, NJ, USA

Abhijeet Singh Immuno Engineering and Therapeutics Laboratory, Department of Pharmaceutical Engineering and Technology, Indian Institute of Technology (Banaras Hindu University) Varanasi, Varanasi, India

Ankita Singh Department of Pathology, Sanjay Gandhi Post Graduate Institute of Medical Sciences, Lucknow, India

Dixa A. Vaghela Department of Pharmaceutics and Pharmaceutical Technology, L.M. College of Pharmacy, Ahmedabad, Gujarat, India

Vaishali Department of Pharmaceutical Engineering and Technology, Indian Institute of Technology (IIT) BHU, Varanasi, Uttar Pradesh, India

Nidhi Verma Department of Pharmaceutical Engineering and Technology, Indian Institute of Technology (IIT) BHU, Varanasi, Uttar Pradesh, India

Rahul Kumar Verma Pharmaceutical Nanotechnology Lab, Institute of Nano Science and Technology, Mohali, Punjab, India

Khushboo Warjurkar Department of Biosciences and Bioengineering, Indian Institute of Technology Jammu, Jagti, Jammu, India

Conventional and Modern Theranostics Approaches

1

Dibyani Nanda, Deeksha Hattila, and Avtar Singh Meena

Abstract

Theranostics is revolutionizing medical science by seamlessly combining diagnostics with targeted therapies to advance precision medicine. This integrated approach uses sophisticated tools like biological pathways, nanotechnology, and molecular engineering to accurately diagnose diseases and deliver personalized treatments. Traditional methods such as radiopharmaceuticals, biomarker analysis, and transarterial chemoembolization have provided the groundwork for aligning therapy with diagnostic insights. However, recent innovations, including nanotheranostics, artificial intelligence, and genomic technologies, have taken personalized medicine to new heights. These breakthroughs utilize nanoscale carriers, advanced imaging techniques, and cutting-edge therapies like CAR-T cells and liquid biopsies to improve effectiveness while reducing side effects. Additionally, artificial intelligence and next-generation sequencing enable clinicians to analyze complex data and forecast treatment responses, paving the way for a more precise and patient-focused approach to healthcare. This chapter delves into the journey of theranostics, examining the interplay between traditional practices and modern advancements and their combined role in addressing challenging diseases such as cancer.

Keywords

Chemotherapy · Cancer · Liquid biopsy · Theranostics · Nanomedicine

Dibyani Nanda and Deeksha Hattila contributed equally with all other contributors.

D. Nanda · D. Hattila · A. S. Meena (✉)
Intestinal Biology and Liver Disease Lab (IBLD), Department of Biotechnology, All India Institute of Medical Sciences (AIIMS), New Delhi, India
e-mail: avtarjeph@aiims.edu

R. Prasad, M. Mohan Yallapu (eds.), *Advancements in Cancer Theranostics*,
https://doi.org/10.1007/978-981-95-0938-6_1

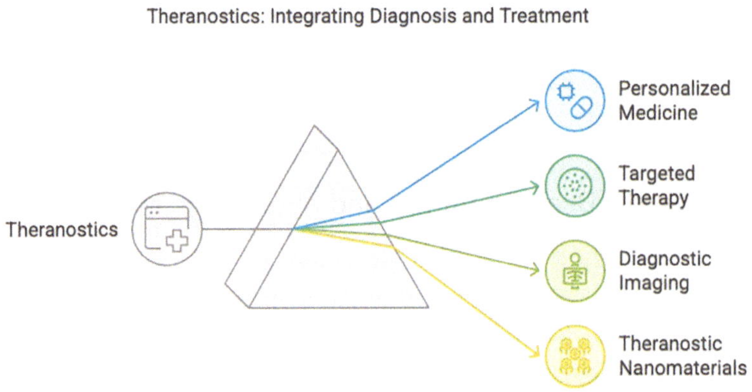

Fig. 1.1 Illustration of theranostics importance for human health

1.1 Introduction

Theranostics is a new area in which diagnostics and targeted therapy are integrated to provide a patient-specific therapeutic strategy. Theranostics is transitioning from trial and error to predictive, preventative, and personalized medicine, resulting in higher quality pharmaceutical treatment. The technique utilizes biological pathways in the body to get diagnostic pictures and provide therapeutic radiation to patients. A diagnostic test identifies a single molecular target on a tumor, allowing therapeutic agents to focus on that receptor rather than the overall illness and location. Furthermore, this advanced technique provides a broad vision to treat complicated and challenging diseases such as cancer, where a specific biological pathway could be targeted. Therapy and diagnostics are the core ingredients of theranostics. It utilizes a single entity to treat and diagnose the disease (or tumor), while monitoring treatment response. The approach of targeting the specific site increases the likelihood that the therapeutic radiation dose will effectively reach the disease site while minimizing damage to surrounding healthy tissues. The theranostic nanomaterials are none other than macromolecular materials or polymers (10–1000 nm) that can adsorb, conjugate, trap, or encapsulate diagnostic and therapeutic agents, enabling simultaneous diagnosis and treatment at the cellular and molecular levels (Fig. 1.1).

1.2 Conventional Theranostics Approaches

These approaches often rely on well-established diagnostic and treatment modalities, such as traditional imaging techniques (e.g., X-ray, CT, and MRI), biomarker detection methods, and standard therapeutic options like chemotherapy, radiation therapy, or surgery (Fig. 1.2). In many cases, these diagnostic and therapeutic methods are used in tandem to enhance the effectiveness of treatment, monitor disease progression, and guide clinical decision-making. For example, diagnostic

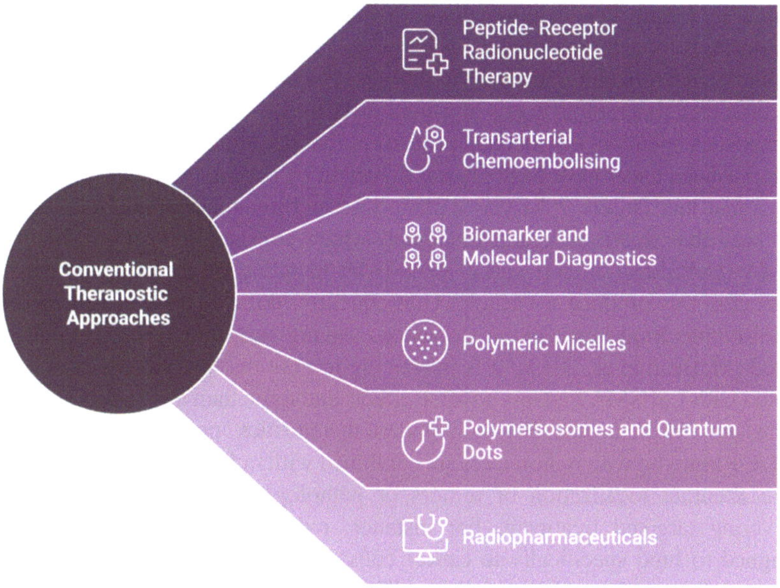

Fig. 1.2 Techniques utilized as conventional theranostics for the treatment and diagnosis of disease

imaging may be utilized to pinpoint the location of a tumor, which is then targeted using radiation therapy or surgically removed. By combining these complementary modalities, conventional theranostic approaches aim to improve patient outcomes, though they may lack the precision and personalization seen in modern advancements.

1.2.1 Radiopharmaceuticals

Radiopharmaceuticals are a cornerstone of theranostic medicine, utilizing radioactive compounds for both diagnostic imaging and therapeutic applications. These compounds, typically radioisotopes, emit radiation that can be detected by various ways of imaging such as positron emission tomography (PET) or single-photon emission computed tomography (SPECT), allowing clinicians to visualize and assess the location and activity of disease within the body. One of the most commonly used radioisotopes in this context is iodine-131, which has been widely applied in the diagnosis and treatment of thyroid disorders, particularly hyperthyroidism and thyroid cancer. Iodine-131 was recognized as the first theranostic agent. This isotope not only helps identify benign thyroid conditions but also plays a crucial role in the treatment of thyroid cancer as the beta-emitting radioactive iodine selectively targets and destroys thyroid cancer cells, while also accumulating in these cells. The gamma camera allows precise imaging by detecting gamma rays emitted by iodine-131. Additionally, Iodine-131 helps to detect early recurrence by monitoring blood thyroglobulin levels or by radioiodine imaging, facilitating the elimination of remnant normal thyroid tissue. For many years, it has been suggested and predicted that radioisotopes like lutetium (Lu) and terbium (Tb) could be used as theranostics agents in the medical profession (Muller et al. 2019). Recently, people with prostate cancer have been proven to benefit by combining radioisotope with an antagonist that targets PSMA as a diagnostic and treatment option (Chang 2004). The way that this radioligand treatment, which consists of lutetium177 (^{177}Lu)-PSMA (LuPSMA), affects cancer cells is by invading their membrane and using endocytosis to produce a vacuole. Consequently, the exposure of cancer cells to radioisotopes limits their cytotoxic action, leaving surrounding normal cells unaffected (McBean et al. 2019). It is interesting to note that individuals with mCRPC treated with LuPSMA had less adverse effects and more than a 50% decrease in PSA expression (McBean et al. 2019). In modern theranostics approaches, techniques like PET/CT imaging with radiotracers are combined with radionuclide therapy, enabling more accurate localization of tumors or pathological tissue while concurrently delivering targeted treatment. For instance, radiolabeled drugs or peptides are designed to bind specifically to cancer cells, allowing for both imaging of tumor sites and therapeutic radiation delivery. The use of PET/CT scans allows clinicians to monitor the biodistribution of the radiopharmaceutical in real-time, guiding the administration of therapy while assessing treatment response. By combining these diagnostic and therapeutic functions, radiopharmaceuticals not only enhance precision in diagnosis but also improve treatment outcomes through personalized, targeted interventions that are monitored throughout the treatment process. This integration of imaging and therapy is a hallmark of modern theranostic strategies, allowing for better prediction, monitoring, and treatment of diseases like cancer, neurological disorders, and cardiovascular diseases.

1.2.2 Biomarkers and Molecular Diagnostics

Biomarkers and molecular diagnostics form a cornerstone of conventional theranostics, offering significant value in both disease monitoring and therapeutic decision-making. Biomarker testing involves the measurement of specific biological markers—substances or indicators present in the blood, tissues, or other bodily fluids—that are associated with a disease or the body's response to treatment. These biomarkers are crucial in identifying the presence and stage of a disease, determining prognosis, and assessing how well a treatment is working. For instance, the Prostate-Specific Antigen (PSA) test is commonly used to monitor prostate cancer. Elevated PSA levels often signal the presence of prostate cancer or its recurrence, allowing for timely intervention and adjusting the treatment course as needed. In breast cancer, testing for HER2 (Human Epidermal growth factor Receptor 2) expression is a critical diagnostic tool. HER2-positive tumors tend to be more aggressive, and knowing a patient's HER2 status helps in deciding whether they would benefit from targeted therapies like trastuzumab (Herceptin), a monoclonal antibody that specifically targets the HER2 receptor to slow down or stop the growth of the cancer. These molecular diagnostics allow for a more precise approach to treatment, minimizing unnecessary interventions and improving patient outcomes by tailoring therapies based on individual biomarkers.

1.2.3 Transarterial Chemoembolization

Transarterial chemoembolization (TACE) approach is one of the earliest theranostics methods used in interventional radiology. According to the Barcelona Clinic Liver Cancer (BCLC) system, TACE is suggested for asymptomatic persons with few unresectable multinodular tumors, no vascular invasion or extrahepatic spread, and normal liver function at an intermediate stage. It combines liver cancer screening, tumor treatment, and real-time therapy evaluation. This approach consists of two major steps: cytotoxic medicines and embolization particles are delivered into the branches of the artery supplying the tumor, resulting in ischemic tumor necrosis. Blood flow through the portal vein is severely limited and dominant; thus, blood supply to normal liver regions is maintained. Conventional TACE (cTACE) and drug-eluting bead TACE (DEB-TACE) are the two types of TACE methods used at present. When it comes to cTACE, the artery, which primarily supplies blood to the tumor, is injected with water-in-oil emulsion of cytotoxic and ethiodized oil such as Lipidol, followed by the tumor blood vessels with embolic agents. While in DEB-TACE, chemotherapeutic loadable microspheres are injected, which selectively clog the vessels and allow the gradual release of cytotoxic chemotherapy agents such as doxorubicin, epirubicin, and cisplatin. Lipiodol has various theranostic properties besides acting as a carrier for cytotoxic medicines. Its radiopaque characteristic allows for accurate real-time monitoring during emulsion distribution. Furthermore, Lipiodol has the unique virtue of being preferentially taken up by hyper-arterialized liver tumors, including HCC, and being effective for a long period. However, cTACE has various drawbacks, which include off-target release, the need for further embolizing the targeted arteries, nonstandardized emulsion preparation, and drug release pharmacokinetics based on emulsion type. To circumvent the aforesaid limitations, nonembolic microspheres (drug-eluting beads or DEBs) with the potential to transport medications have been created over the last decades.

1.2.4 Quantum Dots

Group II, III, IV, V, and VI elements of the periodic table are used to create quantum dots (QDs), which are nanoscale semiconducting nanocrystals. These semiconducting nanocrystals are promising candidates for bio-imaging agents because of their propensity to glow when excited (Kang et al. 2018; Zhang et al. 2008). Moreover, QDs function as efficient drug carriers for cancer-targeted therapy (Badrigilan et al. 2019; Wang et al. 2017). Thus, QDs combined with certain targeted agents and anticancer medications, together with the right imaging methods like CT and MRI, may open the door to the creation of concurrent cancer detection and therapy (Bansal et al. 2019; Brunetti et al. 2018). Brunetti et al. provided a classic example of cancer theranostics employing QDs conjugated with tetra-branched peptides as a targeting agent, as well as an anticancer medication.

1.2.5 Polymeric Micelles

Polymeric micelles are spherical, nanosized supramolecular structures with a diameter of 10–100 nm that are generated by self-assembly of amphiphilic copolymers in an aqueous environment (Zhang and Mi 2019). The hydrophobic core of the micelle serves as a reservoir for a variety of bioactive compounds, including anticancer drugs, antibodies, small molecules, and peptides for cancer treatment (Kumar et al. 2012). Alternatively, due to their hydrophobic surface, these micelles are easily conjugated with targeting agents, such as growth factors and antibodies (Blanco et al. 2009). In in vivo conditions, the hydrophilic PEG chain prevents the binding of serum proteins, hence preventing polymeric micelles from being rapidly cleared by macrophages in the blood. The drug's bioavailability is increased by the micelles and prolonged circulation in the blood. Due to their tiny size, micelles accumulate at tumor sites because of increased permeability and retention effects caused by prolonged blood circulation times. A radiolabeled peptide that is specifically intended to target receptors that are overexpressed on cancer cells is administered systemically as part of peptide receptor radionuclide therapy (PRRT), a molecularly targeted therapeutic approach (Bodei et al. 2013). Lutetium-177 or yttrium-90 beta particles primarily trigger DNA single-strand breaks, which ultimately cause cell apoptosis. Furthermore, in addition to the direct consequences of radiation on treated cells, beta particles could influence neighboring cells via the crossfire effect and bystander effect, boosting the efficacy of PRRT. The former impact is attributed to the longer range of beta-particles relative to the cell diameter, while the latter pertains to the creation of biological effects in cells around the targeted cells as if they were directly attacked. The drug administration of PRRT patients can be seen by performing gamma-emitting imaging. For instance, patients with metastatic gastroenteropancreatic neuroendocrine tumors (GEPNETs) find the use of PRRT especially attractive. GEP-NETs often possess the molecular trait of overexpressing somatostatin receptors. These receptors can be targeted with radiolabeled somatostatin analogues, including [^{90}Y-DOTA0, Tyr3] octreotide (^{90}Y-DOTATOC) or [lutetium-177 DOTA0, Tyr3] octreotate (^{177}Lu-DOTATATE). For instance, ^{177}Lu-DOTATATE has been utilized more often to treat GEP-NETs, showing positive outcomes in terms of overall survival and tumor response (Brabander et al. 2017). Complete responses, on the other hand, are still uncommon and can result from inadequate radiopeptide tumor absorption. Because of this, a growing number of radioisotope medications that were once intended for systemic use are now being reconsidered for locoregional distribution. Even though systemic effects can occur after intra-arterial locoregional distribution, selective intra-arterial injection of radiolabeled peptides is especially intriguing for the treatment of cancer patients with liver-only or liver-predominant sickness. Specifically, compared to systemic administration, intra-arterial delivery resulted in greater tumor absorbed radiation doses and improved tumor response rates, as outlined in a recent review of seminal publications (Feuerecker et al. 2021). Furthermore, the primary restriction of PRRT may be mitigated by intra-arterial administration by reducing the absorbed kidney dosage. When considered collectively, these first results emphasize that, apart from precisely directing radiopeptides toward tumors, enhanced therapeutic effectiveness and specificity can be achieved by targeted intra-arterial administration. In order to challenge the true predominance of systemic PRRT, more research is required to validate these encouraging results obtained by locoregional intra-arterial administration of radiolabeled peptides.

1.2.6 Polymersomes

Polymersomes are spherical nanostructured entities having a hollow aqueous compartment surrounded by a bilayer membrane composed of synthetic amphiphilic block copolymers. Polymersomes are synthetic polymers that self-assemble and are presently being developed to accomplish some of the same tasks as strong viral capsids, such as drug transport, targeting, and release. Polymersomes have more substantial biomedical uses than liposomes. Their high stability and chemical variety allow them to modify the characteristics of their membranes (for example, selective permeability). The enhanced mechanical stability is due to the lower critical aggregation concentration of the constituent amphiphilic block copolymers as a result of the amphiphiles' greater molecular weight and the entangled nature of the membranes. However, there are limitations that can be overcome by reducing polydispersity, improving reproducibility, and optimizing the composition of the nanocarriers for clinical translation and regulatory approval. It also discusses specific examples and approaches to synthesize polymers, functionalize them for theranostic applications, and assemble them into nanoparticles, supporting the modular approach exemplified in the document. For instance, lower doses of PNP-embedded dexamethasone made with PEG are used to treat arthritis. Because of its prolonged availability, this micelle efficiently lowers bone degradation and edema at joints (Wang et al. 2016). Polymersomes serve as multifunctional platforms for theranostics, which combines therapeutic and diagnostic functions in a single system, thereby enhancing personalized medicine. Polymersomes are unique polymeric vesicles that offer high loading capacity, stability, and ease of functionalization, making them ideal for drug delivery and medical imaging applications. They significantly improve imaging techniques, particularly magnetic resonance imaging (MRI), by incorporating superparamagnetic iron oxide nanoparticles (SPIONs), which enhance contrast and imaging quality.

1.3 Modern Theranostics Approaches

Modern approaches in theranostics are centered on the principles of precision medicine, which aim to customize treatment strategies based on the unique genetic, molecular, and clinical characteristics of individual patients. These approaches harness the potential of nanotechnology, enabling the development of nanoscale materials that can simultaneously deliver therapeutic agents and facilitate real-time imaging or monitoring. Additionally, molecular engineering plays a pivotal role in designing highly specific therapeutic agents, such as targeted drug delivery systems, engineered antibodies, or gene-editing tools like CRISPR, which can precisely modulate disease-related pathways (Fig. 1.3). By integrating these advanced technologies, modern theranostic strategies prioritize targeted therapies, ensuring that treatments act selectively on diseased cells while sparing healthy tissues, thereby minimizing side effects. Furthermore, these techniques emphasize minimally invasive interventions, allowing for reduced recovery times, lower risks, and enhanced patient outcomes. Together, these innovations pave the way for more effective, patient-centric medical solutions that combine diagnosis, treatment, and monitoring into a unified platform.

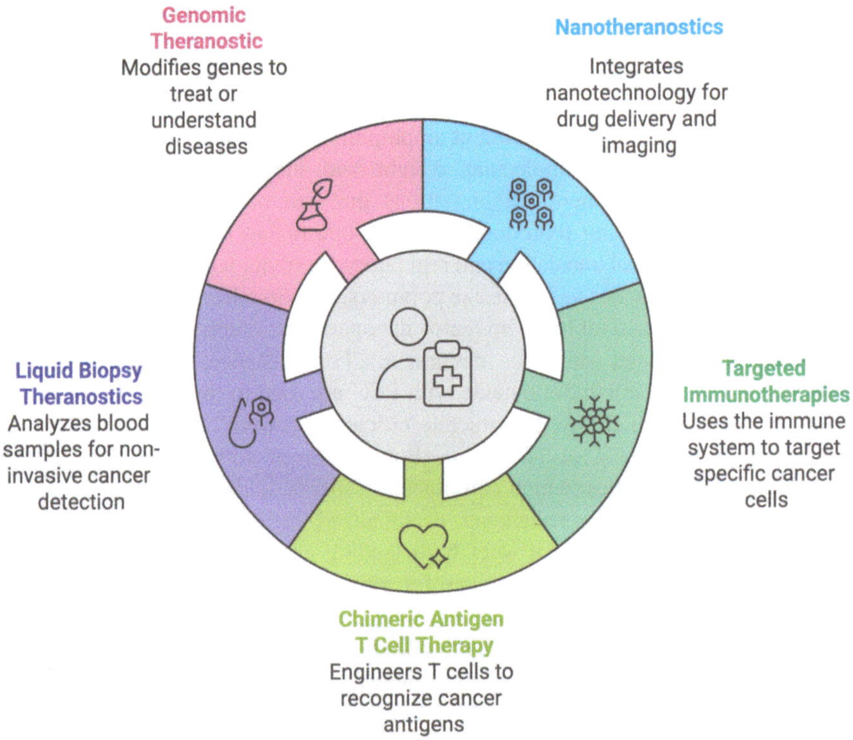

Fig. 1.3 Techniques utilized as modern theranostics for the treatment and diagnosis of disease

1.3.1 Nanotheranostics

Nanotheranostics mainly focuses on strategies for advanced theranostics, i.e., to apply and further develop the various nanocarriers such as polymer conjugations, dendrimers, micelles, liposomes, metal and inorganic nanoparticles, carbon nanotubes, nanoparticles of biodegradable polymers for sustained, controlled and targeted co-delivery of diagnostic and therapeutic agents for better theranostic effects, targeted delivery, fewer side effects, and real-time monitoring (Muthu et al. 2014). The core component includes (a) nanocarriers for carrying drugs, imaging agents, targeting ligands, etc. that include liposomes, dendrimers, quantum dots, gold nanoparticles, magnetic nanoparticles, and polymeric micelles; (b) therapeutic agents like doxorubicin, siRNA, antibodies, and photothermal agents like gold nanoshells to treat the particular disease; (c) diagnostic agents like fluorescent dyes, radioisotopes, and magnetic resonance contrast agents to improve imaging; and (d) targeting ligands to identify special cell types like breast cancer cells using HER2 (Siafaka et al. 2021; Akpa et al. 2023).

Different types of nanoparticles have different applications, like fluorescence imaging using quantum dots, PET scans using radiolabeled nanoparticles, and detection of superparamagnetic iron oxide nanoparticles (SPIONs) using MRI to diagnose cancer. Photothermal Therapy (PTT): Gold nanoshells absorb light and generate heat to kill cancer cells. Photodynamic Therapy (PDT): Light-activated nanoparticles produce reactive oxygen species. Liposomal doxorubicin is used for localized chemotherapy. Quantum Dots conjugated with drugs or siRNA are used for fluorescence imaging (Yasir et al. 2024). Magnetic drug delivery methods employ therapeutic components that are enclosed in polymer or superficially connected to MNPs, as well as MNP composites. The ability to deliver drugs to the exact location of the disease in a magnetic field outside the body while functioning in a range of environments, including light, electric current, temperature, magnetic fields, and ultrasound imaging, is one of the reasons why site-specific drug delivery is so essential (Chen and Hou 2022).

1.3.2 Theranostics Nanocarriers

Theranostic nanomaterials or nanocarriers are manufactured nanoparticles that are utilized for both diagnostic and therapeutic purposes. These have been engineered in such a way that they could be able to contain various components and imaging agents in addition to their main therapeutic function, which enables simultaneous and synergistic diagnosis and therapy. Nanomaterials such as polymer NPs, inorganic NPs, and organic/inorganic hybrid nanomaterials can be used as carriers and improved theranostic agents. This application may lead to a reduction in harmful effects and an increase in positive synergistic effects (Rong et al. 2020; Qian et al. 2017; Puja et al. 2023). Nanomaterials may be tailored for specific activities by adjusting controlled physical and chemical properties such as size, shape, surface ionization, solubility, and pH to precisely govern the pharmacokinetics of medications and imaging agents (Caldorera-Moore et al. 2011). Carriers of nanomaterials employed in theranostics are thoroughly explored before being delivered to the body in order to increase therapeutic efficiency and potency during treatment (Lim et al. 2021). The beneficial characteristics of nanoparticles are equal strength, an active surface, and distinct energy levels. The suitability of NPs for medication delivery and targeting improves with decreasing size. Because of their small dimensions and mobility, NPs have a greater cellular absorption rate than microparticles, allowing them to target a range of intracellular and cellular targets. Because the targeted cells must internalize the NPs, particle shape is also critical for biodistribution. Furthermore, immune system cells may recognize rod-shaped cationic NPs as rod-shaped bacteria, making them better candidates for endosomal absorption (Singh and Lillard 2009).

1.3.3 Targeted Immunotheranostics

Immune Theranostics is the synergy of theranostic procedures in nuclear medicine and immune treatment that leverages the combined power of molecular imaging and targeted modulation of the typically immunosuppressive tumor microenvironment (TME). This approach employs diagnostic and therapeutic radioisotopes to enhance precision and effectiveness in addressing the TME while supporting immunotherapy (Schaefer et al. 2020). It is composed of: (a) immunotherapeutic agents like Immune Checkpoint Inhibitors (anti-PD-1 and anti-CTLA-4 antibodies), cytokines such as IL-2 and interferons, and adoptive cancer therapies such as CAR-T cells; (b) imaging agents like radiolabeled antibodies to target specific T-cell receptors or fluorescent tracers to track immune cell infiltration.; and (c) nanocarriers such as liposomes, dendrimers, or polymeric nanoparticles, to carry the above components (Schaefer et al. 2020; Kim et al. 2021; Weber et al. 2022). One such example is PDD and PTT—photodynamic therapy (PDT) and photothermal therapy (PTT)—which are innovative cancer therapies that could induce immunogenic cell death (ICD) through the generation of reactive oxygen species (ROS) or localized heat for the treatment site, respectively. Both therapies use nanotechnology to improve targeting, efficacy, and immune responses. While PDT relies on light-sensitive photosensitizers (PS) drugs like verteporfin and chlorin e6, PTT employs heat-sensitive agents, such as gold nanorods and graphene oxide, activated by near-infrared (NIR) light.

1.3.4 FAP-Based CAR-T-Cell Therapy

Fibroblast activation protein α (FAP) acts as a target for cancer therapies, specifically CAR-T-cell therapy. When FAP is expressed in normal tissues, it is limited to activated fibroblasts during tissue remodeling, whereas it is highly expressed in cancer-associated fibroblasts (CAF) in the stroma of many carcinomas. FAP is involved in the remodeling of the extracellular matrix, migration of tumor cells, and the formation of an immunosuppressive microenvironment, making FAP an attractive but complex therapeutic target. Despite this promise, previous clinical experiences with FAP-targeting antibodies and inhibitors (e.g., Sibrotuzumab and Talabostat) demonstrated limited clinical efficacy. During the past few years, novel ^{68}Ga-FAPi-04 and ^{18}F-FAPi-74 have made important progress, which could be used not only to detect tumors with high FAP expression but also to monitor the efficacy of FAP-targeted therapy. Lee et al. showed that ^{18}F-FAPi-74 PET imaging can be used to monitor tumor FAP expression during CAR-T therapy, tracking a time course of FAP loss that occurs prior to observable tumor shrinkage. This noninvasive imaging can facilitate patient stratification, monitor therapy response, and overcome problems like antigen escape. The theranostic potentials of molecular imaging with CAR-T therapy facilitate early assessment of target loss or recurrence, provide a more accurate evaluation of treatment, limit radiation exposure compared to CT imaging, and open the door to precise monitoring and adaptation of CAR-T therapy (Weber et al. 2022).

1.3.5 Liquid Biopsy Theranostics

Theranostics, merging precision diagnosis with therapy, shows immense potential in cancer treatment. Liquid biopsies, especially via the analysis of circulating tumor cells (CTCs), facilitate real-time cancer monitoring. A liquid biopsy-guided drug release (LBDR) device integrates cancer detection and therapy on a microfluidic substrate. The device makes use of two types of materials: magnetic nanospheres (MNs) for CTC detection and drug-loaded MNs for treatment. When CTCs connect to aptamers on the recognition MNs, complementary DNA strands (cDNAs) are released, activating the drug-loaded MNs to deliver anticancer medicines. The drug release adjusts proportionally to the detected CTC count, ensuring a personalized approach. This cutting-edge system demonstrates a robust tool for customized cancer therapy, integrating precise diagnostics with adaptive drug delivery. Different biomarkers can be used in this technique, like Circulating Tumor Cells (CTCs), Circulating Tumor DNA (ctDNA), Extracellular Vesicles (EVs) and Exosomes, MicroRNAs (miRNAs), and Proteins and Metabolites (Xu et al. 2020). Current methods of cancer screening are not very effective at diagnosing cancer at early stages. Due to this limitation, the advancement of exosomes as a screening tool for cancer through liquid biopsies based on biological fluids, namely blood, plasma, serum, and urine, has been facilitated. These samples have exosomes sourced from normal and tumor cells, where the latter possess essential information for early-stage cancer diagnosis. To enrich this approach, exosome barcoding has been developed as a novel tool to distinguish tumor-derived exosomes or TEXs (Nyahatkar and Kalele 2024). TEX differentiability in exosome barcoding is achieved through the use of different methods. Chemical permits labeling through fluorescent dyes, Raman dyes, and redox probes for identification of exosomes with high specificity. Some examples of redox probes include using nanoelectrochemistry sensors to detect the biomarkers of cancer in the early stage. Another essential aspect of work is the nanoparticle coding, where quantum dots and plasmonic nanoparticles are utilized for the capture of TEXs based on the surface markers like CD63 or CD9. Additional improvements, such as incorporating magnetic particles, also elevate overall clinical sample detection sensitivity. Molecular tagging techniques, such as those that include the use of antibody-DNA conjugates and CRISPR-associated technologies, provide a better understanding of the additional roles of exosomes in gene silencing or mutagenesis (Boriachek et al. 2017). New methods of physical coding have also been developed, such as optical assays and electrochemical nanosensors. These approaches provide rapid, label-free detection of exosomal biomarkers, which have greater sensitivity than current techniques such as ELISA and Western blotting procedures. Highly sensitive methods, such as nano-plasmonic sensors, and advanced technologies like droplet digital PCR—which surpasses conventional PCR—have significantly improved the detection of low-abundance biomarkers in liquid biopsy applications. PCR, together with antibody-DNA conjugates and microfluidics, again enhances the analysis of exosomal contents that is achieved by PCR, antibody-DNA conjugates, and microfluidic devices (Nyahatkar and Kalele 2024). Liquid biopsy-based theranostics are used in various fields, including oncology, for the early detection of mutations like KRAS, TP53, or EGFR in ctDNA, therapeutic guidance for targeted therapies in identifying ALK rearrangements or BRAF mutations, and monitoring disease progression by longitudinal tracking of ctDNA for metastatic burden. They are also used in noncancerous diseases like cardiovascular diseases, neurological disorders, autoimmune diseases, and infectious diseases, detecting miRNAs and EVs related to myocardial infarction or heart failure, and monitoring inflammatory markers in liquid biopsy samples (Lone et al. 2022).

1.3.6 Genomic Theranostics

Another treatment option is known as genomics theranostics, an innovative method of attacking cancer and other diseases by diagnosing the disease then following up with treatment based on possible cure on the disease based on the disease's genetics. Next-generation sequencing is a breakthrough technology in cancer genomic research that has empowered rapid and genuine genome sequencing. First-generation Sanger sequencing is not parallel as it uses a chromatographic separation process; NGS, on the other hand, is massively parallel and thus fast. This technique is now essential in detecting somatic alterations in cancer subtypes, providing diagnostic, prognostic, and therapeutic avenues. Present technologies are capable of sequencing many samples simultaneously within weeks and further refine the concept of individual targeted sequencing at the single cell level to discover exclusive driver modifications (Yin et al. 2014). The cross-boundary projects, such as ICGC and TCGA, also use NGS to identify genomic patterns in various types of cancer. NGS in combination with clinical workflow could mean that patients would get constitutional and tumor sequencing at different time points in order to dissect and capture intra-tumor heterogeneity and disease evolutionary trends.

New generations of sequencing help improve cancer theranostics due to genome maps of single cells screened in tumors. Altogether, these enable molecular as well as medical imaging, drug targeting, novel drug design, and even the molecular analysis of drugs' action profiles. Moreover, genomic maps direct gene modulation methods (e.g., RNA interference) toward mimicking such states as drug resistance or metastasis in cell cultures or animals and offer a search for solutions to these problems (Turner 2018). With the launch of the Human Genome Project (HGP) in 1990 and its completion in 2003, genetics and medicine have been changed radically and have definitively advanced the field of personalized medicine and theranostics. The HGP offered a guide to where people could begin to look for differences in the human genome and how these differences affect sickness susceptibility, disease development, and therapy outcomes. This knowledge was the key to the development of targeted therapies, especially in oncology's moving from broad methods to molecular methods based on the individual genetics of the patient. HGP also helped the technologies being developed for genetic testing to ascertain the accurate position of the mutations and biomarkers in the human body. These tools comprise the diagnostic application of theranostics and help predict treatment responses and augment the therapeutic results. Through a timely and individualized approach to patient care, the HGP has enhanced positive treatment outcomes while ensuring minimal side effects and enhancing overall patient quality of life. The science of genomics, since the completion of the HGP, continues to grow rapidly as researchers explore the detailed molecular pathways linking genotype to phenotype, disease processes, and treatment strategies. This research goes on to create new applications across the numerous medical fields, extending the scope for better treatment based on theranostics for individual patients (Turner 2018).

1.3.7 AI-Based Theranostics

Modern molecular imaging generates vast amounts of data, creating a need for advanced automated tools and software systems to analyze them effectively. Since the approval of the first AI-powered medical device by the FDA in 1995, artificial intelligence has become a critical part of healthcare, with more than 520 AI and machine learning algorithms approved to date. Most of these are utilized in radiology and oncology. Because every patient and tumor are unique, identifying the most effective treatment requires a comprehensive analysis that considers both individual patient characteristics and tumor-specific details. AI offers tremendous potential in the field of theranostics, as it can integrate and interpret diverse data types, including multi-omics data like genomics and proteomics (Belge Bilgin et al. 2024). There are certain tools in artificial intelligence that are used in theranostic approaches. These tools are as follows: (1) Machine Learning (ML) Algorithms, which analyze vast datasets to detect patterns and correlations. In theranostics, ML is widely used in identifying biomarkers from genomic data, predicting disease progression and treatment response, and also in classifying medical images (e.g., MRI, CT scans). (2) Natural Language Processing (NLP) processes and extracts information from unstructured medical records, research articles, and clinical notes by extracting actionable insights from electronic health records (EHRs) and mining literature for novel biomarker discovery. (3) Deep Learning (DL) Models use artificial neural networks to perform complex data analysis. These models excel in image analysis, natural language processing, and signal processing (Currie and Rohren 2021).

For example, Radiomics and AI methodologies are gaining significant traction within the scientific community for their numerous potential applications in patients with prostate cancer (PCa). These applications range from technical improvements in image reconstruction and segmentation, which enhance workflow efficiency, to clinical applications like lesion classification and the assessment of images for predictive and prognostic purposes. Despite their promise, extensive validation studies and substantial efforts are still required to integrate these approaches into routine clinical practice effectively (Aerts 2016). Emerging AI techniques are transforming our comprehension of protein–ligand interactions. These innovative models can utilize new hit molecules, derived from high-throughput screenings or existing literature, as test sets to accelerate the process of hit-to-lead optimization. Once lead compounds are identified, AI can further assist by predicting the most suitable labeling precursors and efficient synthesis pathways, streamlining the rapid and effective development of theranostic agents (Saboury et al. 2023).

1.4 Conclusion and Future Perspectives

Theranostics, as a multidisciplinary field, bridges the gap between diagnostics and therapeutics to enable personalized, precise, and efficient medical care. This approach has evolved from conventional methods utilizing radiopharmaceuticals, biomarkers, and chemoembolization to modern techniques employing nanotechnology, immunotherapy, and artificial intelligence. These advancements have not only enhanced disease targeting and therapeutic outcomes but have also minimized collateral damage to healthy tissues. By integrating diagnosis, therapy, and real-time monitoring, theranostics embodies the future of patient-centric healthcare, particularly in oncology, cardiovascular diseases, and other complex medical conditions. The synergy of conventional methods, such as biomarker-driven diagnostics and targeted radiation, with cutting-edge technologies like nanotheranostics, genomic sequencing, and AI-powered imaging, underscores a paradigm shift in medical practice. These approaches ensure the delivery of treatments tailored to individual patient profiles, fostering better outcomes and improved quality of life. The future of theranostics is positioned to achieve groundbreaking advancements through the integration of innovative technologies and collaborative efforts across disciplines. A significant area of progress lies in nanotechnology, where the development of multifunctional nanocarriers with enhanced specificity, controlled drug release, and biocompatibility is expected to revolutionize both diagnostic and therapeutic capabilities. Platforms like polymeric micelles, quantum dots, and polymersomes hold immense potential for refinement and clinical applications. Simultaneously, the merging of genomic insights with artificial intelligence offers transformative opportunities for precision medicine. By identifying novel biomarkers, predicting treatment responses, and optimizing therapeutic regimens, AI-driven genomic integration will personalize theranostic solutions like never before. Immune-based theranostics also presents a promising frontier, particularly with the integration of immunotherapy strategies such as immune checkpoint inhibitors and CAR-T-cell therapy. These approaches, combined with advanced techniques like photodynamic and photothermal therapies, promise to deliver highly targeted and effective cancer treatments. Furthermore, advancements in liquid biopsy technologies, which enable real-time monitoring of circulating tumor cells, exosomes, and cell-free DNA, will facilitate early detection, treatment customization, and continuous disease management. These noninvasive approaches are poised to redefine patient care by offering minimally invasive yet highly informative diagnostic options. To ensure widespread impact, future efforts must address the sustainability and accessibility of theranostic technologies. Additionally, streamlined regulatory pathways and robust clinical trials are critical to overcoming translational challenges and integrating new modalities into standard clinical care. By addressing existing limitations and harnessing the potential of emerging technologies, the field is set to revolutionize diagnostics and therapeutics, ultimately improving patient outcomes and reshaping healthcare delivery in the years to come.

Author Contribution D.N. and D.H. have done the literature and prepared the first draft. A.S.M. has done final editing and revision.

References

Aerts HJ (2016) The potential of radiomic-based phenotyping in precision medicine: a review. JAMA Oncol 2(12):1636–1642

Akpa PA et al (2023) Nanotheranostics: platforms, current applications, and mechanisms of targeting in breast and prostate cancers. J Nanotheranostics 4(3):346–383

Badrigilan S et al (2019) Iron oxide/bismuth oxide nanocomposites coated by graphene quantum dots: "three-in-one" theranostic agents for simultaneous CT/MR imaging-guided in vitro photothermal therapy. Photodiagn Photodyn Ther 25:504–514

Bansal S et al (2019) Development of biosurfactant-based graphene quantum dot conjugate as a novel and fluorescent theranostic tool for cancer. Int J Nanomedicine 14:809–818

Belge Bilgin G et al (2024) Theranostics and artificial intelligence: new frontiers in personalized medicine. Theranostics 14(6):2367–2378

Blanco E et al (2009) Multifunctional micellar nanomedicine for cancer therapy. Exp Biol Med (Maywood) 234(2):123–131

Bodei L et al (2013) The joint IAEA, EANM, and SNMMI practical guidance on peptide receptor radionuclide therapy (PRRNT) in neuroendocrine tumours. Eur J Nucl Med Mol Imaging 40(5):800–816

Boriachek K et al (2017) Quantum dot-based sensitive detection of disease specific exosome in serum. Analyst 142(12):2211–2219

Brabander T et al (2017) Long-term efficacy, survival, and safety of [(177)Lu-DOTA(0),Tyr(3)] octreotate in patients with gastroenteropancreatic and bronchial neuroendocrine tumors. Clin Cancer Res 23(16):4617–4624

Brunetti J et al (2018) Near-infrared quantum dots labelled with a tumor selective tetrabranched peptide for in vivo imaging. J Nanobiotechnol 16(1):21

Caldorera-Moore ME, Liechty WB, Peppas NA (2011) Responsive theranostic systems: integration of diagnostic imaging agents and responsive controlled release drug delivery carriers. Acc Chem Res 44(10):1061–1070

Chang SS (2004) Overview of prostate-specific membrane antigen. Rev Urol 6(Suppl 10):S13–S18

Chen Y, Hou S (2022) Application of magnetic nanoparticles in cell therapy. Stem Cell Res Ther 13(1):135

Currie G, Rohren E (2021) Intelligent imaging in nuclear medicine: the principles of artificial intelligence, machine learning and deep learning. Semin Nucl Med 51(2):102–111

Feuerecker B et al (2021) Activity and adverse events of actinium-225-PSMA-617 in advanced metastatic castration-resistant prostate cancer after failure of lutetium-177-PSMA. Eur Urol 79(3):343–350

Kang SJ et al (2018) Anti-EGFR lipid micellar nanoparticles co-encapsulating quantum dots and paclitaxel for tumor-targeted theranosis. Nanoscale 10(41):19338–19350

Kim J et al (2021) Recent advances in tumor microenvironment-targeted nanomedicine delivery approaches to overcome limitations of immune checkpoint blockade-based immunotherapy. J Control Release 332:109–126

Kumar R et al (2012) In vitro evaluation of theranostic polymeric micelles for imaging and drug delivery in cancer. Theranostics 2(7):714–722

Lim J-W et al (2021) Application of nanomaterials as an advanced strategy for the diagnosis, prevention, and treatment of viral diseases. Pharmaceutics 13(10):1570

Lone SN et al (2022) Liquid biopsy: a step closer to transform diagnosis, prognosis and future of cancer treatments. Mol Cancer 21(1):79

McBean R et al (2019) Lu177-PSMA therapy for men with advanced prostate cancer: initial 18 months experience at a single Australian tertiary institution. J Med Imaging Radiat Oncol 63(4):538–545

Muller C et al (2019) Preclinical investigations and first-in-human application of (152)Tb-PSMA-617 for PET/CT imaging of prostate cancer. EJNMMI Res 9(1):68

Muthu MS et al (2014) Nanotheranostics—application and further development of nanomedicine strategies for advanced theranostics. Theranostics 4(6):660–677

Nyahatkar S, Kalele K (2024) Exosomes-based liquid biopsy: an effective approach for drugs and therapeutics resistance screening in cancer. J Liq Biopsy 4:100144

Puja AM et al (2023) Medicinal plant enriched metal nanoparticles and nanoemulsion for inflammation treatment: a narrative review on current status and future perspective. Immuno 3(2): 182–194

Qian CG et al (2017) Conjugated polymer nanomaterials for theranostics. Acta Pharmacol Sin 38(6):764–781

Rong L, Lei Q, Zhang X-Z (2020) Recent advances on peptide-based theranostic nanomaterials. View 1(4):20200050

Saboury B et al (2023) Artificial intelligence in nuclear medicine: opportunities, challenges, and responsibilities toward a trustworthy ecosystem. J Nucl Med 64(2):188–196

Schaefer N, Prior JO, Schottelius M (2020) From theranostics to immunotheranostics: the concept. Nucl Med Mol Imaging 54(2):81–85

Siafaka PI et al (2021) Current update on nanoplatforms as therapeutic and diagnostic tools: a review for the materials used as nanotheranostics and imaging modalities. Asian J Pharm Sci 16(1):24–46

Singh R, Lillard JW Jr (2009) Nanoparticle-based targeted drug delivery. Exp Mol Pathol 86(3): 215–223

Turner JH (2018) Recent advances in theranostics and challenges for the future. Br J Radiol 91(1091):20170893

Wang Q et al (2016) Targeted delivery of low-dose dexamethasone using PCL–PEG micelles for effective treatment of rheumatoid arthritis. J Control Release 230:64–72

Wang Y et al (2017) Quantum-dot-based theranostic micelles conjugated with an anti-EGFR nanobody for triple-negative breast cancer therapy. ACS Appl Mater Interfaces 9(36): 30297–30305

Weber WA et al (2022) A theranostic approach for CAR-T cell therapy. Clin Cancer Res 28(24): 5241–5243

Xu CM et al (2020) A liquid biopsy-guided drug release system for cancer theranostics: integrating rapid circulating tumor cell detection and precision tumor therapy. Lab Chip 20(8):1418–1425

Yasir M et al (2024) Theranostics: a multifaceted approach utilizing nano-biomaterials. Discov Nano 19(1):35

Yin Z et al (2014) Chapter 2: Genomics-based cancer theranostics. In: Chen X, Wong S (eds) Cancer theranostics. Academic Press, Oxford, pp 9–20

Zhang H, Mi P (2019) Chapter 12: Polymeric micelles for tumor theranostics. In: Cui W, Zhao X (eds) Theranostic bionanomaterials. Elsevier, pp 289–302

Zhang H, Yee D, Wang C (2008) Quantum dots for cancer diagnosis and therapy: biological and clinical perspectives. Nanomedicine (Lond) 3(1):83–91

Functional Nanohybrids for Solid Tumor Theranostics

2

Vivek P. Chavda, Surbhi Koul, Dixa A. Vaghela, Anjali Bedse, and Rohankumar Patel

Abstract

Solid tumors pose significant challenges in cancer diagnosis and treatment due to their complex microenvironment and heterogeneity. Functional nanohybrids, integrating multiple components for targeted imaging and therapy, have emerged as a promising approach for solid tumor theranostics. This chapter explores the design and development of nanohybrids, such as lipid-polymer, metal-organic frameworks, and exosome-based systems, for enhanced tumor accumulation, controlled drug release, and synergistic therapeutic effects. The chapter discusses strategies to overcome biological barriers, optimize pharmacokinetics, and improve tumor penetration. Case studies demonstrating the efficacy of nanohybrids in preclinical models of solid tumors are presented. The chapter concludes by highlighting the potential of functional nanohybrids in advancing personalized cancer management.

Keywords

Nanohybrids · Solid tumors · Theranostics · Targeted therapy · Drug delivery · Imaging techniques · Cancer treatment

V. P. Chavda (✉) · S. Koul · D. A. Vaghela
Department of Pharmaceutics and Pharmaceutical Technology, L.M. College of Pharmacy, Ahmedabad, Gujarat, India
e-mail: Vivek.chavda@lmcp.ac.in

A. Bedse
Department of Pharmaceutics, K. K. Wagh College of Pharmacy, Nashik, Maharashtra, India

R. Patel
Analytical Development, Amneal Pharmaceuticals, Bridgewater, NJ, USA

R. Prasad, M. Mohan Yallapu (eds.), *Advancements in Cancer Theranostics*,
https://doi.org/10.1007/978-981-95-0938-6_2

2.1 Introduction

Theranostics is an advanced technique in oncology that combines diagnostic and therapeutic functions. It aims to customize cancer treatment, improving the effectiveness and safety of therapeutic approaches. Theranostic drugs detect biomarkers or molecular targets linked to cancer cells, enabling clinicians to obtain accurate images of tumors (Burkett et al. 2023). After cancer cells are detected, specific substances can administer precise treatment, ensuring that the therapeutic effect is properly targeted toward cancer cells, reducing harm to healthy tissues. Theranostic agents often consist of targeting moieties that selectively bind to cancer cells, and tumor imaging is facilitated by radiolabeled or fluorescent markers (Barkat et al. 2024; Burkett et al. 2023; Aboagye et al. 2023).

Nanohybrids are composite materials that integrate nanoparticles with additional materials, typically at the molecular or atomic scale, in order to generate a novel material with improved or distinctive characteristics. These materials are usually made up of two or more constituents, with at least one of them being in the nanoscale range (1–100 nm) (Harish et al. 2022). The hybridization process frequently entails the amalgamation of organic and inorganic materials, resulting in structures that can demonstrate the collective benefits of both components, such as higher surface area, quantum effects, or improved mechanical strength (Gomez-Romero et al. 2024). Nanohybrids are employed in the medical domain for purposes like medication administration, imaging, and tissue engineering. Nanohybrid materials can be designed to transport medications to targeted locations within the body, thereby enhancing therapeutic effectiveness and minimizing adverse reactions (Bala et al. 2020; Mujahid et al. 2022).

The objective of this chapter is to provide a comprehensive exploration of the function of nanohybrids in theranostics, an innovative method for treating cancer. The chapter elucidates the essential principles of theranostics, encompassing their various classifications, functionalization, and the processes by which they selectively target tumors (Ghobadinezhad et al. 2022). Furthermore, it addresses the issues pertaining to safety and the process of translating research findings into clinical practice. Particular emphasis will be put on the use of nanohybrids in distinct types of cancer, such as breast, prostate, and lung cancer, through the analysis of case studies (Ghobadinezhad et al. 2022). It examines issues related to safety and toxicity, and emphasizes the significance of biocompatibility and regulatory standards. It also explores upcoming developments, such as the fusion of nanotechnology with AI and machine learning, and its potential for use in clinical settings (Kyriakides et al. 2021). This chapter is meant for researchers, medical practitioners, and professionals who have a keen interest in the future of cancer diagnostics and therapeutics.

2.2 Basics of Theranostics

Theranostics is an advanced medical method that integrates diagnostic testing with focused therapeutic interventions, integrating diagnosis and therapy. This integration enables enhanced accuracy and customized therapy, especially in complicated conditions such as cancer (Jeelani et al. 2014). The amalgamation of diagnosis and therapy is a fundamental element of customized medicine. Through the utilization of diagnostic techniques, it is possible to detect the precise molecular markers or genetic alterations that are present in a patient's disease (Salih et al. 2023). This enables the customization of therapy to match the individual's own biological profile (Fig. 2.1). This results in more efficient therapies with reduced adverse reactions in contrast to conventional treatment methods (Eberlein et al. 2017; Nunes et al. 2021).

Theranostics enables the personalization of treatment based on the patient's distinct tumor profile, resulting in improved outcomes and less side effects. Real-time monitoring allows for the simultaneous monitoring of treatment effectiveness and illness development, enabling dynamic modifications to therapy (Anitha et al. 2024). Theranostic methods specifically focus on malignant tissues, thereby minimizing harm to healthy cells and enhancing the patient's well-being throughout and following therapy (Yao et al. 2020).

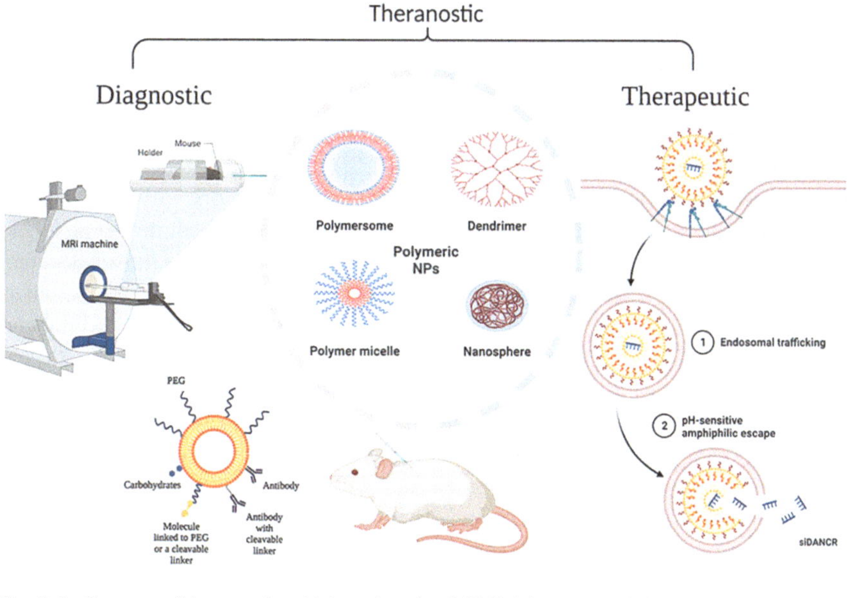

Fig. 2.1 Concept of theranostics. (Adapted under CCBY 4 from Hosseini et al. 2023)

Theranostics is transforming the field of clinical practice, namely in oncology, by providing a customized approach to both diagnosis and therapy. The clinical significance of this rests in the capacity to accurately target and treat diseases based on distinct molecular traits. This level of precision minimizes the need for trial-and-error in conventional treatments, resulting in enhanced efficacy and personalized care (Turner 2018). Theranostics has a significant influence as it improves treatment results by facilitating the real-time monitoring of the effectiveness of therapy, which enables the ability to make dynamic alterations in treatment (Wang et al. 2023). Consequently, this leads to enhanced disease control and increased patient survival rates. Furthermore, the precise targeting of theranostic drugs reduces the occurrence of undesirable side effects, hence enhancing the patient's quality of life during the treatment process (Yordanova et al. 2017).

Theranostics has demonstrated exceptional treatment effectiveness for solid tumors, namely in prostate and neuroendocrine malignancies. Prostate cancer: An exemplary instance involves the utilization of Prostate-Specific Membrane Antigen (PSMA)-targeted theranostics (Uemura et al. 2023). Prostate-specific membrane antigen (PSMA) is highly expressed in cells of prostate cancer. Clinicians can utilize a PSMA ligand that is labeled with a radioactive isotope to precisely visualize tumors through PET scans and administer focused radiation at the same time. This method has greatly enhanced results in individuals with advanced, metastatic prostate cancer, providing a novel therapy option when traditional medicines are ineffective (Czarniecki et al. 2018).

Another instance is Peptide Receptor Radionuclide Therapy (PRRT) for NETs, in which a somatostatin analogue is employed to specifically bind to somatostatin receptors on the tumor (Loharkar and Basu 2023). The radiolabeled counterpart not only allows for the visualization of the tumor but also administers targeted radiation, resulting in a decrease in tumor size and an extended lifespan for patients with inoperable or metastatic NETs (Bodei et al. 2022). These instances highlight the effectiveness of theranostics in the management of solid tumors, providing tailored and focused therapy options that lead to better clinical results (Vahidfar et al. 2021).

The development of theranostic agents that possess both high diagnostic accuracy and therapeutic efficacy is a challenging task that necessitates the utilization of modern technologies and the collaboration of multiple disciplines (Al-Ibraheem 2023). Theranostic agents, which have the ability to perform both diagnostic and therapeutic functions, face specific difficulties in obtaining regulatory approval. This is because they need to satisfy the requirements for both diagnostic equipment and therapeutic medications (Vaghela et al. 2024). Theranostics may lower treatment costs overall, but the early development and implementation costs may be high, which may restrict accessibility, especially in environments with limited resources. The cost of developing and implementing theranostic agents may prevent some patients from accessing these cutting-edge therapies. Clinical implementation is challenged by the need for advanced technologies and competence to integrate therapy and diagnostics (Whitlow et al. 2016).

2.3 Nanohybrid: Fundamentals and Classifications

A nanohybrid refers to a composite material composed of nanoscale components that are combined to enhance the properties and functionality of drug delivery systems. It is a particle that is both organic and inorganic, joined at the nanoscale by noncovalent connections (Choi et al. 2021).

The properties of nanohybrids can be tailored by selecting and combining different materials, such as metals, polymers, or ceramics. Surface modifications can be made to enhance stability. Functionalization of nanohybrids with ligands or antibodies allows for targeting specific cells or tissues, reducing off-target effects (Chavda et al. 2023). Nanohybrids can be engineered to release drugs in a controlled manner, either in response to specific triggers or over extended periods. The encapsulation of drugs within nanohybrids can protect them from degradation, improving stability and bioavailability. Nanohybrids can be constructed in various forms, such as core-shell structures, hollow spheres, or layered composites, depending on the desired application (Abdelkawi et al. 2023). Types of nanohybrids are described in Table 2.1.

Because of the improved qualities at the surface and interface, as well as the synergistic effect among components, composition and structurally controlled hybridization have emerged as a viable strategy to attaining multifunctionality. The microfluidic approaches are gaining more and more attention in the synthesis of nanoparticles (Díez et al. 2022).

2.4 Top-Down Approach and Bottom-Up Approach

2.4.1 Top-Down Approach

It includes the reduction of bulk materials to the nanometer scale through physico-chemical methods. This approach typically begins with a larger material and progressively reduces its size. Nanomaterials are synthesized by breaking down bulk solids into nanosized particles and providing the desired shape and order (Arole and Munde 2014).

There are many techniques involved in the top-down approach.

Lithography: This technique uses light to create patterns in a material, creating nanoscale features. There are two main types of lithography, masked lithography and maskless lithography. In masked nanolithography, a particular mask or template is used to transfer nanopatterns across a sizable surface area. Electron beam lithography, focused ion beam lithography, and scanning probe lithography are examples of maskless lithography. Maskless lithography is a technique that does arbitrary nanopattern writing without any requirement for a mask (Baig et al. 2021).

Table 2.1 Tabulated summary of nanohybrids

Category-subtype	Features	Definition	Examples	Application	References
Core–shell nanohybrid					
Metal–metal	• Enhanced catalytic properties • Plasmonic resonance • Improved stability and catalytic efficiency • Improve targeting at a specific site	Small, spherical, nanosized particles of metal are the building blocks of metallic nanohybrids (MHNs) and nanocomposites, which are covered in a dielectric substance	Gold-silver nanoparticle	• A nanohybrid derived from gold or silver can be used to target tumor • They are also used in multi-image-guided combined therapy for tumors • It has the capacity to accumulate in the tumor after injection in preclinical model • Nanometallic hybrids are widely used in removing bacteria and viruses from water bodies	Cheng et al. (2019) and Singh and Bhateria (2021)
Metal oxide	• Magnetic and optical properties • Enhanced optical and magnetic behavior • High surface area and more catalytic efficiency • Good material for environmental monitoring	A composite material that combines metal oxides with other materials to enhance their functionalities	Fe_3O_4-SiO_2 CuO-NPs	• Metal oxide nanohybrids are used in the manufacturing of electronic sensors and items • They are also key components in magnetic imaging techniques • Among many types, tungsten oxide exhibits electronic, structural, and physical and mechanical properties	Cheng et al. (2019), Liu et al. (2017) and Mujahid et al. (2022)

Polymeric-inorganic	• Controlled drug release • Biocompatibility • Sustained and targeted	They are a class of advanced materials that combine organic polymers with inorganic components, resulting in unique properties	Chitosan-silica nanoparticle	• They are used for targeted drug delivery. For example, hybrid nanomaterials can incorporate stabilizing agents and targeting ligands to enhance their therapeutic efficacy while reducing toxicity • They are utilized in bioimaging techniques, photothermal, and photodynamic therapy	Sun et al. (2021) and Kim et al. (2013)
Composite nanohybrid					
Polymer-matrix composite	• High strength, light weight, corrosion resistance • Enhanced mechanical properties • Better conductivity	It is a composite material made up of different lengths of continuous or short fibers joined by an organic polymer matrix	Epoxy/carbon nanotube composite	• Polymer nanohybrids used in the production of biosensors, and they possess an advantage to electric field • In pharma, they are utilized in biosensor development for the detection of pathogenic interactions and their DNA through electrochemical methods • They are also helpful in immunoassay and immunological detection	Kavitha et al. (2022)

(continued)

Table 2.1 (continued)

Category-subtype	Features	Definition	Examples	Application	References
Ceramic matrix composites (CMCs)	• High temperature stability, wear resistance • Improved thermal and mechanical stability • Ease of handling and positioning • Nontoxic • Possibility of sterilization	CMCs are advanced materials that consist of a ceramic matrix reinforced with ceramic fibers or particles	Alumina-silicon carbon composite	• These nanohybrids are employed to repair and reconstruct damaged parts of musculoskeletal system by integration of customized structures • The ability to incorporate various biochemicals and drugs, which can be used to deliver therapeutics over a controlled release rate	Vaiani et al. (2023)
Metal matrix composite (MMC)	• High thermal conductivity • Ductility • Improved strength and thermal properties	It is a material that combines a metal with other materials, such as particles or fibers to create lightweight structured material	Aluminum carbon fiber composite	• Metal oxide is used to remove heavy metal contaminants and microbes from water bodies • Applicable in the manufacturing of electronic devices in healthcare industry • The metal ion combination creates a self-assembly of functional proteins	Mujahid et al. (2022)

Hybrid organic-inorganic nanohybrid					
Silica-based hybrid	• High surface area, tunable pore size • Biocompatibility • Enhanced drug loading and release profiles	A silica-based nanohybrid is a complex material that is made of silica nanoparticles and organic compounds, and is created through a process called nanohybrid technology	Silica-polymer hybrid	• Silica-based nanohybrids are classified as solid lipid nanoparticles, and mesoporous silica nanoparticles have large surface area and specific pore volumes for drug loading • A drug release rate can be modified by editing the mesopore size • This type of nanoporous system is useful to treat many complicated disorders by giving controlled release formulation	Zhao et al. (2019) and Ilhan-Ayisigi and Yesil-Celiktas (2018)
Metal-organic frameworks (MOFs)	• High porosity, tunable structure, diverse functionalities • High surface area	A metal-organic nanohybrid is a nanosized particle that is made up of organic and inorganic elements linked together at the nanoscale	MOF nanoparticle	• They are used as potential drug delivery system due to their structural properties • By modifying them physically and chemically, a targeted delivery can be achieved	Mhettar et al. (2023)
Organic-inorganic nanohybrids	• Versatile properties • Enhanced functionalities • Improved mechanical and thermal properties	An organic–inorganic nanohybrid is a synthetic material made of organic and inorganic components that are linked together by covalent or noncovalent interactions	Organosilica hybrids	• Drug delivery can be achieved by modifying the properties of this nanohybrid • By using their chemical structure, we can fit our drug molecule into hybrids for controlled and targeted drug delivery • They are useful in environment remediation	Zhao et al. (2019)

Mechanical milling: It is a cost-effective method; in order to create blends of various phases and to produce nanocomposites, mechanical milling is a useful technique. It is useful to produce nanohybrid composites from metal oxide and carbide-based aluminum nanohybrids (Yadav et al. 2012).

Nanoimprint etching: Methods such as reactive ion etching (RIE) can shape materials into nanoscale structures. It is used in the fabrication of semiconductor devices and microelectromechanical systems (Huff 2021).

2.4.2 Bottom-Up Approach

It includes the assembly of nanoscale materials from smaller units such as atoms. This method builds up structures from the atomic or molecular level.

Techniques involved in the bottom-up approach are as follows:

Chemical vapor deposition (CVD): This technique involves the chemical reaction of gaseous precursors to form solid nanomaterials. The substrate is exposed to one or more volatile precursors during the process. The substrate is broken down and reacts with these precursors to create the desired deposit. During the process, vaporized precursors are initially adsorbed at a high temperature onto a substrate, where they either break down or react to form crystals (Kumar et al. 2017).

Sol-gel process: This involves the transition of a solution (sol with a diameter of 1–100 nm) into a homogeneous liquid medium, followed by agglomeration to form a continuous 3D network with a pore diameter. Three methods have generally been used to create sol–gel films: (1) gelation of a solid colloidal particle solution; (2) alkoxide hydrolysis and polycondensation followed by hypercritical gel drying; and (3) alkoxide hydrolysis and polycondensation followed by aging and drying in ambient conditions (Hench and West 1990).

Self-assembly: Molecules spontaneously organize into nanoscale structures due to intermolecular interactions. This approach is used in creating nanostructured materials such as quantum dots, nanotubes, and nanowires. It allows for the creation of highly uniform nanostructures and can be more cost-effective than top-down methods. In Fig. 2.2, achieving precise control over the size and shape can be challenging, and scalability can be an issue.

2.5 Self-Assembling Methods

Self-assembly is a process where nanoscale components spontaneously form organized structures through noncovalent interactions such as van der Waals forces, hydrogen bonding, or electrostatic interactions (Yadav et al. 2020).

Molecular self-assembly: Molecules or nanoparticles self-organize into specific structures due to designed interactions. This phenomenon is driven by noncovalent interactions, such as hydrogen bonding, hydrophobic forces, and

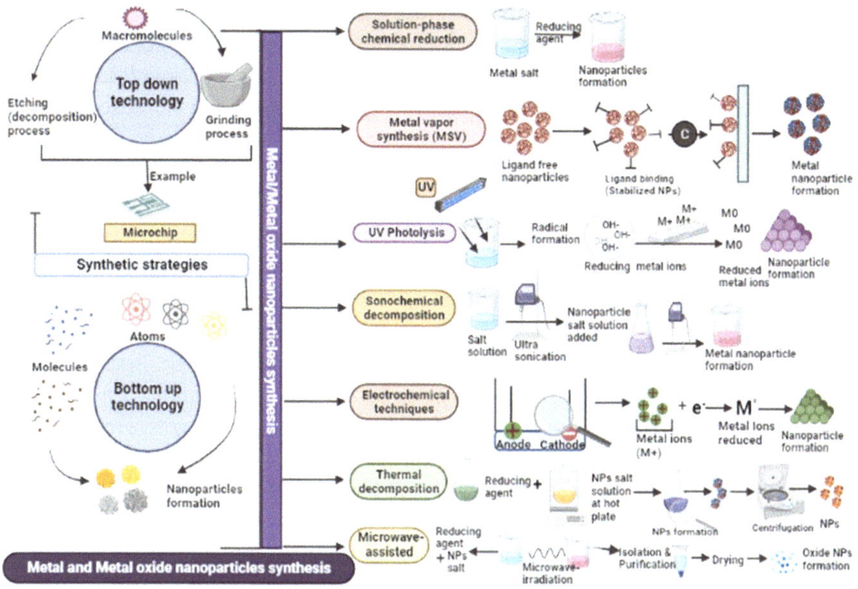

Fig. 2.2 A detailed understanding of top-down and bottom-up approaches. (Adapted under CCBY 4 from Mujahid et al. 2022)

van der Waals forces, allowing for the formation of complex nanostructures at the nanoscale, typically ranging from 1 to 100 nm (Luo et al. 2015). *Langmuir-Blodgett films:* A technique in which monolayers of molecules are transferred onto substrates to create thin films with controlled properties. *Dip-Pen nanolithography:* A technique that uses an AFM tip to deposit molecules onto a surface to create nanoscale patterns (Dhotel et al. 2013).

Self-assembly is used to fabricate nanowires, nanoparticles, and other nanostructures with precise control over size and shape. Self-assembly is a cost-effective method that can produce highly ordered structures with minimal external energy. On the other hand, the self-assembly process can be sensitive to environmental conditions and may require precise control over conditions to achieve the desired structure. Functionalization involves the addition of specific chemical groups or molecules to nanoparticles to alter their surface properties and improve their performance in various applications. This process enhances the interaction between nanoparticles and their environment, leading to improved functionality in areas such as sensing, imaging, and drug delivery (Lombardo et al. 2020).

Chemical functionalization involves *covalent bonding*: Attaching functional groups directly to the nanoparticle surface through covalent bonds. This method provides stable and durable modifications. *Coordination chemistry*: Using metal–ligand interactions to attach functional molecules to the surface of nanoparticles (Sanità et al. 2020).

Physical adsorption includes *electrostatic adsorption:* Utilizing electrostatic interactions to attach molecules to nanoparticles. This method is generally reversible and depends on the charge properties of both the nanoparticles and the functional molecules. *Van der Waals forces*: Molecules can be adsorbed onto nanoparticles through weak van der Waals interactions (Agboola and Benson 2021).

Self-assembled monolayers: Self-assembled monolayers (SAMs) involve the spontaneous formation of a monolayer of molecules on the nanoparticle surface. This technique can provide precise control over the surface chemistry of nanoparticles (Ahmad et al. 2023). Functionalized nanoparticles are used in targeted drug delivery, imaging, and diagnostics. The surface-modified gold nanoparticles can enhance imaging contrast or deliver therapeutic agents to specific cells. Functionalization can tailor nanoparticles for specific applications, enhance stability, and improve interaction with biological or chemical targets. However, the functionalization process can be complex and may require careful optimization of conditions to achieve the desired results (Chavda et al. 2024b).

2.6 Functionalization for Targeted Theranostics

2.6.1 Surface Modification and Ligand Binding

In the field of health, nanotechnology has a significant influence on the creation of diagnostic and therapeutic agents. These days, the ability of nanohybrids to act as a carrier for imaging different biological systems or to give image-guided therapeutic alternatives in the treatment of numerous fatal diseases has led to an increase in their popularity (Sim and Wong 2021). Table 2.2 summarizes ligand binding of nanohybrids.

One tactic to alter nanoparticle's physicochemical characteristics is functionalization. Clinically effective formulations that combine treatment and diagnostics are termed as nanotheranostics. Biomedical and environmental applications such as MRI, drug administration, catalysis, solar hydrogen production, sensors, water and air purification, and catalysis have all been investigated as potential uses for functionalized magnetic nanoparticles (MNM) (Pusta et al. 2023).

2.6.2 Imaging Agents and Diagnostics

An interdisciplinary field known as "nanoparticle-based medicine" has arisen with the potential to comprehend the molecular causes, dynamics, components, and treatments of disease. These include noninvasive cellular and molecular-imaging techniques, including ultrasound, optical, PET, computed tomography, and MRI.

Table 2.2 Ligand binding of nanohybrids

Nanocomposition	Size	Ligand	Target	Release kinetics	Applications	References
PEGylated PLG-co-hydroxymethyl GA-based PNP	100–200 nm	Anti-HER2 nanobody (1A4)	HER2	Sustained release profile	• To deliver drug at target site • Able to provide controlled release drug delivery for anti-cancer and inflammatory diseases-related drugs	Ahmad et al. (2023) and Perinelli et al. (2019)
PLGA-based PNP	100–500 nm	FA	FR	Biphasic release pattern	• To deliver drug at target site in vivo and in vitro • Compatible with various receptor binding and with cells • Has anti-microbial properties, so prevent contamination at laboratory level	Shakya et al. (2023)
HA-b-PLGA-based PNP	100–500 nm	HA	CD44	Biphasic release pattern	• HA-b-PLGA-based PNP nanohybrids can encapsulate therapeutic agents, allowing their controlled release • They are capable of reducing systemic toxicity of drug molecules	Guo et al. (2023)
(PLGA) and carboxymethyl chitosan (CMC)-based PNP	100–500 nm	Tf	Tf receptor	Biphasic release pattern	• It is known for its adjustable degradation rates, which can be tailored by altering the ratio of lactic to glycolic acid in its composition • Combination of PLGA and CMC in nanoparticle formulation has shown promising results in enhancing drug delivery	Herdiana et al. (2021)
PEG-PCL-based polymeric micelles	10–50 nm	GE-11 peptide	EGFR	Controlled release pattern	• They are able to enhance the solubility and bioavailability of hydrophobic drugs • This system is used in ocular drug delivery, cancer therapy, and to modulate the interaction of proteins	Ahmad et al. (2023)

2.6.2.1 Optical Imaging

The development of optical microscopy techniques has made it possible to image single-entity nanoscale objects, such as biological molecules, virions, nanomaterials, and organelles. The basic principles involve the emission, transmission, reflection, and absorption of light, typically within the visible, ultraviolet (UV), and near-infrared (NIR) spectra (Wang et al. 2022).

Fluorescence imaging: This technique uses fluorophores, which are molecules that emit light upon excitation by a specific wavelength.

Confocal microscopy: Confocal microscopy improves the resolution and contrast of fluorescence imaging by using spatial pinholes to eliminate out-of-focus light (Zhou et al. 2024).

2.6.2.2 Magnetic Resonance Imaging (MRI)

MRI is a revolutionary imaging technique that has transformed medical diagnostics and research. It employs powerful magnetic fields and radio waves to generate detailed images of the internal structures of the body (Romero et al. 2023). MRI is based on the principles of nuclear magnetic resonance (NMR), where nuclei of certain atoms, when placed in a strong magnetic field, absorb and re-emit electromagnetic radiation. This phenomenon is particularly prominent in hydrogen atoms, which are abundant in the human body due to the high water content in tissues. The MRI scanner consists of a powerful magnet that creates a strong and uniform magnetic field. Radiofrequency (RF) coils are used to transmit RF pulses and receive the resulting signals (Hosadurg and Kramer 2023). An in vivo study was carried out to check the efficacy of this nanomaterial function. In this study, iron oxide particles conjugated with near-infrared (NIR) dye labeled HER2 and chemotherapeutic drug cisplatin. In an orthotopic human ovarian cancer xenograft model with high levels of HER2 expression, MRI-guided therapy and optical imaging detection of the therapy-resistant tumor were investigated. The outcome demonstrates that in the mouse model of ovarian cancer, it significantly reduced the main tumor as well as the peritoneal and lung metastases (Satpathy et al. 2019).

2.6.2.3 Positron Emission Tomography (PET)

Positron Emission Tomography (PET) is a cutting-edge imaging technique that has revolutionized the fields of medical diagnostics and research. PET provides detailed images of metabolic and physiological processes in the body, enabling clinicians and researchers to detect and monitor various diseases with high precision. PET imaging is based on the detection of positrons emitted from radioactive tracers injected into the body. When a positron encounters an electron, they annihilate each other, producing two gamma photons that travel in opposite directions. These photons are detected by the PET scanner, which uses this information to construct detailed images of the tracer distribution within the body (Vaquero and Kinahan 2015).

Radioactive tracers: PET tracers are biologically active molecules labeled with positron-emitting isotopes. The most commonly used tracer is fluorodeoxyglucose (FDG), which mimics glucose and accumulates in metabolically active tissues.

Detection system: It consists of detectors in the form of a ring that surround the patient. These detectors capture the gamma photons produced by positron-electron annihilation and convert them into electrical signals (Wang et al. 2017).

2.7 Therapeutic Agents

These hybrid nanomaterials are designed to target solid tumors with high specificity and efficacy, offering a promising strategy for improving cancer treatment outcomes. Targeted Delivery-Functional nanohybrids are engineered to recognize and bind to tumor markers, ensuring that therapeutic agents are delivered directly to cancer cells while minimizing damage to healthy tissues (Chavda et al. 2024a). Functional nanohybrids are at the forefront of innovative cancer treatment strategies, especially in the theranostics of solid tumors (Chehelgerdi et al. 2023). These nanohybrids combine diagnostic and therapeutic functions, allowing for precise targeting. Doxorubicin is widely used in chemotherapy and has been effectively integrated into functional nanohybrids. These nanohybrids improve the solubility and bioavailability of doxorubicin while reducing its cardiotoxicity. Research shows that doxorubicin-loaded nanohybrids exhibit enhanced antitumor activity and reduced side effects (Lee et al. 2023). Cisplatin forms DNA crosslinks, inhibiting DNA synthesis and triggering cell death. Nanohybrids designed for cisplatin delivery, such as dendrimers and mesoporous silica nanoparticles, offer controlled release and reduced nephrotoxicity (Cheng et al. 2021).

2.7.1 Gene Therapy Components

Gene therapy represents a cutting-edge approach to treat various genetic disorders and cancers, including solid tumors. The advent of functional nanohybrids has revolutionized gene therapy by enhancing the stability and efficacy of therapeutic genes, particularly in the challenging environment of solid tumors (DAS et al. 2015). Nanoparticle vectors are pivotal in gene therapy, serving as carriers that protect and deliver genetic material into target cells. Lipid nanoparticles, such as those used in mRNA vaccines, encapsulate nucleic acids, protecting them from degradation and facilitating cellular uptake. Therapeutic genes are the core components of gene therapy, designed to replace defective genes, silence harmful genes, or express beneficial proteins. For instance, mRNA-based therapies enable transient expression of therapeutic proteins, offering a safer alternative to permanent genetic modification (Hou et al. 2021).

2.7.2 RNA-Based Therapeutics

RNA-based therapeutics represents a cutting-edge approach in modern medicine, particularly for the treatment of solid tumors. Functional nanohybrids enhance the delivery, stability, and therapeutic efficacy of RNA molecules, making them an integral part of solid tumor theranostics. Some lipid-based nanocarriers can fuse directly with the cell membrane, facilitating the release of RNA into the cytoplasm without the need for endosomal processing. This direct fusion mechanism enhances the efficiency of RNA delivery and reduces potential degradation by endosomal enzymes (Luo et al. 2023). Nanocarriers are critical for the effective delivery of RNA therapeutics. Lipid-based nanoparticles (LNPs), polymeric nanoparticles, and inorganic nanoparticles are commonly used to encapsulate and protect RNA molecules. Polymeric nanoparticles provide versatility and can be tailored for controlled release and targeted delivery. Therapeutic RNA molecules include messenger RNA, small interfering RNA, and microRNA. Each type of RNA serves a specific function: mRNA is used for protein expression, siRNA for gene silencing, and miRNA for gene regulation. siRNA can selectively silence oncogenes, while miRNA can restore normal gene regulatory networks disrupted in cancer (Lam et al. 2015).

Some lipid-based nanocarriers can fuse directly with the cell membrane, facilitating the release of RNA into the cytoplasm without the need for endosomal processing. This direct fusion mechanism enhances the efficiency of RNA delivery and reduces potential degradation by endosomal enzymes (Luo et al. 2023).

2.8 Mechanism of Tumor Targeting

2.8.1 Tumor Microenvironment and Cellular Uptake

The tumor microenvironment (TME) refers to the complex and dynamic environment surrounding a tumor, which incorporates various types of cells, signaling molecules, extracellular matrix, immune cells like Tumor-Associated Macrophage and Dendritic cells, Fibroblasts, Endothelial cells, Mesenchymal stem cells, and Adipocytes. TME performs an important role in tumor development, progression, metastasis, and response to treatment. Key components in TME include diverse signaling molecules like cytokines and growth factors, cancer cells, and immune cells. Various types of mechanisms are involved in that the TME can convert healthy noncancerous cells to tumor cells. Basic characteristics of the tumor microenvironment are hypoxia, hypoglycemia, acidosis, and inflammation, as shown in Fig. 2.3.

The approach to tumor targeting involves disrupting blood vessels that deliver nutrients and oxygen to tumors, preventing the growth of new cells. This is achieved through the enhanced permeability and retention (EPR) effect, which allows large molecules like nanoparticles to accumulate more in tumor tissues. Hypoxia, a common issue in tumor regions, can lead to resistance to certain therapies. To address this, drugs that activate under low oxygen conditions can normalize abnormal blood vessels and improve oxygenation. Targeting macrophages associated with

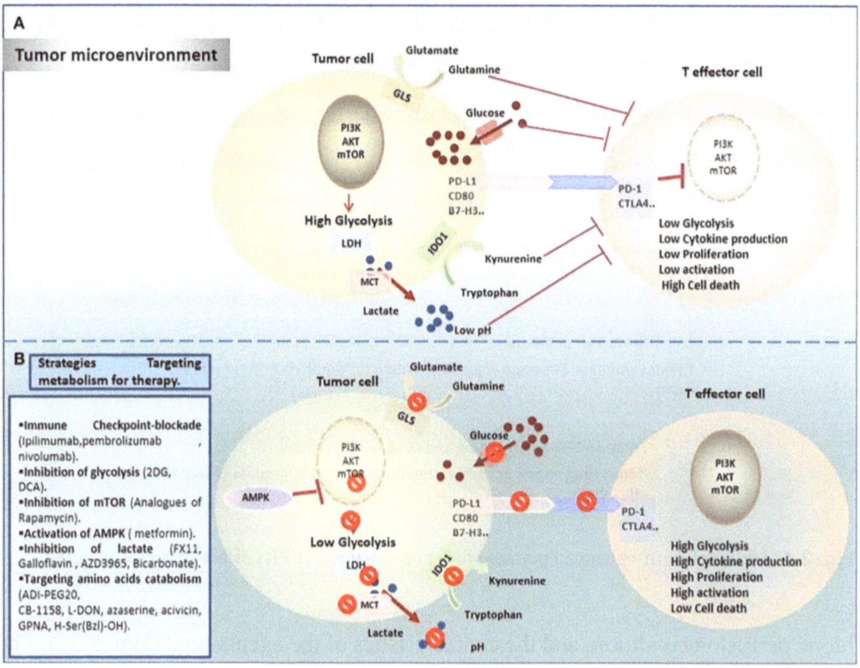

Fig. 2.3 Mechanism of tumor targeting. (Adapted from CCBY 4 fromZhen)

tumors (TAMs) can slow down tumor cell growth. The extracellular matrix (ECM) in the tumor microenvironment acts as a barrier to drug delivery, and improving the penetration of therapeutic agents can improve treatment efficacy. Immune checkpoints, such as PD-1/PD-L1 and CTLA-4 pathways, can also be used to modulate the immune system's ability to attack tumor cells.

2.8.2 Enhanced Permeability and Retention (EPR) Effect

The EPR effect is a significant pathophysiological instance in solid tumors, directing the advancement of nanomedicine for targeted and effective anticancer effects. Nonetheless, its heterogeneity across diverse microenvironments has posed challenges to its relevance in the design and use of nanomedicine. Strategies to enhance delivery and accumulation efficiencies encompass adjusting particle size and surface charge, implementing advanced drug release mechanisms, and refining administration kinetics (Wu 2021). The enhanced permeability and retention effect (EPR effect) facilitates the accumulation of macromolecular substances, such as albumin, in tumor vascularized regions, hence facilitating targeted administration and retention. The efficacy of EPR therapy is dependent upon tumor type, location,

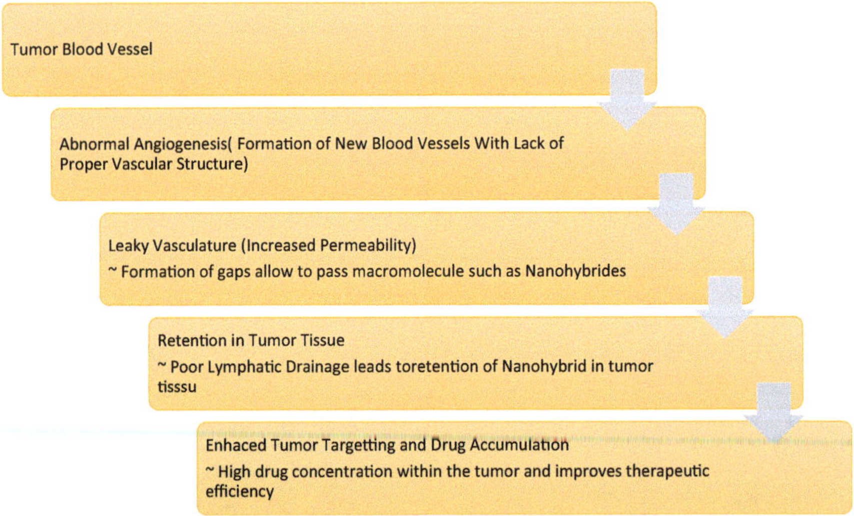

Fig. 2.4 Mechanism of enhanced permeability and retention (EPR) effect in tumor targeting

blood perfusion condition, and the characteristics of the agents used (Wu 2021). The tumor targeting and EPR mechanism have been given in Figs. 2.3 and 2.4.

2.8.3 Receptor-Mediated Endocytosis

The application of receptor targeting for selective drug binding and internalization has received attention since early developments in cell biology identified inherent internalization pathways for the uptake of vital components from the external environment. This hypothesis was studied in the 1980s and 1990s, with results indicating rapidly proliferating tumor cells exhibited overexpressed receptors. The transferrin receptor (TfR) was among the initial methods, binding the glycoprotein transferrin (Tf) and enabling its cellular entry. TfR continues to be a significant target for drug delivery in malignancies, facilitating the transcytosis of conjugates through barrier epithelial and endothelial cells (Xu et al. 2013).

2.9 Passive vs. Active Targeting Strategies

The higher permeability and retention effect influence passive targeting. Active targeting depends on ligand-receptor binding, enhancing selective accumulation at targeted areas and differentiating between tumor and healthy tissues. This might be accomplished by focusing on endothelial cells, tumor cells, the acidic milieu of malignancies, and the nucleus (Attia et al. 2019). Altering polymeric drug carriers with RGD- and NGR-derived oligopeptides facilitates active targeting of tumor vasculature, resulting in effective in vivo accumulation at target sites and ex vivo colocalization with angiogenic endothelial cells. Nonetheless, the prolonged concentrations of modified carrier materials are inferior to those of unmodified control copolymers, indicating that the advantages of active tumor targeting over passive targeting should not be overestimated.

2.9.1 Passive Targeting Approaches

Passively targeted cancer drugs accumulate in tumors due to leaky vasculature and defective lymphatic drainage, leading to a 70-fold accumulation of macromolecules in the tumor microenvironment (TME). The EPR effect is influenced by factors such as circulation time, targeting, and the ability to overcome barriers. Size, shape, and surface properties of the drug particles play a crucial role in the EPR-dependent delivery process.

Major drawbacks of passive targeting include the inability to distinguish between healthy and diseased tissues, inadequate tumor accumulation, and intra- and inter-individual tumor heterogeneity (Subhan et al. 2021).

2.9.2 Active Targeting Approaches

Active targeting allows for identifying precise, high-affinity regions of the tumor for binding, distributing the therapeutic agent selectively across different tumors, and circumventing the challenges related to multidrug resistance. Moreover, several moieties exhibit overlapped selectivity for targets that are positively utilized to produce a synergistic anticancer impact. The disease-oriented approach that takes into account pathophysiological changes is beneficial for formulating and advancing active targeted techniques (Vyas et al. 2022).

2.10 Challenges in Tumor Targeting

Various challenges observed during drug targeting are followed in Fig. 2.5.

2.10.1 Heterogeneity of Tumor Types

Heterogeneity denotes the variability in cancer cells both within and across tumors, encompassing changes in shape, transcriptional patterns, metabolism, and metastatic capability. It poses challenges in cancer ecosystems, resulting in tumor resistance, aggressive metastasis, and recurrence. Comprehending tumor heterogeneity is essential for formulating successful cancer treatments. Emerging single-cell and spatial genomic technologies elucidate its role in fatal consequences and physiological difficulties. A multidisciplinary strategy employing bioinformatics and

Fig. 2.5 Challenges in tumor targeting

computational technologies can facilitate the acquisition of integrated, multilayered knowledge of tumor heterogeneity necessary for the implementation of individualized, more effective medicines for cancer patients (Sutherland and Visvader 2015).

(a) Intertumoral heterogeneity: Intertumoral heterogeneity in cancer arises from intrinsic and extrinsic factors. Intrinsic variables encompass cellular mutations and the "cell of origin," whereas external factors, such as the microenvironment, affect cellular dedifferentiation. Tissue-specific stem and progenitor cells serve as the principal targets for tumor initiation. Identifying these target cell types is essential for early detection, finding premalignant clones, and developing preventive medicines for vulnerable cancer populations (Sutherland and Visvader 2015).

(b) Intratumoral heterogeneity: Intratumor heterogeneity refers to the variability inherent within a single tumor. It involves phenotypic variety, including apoptosis, cell surface indications, growth rate, (epi)genetic abnormalities, and other cancer hallmarks that ultimately contribute to disease progression and treatment failure (Prasetyanti and Medema 2017).

(c) Temporal heterogeneity: The presence of intratumoral driver gene heterogeneity and subcloning can induce clonal evolution, leading to genotypic alterations and temporal heterogeneity. This can profoundly affect therapy outcomes, especially in recurring NSCLC cancers. Consequently, genetic sequencing should not serve as the exclusive criterion for targeted therapy; ctDNA analysis can be employed to evaluate tumor heterogeneity and enable customized treatment (Fang et al. 2023).

2.11 Case Studies and Application

2.11.1 Nanohybrid in Breast Cancer Therapy

Nanohybrids are increasingly used in breast cancer treatment due to their effectiveness in multimodal combination therapies. One such nanohybrid is made of gold nanorods coated with degradable mesoporous silica and contains a photosensitizer (IR820) and chemotherapy drug doxorubicin. The inclusion of hyaluronic acid enhances biocompatibility and allows targeted drug delivery to mammalian carcinomas. The nanohybrid's biodegradable nature ensures biosafety. It accumulates in 4T1 tumor cells by interacting with hyaluronic acid and CD44+ receptors. Once inside the 4T1 cells, hyaluronic acid and organosilicon are degraded by hyaluronidase and glutathione, releasing doxorubicin and IR820. The release of doxorubicin and IR820 produces heat and singlet oxygen, enhancing the therapeutic effect (Cheng et al. 2021).

2.11.2 Diagnostic and Therapeutic Application

Nanohybrids, like iron oxide gold nanoparticles, enhance imaging techniques like MRI and CT by improving contrast and enabling targeted imaging. They are also used in biomarker detection, such as fluorescence detection for breast cancer, and biosensing for detecting cancer-related proteins and genetic markers. Nanohybrids can directly deliver chemotherapeutic drugs to cancer cells, increasing treatment efficacy. Liposome-gold nanohybrids are more effective at targeting tumor cells while minimizing the impact on normal cells, leading to more precise and less toxic cancer therapy. Gold nanorod-based nanohybrids are widely used in photothermal and photodynamic therapy, where they absorb infrared light and convert it into heat, effectively killing cancer cells. These can also overcome drug resistance by co-delivering chemotherapeutic agents and multidrug resistance inhibitors, increasing therapy efficacy at the targeted site (Table 2.3).

Table 2.3 Diagnostic and therapeutic application of nanohybrid

Application	Description	Example	Reference
Targeted drug delivery	Nanohybrids can deliver drugs specifically to breast cancer cells, reducing side effects	Gold nanoparticles functionalized with HER2 Ligand: Target HER2-positive breast cancer cells	
Combination therapy	It can deliver multiple therapeutic agents, like chemotherapeutic agents and gene therapies	Silica-based nanohybrid delivers both paclitaxel and siRNA	
Imaging diagnosis	Improve imaging techniques for better detection and monitoring of breast cancer	Iron oxide nanoparticles enhance MRI imaging for the detection of breast cancer	
Gene silencing	Deliver RNA interference agent to silence genes involved in cancer progression	Liposome nanoparticles hybrid deliver RNAi to targeted genes responsible for drug resistance	

2.11.3 Clinical Trials and Outcomes

Preclinical studies have evaluated the effectiveness and safety of gold nanorod-based nanohybrids in mice. The cytotoxicity of these nanohybrids was evaluated using an MTT assay on 4T1 and HepG2 cells, both with and without near-infrared (NIR) laser exposure. The IR&DOX@NC nanohybrid demonstrated greater phototoxicity in 4T1 cells compared to HepG2 cells under laser exposure. Live/dead assays showed extensive cell death when exposed to NIR laser irradiation. The in vivo distribution of Cy5.5-loaded nanohybrids was examined in mice bearing 4T1 tumors. The nanohybrids initially accumulated in the liver, with fluorescence increasing at the tumor site and peaking 18 h after injection. Ex vivo imaging and ICP-OES analysis confirmed significant tumor accumulation and partial excretion of the nanohybrids. The in vivo tumor growth inhibition effectiveness of IR&DOX@NC was also evaluated in 4T1 tumor-bearing mice. The results showed potent antitumor effectiveness of the IR&DOX@NC in combination with laser irradiation, while also showing minimal skin damage (Cheng et al. 2021).

2.12 Nanohybrid in Prostate Cancer Management

Nanohybrids, a category of nanomaterials, integrate features to improve efficiency, significantly contributing to cancer diagnosis and therapy.

2.12.1 Imaging

Imaging is vital for the scanning and identification of tumor cells, offering critical information for therapy decisions and patient management. Diverse imaging modalities may predict malignancy, necessitate biopsies, determine cancer stage, and assess metastasis. Effective diagnostic imaging depends on spatial resolution and the contrast-to-noise ratio (CNR) of images. Clinical imaging techniques are crucial for the detection and localization of prostate cancer, determining its stage and dissemination, and informing suitable treatment options. Current techniques encompass transrectal ultrasound-guided (TRUS) biopsy and local anesthetic transperineal (LATP) biopsy, with TRUS being the recommended approach for prostate cancer diagnosis (Calatayud et al. 2022). The study explores the interaction between fluorescent NDIs and planar aromatic carbon materials, revealing that peripheral substituents affect surface binding. It suggests that these systems, including NDIs and TRGO (thermally reduced graphene oxide), could be used as bioimaging agents for prostate cancer and novel fluorogenic composite materials (Tyson et al. 2016).

2.12.2 Treatment and Efficacy

The study developed stable upconversion immune-nanohybrids (UINBs) using a one-step ligand exchange method. The biotinylated antibody nanohybrid effectively identified prostate cancer cells with persistent, background-free luminous signals, demonstrating the potential of high-specificity diagnostic detection for prostate cancer cells using unique background-free and photostable features of UCNP.

2.13 Nanohybrid for Lung Cancer Theragnostic

Innovations in nanoparticle design and compositions have proved crucial for the diagnosis and treatment of malignant tumors. Theranostics have been employed in lung cancer imaging and treatment, with porphyrins demonstrating notable efficacy owing to their advantageous photophysical characteristics. Hematoporphyrin derivative (HpD) has received approval in more than 120 countries for cancer detection and photodynamic therapy. Positron emission tomography (PET) is utilized with fluorodeoxyglucose (FDG) for the purposes of cancer staging and restaging. Peptidomimetics, including EDTMP and ^{99m}Tc-MDP, are utilized for the palliative management of bone metastases originating from lung cancer. The advancement of lung cancer nanotheranostics encounters numerous hurdles, such as repeatable synthesis, scalability, sterility, nanoparticle-induced cytotoxicity, genotoxicity, and immunotoxicity. The FDA is implementing regulatory modifications to expedite the development of theranostics; however, the prospective advantages of these technologies must be evaluated based on scientifically rigorous, evidence-based, and well-controlled clinical trials (Mukherjee et al. 2019).

2.13.1 Innovative Approaches

Graphene oxide-DNA-based sensors exhibit potential for cost-effectiveness, heightened sensitivity, and minimal detection thresholds. Nonetheless, obstacles persist, including the inability of all dsDNA to dissociate from the graphene oxide surface following the binding of complementary molecules. At present, a single target can be identified by one DNA-based sensor utilizing GO in the liquid phase. Graphene oxide in the solid phase can be utilized for sensing without chemical modification. This study seeks to develop a nanobiosensor utilizing the DNA-GO nanohybrid for the detection of lung cancer and deletion mutations in cancer cells. The proposed method has benefits including rapidity, simplicity, cost-effectiveness, and early disease identification. Biosensors are potent analytical instruments for the swift analysis of chemical molecules (AL-Salman et al. 2024). Liposomal formulations have been employed for drug delivery, with Irinotecan liposome injection under investigation for patients with small cell lung cancer. Exosomes, nanoscale vesicles released by various cell types, are appealing for medication delivery and molecular diagnostics. They are utilized in liquid biopsy evaluations and are being engineered for pharmaceutical delivery and other diagnostic techniques (Mukherjee et al. 2019). Multifunctional nanohybrids are transforming cancer therapy by combining diagnostic and therapeutic capabilities for real-time imaging. Gold-based nanohybrids can be integrated with fluorescent dyes and chemotherapeutic drugs for targeted drug delivery, reducing side effects and enhancing chemotherapy effectiveness. Stimuli-responsive nanohybrids react to environmental factors for site-specific drug delivery.

2.13.2 Success Stories and Ongoing Research

Nanohybrids have shown success in lung cancer research, with systems combining magnetic nanoparticles with chemotherapeutic agents and fluorescent dyes reducing tumors and inhibiting metastasis. Liposomal nanohybrids deliver siRNA molecules, reducing cancer cell proliferation. Further research focuses on improving targeting specificity and therapeutic efficacy, exploring biodegradable nanohybrids, and developing personalized nanohybrids tailored to individual tumor characteristics.

2.14 Safety and Toxicity Concerns

Cytotoxicity studies use cell-based assays such as MTT, CytoTox, and LDH release to evaluate the toxicity of nanohybrids used in cancer treatment. Finding the best nanohybrid formulation and dose to reduce toxicity and increase the death of cancer cells is the objective. This facilitates the development of targeted medications that specifically destroy cancer cells while protecting healthy tissues, hence improving patient outcomes (Emam et al. 2017; Egbuna et al. 2021). In vivo toxicity studies included a hemolysis assay, which requires a modest hemolytic effect (less than 10%), and LAL assays to confirm the absence of endotoxins. The safety and toxicity can be assessed in Swiss mice by several bioassays, including hematological, biochemical, histopathological, and DNA fragmentation assays. The nanomaterials should exhibit little toxicity concerning hematological, biochemical, histological, and DNA damage criteria (Bhaumik et al. 2015).

2.15 Regulatory Considerations and Future Directions

The FDA and EMA have established safety and efficacy standards for nanohybrids in medical applications, requiring thorough preclinical investigations including toxicological, pharmacokinetic, and immunogenicity assessments. Recent radiopharmaceuticals, ^{223}Ra and ^{177}Lu-PSMA617, have been approved for advanced prostate cancer (PCa) due to their ability to prolong overall survival. ^{223}Ra, an α-emitting radionuclide, targets the bone microenvironment and has been approved for patients with CRPC and bone metastases. The European Medicines Agency recommended restricting ^{223}Ra to patients with two previous PCa treatments and not using it with abiraterone acetate plus prednisolone (Perera and Morris 2022). The researchers are advancing in many directions to advance the design of nanohybrids, such as targeted drug delivery, multifunctional hybrids, and surface modifications to achieve the desired therapeutic effect with less side effects and cost control (Thorat and Bauer 2020). Nanohybrids with multiple functions, such as drug delivery with imaging of the affected area or drug delivery with enhanced bioavailability, have been an intense area of research. They can be personalized for the given solid tumors, for example, Folate receptors (FRs) are proteins found in many types of epithelial cancers, including breast, lung, and ovarian cancers. FRs are overexpressed in cancer cells, sometimes 100–300 times more than in normal tissue, and are a recognized biomarker for tumor cells (Fernández et al. 2017). Nanohybrids containing folic acid and theranostic agent target solid tumors 100–300 times more than normal healthy cells, with less side effects to healthy cells. Moreover, they are being evaluated to be integrated with evolving technologies such as artificial intelligence, machine learning, nanorobots, and smart delivery systems to further improve the solid tumor theranostics.

Artificial intelligence and modeling can help predict process parameters to achieve higher yield and quality. It can also predict data trends that can be easily neglected by traditional data evaluation methods, while machine learning predicts how changing the structural property impacts the performance of nanohybrids (Barbosa et al. 2022). Integrating nanorobots is another great tool for target drug delivery systems (Patil and Lim 2023). They can travel to the affected site and deliver the drug through a smart delivery system, which gets activated when it finds a proper environment or other stimuli. The nanohybrids have a great potential for clinical translation with recent advances in nanohybrid technology and continued ongoing success in identifying target-specific characteristics. The main challenges that need to be overcome in the development are complexity in design, cost, regulatory, safety/efficacy, and quality-related issues (Gaber et al. 2017). Nanohybrids can be researched in the future for areas of imaging and diagnostics where they can deliver specific contrast agents to the target site, and that can produce more clearer image through MRI or other imaging technologies (Choi et al. 2021). It can also be researched in the areas of photothermal therapy where the heat generated by the nanoparticles raises the temperature of the tumor tissue, leading to cell death through hyperthermia, which can destroy cancer cells with the minimum harm to adjacent healthy tissue.

2.16 Conclusion

The functional nanohybrids present a holistic approach to cancer treatment and diagnosis. Their distinctive features, such as superior targeting abilities, enhanced imaging contrast, and controlled drug release, allow for more accurate and effective management of solid tumors. By combining diagnostic and therapeutic functions in one platform, these could transform personalized medicine, enabling treatment strategies that reduce side effects and boost effectiveness. Continued research and development are needed to address recent limitations and completely understand the potential of nanohybrids in clinical applications. As these technologies advance, they are expected to considerably influence cancer care, improving patient outcomes and a better quality of life. Overall, functional nanohybrids hold great potential in the field of target-specific drug delivery systems where theranostics can be performed with minimal damage to the healthy cells. By localizing therapeutic agents within the tumor site, they can help in reducing systemic toxicity. The treatment plans can be tailored based on individual patient tumor profiles and responses. This can lead to more effective and customized theranostics strategies.

Author Contribution S.K. and D.A.V. have done the primary literature survey. A.B. and R.P. have prepared the outlines and the draft. V.P.C. has done the final revision and editing in the chapter along with supervision of the project.

References

Abdelkawi A, Slim A, Zinoune Z, Pathak Y (2023) Surface modification of metallic nanoparticles for targeting drugs. Coatings 13(9):1660. https://doi.org/10.3390/coatings13091660

Aboagye EO, Barwick TD, Haberkorn U (2023) Radiotheranostics in oncology: making precision medicine possible. CA Cancer J Clin 73(3):255–274. https://doi.org/10.3322/caac.21768

Agboola OD, Benson NU (2021) Physisorption and chemisorption mechanisms influencing micro (nano) plastics-organic chemical contaminants interactions: a review. Front Environ Sci 9. https://doi.org/10.3389/fenvs.2021.678574

Ahmad MZ, Pathak K, Ahmad J, Aslam M, Bagre A, Patel P, Jain K (2023) Functionalized targeted theranostic nanomedicines. In: Jain K, Jain NK (eds) Multifunctional and targeted theranostic nanomedicines: formulation, design and applications. Springer Nature, Singapore, pp 1–26. https://doi.org/10.1007/978-981-99-0538-6_1

Al-Ibraheem A (2023) Theranostics in developing countries: addressing challenges and potentials from training to practice. World J Nucl Med 22(3):171–173. https://doi.org/10.1055/s-0043-1774733

AL-Salman HNK, Hsu C-Y, Jawad ZN, Mahmoud ZH, Mohammed F, Saud A, Al-Mashhadani ZI, Hadal LSA, Kianfar E (2024) Graphene oxide-based biosensors for detection of lung cancer: a review. Results Chem 7:101300. https://doi.org/10.1016/j.rechem.2023.101300

Anitha K, Chenchula S, Surendran V, Shvetank B, Ravula P, Milan R, Chikatipalli R, Padmavathi R (2024) Advancing cancer theranostics through biomimetics: a comprehensive review. Heliyon 10(6):e27692. https://doi.org/10.1016/j.heliyon.2024.e27692

Arole DV, Munde SV (2014) Fabrication of nanomaterials by top-down and bottom-up approaches—an overview. https://www.semanticscholar.org/paper/FABRICATION-OF-NANOMATERIALS-BY-TOP-DOWN-AND-%E2%80%93-AN-Arole-Munde/34f8921434fb256c9c8cca886722b5c920a1e4d2

Attia MF, Anton N, Wallyn J, Omran Z, Vandamme TF (2019) An overview of active and passive targeting strategies to improve the nanocarriers efficiency to tumour sites. J Pharm Pharmacol 71(8):1185–1198. https://doi.org/10.1111/jphp.13098

Baig N, Kammakakam I, Falath W (2021) Nanomaterials: a review of synthesis methods, properties, recent progress, and challenges. Mater Adv 2(6):1821–1871. https://doi.org/10.1039/D0MA00807A

Bala R, Badjatya S, Madaan R (2020) Strategies practiced to perk up oral palatability and acceptance of bitter drugs. J Drug Deliv Sci Technol 56:101580. https://doi.org/10.1016/j.jddst.2020.101580

Barbosa RDM, Lima CC, de Oliveira FF, Câmara GBM, Viseras C, de Lima e Moura TFA, Souto EB, Severino P, Raffin FN, Fernandes MAC (2022) New machine learning approach for the optimization of nano-hybrid formulations. Nanomanufacturing 2(3):82–97. https://doi.org/10.3390/nanomanufacturing2030007

Barkat MA, Rahman MA, Ansari MA, Ahmad FJ (2024) Introduction to nanofabrication for theranostics application. In: Barkat MA, Ahmad FJ, Rahman MA, Ansari MA (eds) Nanotheranostics for diagnosis and therapy. Springer Nature, Singapore, pp 1–13. https://doi.org/10.1007/978-981-97-3115-2_1

Bhaumik J, Thakur NS, Aili PK, Ghanghoriya A, Mittal AK, Banerjee UC (2015) Bioinspired nanotheranostic agents: synthesis, surface functionalization, and antioxidant potential. ACS Biomater Sci Eng 1(6):382–392. https://doi.org/10.1021/ab500171a

Bodei L, Herrmann K, Schöder H, Scott AM, Lewis JS (2022) Radiotheranostics in oncology: current challenges and emerging opportunities. Nat Rev Clin Oncol 19(8):534–550. https://doi.org/10.1038/s41571-022-00652-y

Burkett BJ, Bartlett DJ, McGarrah PW, Lewis AR, Johnson DR, Berberoğlu K, Pandey MK et al (2023) A review of theranostics: perspectives on emerging approaches and clinical advancements. Radiol Imaging Cancer 5(4):e220157. https://doi.org/10.1148/rycan.220157

Calatayud DG, Neophytou S, Eleni Nicodemou S, Giuffrida G, Ge H, Pascu SI (2022) Nanotheranostics for the sensing, imaging and therapy of prostate cancers. Front Chem 10:830133. https://doi.org/10.3389/fchem.2022.830133

Chavda VP, Gogoi NR, Vaghela DA, Balar PC, Dawre S, Dave DJ (2023) Parenteral microemulsions for drug delivery: advances and update. J Drug Deliv Sci Technol 89:104991. https://doi.org/10.1016/j.jddst.2023.104991

Chavda VP, Balar PC, Bezbaruah R, Vaghela DA, Rynjah D, Bhattacharjee B, Sugandhi VV, Paiva-Santos AC (2024a) Nanoemulsions: summary of a decade of research and recent advances. Nanomedicine 19(6):519–536. https://doi.org/10.2217/nnm-2023-0199

Chavda VP, Teli D, Balar PC, Davidson M, Bojarska J, Vaghela DA, Apostolopoulos V (2024b) Self-assembled peptide hydrogels for the treatment of diabetes and associated complications. Colloids Surf B Biointerfaces 235:113761. https://doi.org/10.1016/j.colsurfb.2024.113761

Chehelgerdi M, Chehelgerdi M, Omer QB, Allela RD, Pecho C, Jayasankar N, Rao DP, Thamaraikani T et al (2023) Progressing nanotechnology to improve targeted cancer treatment: overcoming hurdles in its clinical implementation. Mol Cancer 22:169. https://doi.org/10.1186/s12943-023-01865-0

Cheng Q, Li Z-H, Sun Y-X, Zhang X-Z (2019) Controlled synthesis of a core-shell nanohybrid for effective multimodal image-guided combined photothermal/photodynamic therapy of tumors. NPG Asia Mater 11(1):1–15. https://doi.org/10.1038/s41427-019-0164-4

Cheng D, Ji Y, Wang B, Wang Y, Tang Y, Yun F, Yufang X, Qian X, Zhu W (2021) Dual-responsive nanohybrid based on degradable silica-coated gold nanorods for triple-combination therapy for breast cancer. Acta Biomater 128:435–446. https://doi.org/10.1016/j.actbio.2021.04.006

Choi G, Sanoj Rejinold N, Piao H, Choy J-H (2021) Inorganic–inorganic nanohybrids for drug delivery, imaging and photo-therapy: recent developments and future scope. Chem Sci 12(14):5044–5063. https://doi.org/10.1039/d0sc06724e

Czarniecki M, Mena E, Lindenberg L, Cacko M, Harmon S, Radtke JP, Giesel F, Turkbey B, Choyke PL (2018) Keeping up with the prostate-specific membrane antigens (PSMAs): an introduction to a new class of positron emission tomography (PET) imaging agents. Transl Androl Urol 7(5):831–843. https://doi.org/10.21037/tau.2018.08.03

Das SK, Menezes ME, Bhatia S, Wang X-Y, Emdad L, Sarkar D, Fisher PB (2015) Gene therapies for cancer: strategies, challenges and successes. J Cell Physiol 230(2):259–271. https://doi.org/10.1002/jcp.24791

Dhotel A, Chen Z, Delbreilh L, Youssef B, Saiter J-M, Tan L (2013) Molecular motions in functional self-assembled nanostructures. Int J Mol Sci 14(2):2303–2333. https://doi.org/10.3390/ijms14022303

Díez AG, Rincón-Iglesias M, Lanceros-Méndez S, Reguera J, Lizundia E (2022) Multicomponent magnetic nanoparticle engineering: the role of structure-property relationship in advanced applications. Mater Today Chem 26:101220. https://doi.org/10.1016/j.mtchem.2022.101220

Eberlein U, Cremonesi M, Lassmann M (2017) Individualized dosimetry for theranostics: necessary, nice to have, or counterproductive? J Nucl Med 58(Suppl. 2):97S–103S. https://doi.org/10.2967/jnumed.116.186841

Egbuna C, Parmar VK, Jeevanandam J, Ezzat SM, Patrick-Iwuanyanwu KC, Adetunji CO, Khan J et al (2021) Toxicity of nanoparticles in biomedical application: nanotoxicology. J Toxicol 2021(1):9954443. https://doi.org/10.1155/2021/9954443

Emam AN, Loutfy SA, Mostafa AA, Awad H, Mohamed MB (2017) Cyto-toxicity, biocompatibility and cellular response of carbon dots–plasmonic based nano-hybrids for bioimaging. RSC Adv 7(38):23502–23514. https://doi.org/10.1039/C7RA01423F

Fang Q, Wan X, D'Aiello A, Sun H, Weiquing G, Li Y, Zhou C et al (2023) Temporal genomic heterogeneity guiding individualized therapy in recurrent non-small cell lung cancer. Front Oncol 13:1116809. https://doi.org/10.3389/fonc.2023.1116809

Fernández M, Javaid F, Chudasama V (2017) Advances in targeting the folate receptor in the treatment/imaging of cancers. Chem Sci 9(4):790–810. https://doi.org/10.1039/c7sc04004k

Gaber M, Medhat W, Hany M, Saher N, Fang J-Y, Elzoghby A (2017) Protein-lipid nanohybrids as emerging platforms for drug and gene delivery: challenges and outcomes. J Control Release 254:75–91. https://doi.org/10.1016/j.jconrel.2017.03.392

Ghobadinezhad F, Ebrahimi N, Mozaffari F, Moradi N, Beiranvand S, Pournazari M, Rezaei-Tazangi F et al (2022) The emerging role of regulatory cell-based therapy in autoimmune disease. Front Immunol 13:1075813. https://doi.org/10.3389/fimmu.2022.1075813

Gomez-Romero P, Pokhriyal A, Rueda-García D, Bengoa LN, González-Gil RM (2024) Hybrid materials: a metareview. Chem Mater 36(1):8–27. https://doi.org/10.1021/acs.chemmater.3c01878

Guo X, Zuo X, Zhou Z, Yinuo G, Zheng H, Wang X, Wang G, Caina X, Wang F (2023) PLGA-based micro/nanoparticles: an overview of their applications in respiratory diseases. Int J Mol Sci 24(5):4333. https://doi.org/10.3390/ijms24054333

Harish V, Ansari MM, Tewari D, Gaur M, Yadav AB, García-Betancourt M-L, Abdel-Haleem FM, Bechelany M, Barhoum A (2022) Nanoparticle and nanostructure synthesis and controlled growth methods. Nanomaterials 12(18):3226. https://doi.org/10.3390/nano12183226

Hench LL, West JK (1990) The sol-gel process. Chem Rev 90(1):33–72. https://doi.org/10.1021/cr00099a003

Herdiana Y, Wathoni N, Shamsuddin S, Muchtaridi M (2021) Drug release study of the chitosan-based nanoparticles. Heliyon 8(1):e08674. https://doi.org/10.1016/j.heliyon.2021.e08674

Hosadurg N, Kramer CM (2023) Magnetic resonance imaging techniques in peripheral arterial disease. Adv Wound Care 12(11):611–625. https://doi.org/10.1089/wound.2022.0161

Hosseini SM, Mohammadnejad J, Salamat S, Beiram Zadeh Z, Tanhaei M, Ramakrishna S (2023) Theranostic polymeric nanoparticles as a new approach in cancer therapy and diagnosis: a review. Mater Today Chem 29:101400. https://doi.org/10.1016/j.mtchem.2023.101400

Hou X, Zaks T, Langer R, Dong Y (2021) Lipid nanoparticles for mRNA delivery. Nat Rev Mater 6(12):1078–1094. https://doi.org/10.1038/s41578-021-00358-0

Huff M (2021) Recent advances in reactive ion etching and applications of high-aspect-ratio microfabrication. Micromachines 12(8):991. https://doi.org/10.3390/mi12080991

Ilhan-Ayisigi E, Yesil-Celiktas O (2018) Silica-based organic-inorganic hybrid nanoparticles and nanoconjugates for improved anticancer drug delivery. Eng Life Sci 18(12):882–892. https://doi.org/10.1002/elsc.201800038

Jeelani S, Reddy RCJ, Maheswaran T, Asokan GS, Dany A, Anand B (2014) Theranostics: a treasured tailor for tomorrow. J Pharm Bioallied Sci 6(Suppl 1):S6–S8. https://doi.org/10.4103/0975-7406.137249

Kavitha N, Elavarasan M, Ramachandran R, Uthayakumar S, Chandramohan A, Dinakaran K (2022) Polymer nanohybrid composites as conductive platform for the electrochemical sensing of pathogens. Curr Res Green Sustain Chem 5:100316. https://doi.org/10.1016/j.crgsc.2022.100316

Kim MH, Park D-H, Yang J-H, Choy YB, Choy J-H (2013) Drug-inorganic-polymer nanohybrid for transdermal delivery. Int J Pharm 444(1):120–127. https://doi.org/10.1016/j.ijpharm.2012.12.043

Kumar S, Bhushan P, Bhattacharya S (2017) Fabrication of nanostructures with bottom-up approach and their utility in diagnostics, therapeutics, and others. In: Environmental, chemical and medical sensors. Springer, pp 167–198. https://doi.org/10.1007/978-981-10-7751-7_8

Kyriakides TR, Raj A, Tseng TH, Xiao H, Nguyen R, Mohammed FS, Halder S et al (2021) Biocompatibility of nanomaterials and their immunological properties. Biomed Mater (Bristol, England) 16(4). https://doi.org/10.1088/1748-605X/abe5fa

Lam JKW, Chow MYT, Zhang Y, Leung SWS (2015) siRNA versus miRNA as therapeutics for gene silencing. Mol Ther Nucleic Acids 4(9):e252. https://doi.org/10.1038/mtna.2015.23

Lee J, Choi M-K, Song I-S (2023) Recent advances in doxorubicin formulation to enhance pharmacokinetics and tumor targeting. Pharmaceuticals 16(6):802. https://doi.org/10.3390/ph16060802

Liu X, Iocozzia J, Wang Y, Cui X, Chen Y, Zhao S, Li Z, Lin Z (2017) Noble metal–metal oxide nanohybrids with tailored nanostructures for efficient solar energy conversion, photocatalysis and environmental remediation. Energy Environ Sci 10(2):402–434. https://doi.org/10.1039/C6EE02265K

Loharkar S, Basu S (2023) Peptide receptor radionuclide therapy in neuroendocrine neoplasms and related tumors: from fundamentals to personalization and the newer experimental approaches. Expert Rev Precis Med Drug Dev 8(1):1–32. https://doi.org/10.1080/23808993.2023.2211090

Lombardo D, Calandra P, Pasqua L, Magazù S (2020) Self-assembly of organic nanomaterials and biomaterials: the bottom-up approach for functional nanostructures formation and advanced applications. Materials 13(5):1048. https://doi.org/10.3390/ma13051048

Luo D, Yan C, Wang T (2015) Interparticle forces underlying nanoparticle self-assemblies. Small 11(45):5984–6008. https://doi.org/10.1002/smll.201501783

Luo T, Huo C, Zhou T, Xie S (2023) Progress on RNA-based therapeutics for genetic diseases. J Zhejiang Univ Med Sci (Zhejiang Da Xue Xue Bao Yi Xue Ban) 52(4):406–416. https://doi.org/10.3724/zdxbyxb-2023-0190

Mhettar P, Kale N, Pantwalawalkar J, Nangare S, Jadhav N (2023) Metal-organic frameworks: drug delivery applications and future prospects. ADMET DMPK 12(1):27–62. https://doi.org/10.5599/admet.2057

Mujahid MH, Upadhyay TK, Khan F, Pandey P, Park MN, Sharangi AB, Saeed M, Upadhye VJ, Kim B (2022) Metallic and metal oxide-derived nanohybrid as a tool for biomedical applications. Biomed Pharmacother 155:113791. https://doi.org/10.1016/j.biopha.2022.113791

Mukherjee A, Paul M, Mukherjee S (2019) Recent progress in the theranostics application of nanomedicine in lung cancer. Cancer 11(5):597. https://doi.org/10.3390/cancers11050597

Nunes RF, Zuppani RMF, Coutinho AM, Barbosa FG, Sapienza MT, Marin JFG, Buchpiguel CA (2021) General concepts in theranostics. PET Clin 16(3):313–326. https://doi.org/10.1016/j.cpet.2021.03.010

Patil TV, Lim K-T (2023) Fundamental in polymer-/nanohybrid-based nanorobotics for theranostics. In: Lim K-T, Abd-Elsalam KA (eds) Nanorobotics and nanodiagnostics in integrative biology and biomedicine. Springer International Publishing, Cham, pp 79–108. https://doi.org/10.1007/978-3-031-16084-4_5

Perera M, Morris MJ (2022) From concept to regulatory drug approval: lessons for theranostics. J Nucl Med 63(12):1793–1801. https://doi.org/10.2967/jnumed.121.263301

Perinelli D, Cespi M, Bonacucina G, Palmieri G (2019) PEGylated polylactide (PLA) and poly (lactic-co-glycolic acid) (PLGA) copolymers for the design of drug delivery systems. J Pharm Investig 49. https://doi.org/10.1007/s40005-019-00442-2

Prasetyanti PR, Medema JP (2017) Intra-tumor heterogeneity from a cancer stem cell perspective. Mol Cancer 16(1):41. https://doi.org/10.1186/s12943-017-0600-4

Pusta A, Tertis M, Crăciunescu I, Turcu R, Mirel S, Cristea C (2023) Recent advances in the development of drug delivery applications of magnetic nanomaterials. Pharmaceutics 15(7): 1872. https://doi.org/10.3390/pharmaceutics15071872

Romero ÁB, Furtado FS, Sertic M, Goiffon RJ, Mahmood U, Catalano OA (2023) Abdominal positron emission tomography/magnetic resonance imaging. Magn Reson Imaging Clin N Am 31(4):579–589. https://doi.org/10.1016/j.mric.2023.06.003

Salih S, Elliyanti A, Alkatheeri A, AlYafei F, Almarri B, Khan H (2023) The role of molecular imaging in personalized medicine. J Pers Med 13(2):369. https://doi.org/10.3390/jpm13020369

Sanità G, Carrese B, Lamberti A (2020) Nanoparticle surface functionalization: how to improve biocompatibility and cellular internalization. Front Mol Biosci 7:587012. https://doi.org/10.3389/fmolb.2020.587012

Satpathy M, Wang L, Zielinski RJ, Weiping Qian Y, Wang A, Mohs AM, Kairdolf BA et al (2019) Targeted drug delivery and image-guided therapy of heterogeneous ovarian cancer using HER2-targeted theranostic nanoparticles. Theranostics 9(3):778–795. https://doi.org/10.7150/thno.29964

Shakya AK, Al-Sulaibi M, Naik RR, Nsairat H, Suboh S, Abulaila A (2023) Review on PLGA polymer based nanoparticles with antimicrobial properties and their application in various medical conditions or infections. Polymers 15(17):3597. https://doi.org/10.3390/polym15173597

Sim S, Wong NK (2021) Nanotechnology and its use in imaging and drug delivery (review). Biomed Rep 14(5):42. https://doi.org/10.3892/br.2021.1418

Singh R, Bhateria R (2021) Core-shell nanostructures: a simplest two-component system with enhanced properties and multiple applications. Environ Geochem Health 43(7):2459–2482. https://doi.org/10.1007/s10653-020-00766-1

Subhan MA, Yalamarty SSK, Filipczak N, Parveen F, Torchilin VP (2021) Recent advances in tumor targeting via EPR effect for cancer treatment. J Pers Med 11(6):571. https://doi.org/10.3390/jpm11060571

Sun J, Ma Q, Xue D, Shan W, Liu R, Dong B, Zhang J, Wang Z, Shao B (2021) Polymer/inorganic nanohybrids: an attractive materials for analysis and sensing. TrAC Trends Anal Chem 140: 116273. https://doi.org/10.1016/j.trac.2021.116273

Sutherland KD, Visvader JE (2015) Cellular mechanisms underlying intertumoral heterogeneity. Trends Cancer 1(1):15–23. https://doi.org/10.1016/j.trecan.2015.07.003

Thorat ND, Bauer J (2020) Functional smart hybrid nanostructures based nanotheranostic approach for advanced cancer treatment. Appl Surf Sci 527:146809. https://doi.org/10.1016/j.apsusc.2020.146809

Turner JH (2018) An introduction to the clinical practice of theranostics in oncology. Br J Radiol 91(1091):20180440. https://doi.org/10.1259/bjr.20180440

Tyson JA, Mirabello V, Calatayud DG, Ge H, Kociok-Köhn G, Botchway SW, Dan Pantoş G, Pascu SI (2016) Thermally reduced graphene oxide nanohybrids of chiral functional naphthalenediimides for prostate cancer cells bioimaging. Adv Funct Mater 26(31): 5641–5657. https://doi.org/10.1002/adfm.201601123

Uemura M, Watabe T, Hoshi S, Tanji R, Yaginuma K, Kojima Y (2023) The current status of prostate cancer treatment and PSMA theranostics. Therap Adv Med Oncol 15: 17588359231182293. https://doi.org/10.1177/17588359231182293

Vaghela DA, Pandya MB, Parmar PM, Chauhan SP, Vaishnav A, Patel RK, Patel DM, Modh VA, Raval MK (2024) Nanovaccine via intramuscular, subcutaneous, and intradermal routes. In: Nanocarrier vaccines. John Wiley & Sons Ltd., pp 405–422. https://doi.org/10.1002/9781394175482.ch12

Vahidfar N, Eppard E, Farzanehfar S, Yordanova A, Fallahpoor M, Ahmadzadehfar H (2021) An impressive approach in nuclear medicine. PET Clin 16(3):327–340. https://doi.org/10.1016/j.cpet.2021.03.011

Vaiani L, Boccaccio A, Uva AE, Palumbo G, Piccininni A, Guglielmi P, Cantore S, Santacroce L, Charitos IA, Ballini A (2023) Ceramic materials for biomedical applications: an overview on properties and fabrication processes. J Funct Biomater 14(3):146. https://doi.org/10.3390/jfb14030146

Vaquero JJ, Kinahan P (2015) Positron emission tomography: current challenges and opportunities for technological advances in clinical and preclinical imaging systems. Annu Rev Biomed Eng 17:385–414. https://doi.org/10.1146/annurev-bioeng-071114-040723

Vyas D, Patel M, Wairkar S (2022) Strategies for active tumor targeting-an update. Eur J Pharmacol 915:174512. https://doi.org/10.1016/j.ejphar.2021.174512

Wang Y, Li Y, Wei F, Duan Y (2017) Optical imaging paves the way for autophagy research. Trends Biotechnol 35(12):1181–1193. https://doi.org/10.1016/j.tibtech.2017.08.006

Wang Y, Zhang X, Jing X, Sun X, Zhao X, Li H, Liu Y et al (2022) The development of microscopic imaging technology and its application in micro- and nanotechnology. Front Chem 10:931169. https://doi.org/10.3389/fchem.2022.931169

Wang Y, Staudinger JN, Mindt TL, Gasser G (2023) Theranostics with photodynamic therapy for personalized medicine: to see and to treat. Theranostics 13(15):5501–5544. https://doi.org/10.7150/thno.87363

Whitlow J, Pacelli S, Paul A (2016) Polymeric nanohybrids as a new class of therapeutic biotransporters. Macromol Chem Phys 217(11):1245–1259. https://doi.org/10.1002/macp.201500464

Wu J (2021) The enhanced permeability and retention (EPR) effect: the significance of the concept and methods to enhance its application. J Pers Med 11(8):771. https://doi.org/10.3390/jpm11080771

Xu S, Olenyuk BZ, Okamoto CT, Hamm-Alvarez SF (2013) Targeting receptor-mediated endocytotic pathways with nanoparticles: rationale and advances. Adv Drug Deliv Rev 65(1):121–138. https://doi.org/10.1016/j.addr.2012.09.041

Yadav TP, Yadav RM, Singh DP (2012) Mechanical milling: a top down approach for the synthesis of nanomaterials and nanocomposites. Nanosci Nanotechnol 2(3):22–48

Yadav S, Sharma AK, Kumar P (2020) Nanoscale self-assembly for therapeutic delivery. Front Bioeng Biotechnol 8:127. https://doi.org/10.3389/fbioe.2020.00127

Yao Y, Zhou Y, Liu L, Yanyan X, Chen Q, Wang Y, Shijie W, Deng Y, Zhang J, Shao A (2020) Nanoparticle-based drug delivery in cancer therapy and its role in overcoming drug resistance. Front Mol Biosci 7:193. https://doi.org/10.3389/fmolb.2020.00193

Yordanova A, Eppard E, Kürpig S, Bundschuh RA, Schönberger S, Gonzalez-Carmona M, Feldmann G, Ahmadzadehfar H, Essler M (2017) Theranostics in nuclear medicine practice. Onco Targets Ther 10:4821–4828. https://doi.org/10.2147/OTT.S140671

Zhao N, Yan L, Zhao X, Chen X, Li A, Zheng D, Zhou X, Dai X, Fu-Jian X (2019) Versatile types of organic/inorganic nanohybrids: from strategic design to biomedical applications. Chem Rev 119(3):1666–1762. https://doi.org/10.1021/acs.chemrev.8b00401

Zhou X, Chieng A, Wang S (2024) Label-free optical imaging of nanoscale single entities. ACS Sensors 9(2):543–554. https://doi.org/10.1021/acssensors.3c02526

Optically Activatable Photosensitizers for Imaging and Therapy

3

Sathyaprakash Kumarasamy, Khushboo Warjurkar, Lipy Sharma, and Vinay Sharma

Abstract

Photosensitizers (PSs) are light-responsive compounds that are significant in therapies and imaging. They are at the forefront of advancements in phototheranostics. They generate reactive oxygen species (ROS) upon exposure to specific wavelengths of light. This chapter provides a brief overview of their significant role, molecular structure, and various applications in phototheranostics. The chapter also examines the mechanisms of action and their applications in imaging and therapeutics. The therapeutic use of these PSs is further explored through examples in photodynamic therapy (PDT), photothermal therapy (PTT), and combination therapies. Finally, we discuss the ongoing challenges in selecting these PSs, highlighting how innovative strategies and emerging advancements in phototheranostics could provide effective solutions.

Keywords

Photosensitizers · Reactive oxygen species · Photothermal therapy · Photodynamic therapy

S. Kumarasamy · K. Warjurkar · V. Sharma (✉)
Department of Biosciences and Bioengineering, Indian Institute of Technology Jammu, Jagti, Jammu, India
e-mail: vinay.sharma@iitjammu.ac.in

L. Sharma
Department of Zoology, Central University of Punjab, Bathinda, Punjab, India

R. Prasad, M. Mohan Yallapu (eds.), *Advancements in Cancer Theranostics*,
https://doi.org/10.1007/978-981-95-0938-6_3

3.1 Introduction

The photosensitizers (PSs) are special photochemicals that undergo chemical transformation upon light energy absorption and transfer it to surrounding non-reactive oxygen molecules, converting them to reactive oxygen species (ROS) that can target and kill cancer cells and pathogens. These highly reactive ROS molecules play a crucial role in biomedical applications, such as photodynamic therapy (PDT) and photothermal therapy (PTT). The unanticipated discovery by Raab in the 1900s on the lethal effects of sunlight on the protozoan *Paramecium* triggered by acridine marked the inception of PDT. The PDT is a type of treatment in which PSs get activated in the presence of a specific wavelength of light to produce ROS. These ROS are capable of destroying targeted cells like cancer and microbial cells through non-invasive medical techniques in cancer and antimicrobial therapies, respectively. In PDT, the PSs produce various ROS such as singlet oxygen ($^1O_2$), superoxide radicals ($O_2^{\cdot-}$), and hydrogen peroxide (H_2O_2), leading to cellular damage and cell death. Whereas, in PTT, PSs convert absorbed light into heat, causing localized thermal destruction of the target cells (Celli et al. 2010; Heinemann et al. 2017; Raab 1900; Spikes 1991). Thus, the PSs have emerged as key players in theranostics, which integrates diagnostic and therapeutic capabilities for real-time disease tracking and treatment. Furthermore, for accurate bioimaging and treatment, these PSs can initiate photochemical reactions upon light absorption, enabling the targeting of diseased cells for imaging and therapy simultaneously (Celli et al. 2010; Lovell et al. 2010; Kwiatkowski et al. 2018; Allison and Sibata 2010). The PSs are also employed as enhanced contrast agents in various imaging techniques such as fluorescence imaging, photoacoustic imaging (PAI), and magnetic resonance imaging (MRI), making them useful for real-time diagnostics. The characteristics, such as light absorption, tissue selectivity, and bioavailability, can be improved by modifying PS's chemical structure. These versatilities of PSs signify their importance in personalized medicine preparation, where treatments are tailored as per the unique biological profile of each patient (Yang et al. 2021a; Hester et al. 2020; Tang et al. 2017; Aksel et al. 2021; Patyal et al. 2024). This chapter explores their underlying mechanisms and diverse applications in imaging and therapy and discusses future directions for advancing their use in medicine.

3.2 Mechanism of Optical Activation of Photosensitizers

The optical activation of PSs occurs when they absorb photons at a specific wavelength of light. This process excites the PS molecules and leads to ROS generation (Fig. 3.1). This process typically involves three key steps:

1. Photon absorption: The PS molecules absorb light energy, usually in the visible, near infrared (NIR), or ultraviolet (UV) spectrum, transitioning from a ground state to an excited singlet state.

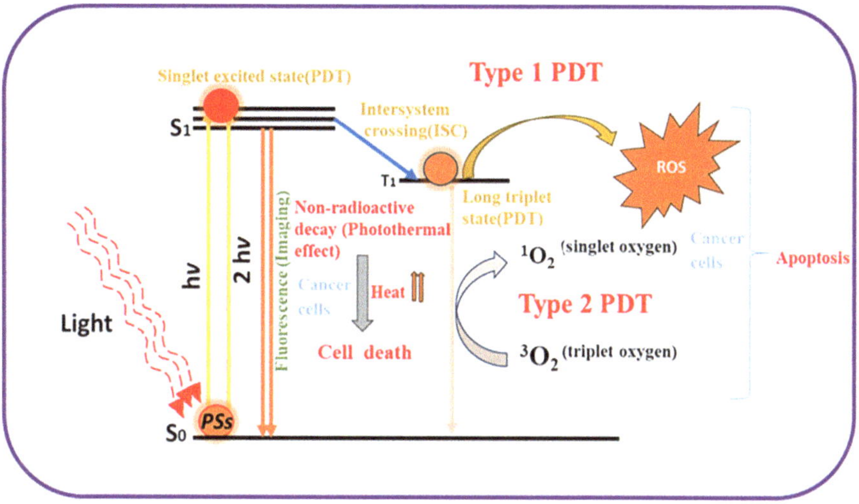

Fig. 3.1 Simplified Jablonski energy level diagram illustrating fluorescence imaging, PDT, and PTT

2. Energy transfer: Following photon absorption, the excited singlet state can undergo a transition to a triplet state via intersystem crossing (ISC). This step becomes significant due to the longer lifetime of the triplet state than the singlet state, providing more time for interactions with molecular oxygen.
3. ROS generation: The triplet state of the PS molecules interacts with ground-state molecular oxygen (non-reactive), leading to the production of $_1O^2$ and other reactive species.

3.3 Theranostic Applications of Photosensitizers

In therapeutic applications, PSs show dual function by providing real-time imaging capabilities to visualize disease progression while delivering the therapeutic agent for treatment through ROS generation. In PDT, for the generation of ROS, two types of reactions occur, named as Type I and Type II, as shown in Fig. 3.1. In **Type I reaction**, the excited PSs transfer electrons directly to biological molecules, such as proteins or lipids, generating radical species. These radicals subsequently react with oxygen to generate ROS, leading to oxidative damage in cells and tissues. This oxidative process is essential in PDT, as it disrupts cellular functions and induces cell death. In **Type II reaction**, the excited PSs transfer their energy directly to O_2, converting it into $^1O_2$. The $^1O_2$ is a highly reactive form of oxygen that can oxidize various cellular components, such as lipids, proteins, and nucleic acids, leading to cell damage or death. This mechanism is a key aspect of PDT, as $^1O_2$ is highly effective in disrupting tumor cells and other targeted cells. Similarly, in PTT, the temperature within the cells rises, enhancing treatment efficacy. This selective vulnerability makes PDT and PTT effective for treating solid tumors, especially when combined with targeting strategies that direct the PSs to tumor-specific markers.

3.3.1 Classification of Photosensitizers by Structure and Their Therapeutic Application

PSs are classified into different types based on their structures, such as porphyrins, chlorins, phthalocyanines, dye-based molecules, inorganic nanoparticles, and organometallic complexes, as mentioned in Table 3.1. Each type has unique properties that make it suitable for specific therapeutic or imaging applications. Their effectiveness mostly depends on factors like light absorption, tissue penetration, and selectivity

3.3.1.1 Porphyrin-Based Photosensitizers

Porphyrin-based PSs (naturally occurring and synthetic) are essential molecules that play a vital role in PDT. The distinctive tetrapyrrole macrocycle structure of porphyrins allows them to efficiently absorb light, typically in the visible and NIR spectrum, making them ideal for PDT applications. Advancements in porphyrin-based PSs over time have resulted in the development of three generations, with each generation addressing the limitations of the previous ones. The first-generation PSs, such as hematoporphyrin derivative and photofrin, introduced in PDT encountered several challenges such as poor selectivity, limited tissue penetration, and prolonged photosensitivity in patients (Allison and Sibata 2010; Tian et al. 1993). Second-generation PSs, such as temoporfin and benzoporphyrin derivatives, have refined chemical structures that enable the absorption of light at longer wavelengths, allowing deeper tissue penetration and enhanced tumor selectivity. They could generate ROS more efficiently, enhancing PDT efficacy while reducing toxicity in normal cells and shortening the duration of photosensitivity (Levy et al. 1996; Lin et al. 1994; Richter et al. 1989; Glidden et al. 2012). Another example of second-generation PSs is verteporfin that absorbs light at 690–700 nm, extending tissue penetration and proving effective for deep-seated tumors and nononcological conditions such as age-related macular degeneration (AMD). It also offers improved pharmacokinetics with faster clearance from the bloodstream, reducing prolonged photosensitivity (Jeising et al. 2022; Mellish and Brown 2001; Ziemssen and Heimann 2012). The third generation of PSs marks a significant advancement by integrating advanced targeting strategies and nanoparticle-based delivery systems, which enhances tumor specificity and minimizes various side effects. Innovations such as porphysome nanoparticles, antibody–porphyrin conjugates, polymer-porphyrin complexes, and gold nanoparticle conjugates have further improved solubility, stability, and precision in treatment. These advancements lead to an effective targeting of cancer cells and improve patient treatment outcomes (Kennedy and Pottier 1992; Colditz and Jeffree 2012; Ibarra et al. 2018; Azarian et al. 2024; Rabiee et al. 2020).

Table 3.1 Summarizes various types of PSs along with their theranostic properties, advantages, and disadvantages

PSs	Class	Generation	Therapeutic applications	Fluorescence imaging applications	Advantages	Disadvantages	Ref.
Hematoporphyrin derivative (HpD)	Porphyrin	First	Esophageal and lung cancers	N/A	Established clinical use, effective for superficial tumors	Non-specific uptake, prolonged skin photosensitivity	Tian et al. (1993)
Photofrin (Porfimer sodium)	Porphyrin	First	Esophageal, non-small cell lung cancer, and brain tumors	Fluorescence-guided surgery	FDA-approved, widely used, good for superficial cancers	Limited tissue selectivity, requires photoprotection	Kwiatkowski et al. (2018), Allison and Sibata (2010), Tian et al. (1993), Bartusik-Aebisher et al. (2022)
Temoporfin	Chlorin	Second	Head and neck cancers	N/A	Deeper tissue penetration, reduced side effects	Slow release and poor water solubility	Ronn et al. (1996), Tan et al. (2010), Yakavets et al. (2019)
Benzoporphyrin derivative	Porphyrin	Second	Cancer therapy; age-related macular degeneration (AMD)	Tumor imaging	Rapid clearance; effective in multiple applications	Moderate tissue penetration, potential phototoxicity	Andersson-Engels et al. (1993), Levy et al. (1996), Lin et al. (1994), Richter et al. (1989)
Protoporphyrin IX (PpIX)	Porphyrin	Third	PDT for various tumors, used in combination with ALA	N/A	Generated from Amino Levulinic Acid (ALA); effective in tumor imaging	Limited stability in circulation and rapid clearance may limit efficacy	Kennedy and Pottier (1992), Krammer and Überriegler (1996), Colditz and Jeffree (2012)

(continued)

Table 3.1 (continued)

PSs	Class	Generation	Therapeutic applications	Fluorescence imaging applications	Advantages	Disadvantages	Ref.
Chlorin e6	Chlorin	Second	Solid tumors therapy	Used for imaging in PDT	High $_1O^2$ yield, good tissue penetration	Photosensitivity issues, limited systemic availability	Akhlynina et al. (1997), Moon et al. (2009), Hak et al. (2023), Spikes and Bommer (1993a)
Chlorin e6 conjugates	Chlorin	Second	Targeted PDT for solid tumors	Organelle-specific imaging	Improved tumor selectivity and efficacy	Conjugation process may affect photophysical properties	Akhlynina et al. (1997), Hak et al. (2023), Spikes and Bommer (1993a, b), Hamblin et al. (2001)
Phthalocyanine	Phthalocyanine	Second	PDT for various cancers	Organelle-specific imaging	High photostability, good tissue penetration	Phototoxicity and photobleaching	Miller et al. (2007), Nyokong (2011), Whitacre et al. (2000), Jung et al. (2017)
Silicon phthalocyanine	Phthalocyanine	Second	Tumor treatment, especially in animal models	N/A	High $_1O^2$ production, suitable for deep tissue penetration	Poor solubility and hydrophobicity	Lam et al. (2011), Gümrükçü Köse and Keser Karaoğlan (2023), Leung et al. (2008)
Indocyanine Green	Dye	–	Vascular imaging; tumor detection	Strong NIR absorption, used in sentinel node mapping	Excellent tissue penetration, minimal background fluorescence	Short circulation time, low photostability	Meister et al. (2014), Shemesh et al. (2014), Nagahara et al. (2013), Yan et al. (2016)

1. BODIPY (boron-dipyrromethene) 2. BODIPY 630/650 3. 4,4-Difluoro-4-bora-3a,4a-diaza-s-indacene	Dye	–	Potential cancer therapy, imaging of biological processes	Organelle-specific imaging, high photostability	Versatile with tunable properties, useful for various applications	Low $_1O^2$ generation	Wang et al. (2018), Sun et al. (2019), Awuah and You (2012), Kamkaew et al. (2013), Turksoy et al. (2019), Chen et al. (2018), Wen et al. (2022), Dale et al. (2012), Li et al. (2020)
Rose Bengal	Dye	–	Treatment of skin cancers; potential antiviral activity	Organelle targeting and imaging	High $_1O^2$ yield, good visibility in imaging	Potential toxicity and photobleaching	Guo et al. (2010), Dhaini et al. (2022), Hua et al. (2017)
Gold nanoparticles	Inorganic nanoparticle	–	Enhances ROS production, PTT	N/A	High biocompatibility; can improve light absorption	Potential systemic toxicity, requires careful dosing	Hirsch et al. (2003), Loo et al. (2004), Huff et al. (2007)
Titanium dioxide	Inorganic	–	Photocatalytic applications, antimicrobial activity	N/A	Strong oxidizing agent, useful for various applications	Low solubility, potential environmental concerns	Aksel et al. (2021), Behnam et al. (2018), Cheng et al. (2017)
Iridium complexes	Organometallic	–	Cancer therapy, PDT, and potential in targeted therapies	Fluorescent probes for bioimaging	High photostability, tunable luminescence properties	Biocompatibility	Cai et al. (2021), Guo et al. (2019), Yang et al. (2021b), Yuan et al. (2019), Zhong et al. (2023), Liu et al. (2024)

3.3.1.2 Chlorin-Based Photosensitizers

Chlorin-based PSs represent a significant advancement in the field of PDT. These agents are derived from the structural modification of porphyrins and possess improved photophysical and chemical properties that enhance their utility in clinical applications. Chlorins have the ability to absorb light at longer wavelengths, enabling deeper tissue penetration while generating ROS that increases therapeutic effectiveness. Due to these advantages, chlorin-based PSs have become an essential in the field of PDT. Chlorins are structurally similar to porphyrins except for one double bond in the pyrrole ring, which is saturated. This modification shifts their absorption spectrum toward the red to NIR, typically around 650–700 nm, facilitating deeper tissue penetration (Akhlynina et al. 1997; Moon et al. 2009; Hak et al. 2023; Spikes and Bommer 1993a). The altered structure leads to a higher molar absorption coefficient, resulting in more efficient light absorption and greater $^1O_2$ production upon photoactivation. This enhanced ROS generation makes chlorins highly cytotoxic to malignant cells. Chlorin-based PSs began with first-generation agents such as Chlorin e6, which absorbs light at around 660 nm, allowing for deeper tissue penetration compared to porphyrin-based compounds. However, these early agents faced some limitations, such as poor tumor selectivity and prolonged photosensitivity (Akhlynina et al. 1997; Hak et al. 2023; Gilyadova et al. 2022). Second-generation chlorin PSs, like temoporfin (mTHPC, Foscan), developed stronger absorption at approximately 650–700 nm. Temoporfin provides greater tumor selectivity and deeper penetration, making it particularly useful for deep-rooted tumors (Ronn et al. 1996; Tan et al. 2010; Yakavets et al. 2019). The third generation of chlorin-based PSs has advanced targeting strategies and delivery systems to enhance tumor selectivity and biodistribution with minimum side effects. These agents utilize targeting ligands or antibodies to direct the PSs toward cancer cells, reducing accumulation in normal tissues and improving the precision of the PDT treatment. Nanoparticle-based delivery systems, including liposomes and gold nanoparticles, encapsulate chlorins, improving their solubility, stability, and circulation time. Liposome-loaded Chlorin e6 has shown promise in preclinical studies, allowing for sustained therapeutic effects and precise tumor targeting. Chlorin-gold nanoparticle PSs complexes have unique optical properties of gold to enhance ROS generation and improve targeting, which represents a promising frontier in third-generation PDT agents. These developments mark significant progress in the precision and effectiveness of chlorin-based PSs in cancer treatment (Hak et al. 2023; Spikes and Bommer 1993a; Hamblin et al. 2001).

3.3.1.3 Phthalocyanine-Based Photosensitizers

Phthalocyanine-based PSs are an important class of compounds utilized in PDT due to their excellent photophysical properties, such as strong absorption in the NIR region, high photostability, and efficient generation of ROS. Phthalocyanines are macrocyclic compounds made up of four isoindole units linked by nitrogen atoms, forming a highly conjugated 18-π electron system. This structure is similar to porphyrins, except that phthalocyanines also include benzene rings attached to the isoindole units, extending the conjugation and shifting their absorption into the NIR range between 600 and 800 nm. This shift into the NIR spectrum allows for deeper tissue penetration, making phthalocyanines effective for treating deep-seated tumors within the body (van Lier and Spikes 2007; Rosenthal and Ben-Hur 1995; Ishii 2012; McKeown 2003; Li et al. 2019). Phthalocyanines can also form stable complexes with metal ions at their central nitrogen core, enhancing their photodynamic activity. Metals like zinc, aluminum, and silicon increase the production of $^1O_2$, and improve the overall therapeutic efficacy (Ding and Han 2015; Roguin et al. 2019; Egorin et al. 1999; Mohammed et al. 2019; Brasseur et al. 1999). Zinc phthalocyanine, a first-generation PS, showed strong absorption around 680–690 nm and efficiently generated $^1O_2$. However, some limitation, such as poor solubility and non-specific tissue distribution, limits its clinical use. Therefore, researchers developed sulfonated derivatives of zinc phthalocyanine tetrasulfonate to improve solubility and tumor selectivity. However, these agents still have some limitations, including long retention times in the body and poor targeting precision (Roguin et al. 2019; Egorin et al. 1999; Brasseur et al. 1999). Second-generation phthalocyanine PSs, such as aluminum phthalocyanine chloride and silicon phthalocyanine, were developed to address these limitations. Both of these PSs absorbed light in the NIR region, allowing deeper tissue penetration for effective phototherapy. These compounds exhibited increased phototoxicity and reduced systemic toxicity, generating high levels of ROS to damage tumor cells more efficiently. Some second-generation agents were conjugated to tumor-targeting ligands or incorporated into nanoparticle delivery systems, improving selective targeting of cancerous tissues (Egorin et al. 1999; Mohammed et al. 2019; Obaid et al. 2015). Third-generation phthalocyanine PSs focused on enhancing tumor specificity and optimizing delivery mechanisms. Their strategies included loading phthalocyanines into nanoparticles to improve solubility, protect them from degradation, and achieve improved therapeutics. These nanoparticles release the PS in response to tumor-specific stimuli and accumulate in tumors due to the enhanced permeability and retention effect (EPR). Phthalocyanines are conjugated to target ligands like monoclonal antibodies or peptides that recognize cancer cell receptors, ensuring the PSs light-activated cytotoxic effects are confined to tumors. Metallo-phthalocyanines PSs remain crucial in these advanced strategies, and combining phthalocyanines with metal ions enhances their ROS production, stability, and light absorption properties, making them more effective in clinical applications (Muehlmann et al. 2014; Fauiod et al. 2024; Stuchinskaya et al. 2011; Sato et al. 2018).

3.3.1.4 Dye-Based and Inorganic Photosensitizers

Dye-based PSs are synthetic organic molecules widely used in therapeutic applications and imaging. They play a crucial role in PDT due to their chemical diversity, ease of synthesis, and tunable photophysical properties. These dyes include xanthene, cyanine, phenothiazinium, BODIPY, and iridium-based dyes, which have improved light absorption properties, $^1O_2$ production, and tumor selectivity. Xanthene dyes like Rose Bengal are known for their phototoxic effects in treating superficial cancers and microbial infections, but their shorter wavelength absorption limits deep tissue penetration (Guo et al. 2010; Dhaini et al. 2022;

Azevedo et al. 2024). Cyanine dyes such as indocyanine green absorb in the NIR region, allowing deeper tissue penetration, which makes them suitable for treating deep-rooted tumors despite challenges like rapid clearance and low ROS production (Meister et al. 2014; Shemesh et al. 2014; Schaafsma et al. 2011). Phenothiazinium dyes, especially methylene blue, are effective in generating $^1O_2$ and are used in antimicrobial PDT and cancer treatment with modifications that enhance their selectivity and reduce toxicity (Wainwright & Giddens 2003; Wainwright & Schmitt et al. 2017). BODIPY dyes are gaining attention for their tunable optical properties and resistance to photobleaching, which makes them promising compounds for PDT applications (Awuah and You 2012; Kamkaew et al. 2013; Chen et al. 2018). Recent advances in targeting and delivery strategies, such as conjugation with antibodies or nanoparticles, aim to improve selectivity and reduce side effects by enhancing the accumulation of PSs in diseased tissues. Nanoparticle-based systems also protect PSs from premature degradation and enable controlled release, which makes them particularly beneficial for treating solid tumors (Aksel et al. 2021; Huff et al. 2007, Behnam et al. 2018).

Inorganic PSs address some of the limitations associated with traditional organic PSs and find applications in both diagnostic and therapeutic settings. These inorganic compounds have diverse chemical structures that allow tailoring for tunable light absorption properties for specific medical uses. For example, iridium-based dyes exhibit strong absorption in the visible to NIR spectrum and are efficient in generating ROS, making them valuable for PDT through their photostability and targeting capabilities (Cai et al. 2021; Yang et al. 2021b; Zhong et al. 2023). Gold nanoparticles and titanium nanoparticles are well known for their robust absorption capabilities in the NIR region. These nanoparticles play a significant role in PTT, where their ability to absorb light leads to localized heating that effectively destroys cancer cells. Moreover, gold and titanium nanoparticles can be conjugated with traditional PSs, enhancing therapeutic outcomes by synergistically combining the effects of PTT and PDT (Aksel et al. 2021; Hirsch et al. 2003; Loo et al. 2004; Huff et al. 2007; Behnam et al. 2018; Cheng et al. 2017). This innovative approach increases the theranostics potential of these PSs, providing both diagnostic imaging and therapeutic benefits, which greatly enhances their value in the medical field.

3.4 Photosensitizers for Imaging

The PSs not only play an important role in PDT to generate ROS but also help in various imaging modalities (as shown in Table 3.1) due to their inherent photophysical properties. Their ability to show fluorescence upon light activation makes them invaluable for real-time visualization of biological processes. Furthermore, conjugating PSs with other contrast imaging agents can enhance the capabilities of imaging technologies, leading to improved diagnostic accuracy. Among their various applications, PSs are increasingly utilized for organelle-targeted imaging, enabling precise visualization and monitoring of specific cellular compartments.

The PSs serve as bright fluorophores that emit light in the visible and NIR range, making them useful for imaging in living organisms. A major advantage of these fluorescent PSs is their ability to target specific organelles, enabling detailed monitoring of cellular pathways and providing valuable insights into disease processes, especially in cancer and neurodegenerative diseases (Celli et al. 2010; Lovell et al. 2010; Reynolds 1997; Liu et al. 2019; Shuang et al. 2021; Mao et al. 2024; Kumarasamy and Sharma 2024). When these PSs accumulate in organelles such as mitochondria, lysosomes, or the endoplasmic reticulum, they facilitate real-time visualization of critical cellular activities. For instance, PSs that target mitochondria can reveal metabolic changes associated with cancer cell proliferation, while those directed at lysosomes provide insights into processes like autophagy and apoptosis (Jung et al. 2017; Guo et al. 2019; Brillo et al. 2021). Using organelle-targeted PSs for real-time tracking improves treatment precision by an image-guided theranostic approach. The fluorescence emitted by PSs aids in guiding therapeutic interventions and distinguishing malignant cells from normal ones. This capability serves as an optical biopsy, potentially reducing the need for invasive diagnostic procedures. Monitoring this fluorescence offers real-time feedback on treatment efficacy. A reduction in signal indicates successful targeting and destruction of affected cells, serving as a valuable measurement tool for adjusting therapy as necessary (Allison and Sibata 2010; Kennedy and Pottier 1992).

3.4.1 In Vitro Imaging

In vitro organelle-targeted studies using various PSs are essential for examining the unique characteristics of different cellular compartments. These studies facilitate the precise monitoring of the cellular microenvironment, including pH, oxygen levels, and ROS, which is vital for differentiating cancer cells and enhancing the efficacy of targeted therapies. Guo et al. synthesized inorganic-based PSs iridium (Ir) complexes that demonstrated significant activity in targeting mitochondria, achieving a Pearson correlation coefficient of 95.4% with Mito-Tracker Green. This high level of colocalization was confirmed in various tumor cell lines in Fig. 3.2a–c, including A54, HepG2, and RAW 264.7 with Pearson colocalization coefficients of 94.8%, 93.0%, and 90.0%, respectively. The average fluorescence lifetimes increased as oxygen concentrations decreased, ranging from 659.9 ns at 21% to 957.0 ns at 0%. Fixed HeLa cells treated with Ir1 showed significant enhancements in luminescence and lifetimes, indicating that these complexes effectively respond to low-oxygen conditions. This highlights their potential for applications in phototherapy and effective oxygen monitoring in PDT studies (Guo et al. 2019). Similarly, Jung et al. synthesized the mitochondria-targeted NIR-absorbing cryptocyanine PSs. These PSs exhibited strong mitochondrial localization in HeLa cells, with a Pearson correlation coefficient of 0.95 when compared to Mito Tracker Green. This high level of colocalization highlights the ability of the probe to specifically target and visualize mitochondria, facilitating precise real-time imaging. The lower overlap with ER-Tracker and Lyso-Tracker dye revealed the

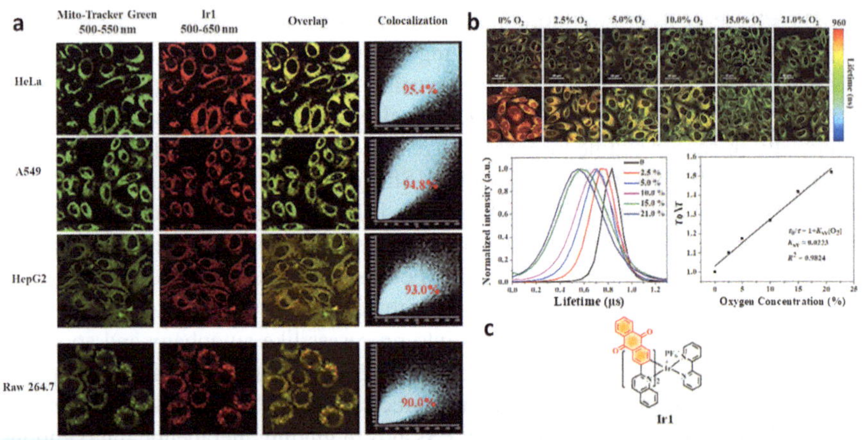

Fig. 3.2 This figure illustrates the behavior of the iridium-based photosensitizer Ir1. (**a**) Mito-Tracker images show its subcellular localization in various tumor cells. (**b**) PLIM images depict Ir1 in living and fixed HeLa cells under different O_2 concentrations, along with luminescence lifetime distributions that provide insights into Ir1's response to varying oxygen levels. (**c**) The chemical structure of the iridium-based photosensitizer is also presented. (Reproduced from Guo et al. 2019 with permission from Springer)

Pearson correlation coefficients of 0.64 and 0.44, respectively. The dual role of PS as both an imaging agent and PTT makes it a promising candidate for monitoring mitochondrial function and targeting cancer cells more precisely (Jung et al. 2017). NIR-based PSs offer excellent tissue penetration, making them highly effective for imaging in PTT. The deep tissue penetration ensures clear, precise imaging, which is crucial for image-guided therapy, enhancing both the accuracy and effectiveness of the treatment. Tu et al. synthesized NIR PSs for lysosome-targeted PDT. These PSs demonstrated specific targeting to lysosomes in cells, with Pearson correlation values of 0.90 for HeLa and 0.81 for A549 cells, confirming their precise localization. Additionally, two-photon fluorescence imaging ensures deep tissue penetration up to 210 μm in vivo, highlighting its potential for both bioimaging and as a therapeutic tool in cancer treatment (Tu et al. 2021). Lipid droplets are critical organelles that store lipids and regulate metabolism, playing a vital role in energy balance and cellular signaling, with their dysfunction linked to diseases such as obesity, diabetes, and cancer. Li et al. synthesized a BODIPY-based fluorogenic PS that demonstrates bright fluorescence in oil, exhibiting significantly enhanced brightness compared to aqueous solutions. This probe specifically stains lipid droplets in live and fixed cells, facilitating high-contrast imaging. When evaluated for their colocalization with Nile Red dye, the Pearson correlation coefficient was found to be 0.89, indicating a strong correlation in staining patterns. Compared to commercial dyes for lipid droplet imaging such as Nile Red, these PSs showed superior photostability. Furthermore, these PSs exhibit less toxicity and compatibility during cell fixation, making them a promising agent for the specific imaging of lipid droplets (Li et al. 2020; Gao et al. 2022). Yuan et al. synthesized iridium-based PSs targeting the endoplasmic reticulum (ER) in cancer cells. Imaging studies demonstrated a high degree of localization within the ER, with Pearson correlation coefficients of 0.81, indicating strong colocalization with the ER-Tracker Green dye. This precise targeting enhances the efficacy of PDT by facilitating the induction of apoptosis through ER stress mechanisms. The imaging results highlight the potential of these PS for improving therapeutic outcomes while minimizing off-target effects (Yuan et al. 2019).

3.4.2 In Vivo Imaging

In cancer diagnosis, PSs can be used to map the distribution of tumors and metastases, providing a detailed visualization that can guide surgical resection or other interventions. The ability to perform both diagnosis and treatment in one step (theranostic) streamlines the process, improving patient outcomes by reducing the need for multiple invasive procedures.

There are various types of PSs used for in vivo imaging. For instance, ICG is an NIR PSs widely used in clinical fluorescence imaging. It penetrates deeper into tissues compared to visible light, making it particularly effective for imaging vascular structures and evaluating deep tissue tumors. This capability allows for real-time feedback during surgical procedures (Schaafsma et al. 2011). Temoporfin chlorin-based PSs emit fluorescence after being excited by light at around 652 nm. Its applications extend to fluorescence-guided surgery, where it assists in identifying tumor margins, thereby improving surgical outcomes (Ronn et al. 1996; Tan et al. 2010; Yakavets et al. 2019). Multi-modal imaging using MRI is accomplished by integrating PSs with gadolinium-based contrast agents, such as diketopyrrolopyrrole-porphyrin. This combination provides complementary insights into the tissue structure through MRI while enabling real-time fluorescence imaging (Schmitt et al. 2016). Another imaging technique, such as PAI, utilizes the unique properties of PSs such as absorption characteristics, high quantum yield, and selective targeting capabilities. In PAI, PSs absorb light and convert it into heat, resulting in localized thermal expansion that generates ultrasound waves detectable for high-resolution imaging. Various types of nanomaterials serve as PSs for PAI, including metal-organic frameworks (MOFs), porphyrin-based nanomaterials such as covalent organic frameworks (COFs), and carbon-based nanomaterials that enhance the imaging performance through their unique structural and optical characteristics. Wang et al. synthesized porphyrin-derived COFs for enhanced PAI in cancer therapy. These frameworks exhibit improved NIR absorbance, facilitating effective visualization of biological tissues as shown in Fig. 3.3a–c. Their structured arrangement minimizes self-aggregation and quenching, resulting in enhanced imaging quality and therapeutic efficacy. This innovative approach streamlines the integration of PDT and PTT while employing PAI to monitor treatment responses in vivo (Wang et al. 2019).

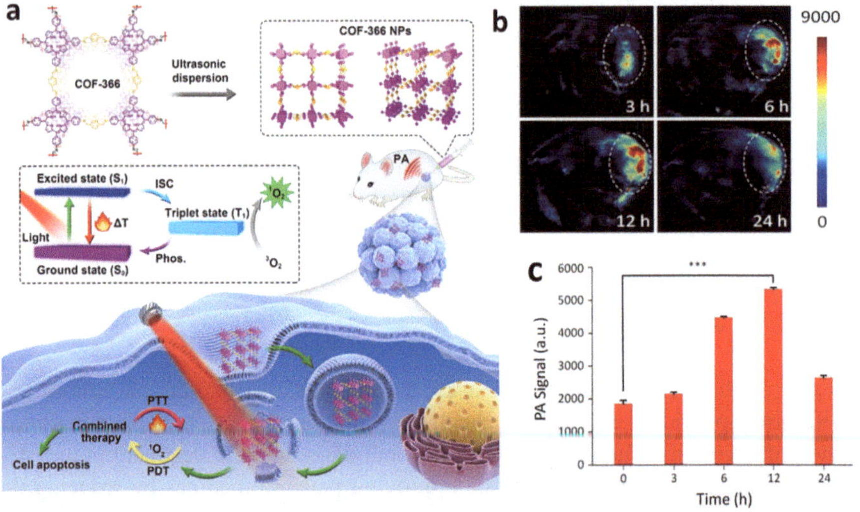

Fig. 3.3 Illustrate the following key components: (**a**) schematic illustrations of the formation of porphyrin-MOF PSs and their role in photoacoustic imaging-guided phototherapy, (**b**) photoacoustic images of tumors after intravenous injection at different time points, and (**c**) quantification of photoacoustic signals in tumors following intravenous injection at various time intervals. (Reproduced from Wang et al. 2019 with permission from Elsevier)

3.5 Challenges and Future Directions

Despite the advancements in PSs development, several challenges still hinder their full potential in clinical applications. One of the major challenges is effective tissue penetration. Although NIR light allows for deeper penetration compared to visible light, it remains inadequate for treating deeply rooted tumors effectively. Another critical challenge is achieving precise targeting specificity. While modern delivery systems have improved, ensuring that PSs selectively accumulate only in diseased tissues without affecting healthy cells is still difficult. Additionally, the translation of PSs from research to clinical use requires significant progress in production scalability, testing, and regulatory approval. To overcome these challenges, optimizing PS design is crucial. Future research should focus on enhancing light absorption properties to maximize tissue penetration, especially in the NIR range. Increasing the efficiency of ROS generation is also essential for improving therapeutic outcomes in treatments like PDT and PTT. Furthermore, PSs must be biocompatible and non-toxic in their inactive state to reduce potential side effects. Incorporating targeting moieties such as antibodies or peptides can enhance their ability to selectively accumulate in target tissues, particularly in cancer therapies. Solubility and pharmacokinetics must also be optimized to ensure efficient delivery and clearance from the body. Looking ahead, future directions in this field include designing PSs with improved quantum yields, biocompatibility, and multifunctional capabilities. The integration of artificial intelligence and machine learning could play a transformative role in refining PSs development, from molecular design to clinical application, paving the way for more personalized and effective therapies.

Acknowledgments The authors wish to acknowledge research support from IIT Jammu and DST-SERB. V.S. would like to acknowledge DST-SERB (Project SRG/2022/001440) and IIT Jammu Seed Grant for financial assistance. S.P. and K.W. would like to acknowledge IIT Jammu for the fellowship. L.S. would like to acknowledge IIT Jammu for the Internship.

Author's Contribution
S.K. performed the data collection and literature review. S.K., K.W., L.S., and V.S. co-wrote the manuscript, and V.S. conceptualized and supervised this study.

Conflict of Interest The authors declare no conflict of interest.

References

Akhlynina TV, Jans DA, Rosenkranz AA, Statsyuk NV, Balashova IY, Toth G, Pavo I, Rubin AB, Sobolev AS (1997) Nuclear targeting of chlorin e6 enhances its photosensitizing activity. J Biol Chem 272:20328–20331

Aksel M, Kesmez Ö, Yavaş A, Bilgin MD (2021) Titaniumdioxide mediated sonophotodynamic therapy against prostate cancer. J Photochem Photobiol B Biol 225:112333

Allison RR, Sibata CH (2010) Oncologic photodynamic therapy photosensitizers: a clinical review. Photodiagn Photodyn Ther 7:61–75

Andersson-Engels S, Ankerst J, Johansson J, Svanberg K, Svanberg S (1993) Laser-induced fluorescence in malignant and normal tissue of rats injected with benzoporphyrin derivative. Photochem Photobiol 57:978–983

Awuah SG, You Y (2012) Boron dipyrromethene (BODIPY)-based photosensitizers for photodynamic therapy. RSC Adv 2:11169

Azarian M, Farani MR, Zare I, Imani M, Kumar K, Huh YS, Mostafavi E (2024) Chapter 15: Functionalized porphysomes and porphyrin-based nanomaterials for cancer therapy. In: Barabadi H, Mostafavi E, Mustansar Hussain C (eds) Functionalized nanomaterials for cancer research. Academic Press, pp 329–344

Azevedo AMO, Seabra CL, Moniz T, Nunes C, Ayala CE, Rangel M, Reis S, Santos JLM, Warner IM, Saraiva MLMFS (2024) An ion metathesis strategy for overcoming therapeutic limitations of rose bengal. Dyes Pigments 225:112050

Bartusik-Aebisher D, Żołyniak A, Barnaś E, Machorowska-Pieniążek A, Oleś P, Kawczyk-Krupka-A, Aebisher D (2022) The use of photodynamic therapy in the treatment of brain tumors—a review of the literature. Molecules 27:6847

Behnam MA, Emami F, Sobhani Z, Dehghanian AR (2018) The application of titanium dioxide (TiO2) nanoparticles in the photo-thermal therapy of melanoma cancer model. Iran J Basic Med Sci. https://doi.org/10.22038/ijbms.2018.30284.7304

Brasseur N, Ouellet R, Madeleine CL, van Lier JE (1999) Water-soluble aluminium phthalocyanine–polymer conjugates for PDT: photodynamic activities and pharmacokinetics in tumour-bearing mice. Br J Cancer 80:1533–1541

Brillo V, Chieregato L, Leanza L, Muccioli S, Costa R (2021) Mitochondrial dynamics, ros, and cell signaling: a blended overview. Life 11. https://doi.org/10.3390/life11040332

Cai X, Wang K-N, Ma W, Yang Y, Chen G, Fu H, Cui C, Yu Z, Wang X (2021) Multifunctional AIE iridium (III) photosensitizer nanoparticles for two-photon-activated imaging and mitochondria targeting photodynamic therapy. J Nanobiotechnol 19:254

Celli JP, Spring BQ, Rizvi I, Evans CL, Samkoe KS, Verma S, Pogue BW, Hasan T (2010) Imaging and photodynamic therapy: mechanisms, monitoring, and optimization. Chem Rev 110:2795–2838

Chen D, Zhang J, Tang Y, Huang X, Shao J, Si W, Ji J, Zhang Q, Huang W, Dong X (2018) A tumor-mitochondria dual targeted aza-BODIPY-based nanotheranostic agent for multimodal imaging-guided phototherapy. J Mater Chem B 6:4522–4530

Cheng Y, Chang Y, Feng Y, Liu N, Sun X, Feng Y, Li X, Zhang H (2017) Simulated sunlight-mediated photodynamic therapy for melanoma skin cancer by titanium-dioxide-nanoparticle–gold-nanocluster–graphene heterogeneous nanocomposites. Small 13:1603935

Colditz MJ, Jeffree RL (2012) Aminolevulinic acid (ALA)–protoporphyrin IX fluorescence guided tumour resection. Part 1: clinical, radiological and pathological studies. J Clin Neurosci 19: 1471–1474

Dale CL, Hill SJ, Kellam B (2012) New potent, short-linker BODIPY-630/650™ labelled fluorescent adenosine receptor agonists. Med Chem Commun 3:333–338

Dhaini B, Wagner L, Moinard M, Daouk J, Arnoux P, Schohn H, Schneller P, Acherar S, Hamieh T, Frochot C (2022) Importance of Rose Bengal loaded with nanoparticles for anti-cancer photodynamic therapy. Pharmaceuticals 15:1093

Ding X, Han B-H (2015) Metallophthalocyanine-based conjugated microporous polymers as highly efficient photosensitizers for singlet oxygen generation. Angew Chem Int Ed 54:6536–6539

Egorin MJ, Zuhowski EG, Sentz DL, Dobson JM, Callery PS, Eiseman JL (1999) Plasma pharmacokinetics and tissue distribution in CD 2 F 1 mice of Pc4 (NSC 676418), a silicone phthalocyanine photodynamic sensitizing agent. Cancer Chemother Pharmacol 44:283–294

Fauiod OG, Fadel M, El-Hussein A, Fadeel DA (2024) Aluminum phthalocyanine tetrasulfonate conjugated to surface-modified iron oxide nanoparticles as a magnetic targeting platform for photodynamic therapy of Ehrlich tumor-bearing mice. Photodiagn Photodyn Ther 50:104356

Gao D, Zhang Y, Wu K, Min H, Wei D, Sun J, Yang H, Fan H (2022) One-step synthesis of ultrabright amphiphilic carbon dots for rapid and precise tracking lipid droplets dynamics in biosystems. Biosens Bioelectron 200:113928

Gilyadova A, Ishchenko A, Shiryaev A, Alekseeva P, Efendiev K, Karpova R, Loshchenov M, Loschenov V, Reshetov I (2022) Phototheranostics of cervical neoplasms with Chlorin e6 photosensitizer. Cancers 14:211

Glidden MD, Celli JP, Massodi I, Rizvi I, Pogue BW, Hasan T (2012) Image-based quantification of benzoporphyrin derivative uptake, localization, and photobleaching in 3D tumor models, for optimization of PDT parameters. Theranostics 2:827–839

Gümrükçü Köse G, Keser Karaoğlan G (2023) Synthesis of a novel axially substituted silicon phthalocyanine sensitizer for efficient singlet oxygen generation by comparing PDT and SPDT studies. Chem Phys 565:111737

Guo Y, Rogelj S, Zhang P (2010) Rose Bengal-decorated silica nanoparticles as photosensitizers for inactivation of gram-positive bacteria. Nanotechnology 21:065102

Guo S, Han M, Chen R, Zhuang Y, Zou L, Liu S, Huang W, Zhao Q (2019) Mitochondria-localized iridium(III) complexes with anthraquinone groups as effective photosensitizers for photodynamic therapy under hypoxia. Sci China Chem 62:1639–1648

Hak A, Ali MS, Sankaranarayanan SA, Shinde VR, Rengan AK (2023) Chlorin e6: a promising photosensitizer in photo-based cancer nanomedicine. ACS Appl Bio Mater 6:349–364

Hamblin MR, Miller JL, Rizvi I, Ortel B, Maytin EV, Hasan T (2001) Pegylation of a Chlorine6 polymer conjugate increases tumor targeting of photosensitizer. Cancer Res 61:7155–7162

Heinemann F, Karges J, Gasser G (2017) Critical overview of the use of Ru(II) polypyridyl complexes as photosensitizers in one-photon and two-photon photodynamic therapy. Acc Chem Res 50:2727–2736

Hester SC, Kuriakose M, Nguyen CD, Mallidi S (2020) Role of ultrasound and photoacoustic imaging in photodynamic therapy for cancer. Photochem Photobiol 96:260–279

Hirsch LR, Stafford RJ, Bankson JA, Sershen SR, Rivera B, Price RE, Hazle JD, Halas NJ, West JL (2003) Nanoshell-mediated near-infrared thermal therapy of tumors under magnetic resonance guidance. Proc Natl Acad Sci USA 100:13549–13554

Hua X-W, Bao Y-W, Chen Z, Wu F-G (2017) Carbon quantum dots with intrinsic mitochondrial targeting ability for mitochondria-based theranostics. Nanoscale 9:10948–10960

Huff TB, Tong L, Zhao Y, Hansen MN, Cheng J-X, Wei A (2007) Hyperthermic effects of gold nanorods on tumor cells. Nanomedicine 2:125–132

Ibarra LE, Porcal GV, Macor LP, Ponzio RA, Spada RM, Lorente C, Chesta CA, Rivarola VA, Palacios RE (2018) Metallated porphyrin-doped conjugated polymer nanoparticles for efficient photodynamic therapy of brain and colorectal tumor cells. Nanomedicine 13:605–624

Ishii K (2012) Functional singlet oxygen generators based on phthalocyanines. Coord Chem Rev 256:1556–1568

Jeising S, Geerling G, Guthoff R, Hänggi D, Sabel M, Rapp M, Nickel A-C (2022) In-vitro use of verteporfin for photodynamic therapy in glioblastoma. Photodiagn Photodyn Ther 40:103049

Jung HS, Lee J-H, Kim K, Koo S, Verwilst P, Sessler JL, Kang C, Kim JS (2017) A mitochondria-targeted cryptocyanine-based photothermogenic photosensitizer. J Am Chem Soc 139:9972–9978

Kamkaew A, Lim SH, Lee HB, Kiew LV, Chung LY, Burgess K (2013) BODIPY dyes in photodynamic therapy. Chem Soc Rev 42:77–88

Kennedy JC, Pottier RH (1992) New trends in photobiology: endogenous protoporphyrin IX, a clinically useful photosensitizer for photodynamic therapy. J Photochem Photobiol B Biol 14: 275–292

Krammer B, Überriegler K (1996) In-vitro investigation of ALA-induced protopoyphyrin IX. J Photochem Photobiol B Biol 36:121–126

Kumarasamy S, Sharma V (2024) Unveiling cellular secrets: illuminating carbon dot lighthouses for improved mitochondrial exploration. Chem Mater 37. https://doi.org/10.1021/acs.chemmater.4c01481

Kwiatkowski S, Knap B, Przystupski D, Saczko J, Kędzierska E, Knap-Czop K, Kotlińska J, Michel O, Kotowski K, Kulbacka J (2018) Photodynamic therapy—mechanisms, photosensitizers and combinations. Biomed Pharmacother 106:1098–1107

Lam M, Hsia AH, Liu Y et al (2011) Successful cutaneous delivery of the photosensitizer silicon phthalocyanine 4 for photodynamic therapy. Clin Exp Dermatol 36:645–651

Leung SCH, Lo P-C, Ng DKP, Liu W-K, Fung K-P, Fong W-P (2008) Photodynamic activity of BAM-SiPc, an unsymmetrical bisamino silicon(IV) phthalocyanine, in tumour-bearing nude mice. Br J Pharmacol 154:4–12

Levy JG, Chan AH, Andrew Strong H (1996) Clinical status of benzoporphyrin derivative, pp 86–95

Li X, Zheng B-D, Peng X-H, Li S-Z, Ying J-W, Zhao Y, Huang J-D, Yoon J (2019) Phthalocyanines as medicinal photosensitizers: developments in the last five years. Coord Chem Rev 379:147–160

Li G, Li J, Otsuka Y, Zhang S, Takahashi M, Yamada K (2020) A BODIPY-based fluorogenic probe for specific imaging of lipid droplets. Materials 13:677

Lin SC, Lin CP, Feld JR, Duker JS, Puliafito CA (1994) The photodynamic occlusion of choroidal vessels using benzoporphyrin derivative. Curr Eye Res 13:513–522

Liu Z, Zou H, Zhao Z, Zhang P, Shan G-G, Kwok RTK, Lam JWY, Zheng L, Tang BZ (2019) Tuning organelle specificity and photodynamic therapy efficiency by molecular function design. ACS Nano 13:11283–11293

Liu J, Yang X, Wu S, Gong P, Pan F, Zhang P, Lee C-S, Liu C, Wong KM-C (2024) Iridium(III) complexes decorated with silicane-modified rhodamine: near-infrared light-initiated photosensitizers for efficient deep-tissue penetration photodynamic therapy. J Mater Chem B 12:3710–3718

Loo C, Lin A, Hirsch L, Lee M-H, Barton J, Halas N, West J, Drezek R (2004) Nanoshell-enabled photonics-based imaging and therapy of cancer. Technol Cancer Res Treat 3:33–40

Lovell JF, Liu TWB, Chen J, Zheng G (2010) Activatable photosensitizers for imaging and therapy. Chem Rev 110:2839–2857

Mao Q, Meng Y, Feng Y, Li H, Ma T (2024) Organelle imaging with carbon dots: strategies, challenges, and perspectives. Inorg Chem Front 11:713–734

McKeown NB (2003) Phthalocyanines. In: McCleverty JA, Meyer TJ (eds) Comprehensive coordination chemistry II. Pergamon, Oxford, pp 507–514

Meister J, Hopp M, Schäfers J, Verbeek J, Kraus D, Frentzen M (2014) Indocyanine green (ICG) as a new adjuvant for the antimicrobial photo-dynamic therapy (aPDT) in dentistry. In: SPIE, pp 205–214

Mellish KJ, Brown SB (2001) Verteporfin: a milestone in ophthalmology and photodynamic therapy. Expert Opin Pharmacother 2:351–361

Miller JD, Baron ED, Scull H, Hsia A, Berlin JC, McCormick T, Colussi V, Kenney ME, Cooper KD, Oleinick NL (2007) Photodynamic therapy with the phthalocyanine photosensitizer Pc 4: the case experience with preclinical mechanistic and early clinical–translational studies. Toxicol Appl Pharmacol 224:290–299

Mohammed I, Oluwole DO, Nemakal M, Sannegowda LK, Nyokong T (2019) Investigation of novel substituted zinc and aluminium phthalocyanines for photodynamic therapy of epithelial breast cancer. Dyes Pigments 170:107592

Moon Y-H, Kwon S-M, Kim H-J et al (2009) Efficient preparation of highly pure chlorin e6 and its photodynamic anti-cancer activity in a rat tumor model. Oncol Rep 22:1085. https://doi.org/10.3892/or_00000540

Muehlmann LA, Ma BC, Longo JPF, de Fátima Menezes Almeida Santos M, Azevedo RB (2014) Aluminum–phthalocyanine chloride associated to poly(methyl vinyl ether-co-maleic anhydride) nanoparticles as a new third-generation photosensitizer for anticancer photodynamic therapy. Int J Nanomedicine 9:1199–1213

Nagahara A, Mitani A, Fukuda M et al (2013) Antimicrobial photodynamic therapy using a diode laser with a potential new photosensitizer, indocyanine green-loaded nanospheres, may be effective for the clearance of Porphyromonas gingivalis. J Periodontal Res 48:591–599

Nyokong T (2011) Desired properties of new phthalocyanines for photodynamic therapy. Pure Appl Chem 83:1763–1779

Obaid G, Chambrier I, Cook MJ, Russell DA (2015) Cancer targeting with biomolecules: a comparative study of photodynamic therapy efficacy using antibody or lectin conjugated phthalocyanine-PEG gold nanoparticles. Photochem Photobiol Sci 14:737–747

Patyal R, Warjurkar K, Sharma V (2024) Advancements in tumor diagnostics through carbon dot-assisted photoacoustic imaging. Adv Opt Mater 13:2402343

Raab O (1900) Uber die Wirkung fluoreszierender Stoffe auf Infusorien. Z Biol 39:524–546

Rabiee N, Yaraki MT, Garakani SM et al (2020) Recent advances in porphyrin-based nanocomposites for effective targeted imaging and therapy. Biomaterials 232:119707

Reynolds T (1997) Photodynamic therapy expands its horizons. JNCI J Natl Cancer Inst 89:112–114

Richter A, Sternberg E, Waterfield E, Dolphin D, Levy JG (1989) Characterization of benzoporphyrin derivative, a new photosensitizer, pp 132–138

Roguin LP, Chiarante N, García Vior MC, Marino J (2019) Zinc(II) phthalocyanines as photosensitizers for antitumor photodynamic therapy. Int J Biochem Cell Biol 114:105575

Ronn AM, Nouri M, Lofgren LA, Steinberg BM, Westerborn A, Windahl T, Shikowitz MJ, Abramson AL (1996) Human tissue levels and plasma pharmacokinetics of temoporfin (Foscan®, mTHPC). Lasers Med Sci 11:267–272

Rosenthal I, Ben-Hur E (1995) Role of oxygen in the phototoxicity of phthalocyanines. Int J Radiat Biol 67:85–91

Sato K, Ando K, Okuyama S et al (2018) Photoinduced ligand release from a silicon phthalocyanine dye conjugated with monoclonal antibodies: a mechanism of cancer cell cytotoxicity after near-infrared photoimmunotherapy. ACS Cent Sci 4:1559–1569

Schaafsma BE, Mieog JSD, Hutteman M, Van Der Vorst JR, Kuppen PJK, Löwik CWGM, Frangioni JV, Van De Velde CJH, Vahrmeijer AL (2011) The clinical use of indocyanine green as a near-infrared fluorescent contrast agent for image-guided oncologic surgery. J Surg Oncol 104:323–332

Schmitt J, Heitz V, Sour A, Bolze F, Kessler P, Flamigni L, Ventura B, Bonnet CS, Tóth É (2016) A theranostic agent combining a two-photon-absorbing photosensitizer for photodynamic therapy and a gadolinium(III) complex for MRI detection. Chem Eur J 22:2775–2786

Shemesh CS, Hardy CW, Yu DS, Fernandez B, Zhang H (2014) Indocyanine green loaded liposome nanocarriers for photodynamic therapy using human triple negative breast cancer cells. Photodiagn Photodyn Ther 11:193–203

Shuang E, He C, Wang JH, Mao Q, Chen X (2021) Tunable organelle imaging by rational design of carbon dots and utilization of uptake pathways. ACS Nano 15:14465–14474

Spikes JD (1991) The origin and meaning of the term "photodynamic" (as used in "photodynamic therapy", for example). J Photochem Photobiol B Biol 9:369–371

Spikes JD, Bommer JC (1993a) Photosensitizing properties of mono-l-aspartyl chlorin e6 (NPe6): a candidate sensitizer for the photodynamic therapy of tumors. J Photochem Photobiol B Biol 17: 135–143

Spikes JD, Bommer JC (1993b) Photobleaching of mono-L-aspartyl chlorin e (NPe6): a candidate sensitizer for the photodynamic therapy of tumors. Photochem Photobiol 58:346–350

Stuchinskaya T, Moreno M, Cook MJ, Edwards DR, Russell DA (2011) Targeted photodynamic therapy of breast cancer cells using antibody-phthalocyanine-gold nanoparticle conjugates. Photochem Photobiol Sci 10:822–831

Sun W, Zhao X, Fan J, Du J, Peng X (2019) Boron dipyrromethene nano-photosensitizers for anticancer phototherapies. Small 15:1804927

Tan IB, Dolivet G, Ceruse P, Poorten VV, Roest G, Rauschning W (2010) Temoporfin-mediated photodynamic therapy in patients with advanced, incurable head and neck cancer: a multicenter study. Head Neck 32:1597–1604

Tang Q, Xiao W, Huang C, Si W, Shao J, Huang W, Chen P, Zhang Q, Dong X (2017) pH-Triggered and enhanced simultaneous photodynamic and photothermal therapy guided by photoacoustic and photothermal imaging. Chem Mater 29:5216–5224

Tian M-E, Liu F-W, Qian J-P, Ji Q, Feng Y-Q (1993) Endoscopic photodynamic therapy with hematoporphyrin derivative in the treatment of malignant tumors: report of 120 cases, pp 108–113

Tu Y, Xia W, Wu X, Wang L (2021) A lysosome-targeted near-infrared photosensitizer for photodynamic therapy and two-photon fluorescence imaging. Org Biomol Chem 19(27): 6098–6107

Turksoy A, Yildiz D, Akkaya EU (2019) Photosensitization and controlled photosensitization with BODIPY dyes. Coord Chem Rev 379:47–64

van Lier JE, Spikes JD (2007) The chemistry, photophysics and photosensitizing properties of phthalocyanines. In: Ciba foundation symposium 146—photosensitizing compounds: their chemistry, biology and clinical use, pp 17–39

Wainwright M, Giddens RM (2003) Phenothiazinium photosensitisers: choices in synthesis and application. Dyes Pigments 57:245–257

Wainwright M, McLean A (2017) Rational design of phenothiazinium derivatives and photoantimicrobial drug discovery. Dyes Pigments 136:590–600

Wang Z-W, Su D, Li X-Q, Cao J-J, Yang D-C, Liu J-Y (2018) A H2O2-responsive boron dipyrromethene-based photosensitizer for imaging-guided photodynamic therapy. Molecules 24:32

Wang D, Zhang Z, Lin L, Liu F, Wang Y, Guo Z, Li Y, Tian H, Chen X (2019) Porphyrin-based covalent organic framework nanoparticles for photoacoustic imaging-guided photodynamic and photothermal combination cancer therapy. Biomaterials 223:119459

Wen H, Wu Q, Li C, Sun T, Xie Z (2022) 4,4-Difluoro-4-bora-3a,4a-diaza-s-indacene (BDPI)-triphenylphosphine nanoparticles as a photodynamic antibacterial agent. ACS Appl Nano Mater 5:1500–1507

Whitacre CM, Feyes DK, Satoh T, Grossmann J, Mulvihill JW, Mukhtar H, Oleinick NL (2000) Photodynamic therapy with the phthalocyanine photosensitizer Pc 4 of SW480 human colon cancer xenografts in athymic mice. Clin Cancer Res 6:2021–2027

Yakavets I, Millard M, Zorin V, Lassalle H-P, Bezdetnaya L (2019) Current state of the nanoscale delivery systems for temoporfin-based photodynamic therapy: advanced delivery strategies. J Control Release 304:268–287

Yan F, Wu H, Liu H, Deng Z, Liu H, Duan W, Liu X, Zheng H (2016) Molecular imaging-guided photothermal/photodynamic therapy against tumor by iRGD-modified indocyanine green nanoparticles. J Control Release 224:217–228

Yang M, Wang X, Pu F, Liu Y, Guo J, Chang S, Sun G, Peng Y (2021a) Engineered exosomes-based photothermal therapy with MRI/CT imaging guidance enhances anticancer efficacy through deep tumor nucleus penetration. Pharmaceutics 13:1593

Yang K, Zhou Y, Wang Y, Zhao S, Wu X, Peng X, Huang L, Jiang L, Lan M, Yi X-Y (2021b) An iridium complex as an AIE active photosensitizer for image-guided photodynamic therapy. Chem Asian J 16:1780–1785

Yuan B, Liu J, Guan R, Jin C, Ji L, Chao H (2019) Endoplasmic reticulum targeted cyclometalated iridium(III) complexes as efficient photodynamic therapy photosensitizers. Dalton Trans 48: 6408–6415

Zhong M, He J, Zhang B, Liu Q, Fang J (2023) Mitochondria-targeted iridium-based photosensitizers enhancing photodynamic therapy effect by disturbing cellular redox balance. Free Radic Biol Med 195:121–131

Ziemssen F, Heimann H (2012) Evaluation of verteporfin pharmakokinetics—redefining the need of photosensitizers in ophthalmology. Expert Opin Drug Metab Toxicol 8:1023–1041

Next-Generation of Multimode Imaging and Synergistic Therapeutics for Solid Tumors

4

Pavithra Kurungottu, A. S. Lakshmi, Agnes Sebastian, Aleena Sabu Koshi, Ashwin T. Shaji, S. Karthik, Anushka Balakrishnan, and Rajendra Kurapati

Abstract

Cancer has been the second leading cause of mortality, which prompted researchers to develop advanced techniques in diagnostic imaging to manage the diagnostic and therapeutic treatment of cancer patients. Various nanoimaging and therapeutics approaches have been applied to diagnose and treat cancer at an early stage. Especially, silica, gold, and lipid nanoparticles-based medicines have gained tremendous attention for localized diagnosis and the treatment of cancer. However, reproducibility, batch-to-batch variability, scalability, colloidal stability, storage stability, etc., are still being questioned. Herein, we have discussed various modalities in nanoparticle-based imaging and synergistic therapeutics for cancer diagnosis and therapy.

Keywords

Multimode imaging · Cancer therapy · Synergistic · Solid tumors · Nanomedicine

4.1 Introduction

In cancer, a second leading cause of mortality, especially solid tumors, is defined as abnormal masses of tissues that can be cancerous (malignant) or noncancerous. In the case of malignant solid tumors, the tumor microenvironment (TME) is composed of various tissues and thus shows significant heterogeneity, such as the pH, various

P. Kurungottu · A. S. Lakshmi · A. Sebastian · A. S. Koshi · A. T. Shaji · S. Karthik
A. Balakrishnan · R. Kurapati (✉)
School of Chemistry, Indian Institute of Science Education and Research Thiruvananthapuram, Thiruvananthapuram, India
e-mail: rkurapati@iisertvm.ac.in

R. Prasad, M. Mohan Yallapu (eds.), *Advancements in Cancer Theranostics*,
https://doi.org/10.1007/978-981-95-0938-6_4

nutrient levels, and oxygen (Yin et al. 2024). The heterogeneity and the flexibility among cancer tissues stand as the main obstacles in cancer treatments (Miranda-Gonçalves et al. 2020). Solid tumors were found in various parts of the body and are divided mainly into three categories, such as carcinoma, sarcoma, and melanoma. In general, solid tumors are formed by embedding the cancer cells in the network of nonmalignant cells and extracellular matrix (tumor stroma) (Schietinger et al. 2013). The different forms of solid tumors include brain cancer, colorectal cancer, prostate cancer, breast cancer, etc. Although surgery is the main resection for the removal of solid tumors, inaccurate detection of solid tumors can lead to the incomplete removal of the cancer cells or excessive removal of healthy tissue (Faucher et al. 2023). As much as adequate treatment is important in cancer, the earlier diagnosis of the disease is also important. In oncology, it is crucial to use imaging to precisely characterize the tumor and its response to treatment and detect metastasis. Therefore, accurate detection and imaging of solid tumors are critical in limiting the excessive excision of healthy tissues or incomplete removal of the cancer tissue, which determines the success of the tumor treatment. The incomplete removal of tumors is followed by chemotherapy, radiation, etc.

Conventional imaging techniques, like ultrasound scanning (US), X-ray computed tomography (CT), molecular resonance imaging (MRI), positron emission tomography (PET), single photon emission computed tomography (SPECT, involves nuclear medicine imaging), photoacoustic imaging (PA), assist in a detailed understanding of tumor size and morphology (Soliman et al. 2020). Further, the increased cases of malignant tumors have led to an increase in demand for standardized imaging tests for patient assessment, which can detect even the slightest tumor growth in the initial stage of cancer. As the risk factors of cancer treatment are increasing day by day, imaging technologies, which greatly leverage the early detection and treatment of cancer, are also getting advanced. In treating cancer, it is crucial to comprehend how to best combine these cutting-edge approaches with tried-and-true conventional ways. Multimodal or multiplexed imaging can be defined as integrating various imaging techniques used to provide characteristic diagnosis about the neoplasm and analysis of the diagnosis. Combining various multimodal imaging techniques, like MRI, CT, PET, and ultrasound, can comprehensively provide better insights into precision medicine (Tempany et al. 2015). For example, a combination of optical, magnetic, and radioactive reporters to be detected by SPECT, MRI, and PET, as shown in Fig. 4.1. The goal of multimodal imaging is to enable early detection and localization of the tumor using medical imaging technology and an image fusion system that provides image information of the tumor from different angles and dimensions, enabling precise diagnosis (Papathanassiou and Liehn 2008). In most cancer treatments, minimally invasive tumors are surgically removed, guided by precise imaging and target-specific drug therapies. Multimodal imaging has become an integral part of precision oncology.

The journey of medical imaging started in 1895 when Wilhelm Conrad Roentgen discovered the X-ray, and its immense applications in visualizing internal structures were revolutionary (Tubiana 1996). However, X-ray imaging is limited due to poor contrast, especially for soft tissues. Later, we witnessed the introduction of

a)

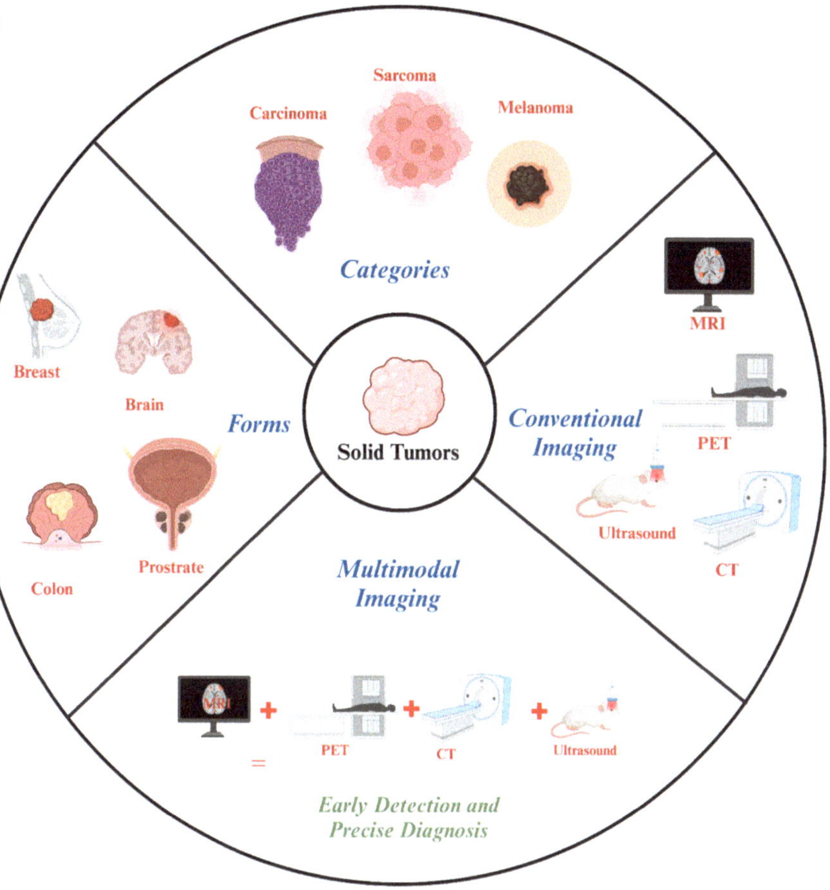

b)

	PET	MRI	US	CT	SPECT	Optical
Strengths	Unlimited penetration depth and high sensitivity	High spatial resolution and unlimited tissue penetration	High sensitivity, real time and low cost	High spatial resolution and penetration depth	Unlimited penetration depth and high sensitivity	High sensitivity, real time and cost effective
Limitations	Poor spatial resolution and radiation risk	Low sensibility, high cost and long imaging time	Poor resolution and operator dependent analysis	Radiation risk	Poor spatial resolution and radiation risk	Low spatial resolution and small penetration depth

Fig. 4.1 (**a**) Scheme illustrating various categories and forms of solid tumors and imaging techniques for solid tumors with an emphasis on multimodal imaging. (**b**) Table listing out the strength and limitations of conventional imaging techniques

ultrasound imaging, followed by CT imaging, which allowed better visualization of tumor morphology, location, and other structures. Later, multimode imaging was brought by combining CT imaging with nuclear medicine detectors (MRI). On the other hand, synergistic therapy (or combination therapy) is the cooperation of several treatments combined into a single therapy with improved therapeutic effects compared to individual treatment (Cazares-Cortes et al. 2019). Nowadays, we have synergistic therapies like radio-sensitizing anticancer drugs that kill not only the cancer cells with chemotherapy but also make the cells more sensitive to ionizing radiation for radiation therapy (Fan et al. 2017). Various combinations of therapies like chemotherapy, radiation therapy, photothermal therapy, immunotherapy, and gene therapy have demonstrated high anticancer efficacy in laboratory and preclinical studies, with a promising future for synergistic therapies. The rationale behind introducing synergistic therapies lies in the complexity of disease models like cancer, as it involves multiple pathways and shows resistance mechanisms against several drugs and our immune cells. Targeting cancer cells can be made easier and more effective by employing synergistic therapies, as they can disrupt cancer proliferation and survival. For example, adjuvant (helper) radiation therapy, when given before surgery for rectal cancer, reduces the risk of recurrence, thus likely curing the patient. Multimodal imaging can detect early signs of resistance and allow synergistic therapy to counteract the resistance mechanisms. It helps in personalized medicine and designing the treatment specific to the patient's condition. Multimodal imaging, along with synergistic therapeutics, continues to advance effective cancer treatment strategies with better precision and efficacy.

4.2 Advances in Imaging Modalities

4.2.1 High-Resolution Imaging Techniques

As discussed in the previous section, there are different high-resolution techniques, such as MRI, CT, PET, SPECT, and PA (see Fig. 4.1). However, CT and MRI are considered very efficient and noninvasive methods for tumor imaging and early-stage detection. CT imaging involves collections of X-rays produced from different angles of the subject being analyzed and the generation of topographical images (Hussain et al. 2022). Further advancements in CT, three areas such as isotropic volume coverage, high temporal resolution, and spectral information for material classification and differentiation, are mainly being improved for better imaging (García-Figueiras et al. 2021). Whereas in MRI, the interaction between the applied radiofrequency pulses and the magnetic moment of hydrogen atoms is transmitted for imaging purposes. In the case of head and neck cancer, a heterogeneous group of cancers, MRI and CT are widely applied for diagnosis and monitoring, as given in Fig. 4.2. However, these two methods mostly rely on the human interpretation of images, which is unpredictable and subjective. To overcome this issue, deep learning, which is a subfield of machine learning, was introduced. Deep learning uses multi-layered artificial neural networks, or deep neural networks, to

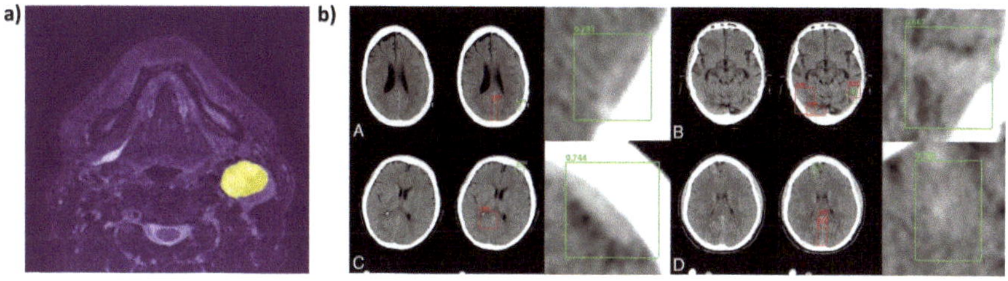

Fig. 4.2 (**a**) MRI image of neck lymph node metastasis (yellow color) and (**b**) CT image of the head to assess the hemorrhage evaluation. (Adapted with permission from Illimoottil and Ginat 2023 and Chang et al. 2018)

automatically build hierarchical representations of input data. Incorporation of deep learning techniques into imaging modalities upgrades the efficiency and accuracy of the data and also makes it compatible (Illimoottil and Ginat 2023; LeCun et al. 2015). Advanced MRI techniques, containing diffusion, perfusion, spectroscopy, tractography, and functional MRI, are significantly applied for brain tumor imaging and evaluation. By examining the Brownian mobility of water molecules, diffusion-weighed imaging (DWI) makes it possible to evaluate postoperative damage, tumor hypoxic areas, etc. Susceptibility-weighed imaging (SWI) techniques and other high-resolution 3-D T2* gradient echo sequences are extremely susceptible to magnetic susceptibility effects caused by mineralization or blood products. With the help of this approach, internal vascular architecture and tumor hemorrhage can be depicted (Villanueva-Meyer et al. 2017; Petrella and Provenzale 2000). Dynamic susceptibility contrast (DSC), dynamic contrast-enhanced (DCE), and arterial spin labeling (ASL) are some of the cerebral perfusion techniques that are being applied nowadays (Villanueva-Meyer et al. 2017). The work done by Chang and co-workers utilized a hybrid 3D/2D convolutional neural network (CNN) in head CT scans to assess hemorrhages (Chang et al. 2018). Further, Xie et al. demonstrated a deep learning multi-view knowledge-based collaborative (MV-KBC) model that can distinguish between malignant and benign nodules using the available data. The model can understand its features by breaking down a 3-D lung nodule into nine fixed images. For every viewpoint, a knowledge-based collaborative (KBC) sub-model was created. These three different kinds of image patches are intended to refine three pretrained ResNet-50 networks that, in turn, describe the features of the nodules (Xie et al. 2019).

4.2.2 Molecular and Functional Imaging

Other commonly used nuclear imaging techniques for tumor detection are PET and SPECT. In nuclear imaging, a radiotracer will be administered to the body, and the 3-D images of the targeted part will be obtained (Zhang et al. 2020). Radio tracers such as ^{18}F-dihydroxyphenylalanine (^{18}F-DOPA), ^{18}F-fluorodeoxyglucose

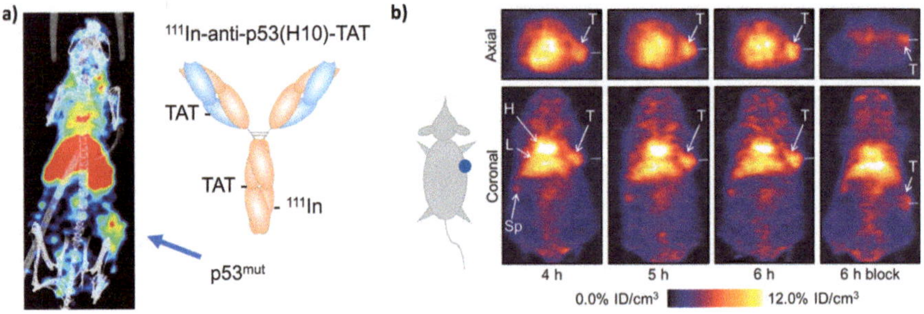

Fig. 4.3 (**a**) In vivo SPECT image showing the p53 inside the tumour with help of radiolabelled TAT-modifed antibodies and (**b**) PET images of a mice with SK-OV-3 tumours on the right flank. (Adapted with permission from 2019 Alakonya et al. 2024 and Patra et al. 2019)

(^{18}F-FDG), and ^{68}Ga-labeled somatostatin analogs (^{68}Ga-DOTA-peptides) are found to give higher spatial resolution and sensitivity in detecting medullary thyroid carcinoma, as shown in Fig. 4.3. Because of the higher half-life shown by the ^{18}F nuclei, it is widely used for clinical applications. Furthermore, the combination of different monoclonal antibodies (mAb) and radionuclides significantly contributed to PET imaging as immune-PET tracers. Patra et al. reported the photochemical conjugation of mAb with ^{68}Ga via an aryl azide-functionalized chelate. This one-pot synthesis method helped retain the parent protein's properties rather than going through complex processes, and showed better in vivo PET imaging ability (Patra et al. 2019). Further, Alakonya et al. reported a noninvasive in vivo visualization of p53 in tumor tissue using SPECT imaging in mouse models. To enable SPECT imaging, antip53 monoclonal antibodies were coupled to a metal ion chelator and the cell-penetrating transactivator of the transcription (TAT) peptide. They were then radiolabeled with ^{111}In. Compared to non-p53-specific cells, conjugates were retained longer in cells overexpressing p53 and showed enhanced uptake in in vivo SPECT imaging (Alakonya et al. 2024). Similar mAb, affibody molecules, which are engineered scaffold proteins consisting of 58 amino acids and peptide molecules, are also used for PET and SPECT imaging after being labeled with different radionuclides, as shown in Fig. 4.3 (Bai et al. 2023). However, the main drawbacks of nuclear imaging techniques are radiation exposure and the high cost and restricted supply of radiopharmaceuticals (Imperiale et al. 2024).

4.2.3 Optical Imaging Technologies

Optical imaging techniques such as near-infrared fluorescence imaging (NIRFI) and bioluminescent imaging (BLI) are promising technologies for in vivo tumor imaging because of their high sensitivity, high spatiotemporal resolution, real-time approach, cost-effectiveness, and especially the nonionizing radiation compared to other noninvasive imaging methods (Zhao et al. 2024a). However, optical imaging techniques are mainly found in applications in preclinical studies. BLI converts chemical energy into light or photons through various enzymatic reactions. Compared to traditional methods, recent developments in BLI have boosted the visibility of individual tumor cells by a ratio of 100:1000 (Mali 2024). The work by Huang et al. put forwarded a novel approach by utilizing an NIR-II fluorescence/dual bioluminescence multiplexed imaging platform, for assessing tumor-promoting/suppression properties of stem cells by simultaneously monitoring the tropism of transplanted stem cells toward the tumor and then evaluating the growth and nuclear factor-kappa B (NF-κB) activity of the tumor following stem cell transplantation. This was achieved by employing a CRISPR-Cas9 gene editing technology to create a genetically modified human primary glioma cell line-U87-MG (GE-U87-MG), with an RFLuc reporter gene driven by the Elongation factor 1-alpha 1 (EF1α) promoter and a GLuc reporter gene driven by the NF-κB promoter (Huang et al. 2024). Similarly, Hersh et al. developed an EGFR-targeting bioluminescent fusion protein that was loaded on the G5-polyamidoamine (G5-PAMAM) dendrimer nanocarrier to identify pancreatic ductal adenocarcinoma. EGFR-binding affibody (ZEGFR) was introduced for PDAC-targeting, and *Gaussia* luciferase (GLuc) was used as the bioluminescent reporter. The fusion protein–dendrimer complex was observed to be localized in PDAC cells both in vitro and in vivo by simultaneously exhibiting BLI (Hersh et al. 2023). In the case of fluorescence imaging, quantum dots (QDs), rare-earth nanoparticles, and NIR-II organic fluorophores are mainly explored as NIR-II imaging probes (Li et al. 2024; Deng et al. 2023).

4.3 Integration of Imaging Techniques

4.3.1 Hybrid Imaging Systems

Nonetheless, most of the imaging techniques have certain limitations. For instance, MRI has a long imaging duration and relatively low specificity, whereas PET/SPECT is limited by poor spatial resolution and radiation exposure hazards. On the other hand, optical imaging has a shallow penetration depth and poor spatial resolution. Therefore, to overcome these limitations, hybrid modalities are developed by combining the different imaging techniques known as multi-mode imaging (Bai et al. 2023). A single patient scan can provide a multitude of molecular, functional, and morphological data when using positron emission tomography (PET) in association with computed tomography (CT) or magnetic resonance imaging (MRI) as a multimodality PET–CT or PET–MRI system. For example, the diagnostic efficacy of hybrid PET/MRI with Ga-68 prostate-specific membrane antigen (PSMA) was developed in order to identify skeletal metastases in patients diagnosed with prostate cancer. The efficiency of the dual imaging modality was compared to the efficacy of stand-alone PSMA PET, accelerating and alone, whole-body (WB) MRI assessment. The PET/MRI system showed more accurate results (Regmi et al 2021). In another work, a noninvasive method was developed combining PET/MRI for the treatment of non-small cell lung cancer (NSCLC).

4.3.2 Multimodal Image Fusion and Analysis

Even though multi-modal imaging techniques are promising areas in the biomedical field, they have certain limitations. One of the main limitations is the difference in image registration, as the organs or tissue groups are imaged using different physical principles. This heterogeneity makes the integration and evaluation of the multi-modal imaging entities complex. Also, the acquisition time might increase due to the combination of more than one technique, and it can be expensive (Alam and Rahman 2019; Kalamkar and Mary 2023). The heterogeneous structure of the data, misalignment and nonlinear correlations between input data, or inadequate data throughout the fusion process make multimodal image fusion difficult. A number of techniques have been introduced in recent years to improve the performance. The fused image is more informative and precise than all single images. Various medical imaging fusion modalities are MRI, SPECT, PET, CT, Structural Positron Emission Tomography (SPET), and magnetic resonance angiography (MRA). These are various algorithms. According to the data set, image fusion is classified into multi-focus, multi-sensor, multi-scale, multi-spectral, and multi-temporal. Multi-focus image fusion techniques combine many input images with various focuses to create a single image. Multi-sensor image fusion in medical diagnosis combines data from different sources to enhance image clarity and reveal details that are not visible to the human eye. Multi-scale fusion integrates differently exposed low dynamic range (LDR) images, multi-spectral fusion merges spectral features for better interpretation, and multi-temporal fusion reduces information loss while capturing all clinical variables. These techniques improve abnormality detection and diagnostic accuracy. Medical imaging fusion is categorized as spatial domain, transform fusion, fuzzy logic, morphological methods, and sparse representation methods. Further fusion quality evaluation metrics can be categorized into three levels: pixel-level, feature-level, and decision-level (Saleh et al. 2023). Each method has its own pros and cons. Ultimately, medical image fusion technologies are constantly evolving according to the need and data accuracy.

4.4 Nanotechnology in Imaging and Therapy

The large volume-to-surface area ratio in nanoparticles employed for cancer imaging aids them in being densely covered with antibodies, peptides, aptamers, and other moieties. These moieties can bind and recognize specific cancer sites. Fluorescence imaging is a powerful method due to its high sensitivity, in situ visualization, and rapid response (Mao et al. 2024). Quantum dots (QDs), gold nanoparticles (AuNPs), and polymer dots (PDs) are some of the commonly used nanoparticle probes in imaging diagnostics. Quantum dots, with their target specificity and excellent fluorescence properties, stand out as an excellent candidate for tumor theranostics and imaging applications. Quantum dots exhibit a high quantum yield (Harun et al. 2013). The biological imaging window ranges from the visible region (400–700 nm) to the first near-infrared region, NIR-II (700–900 nm), to the second near-infrared region, NIR-II (1000–1700 nm) (Hu et al. 2020). The NIR-II light can penetrate deep into the tissues and has less scattering and attenuation, enhancing the imaging activity (Chu et al. 2022). Thus, the NPs or quantum dots which has optical properties in these regions can be applied for optical imaging techniques such as fluorescence imaging. In vivo fluorescence imaging is severely interfered with by tissue absorption, scattering, and spontaneous fluorescence. Specifically, in the near-infrared (NIR)-I and NIR-II windows, the absorption and scattering of oxygenated whole blood, deoxygenated whole blood, skin, and fat in biological tissues are significantly lower than that in the visible region (Smith et al. 2009). Juan Ning et al. synthesized o-CDs@Mn nanomaterials that possessed enzyme-like activity and had the potential for fluorescence imaging and tumor therapeutic applications. It showed a photothermal conversion efficiency of 72.31% and photothermal activity, with a tumor cell apoptosis of 79.39% (Ning et al. 2024).

4.5 Synergistic Therapeutics for Solid Tumors

4.5.1 Combination Therapies

Combination therapy, a treatment modality that combines two or more therapeutic agents, is a cornerstone of cancer therapy (Vafaei et al. 2022). There have been significant advances in cancer treatment over the past several years through the use of chemotherapy, radiation therapy, targeted therapy, including immunotherapy. However, such treatments have limitations (Wargo et al. 2015). Therefore, there is an increasing attention for merging these individual therapies to develop hybrid or synergistic therapies. Preclinical reports, in fact, provide a solid rationale for combining tumor-targeted and immunotherapies in order to enhance tumor clearance, and several ongoing clinical trials are assessing the potential synergistic effects of both approaches (Fang et al. 2021). Understanding how different treatments interact with the immune system and tumor microenvironment will help researchers develop safer and more effective combination therapies. These strategies aim to create long-lasting immune responses that prevent cancer from returning, setting the stage for future breakthroughs in cancer treatment. Immunotherapy enhances the body's natural immune defences to identify and eliminate cancer cells more effectively. A prominent approach involves immune checkpoint inhibitors (ICIs), which block inhibitory proteins, such as PD-1 (Programmed Death-1) and CTLA-4 (Cytotoxic T-Lymphocyte Antigen-4), that cancer cells exploit to evade immune detection (Darvin et al. 2018). By inhibiting these checkpoints, ICIs restore the immune system's ability to recognize and attack tumors.

Chemotherapy and immunotherapy work synergistically to enhance treatment outcomes (Luo et al. 2019). Chemotherapy employs potent drugs to inhibit cancer cell division and disrupt their growth by damaging cellular DNA. However, in some cases, chemotherapy alone proves insufficient (Chidambaram et al. 2011). Greater efficacy is often achieved through combination regimens, where multiple drugs target cancer from various angles, maximizing therapeutic impact. Chemotherapy impacts not only malignant cells but also rapidly proliferating healthy cells, including hematopoietic cells, leading to myelosuppression and immunosuppression (Crawford et al. 2024). This raised initial concerns among clinicians about the potential for chemotherapy to diminish the effectiveness of immunotherapy. Chemotherapy plays a pivotal role in enhancing the immune system's effectiveness by reducing tumor size, a process known as "debulking." By lowering the tumor burden, chemotherapy minimizes the number of cancer cells the immune system must target, making it easier for immune cells to mount an effective response. Additionally, smaller tumors release fewer immunosuppressive signals that cancer cells often use to evade immune detection, further improving immune function. Chemotherapy also addresses a key limitation of immunotherapy by transforming "cold" tumors (those lacking immune cell infiltration) into "hot" tumors, which are more immunologically active and responsive to treatment (Zhang et al. 2022). Certain chemotherapy agents promote immune cell infiltration by releasing tumor antigens and enhancing the expression of immune-stimulating molecules. Furthermore, chemotherapy can increase the expression of immune checkpoint molecules, improving the efficacy of immune checkpoint inhibitors when used in combination therapy.

While chemotherapy is effective at shrinking large tumors, it may not eliminate small, disseminated cancer cells, particularly in metastatic disease. Immunotherapy complements chemotherapy by targeting these scattered cells and preventing relapse. Together, this combination provides a more comprehensive approach, especially for tumors that are resistant to either therapy alone. It also helps counteract the tumor's ability to adapt and develop treatment resistance over time. Incorporating chemo-immunotherapy allows for personalized treatment strategies tailored to the patient's specific cancer type and immune profile. Though challenges remain, this synergistic approach holds significant promise for improving cancer outcomes and offers a more effective path forward, particularly for hard-to-treat malignancies. Targeted therapy is designed to address specific proteins, genes, or molecular pathways that drive cancer growth, offering personalized treatment based on the unique characteristics of a patient's tumor (Saeed et al. 2023). However, challenges such as drug resistance and incomplete tumor elimination can limit its effectiveness, highlighting the need for combination strategies (Boumahdi and de Sauvage 2020).

Immunotherapy, particularly immune checkpoint inhibitors (ICIs), plays a crucial role in enhancing treatment outcomes. These inhibitors reactivate exhausted T cells by blocking inhibitory signals like PD-1, thereby restoring immune function. When used alongside targeted therapy, ICIs improve the immune response by not only attacking the tumor but also countering the mechanisms that cancer cells use to evade detection (Li et al. 2023). This combination approach provides a comprehensive treatment strategy: targeted therapy reduces tumor size and releases tumor antigens, making cancer cells more recognizable to the immune system. In turn, immunotherapy activates T cells to target any surviving cancer cells, reducing the risk of relapse and metastasis. The synergy between targeted therapy and immunotherapy effectively addresses the limitations of each method. Targeted therapy weakens cancer cells by blocking essential growth pathways, while immunotherapy maintains an active immune response to eliminate resistant cells. This dual strategy is particularly effective in cancers with high mutation burdens, where numerous abnormal proteins can serve as immune targets.

There are a lot of benefits to combination therapies. Combination therapies simultaneously target multiple pathways involved in cancer cell growth and survival. This multi-target strategy complicates the ability of cancer cells to adapt and evade treatment. By attacking different aspects of tumor biology, such as growth signaling, apoptosis, and immune evasion, combination therapies create a more comprehensive assault on cancer cells, significantly improving the likelihood of treatment success (Mahoney et al. 2015). Cancer cells frequently develop resistance to single-drug treatments through various mechanisms, such as genetic mutations or activation of alternative signaling pathways (Jin et al. 2019). By using different drugs in combination, the development of resistance is limited. This approach disrupts the selective pressure that leads to resistant cell populations, thereby enhancing the durability of treatment responses. Tumors are often composed of heterogeneous cell populations, each with distinct genetic and phenotypic characteristics (Bhat et al. 2024). Combination therapies can more effectively target these variations by employing drugs that act on different subsets of tumor cells. This is particularly important in advanced cancers, where the genetic diversity of the tumor can make it more challenging to achieve a complete response with monotherapy. Certain chemotherapy or targeted therapies can prime the immune system by reducing tumor burden and removing immune-suppressing signals in the tumor microenvironment (Budczies et al. 2024). This priming effect can enhance the efficacy of subsequent immunotherapy treatments. Combining immunotherapies, such as checkpoint inhibitors, with chemotherapy or targeted therapies can significantly improve treatment outcomes. These combinations can enhance T-cell infiltration into tumors, thereby increasing the effectiveness of immune responses. Additionally, they can disrupt the immunosuppressive tumor microenvironment, making it more conducive for immune cells to attack cancer. The synergistic effects of combining targeted therapies with immunotherapy can produce greater outcomes than either treatment alone. These combinations can boost T-cell activity, augment immune responses, and disrupt tumor vasculature, collectively leading to more substantial tumor regression and improved patient survival rates. Combination therapies are particularly beneficial in managing advanced cancers that have metastasized to distant organs (Mokhtari et al. 2017). By integrating various treatment modalities, these approaches can better control cancer spread, improve overall survival, and enhance the quality of life for patients. Combination therapies can help establish immunological memory (Vilgelm et al. 2016), which allows the immune system to recognize and respond to residual cancer cells effectively. This long-lasting immunity reduces the risk of cancer recurrence, enabling the body to mount an effective response to any future cancerous developments.

While combination therapies hold significant promise in enhancing cancer treatment outcomes, they also present several challenges that must be addressed. Combining therapies can heighten the risk of adverse effects on healthy tissues and organs. Treatments that target both the immune system and cancer cells may lead to inflammation or autoimmune reactions (Zhou et al. 2023). The sequence and timing of administering combination therapies are critical for achieving optimal effectiveness (Kwon et al. 2021). Incorrect administration can lead to unintended consequences, such as the suppression of immune activity rather than enhancement. For instance, prolonged use of sunitinib, a targeted therapy, can increase the number of regulatory T cells (Tregs), which suppress immune responses, thereby undermining the overall treatment strategy. Designing effective combination therapies necessitates meticulous planning to balance the interactions between multiple drugs. This complexity poses challenges for both healthcare providers and patients, as managing drug interactions and ensuring optimal dosing require careful oversight. Combination therapies, particularly those involving innovative immunotherapies or targeted agents, can be prohibitively expensive. This financial burden may limit access to these potentially life-saving treatments for many patients, raising concerns about equity in cancer care. The combination of different therapies can lead to the development of novel side effects that may not be anticipated based on individual treatments. For example, combining a checkpoint inhibitor with chemotherapy might result in unexpected immune overactivation, leading to severe inflammatory responses. These new side effects can significantly impact patients' quality of life and require vigilant management by healthcare providers. Not all patients respond uniformly to combination therapies. Individual variations in genetics, tumor biology, and overall health can complicate the identification of effective combinations for specific patient populations (Krzyszczyk et al. 2018). This variability underscores the need for personalized approaches to treatment and the ongoing research necessary to optimize therapeutic strategies.

4.5.2 Immuno Therapy and Imaging

The immune system is highly complex in its nature and architecture, and the immune responses are highly variable and dynamic (Subramanian et al. 2015). Medical Imaging enables noninvasive monitoring of these processes in vivo. To understand the same, consider one of the most common imaging requirements in clinics—cancer imaging. It has transitioned from solely assessing tumor anatomy to comprehensively evaluating tumor phenotypes and their heterogeneity (Bailly et al. 2019). Traditional cancer imaging does not always work well with the requirements of monitoring precision medicine, as in cancer immunotherapy (van de Donk et al. 2022). There is a clear need for approaches that focus on tracking immune-related activities, like immune cell behavior or checkpoint pathways. From the clinical viewpoint, these elements can be evaluated with different imaging techniques, but in practice, PET and SPECT are the fundamental tools. Functional and molecular imaging techniques that allow evaluation of the molecular and physiologic processes of the tumor in vivo promise to be useful in determining both the immune phenotype and response to treatment.

These techniques can enable the noninvasive evaluation of specific markers that are associated with immune response (e.g., PD-1 and PD-L1) or certain populations of immune cells, and the physiologic and molecular changes associated with their activation and action (Krutzek et al. 2022). Currently, the use of these imaging techniques in clinical practice to monitor dynamic and spatial changes in immunity status or to predict the response to treatment has been limited. There are various techniques available for this purpose. Diffusion-weighted MRI (DW-MRI) is becoming more critical in evaluating how tumors respond to immunotherapy. By measuring water movement, it can detect inflammation and cell density changes caused by immune cells attacking the tumor (Iima et al. 2024). However, interpreting these results is complex because different cancers show different patterns. Advanced techniques are also being explored to better predict treatment success and distinguish between real tumor growth and temporary swelling. Similarly, perfusion imaging measures how well blood flows through a tumor and plays an important role in understanding the immune microenvironment (Foray et al. 2020). It uses noninvasive imaging techniques, such as CT scans or MRI. It is assumed that tumors with better blood flow might be more accessible to immune cells, improving the chances of immunotherapy workup. Poor blood flow can limit immune cells from entering the tumor, making treatment less effective. But the clinical applications are limited to a few types of cancers. Brain tumors like glioblastoma are an exception, where blood flow measurements can help doctors tell the difference between tumor growth and inflammation caused by the immune response. Much research is focused on molecular imaging strategies to develop biomarkers to evaluate the response to immunotherapy. To date, the assessment of patients treated with immunotherapy has been based mainly on the use of fluorine 18 (^{18}F)-FDG PET (Fluorodeoxyglucose Positron Emission Tomography). FDG PET plays a vital role in tracking how well immunotherapy works by measuring metabolic changes in tumors (Léger et al. 2022). Though it is useful, it can also create confusion since both immune cells and cancer cells consume glucose, sometimes mimicking tumor progression. New methods and criteria are being developed to better assess treatment outcomes, but challenges remain in interpreting FDG PET results for different cancers and treatment types. FDG PET can spot the side effects caused by immunotherapy much earlier, even before symptoms appear (Gandy et al. 2020). PET imaging with other tracers is also studied to understand how well they can monitor cancer treatment. Fluorothymidine PET (FLT-PET) helps in assessing how fast cells are proliferating. Unlike FDG, FLT does not light up during inflammation, meaning it could better distinguish between actual tumor growth and pseudoprogression (Nappi et al. 2024). Fluoro-α-methyl-L-tyrosine and fluoroethyl-L-tyrosine track amino acid metabolism in tumors. These tracers may help differentiate between treatment effects and true cancer progression. ^{11}C-methionine PET is useful for tracking response in brain tumors treated with immunotherapy (Debreczeni-Máté et al. 2024). Magnetic resonance spectroscopy (MRS) identifies chemical changes in tumors by detecting specific metabolites (Brindle 2024). This technique is not widely used, but it has shown potential in monitoring immunotherapy.

Combining different imaging methods (like MRI, PET, and spectroscopy) may provide both precise anatomical information and molecular-level data, which can be utilized in cancer treatment. PET/MRI, which combines the detailed imaging of MRI with the molecular insights of PET, is still in the experimental stage. Tumor cells use the natural immune-inhibition mechanisms to avoid attack from the immune system. Drugs that block immune "checkpoints" like PD-1, PD-L1, and CTLA-4 have transformed cancer treatment. Imaging can track these checkpoint molecules. PET scans with PD-L1 tracers can predict whether the treatment is working, sometimes more accurately than a biopsy (Zhao et al. 2024b). ^{89}Zr-labeled tracers can mark immune drugs or checkpoints for repeated scans, helping clinicians to monitor how well the drug is distributed and working over time (Li et al. 2018). Tracking immune cells directly in the body using PET or SPECT is challenging. Studies using tracers that attach to CD8+ T cells show good results in predicting tumors that are being attacked by the immune system (Zhang et al. 2024). PET imaging with granzyme B (molecules released by T cells and natural killer cells to destroy cancer cells) tracers can help to identify how active the immune cells are in attacking the tumor cells. Another area of research involves genetically modifying immune cells to make them easier to track with PET (Kazim and Yoo 2024). This approach has already been used to study CAR-T-cell therapy.

Tumors often have macrophages inside them. Scientists use iron oxide nanoparticles with MRI to study these macrophages. If the MRI shows lots of iron inside the tumor, it means the tumor has more M2 macrophages, which indicates a worse prognosis and poor response to treatment. Although PET and SPECT scans are leading tools to track checkpoint molecules and immune cells, some techniques, such as using nanoparticles and genetic cell modification, are still under experimentation; they are promising tools for better monitoring of immune activity in cancer treatment. Currently, only preclinical studies have explored the use of imaging to study factors such as cytokines or the enzyme indolamine 2,3-dioxygenase in oncology, and practical clinical applications have yet to be found (Batista et al. 2009).

4.5.3 Multiparametric Molecular Assessment of the Immune Microenvironment

Molecular imaging allows noninvasive monitoring of the tumor's immune activity and response to immunotherapy using radiotracers. Studies demonstrated how multiple tracers could measure both glycolysis and immune checkpoint markers like PD-L1 and PD-1, simultaneously in patients, offering a detailed view of the tumor's immune landscape (Liu et al. 2023). However, this study shows that combining multiple tracers with granzyme B-based imaging has limited clinical benefit. Challenges include complex data interpretation, signal interference, and the variability of immune responses. Although this method holds promise for more personalized cancer treatment, further research is needed to optimize tracer combinations and improve imaging protocols for better clinical application.

Immunotherapy works by activating the patient's own immune cells to fight against tumor cells by inducing a functional immune response (Giraldo et al. 2015). This approach has brought a change in the treatment paradigm for cancer and has introduced new challenges in medical imaging (Carter et al. 2018). Conventional imaging techniques (CT and MRI) do not seem to be useful for patients undergoing immunotherapy. Apart from other treatment strategies, people undergoing immunotherapy possess different patterns of response and progression, and are associated with a wide spectrum of toxic effects that are not seen with other treatments (Kennedy and Salama 2020). Classic tumor response criteria are limited for the evaluation of the unusual patterns of response that arise from the specific mechanisms of action of immunotherapy (Hodi et al. 2018), so advanced imaging methods must be developed to overcome these shortcomings. Because not all patients benefit from immunotherapy, pretreatment imaging should be performed to identify not only prognostic factors but also factors that allow prediction of response to immunotherapy. Follow-up studies must allow the detection of nonresponders, without confusion of pseudo-progression with real progression, to prevent premature discontinuation of treatment that can benefit the patient.

There is a clear need to identify biomarkers to guide patient selection, predict outcomes, and help evaluate the tumor response to immunotherapy. It is quite challenging to find reliable biomarkers due to the tumor complexity and typical variations from patient to patient (Purkayastha et al. 2023). Technological advances have facilitated the extraction of analyzable data from medical images (radiomics and radiogenomics) that, when combined with other parameters (e.g., clinical, laboratory, and genetic), can improve decision-making in the treatment of patients with cancer (Zhang et al. 2025). There are limitations in the available traditional biomarkers, such as PD-L1 expression, T-cell infiltration, and tumor mutational burden (Moeckel et al. 2023). These markers are not always reliable. For example, some PD-L1-positive tumors do not respond to treatment, and PD-L1-negative tumors sometimes do. This is where imaging data can help clinicians get a clear picture, and it helps in monitoring tumor progression.

AI is now helping analyze medical images through Machine learning and Deep learning. Sun et al. used machine learning to study imaging data from 135 patients. They combined radiomic features with biopsy data to create a signature that could predict how well patients would respond to PD-1 or PD-L1 therapy. Mu et al. analyzed PET/CT scans of 194 lung cancer patients to find patterns in the scans that were linked to treatment response. The process of tumor imaging itself is complex. It requires precise image collection, segmentation, and feature extraction. Immunotherapy responses are unpredictable: some patients show atypical responses, like hyper progression (sudden tumor growth) or unexpected side effects. This makes it harder to measure success using traditional criteria. Deep learning will likely improve tumor detection and analysis, including across multiple imaging methods like PET/CT or PET/MRI (Mahmud et al. 2023). Liquid biopsies (blood tests that detect cancer markers) could complement imaging, allowing for more precise monitoring of how well immunotherapy is working. In time, better imaging biomarkers will help tailor immunotherapy to individual patients, leading to more effective treatments. Radiomics and radio genomics offer new ways to understand how tumors respond to immunotherapy by using advanced imaging and AI. However, challenges like variability in images and unpredictable immune responses make it hard to create reliable biomarkers. In the future, combining imaging with tools like liquid biopsies could make precision medicine more accessible for cancer patients.

4.6 Advanced Radiotherapy Techniques

PET, CT, and MRI techniques mainly rely on deep penetrating radiation to detect in vivo structures and effectively improve the detection rate of malignant tumor lesions. However, the above-mentioned imaging techniques still have some disadvantages: (1) few imaging approaches can reach a millisecond temporal resolution and the micron-scale spatial resolution, so the ability to monitor the dynamic lesions is poor; (2) there is high risk of radiation during the imaging process, which damages human health; and (3) expensive equipment and high detection costs limit their practicality and popularity (Chen et al. 2022). Proton therapy has enhanced target coverage and dramatically decreased dosage and organ at risk (OAR) dose. Intensity-modulated proton therapy (IMPT) promises better coverage of the clinical target volume and reduces exposure to OAR, reducing the likelihood of long-term adverse effects (Qiao et al. 2024). Stereotactic body radiation therapy (SBRT) or stereotactic ablative radiotherapy (SABR) is a testament to the precision that can be achieved in radiation therapy. Targeting a tumor with several beams of radiation energy and using 3-D or 4-D imaging and high doses of highly focused radiation beams ensure the least amount of damage to the neighboring healthy cells.

4.7 Clinical Applications and Case Studies

4.7.1 Case Studies in Multimodal Imaging and Therapy

Novel innovations in imaging technologies are the need of the hour since doctors can only examine the tumors layer-by-layer using conventional imaging modalities, making it difficult for an accurate detection of tumor tissues, effective pre-operative planning, complete resection of tumors during surgery, and post-operative observations. Failure of complete resection of tumor tissues can result in intratumor heterogeneity (ITH) (variation in the local cancer genome which provides a selective advantage in its environment), which can cause the tumors to be more aggressive and, thereby, develop resistance to multiple therapy modes (Bi et al. 2019). Multimodal imaging helps integrate different techniques for diagnosis, treatment, and even intraoperative procedures. The journey from simple X-ray images unable to detect small tumor masses to swiftly evolving nano theranostics and personalized imaging strategies using machine learning and artificial intelligence is quite insightful. In this section, we look into some of the clinical trial case studies using the emerging next-generation imaging modalities.

The Advanced Multimodality Image-Guided Operating (AMIGO) Suite comprises a hybrid operating setup occupying a 5700-ft^2 area of the Brigham and Women's Hospital in Boston, designed by Drs. Ferenc Jolesz and Clare Tempany. The clinical use of AMIGO was initiated in 2011, with 552 procedures performed by June 2014, including craniotomies, PET/CT and MR-guided cryoablations, gynecologic brachytherapy procedures, video-assisted thoracic surgeries, etc. The setup comprises three sterile procedure rooms: MRI, OR, and PET/CT.

MRI room: An IMRIS system consisting of a 3 T MRI scanner with OR-grade medical gases, an MRI-compatible system to deliver and monitor anesthesia, and other therapeutic devices is installed in the room. A radiofrequency shielded door separates the MR scanner from the OR, which opens when MR is needed in the procedure (e.g., focal ablation therapy, prostate biopsy), and the MR scanner traverses through a railing mounted on the ceiling to the OR.

OR: This room contains the tools and facilities for surgical procedures with imaging modalities like 2D and 3D ultrasound imagers, a fluoroscopy unit, a mounted single-plane angiographic X-ray system, and a surgical microscope with NIR imaging capabilities.

PET/CT: There is a combined device consisting of CT and PET, which offers high-resolution images of CT, enabling the doctor to see even small masses in the lungs, brain, etc., and radioisotopes in PET help to monitor metabolic activity (Tempany et al. 2015).

The main aim of incorporating imaging intraoperative MRI is to maximize tumor resection without affecting the neighboring normal tissues in surgeries and to confirm the needle position in biopsies. CT comes into play in lung cancer detection, cases of bony resection, and post-operative imaging. A case study was conducted from January 2013 to December 2020 in which 11 surgeries were carried out for skull base chordomas and chondrosarcomas using the AMIGO Suite, in which a gross total resection was observed in two patients, and the rest of them had subtotal resections. Surgeries performed on six patients with recurrent lesions yielded two cases of clinical improvement and the other four with stable postoperative symptoms. The only case of death was a patient whose initial diagnosis occurred 22 years before the surgery (Essayed et al. 2022).

4.7.2 Translational Research and Clinical Trials

Translation research is the successful translation of basic laboratory research into research that benefits humans. That is the expected end product of all the clinical research. It will be demonstrated by clinical testing. Artificial intelligence is used mainly for detecting, characterizing, and monitoring tumors. Computer-aided detection (CADe) of early breast carcinoma, locating microcalcification, brain metastases in MRIs, and false-negative results in low-dose CT screening, automated segmentation, calculation of dosage for radiation therapy, categorizing tumors into predefined groups based on their characteristics, and associating imaging features with genomic data are the major milestones achieved using AI (Bi et al. 2019). Computer-Aided Nodule Assessment and Risk Yield (CANARY), a lung cancer imaging software, was clinically trialed on 294 patients in 2015, where the Kaplan–Meier analysis classified them into Good, Intermediate, and Poor CANARY risk groups and produced distinct progression-free survival curves. In 2017, automated lung cancer detection was experimented with conventional neural networks (Maldonado et al. 2015).

Photodynamic therapy (PDT) uses a photosensitizer, which, on excited by a light source of a specific wavelength, transfers energy to oxygen and generates reactive oxygen species (ROS) that causes apoptosis, necrosis, and/or autophagy of tumor cells. Over 250 clinical trials have been carried out for PDT in brain tumors, lung cancer, urological cancer, etc. A study of PDT using photoprint (porfimer sodium) carried out post-resurrection surgery concluded that an increasing dose could lead to a better progressive-free trial. At Royal Melbourne Hospital, hematoporphyrin derivative (HPD) was used as a photosensitizer in 136 patients, where they reported survival of more than 36 months post-therapy in newly diagnosed and recurrent AA and GBM patients. This study recognized that old age at diagnosis contributes to poor prognosis. 5-Aminolevulinic acid (5-ALA) also prolonged high-grade gliomas (HGG) survival time. Photofrin II and temoporfin are the leading candidates that were successful in the clinical trials of postoperative PDT for non-small cell lung cancer (NSCLC). 5-ALA also proved to be successful on an intermediate-to-high-risk level in a clinical trial conducted on 45 patients diagnosed with nonmuscle invasive bladder cancer (Mahmoudi et al. 2019). 2-(1-Hydroxyethyl)-2 -devinylpyropheophorbide-a (HPPH), LS 11 (Taporfin Sodium), and Porfirmer Sodium are the main photosensitizers investigated in the clinical trials for esophageal cancer (Alsaab et al. 2020).

Multimodality imaging with two or more imaging techniques permits us to combine the limitations and overcome the intrinsic limitations of individual modalities. Nanoparticles are promising agents for multimodal imaging because of their special features, including nanometer dimensions (1–100 nm), tunable imaging characteristics, and multifunctionality. Their smaller size enhances the permeability and retention of nanoparticles in tumors. These lead to the increased local concentration of imaging agents in tumors. Another advantage of nanoparticles is their potential for facile functionalization with radioactive isotopes and a range of contrast agents. Tailored nanoparticles can also be loaded with therapeutic agents for chemotherapy and/or radiotherapy. Multipurpose nanoagents also help reduce the toxicity associated with using multiple contrast agents for different modalities. Thus, nanoparticles function as unified platforms of therapeutics and imaging. Based on the chemical compositions of nanoparticles, they are mainly classified into three categories: inorganic, organic, and carbon-based NT (Najdian et al. 2024).

Semiconducting molecules of perylene diimide (PDI) with strong light absorption properties in the near-infrared region and good biocompatibility are examples of organic nanoparticles for theranostics, especially as photoacoustic (PA) imaging agents. These PDI nanoparticles were further labeled with ^{64}Cu for positron emission tomography (PET). The size of PDI nanoparticles was altered by using different initial concentrations of PDI molecules in the self-assembly process, and the size-dependent accumulation of the PDI NPs in the lymphatic system after local administration and in tumors was monitored using PA and PET imaging in mouse models after local administration. The studies revealed that 100 nm is the best size for differentiating popliteal and sciatic LNs since the interval is around 60 min for the NPs to migrate from popliteal LNs to sciatic LNs, which is an ideal time window to facilitate surgical sentinel LN biopsy and pathological examination. The NPs with a size of 60 nm appear to be the best for tumor imaging and photothermal cancer therapy due to the maximum tumor accumulation efficiency. Thus, this series of [^{64}Cu]-labeled-PDI nanoparticles (NPs) of different sizes (30, 60, 100, and 200 nm) act as dual positron emission tomography (PET) and PA imaging probes and photothermal therapy agents.

Quantum dots (QDs) are nanoscale inorganic fluorescent crystals with unique optical properties, including high quantum yield, high brightness, high stability, low photo-bleaching, size-tunable absorption, and emission spectra. Here, we present an example of a quantum dot being a CRET (Cerenkov Resonance Energy Transfer) and PET/NIRF nanoprobe-CdSe/ZnS QD doped with ^{64}Cu via a cation exchange reaction. These intrinsically incorporated ^{64}Cu atoms showed less than 3% dissociation, even in the presence of EDTA. Thus, these ^{64}Cu RQDs showed very high radiochemical stability. To improve their colloidal stability and biocompatibility, these RQDs were coated with PEGylation; PEGylated QDs showed higher tumor uptake in xenografted mice than in modification with glutathione. Upon optimizing the QD levels to be injected and ^{64}Cu radioactivity levels, these ^{64}Cu-doped CdSe/ZnS QDs showed successful utility for PET/CRET luminescence dual-modal imaging of U87MG glioblastoma xenograft mice models (Sun et al. 2014).

Upconversion nanoparticles (UCNPs), especially lanthanide-doped nanocrystals, a sub-class of inorganic nanoparticles, are drawing attention now as promising next-generation fluorescent nanoprobes. These nanoparticles convert near-infrared (NIR) radiation to visible light by a mechanism called "upconversion luminescence." UCNPs exhibit excellent chemical and optical properties, including the absence of autofluorescence and photo blinking, high photostability, narrow emission peaks, and low toxicity compared to common down-conversion fluorescent agents (i.e., fluorescent dyes, QDs). A pentamodal imaging agent with dual-targeting capabilities (passive and active) using fluoromagnetic NaGdF4:Yb:Er upconversion nanoparticles (UCNPs) was developed. To target cancer cells, the nanoparticles were functionalized with apo-human serum transferrin protein (Tf), which can readily bind to transferrin receptors that are often overexpressed in cancer cells. The radiolabel ^{99m}Tc was incorporated into the Tf core, achieving high radiolabeling efficiency (~95%) and stability (~100% over 24 h), making it practical for SPECT (Single Photon Emission Computed Tomography) imaging. The UCNPs were excited at a 980 nm wavelength, enabling deep tissue penetration for fluorescence and photothermal imaging. Gadolinium (Gd) in the UCNPs provided a T1-weighted MRI signal, while ytterbium (Yb) acted as a CT contrast agent. In vitro experiments showed that the UCNP@Tf-99mTc formulation underwent receptor-mediated endocytosis and exhibited minimal cytotoxicity, with cell viability remaining above 75% at concentrations up to 1000 ppm. In vivo studies in a mouse model using SPECT/CT imaging revealed favorable bioavailability and renal clearance. MRI studies on 4T1 tumor-bearing mice confirmed both passive and active tumor targeting. The ability to perform multimodal imaging (MRI/SPECT/thermal/CT) in vivo demonstrates the potential of UCNP@Tf-99mTc as a versatile alternative to multiple imaging agents, positioning it as a strong candidate for advanced multimodal image-guided operating (AMIGO) systems (Ezerskyte et al. 2024).

4.8 Challenges and Future Directions

4.8.1 Technical and Practical Challenges

Although the multimodal-imaging nanoparticle-based systems proved advantageous in the noninvasive visualization of tumors, there are a couple of challenges in establishing the feasibility of this approach. The primary challenge is the absorption of radio-labeled nano formulation by the reticuloendothelial system (RES) organs like the liver and spleen. The physicochemical properties of nanoparticles, like dimensions, adaptability, hydrophobicity, and surface charge, significantly impact their distribution, bioavailability, and excretion from the body. Nanoparticles possessing a rigid, globular shape, and a size smaller than 6 nm can be eliminated through the renal pathway. Exceptionally, nanoparticles of size 4–8 nm are quickly cleared from the bloodstream by the reticuloendothelial system. This leads to faster accumulation in the spleen and liver, followed by their elimination via the hepatobiliary pathway. Thus, it may be inferred that smaller-sized nanoparticles exhibit less absorption of RES, resulting in improved imaging properties.

Moreover, the detachment of the radionuclide (typically present as a metallic ion) from the chelator or the dissociation of the radionuclide-containing polymeric coating from the nanoparticle might result in significant accumulation in healthy organs and decrease the target specificity of the radiolabeled nanoparticles. Considering the cellular toxicity of nanoparticles is crucial while developing theranostic systems for multimodal tumor imaging and diagnosis. Nanoparticles showing significant cytotoxicity are unlikely to be included in therapeutic applications. Specific inorganic nanoparticles containing gadolinium, cadmium, and carbon-based nanoparticles exhibit biological toxicity. Biocompatible materials, especially those derived from the human body, are typically preferred over heavy metals, considering their sustainability in human systems (Najdian et al. 2024).

Complex algorithms are required to combine data from different imaging modalities like PET, CT, MRI, and other clinical data; thus, accurate integration of information still needs to be worked on. Due to the patient demographics and imaging protocol variations, reproducing and validating the results by an AI model are challenging. The clinicians find it difficult to interpret AI languages like deep learning; thus, how much they trust the decisions made by the model is questionable. Since the clinical data is used to train the AI models, we must be careful about adversities like data breaches and misuse of digital footprints (Bi et al. 2019). In the case of photodynamic therapy (PDT), improving the solubility of photosensitizers is crucial for proper drug administration and circulation in the patient's blood, thereby enhancing its biodistribution in the body. The off-target accumulation of the photosensitizers is another serious challenge, resulting in scarring of normal tissues surrounding the solid tumors and prolonged phototoxicity. Additionally, the effectiveness of PDT in treating skin cancer is restricted to superficial tumors due to the limited penetration of light into deeper tissues (Alsaab et al. 2020).

4.8.2 Future Trends and Innovations

Although nanoparticles appear to be promising candidates for multimodal cancer imaging, those containing Cd, Se, Zn, Te, Hg, and Pb can be potentially neurotoxic, and/or teratogenic, depending on the dosage, complexation, and accumulation in our body. Even at very low doses, these metals are bound by metallothionein proteins and may be excreted slowly or sequestered in vivo in adipose and other tissues. Cadmium, with its half-life of about 20 years, is carcinogenic, although its mechanism is not known. A thick coating of ZnS with an efficient surface coating is crucial to avoid leakage of toxic metal ions. Many non-Cd-based nanoparticles with Zn, Cu, Mg, carbon-based graphite, and silicon are reported and shown to have more resistance to thermal, chemical, and photochemical disturbances. Although the inorganic QDs are smaller in size and shape than organic nanoparticles like dendrimers, efforts are being made to reduce the nanoparticles' size to reduce their RES uptake and have better imaging properties. The major roadblocks for clinical translation of QDs are inefficient delivery, toxicity, and lack of quantification. However, with the development of smaller non-Cd-based multifunctional QDs and further improvement on conjugation strategy, it is expected that QDs will achieve optimal tumor targeting efficacy with an acceptable toxicity profile for clinical translation in the near future using either NIRF imaging alone or multimodality imaging (Cai et al. 2007; Liu et al. 2020).

We present a potential multimodal imaging therapeutic system—a "dual-lock-and-key" (DKL) DNA nanodevice which consists of inner Au@Ag@Au DSNPs (Double Shell NanoParticles) for microRNA21 (miRNA-21) detection and treatment. As miR-21 is a well-known biomarker for various cancers (breast, lung, prostate), it is chosen as a target for our diagnosis and treatment. The reported system consists of inner Dox/Hairpin/Aptazyme-Au@Ag@Au probes and the outer metal-organic framework loaded with fuel strand. The pH-responsive MOF shell breaks as the DSNPs enter the cell via endocytosis due to the acidic environment of lysosomes, and Mn^{2+} and tannic acid are removed from the MOF. Mn^{2+} enhances the MRI contrast. This also leads to the release of Fuel DNAs and Dox/HA-DSNPs into the cancer cell. The intracellular miR-21 binds to the hairpin DNA, triggering Catalyzed Hairpin Assembly (CHA) and forming Hairpin/Fuel (HF) hybrids on DSNPs. Excess ATP in the cancer cells triggers the Aptazyme to release HA hybrids. The Fluorescence Signal from the Cy5 fluorophore on fuel DNA is quenched due to its close proximity to miR-21. The SERS (Surface Enhanced Raman Spectroscopy) signal is enhanced by Au@Ag@Au plasmonic effects. Doxorubicin acts as a therapeutic agent, and along with tannic acid, it reduces cancer cell proliferation and migration. This system, integrating dual-mode FL/SERS imaging and MRI-guided therapy for precise miR-21 detection and targeted cancer treatment, on proper optimization of toxicity and bio-compatibility, is a promising candidate for clinical purposes in the future (Yue et al. 2024).

The future promising approach is the integration of machine learning models and multi-modal imaging therapy as described in the enhanced prognostic assessment approach of Glioblastoma Multiforme cancer. ML models are first created by gathering the tumor's MRI, PET, and CT scan images and some clinical data about the patient, including demographics, genetic information, and treatment history. Additional clinical data include therapy data, the degree of resection, radiation dose prescriptions and distributions, and chemotherapy regimens. These data need to be preprocessed, including position scaling of image data, dealing with missing values in the patient record database, and harmonizing different data sets. ML approaches to analyze GBM data are Random Forests (RF), Support Vector Machines (SVM), Neural Networks, and Deep Learning. RFs can effectively handle larger datasets while reducing overfitting. Those are used to determine which features (imaging, genetic, or clinical) have the most predictive value in patient outcomes. SVMs are best suited for classification tasks, especially in higher-dimensional datasets. SVMs can differentiate tumor subtypes or classify patients into treatment response categories. Neural Networks and Deep Learning are potent tools for pattern recognition and complex medical image data processing. Deep learning models analyze MRI and PET scans to detect tumor characteristics, predict progression, and identify treatment-resistant regions. After their building, the models are usually checked as cross-validated or externally validated. Cross-validation involves partitioning our extensive data into training and testing datasets in several ways to avoid having the model perform best on any of the partition methods. External validation involves using an entirely different data set not used in the model development process, making the results more reliable. After their building and validation, these ML models, along with various multimodal theranostic agents as described earlier in this chapter, will revolutionize cancer diagnosis and treatment, serving as the most effective tool to assist doctors in making accurate and informed decisions (Rout et al. 2024).

4.9 Ethical, Legal, and Economic Considerations

4.9.1 Ethical Issues in Advanced Imaging and Therapy

The advancements and improvisations in medical imaging modalities are beneficial for better healthcare and diagnosis. However, it challenges the traditional processes set for different settings. This opens the discussion of the ethical, legal, and socioeconomic effects of these techniques in the present scenario. All the medical data include the patient's complete health details about past, present, and future diagnoses. All these data should be handled carefully since mishandling can cause serious issues for patients and medical professionals. Each patient undergoing the examination should be respected by acknowledging their individual autonomy and choices. Also, it should be beneficial by protecting the patient and by promoting their well-being. Finally, justice should be ensured by giving equal benefits to all involved. Thus, the fundamental ethical principles can be summarized as the principle of autonomy, the principle of equality and justice, the principle of beneficence, the principle of nonmalfeasance, and the principle of impossibility (Duquenoy et al. 2008). Also, the security, integrity, material quality, usability, and accessibility of all the electronic records should be verified (Bankman 2000). The major legal challenges in the processing of medical electronic data include reliability and quality of health data and tort-based liability (Hodge et al. 1999). The privacy of each data set should be ensured since it can be used for malpractices or privacy invasions. Each data piece can be used for medical diagnosis or research purposes. Thus, data handling etiquette should be analyzed in both these fields.

4.9.2 Economic Impact and Accessibility

Cost-effectiveness analyses (CEAs) aim to give more details on the potential for optimizing health outcomes while accounting for scarce medical expenditures demand. Given the unique and pivotal role of diagnostic imaging in treating cancers, there is an increasing demand for developing cost-effectiveness analysis (CEA) in imaging technology (Sailer et al. 2015). Imaging is an expensive medical procedure, and more people are getting examined day by day. Also, there is a lack of trained professionals. The major costs arising from the imaging technique are equipment, operation costs, the corresponding radiation exposure of the modality, and the imaging probe (Kjelle et al. 2024). Among all the imaging techniques, ultrasound and X-ray machines are considered cheaper. PET, MRI, and CT are considered as expensive because of the radiation exposure, etc. Currently, a lot of research is going on to make imaging techniques readily available and cost-effective to normal people, ensuring quality health care for everyone. Research on this topic should be appropriately addressed in an era of growing populations and illnesses.

Author Contribution P.K., L.A.S., A.S., and A.S.K. have reviewed the literature and prepared the first draft. P.K., A.S.K., and A.T.S. have prepared the outline and figures. K.S., A.B., and R.K. have done the final editing and revision of this chapter.

References

Alakonya H, Koustoulidou S, Hopkins SL, Veal M, Ajenjo J, Sneddon D, Dias G, Mosley M, Baguña Torres J, Amoroso F, Anderson A, Banham AH, Cornelissen B (2024) Molecular imaging of p53 in mouse models of cancer using a radiolabeled antibody TAT conjugate with SPECT. J Nucl Med 65(10):1626–1632

Alam F, Rahman SU (2019) Challenges and solutions in multimodal medical image subregion detection and registration. J Med Imaging Radiat Sci 50(1):24–30

Alsaab HO, Alghamdi MS, Alotaibi AS, Alzhrani R, Alwuthaynani F, Althobaiti YS, Almalki AH, Sau S, Iyer AK (2020) Progress in clinical trials of photodynamic therapy for solid tumors and the role of nanomedicine. Cancers (Basel) 12(10):2793

Bai J-W, Qiu S-Q, Zhang G-J (2023) Molecular and functional imaging in cancer-targeted therapy: current applications and future directions. Signal Transduct Target Ther 8(1):89

Bailly C, Bodet-Milin C, Bourgeois M, Gouard S, Ansquer C, Barbaud M, Sébille J-C, Chérel M, Kraeber-Bodéré F, Carlier T (2019) Exploring tumor heterogeneity using PET imaging: the big picture. Cancers (Basel) 11(9):1282

Bankman IN (2000) Introduction to segmentation. In: Bankman IN (ed) Handbook of medical imaging. Academic Press, San Diego, pp 67–68

Batista CEA, Juhász C, Muzik O, Kupsky WJ, Barger G, Chugani HT, Mittal S, Sood S, Chakraborty PK, Chugani DC (2009) Imaging correlates of differential expression of indoleamine 2,3-dioxygenase in human brain tumors. Mol Imaging Biol 11(6):460–466

Bhat GR, Sethi I, Sadida HQ, Rah B, Mir R, Algehainy N, Albalawi IA, Masoodi T, Subbaraj GK, Jamal F, Singh M, Kumar R, Macha MA, Uddin S, Akil ASA-S, Haris M, Bhat AA (2024) Cancer cell plasticity: from cellular, molecular, and genetic mechanisms to tumor heterogeneity and drug resistance. Cancer Metastasis Rev 43(1):197–228

Bi WL, Hosny A, Schabath MB, Giger ML, Birkbak NJ, Mehrtash A, Allison T, Arnaout O, Abbosh C, Dunn IF, Mak RH, Tamimi RM, Tempany CM, Swanton C, Hoffmann U, Schwartz LH, Gillies RJ, Huang RY, Aerts HJWL (2019) Artificial intelligence in cancer imaging: clinical challenges and applications. CA Cancer J Clin 69(2):127–157

Boumahdi S, de Sauvage FJ (2020) The great escape: tumour cell plasticity in resistance to targeted therapy. Nat Rev Drug Discov 19(1):39–56

Brindle KM (2024) Imaging cancer metabolism using magnetic resonance. npj Imaging 2(1):1

Budczies J, Kazdal D, Menzel M, Beck S, Kluck K, Altbürger C, Schwab C, Allgäuer M, Ahadova A, Kloor M, Schirmacher P, Peters S, Krämer A, Christopoulos P, Stenzinger A (2024) Tumour mutational burden: clinical utility, challenges and emerging improvements. Nat Rev Clin Oncol 21(10):725–742

Cai W, Hsu AR, Li Z-B, Chen X (2007) Are quantum dots ready for in vivo imaging in human subjects? Nanoscale Res Lett 2(6):265

Carter BW, Bhosale PR, Yang WT (2018) Immunotherapy and the role of imaging. Cancer 124(14):2906–2922

Cazares-Cortes E, Cabana S, Boitard C, Nehlig E, Griffete N, Fresnais J, Wilhelm C, Abou-Hassan A, Ménager C (2019) Recent insights in magnetic hyperthermia: from the "hot-spot" effect for local delivery to combined magneto-photo-thermia using magneto-plasmonic hybrids. Adv Drug Deliv Rev 138:233–246

Chang PD, Kuoy E, Grinband J, Weinberg BD, Thompson M, Homo R, Chen J, Abcede H, Shafie M, Sugrue L, Filippi CG, Su MY, Yu W, Hess C, Chow D (2018) Hybrid 3D/2D convolutional neural network for hemorrhage evaluation on head CT. AJNR Am J Neuroradiol 39(9):1609–1616

Chen L-L, Zhao L, Wang Z-G, Liu S-L, Pang D-W (2022) Near-infrared-II quantum dots for in vivo imaging and cancer therapy. Small 18(8):2104567

Chidambaram M, Manavalan R, Kathiresan K (2011) Nanotherapeutics to overcome conventional cancer chemotherapy limitations. J Pharm Pharm Sci 14(1):67–77

Chu Y, Xu X-Q, Wang Y (2022) Ultradeep photothermal therapy strategies. J Phys Chem Lett 13(41):9564–9572

Crawford J, Herndon D, Gmitter K, Weiss J (2024) The impact of myelosuppression on quality of life of patients treated with chemotherapy. Future Oncol 20(21):1515–1530

Darvin P, Toor SM, Sasidharan Nair V, Elkord E (2018) Immune checkpoint inhibitors: recent progress and potential biomarkers. Exp Mol Med 50(12):1–11

Debreczeni-Máté Z, Freihat O, Törő I, Simon M, Kovács Á, Sipos D (2024) Value of 11C-methionine PET imaging in high-grade gliomas: a narrative review. Cancers (Basel) 16(18):3200

Deng K, Yu Z-L, Hu X, Liu J, Hong X, Zi GGL, Zhang Z, Tian Z-Q (2023) NIR-II fluorescent Ag2Se polystyrene beads in a lateral flow immunoassay to detect biomarkers for breast cancer. Microchim Acta 190(12):462

Duquenoy P, George C, Solomonides A (2008) Considering something 'ELSE': ethical, legal and socio-economic factors in medical imaging and medical informatics. Comput Methods Programs Biomed 92:227–237

Essayed WI, Juvekar P, Bernstock JD, Rassi MS, Almefty K, Zamani AA, Golby AJ, Al-Mefty O (2022) Multimodal intraoperative image-driven surgery for skull base chordomas and chondrosarcomas. Cancers (Basel) 14(4):966

Ezerskyte E, Morkvenas A, Venius J, Sakirzanovas S, Karabanovas V, Katelnikovas A, Klimkevicius V (2024) Biocompatible upconverting nanoprobes for dual-modal imaging and temperature sensing. ACS Appl Nano Mater 7(6):6185–6195

Fan W, Yung B, Huang P, Chen X (2017) Nanotechnology for multimodal synergistic cancer therapy. Chem Rev 117(22):13566–13638

Fang Y, Yu A, Ye L, Zhai G (2021) Research progress in tumor targeted immunotherapy. Expert Opin Drug Deliv 18(8):1067–1090

Faucher FF, Liu KJ, Cosco ED, Widen JC, Sorger J, Guerra M, Bogyo M (2023) Protease activated probes for real-time ratiometric imaging of solid tumors. ACS Cent Sci 9(5):1059–1069

Foray C, Barca C, Backhaus P, Schelhaas S, Winkeler A, Viel T, Schäfers M, Grauer OM, Jacobs AH, Zinnhardt B (2020) Multimodal molecular imaging of the tumour microenvironment. Adv Exp Med Biol 1225:71–87

Gandy N, Arshad MA, Wallitt KL, Dubash SR, Khan S, Barwick TD (2020) Immunotherapy-related adverse effects on 18F-FDG PET/CT imaging. Br J Radiol 93:20190832

García-Figueiras R, Baleato-González S, Canedo-Antelo M, Alcalá L, Marhuenda A (2021) Imaging advances on CT and MRI in colorectal cancer. Curr Colorectal Cancer Rep 17(6): 113–130

Giraldo NA, Becht E, Vano Y, Sautès-Fridman C, Fridman WH (2015) The immune response in cancer: from immunology to pathology to immunotherapy. Virchows Arch 467(2):127–135

Harun NA, Benning MJ, Horrocks BR, Fulton DA (2013) Gold nanoparticle-enhanced luminescence of silicon quantum dots co-encapsulated in polymer nanoparticles. Nanoscale 5(9): 3817–3827

Hersh J, Yang Y-P, Roberts E, Bilbao D, Tao W, Pollack A, Daunert S, Deo SK (2023) Targeted bioluminescent imaging of pancreatic ductal adenocarcinoma using nanocarrier-complexed EGFR-binding affibody–Gaussia luciferase fusion protein. Pharmaceutics 15(7):1976

Hodge J, James G, Gostin LO, Jacobson PD (1999) Legal issues concerning electronic health information privacy, quality, and liability. JAMA 282(15):1466–1471

Hodi FS, Ballinger M, Lyons B, Soria J-C, Nishino M, Tabernero J, Powles T, Smith D, Hoos A, McKenna C, Beyer U, Rhee I, Fine G, Winslow N, Chen DS, Wolchok JD (2018) Immune-modified response evaluation criteria in solid tumors (imRECIST): refining guidelines to assess the clinical benefit of cancer immunotherapy. J Clin Oncol 36(9):850–858

Hu Z, Fang C, Li B, Zhang Z, Cao C, Cai M, Su S, Sun X, Shi X, Li C, Zhou T, Zhang Y, Chi C, He P, Xia X, Chen Y, Gambhir SS, Cheng Z, Tian J (2020) First-in-human liver-tumour surgery guided by multispectral fluorescence imaging in the visible and near-infrared-I/II windows. Nat Biomed Eng 4(3):259–271

Huang D, Yang X, Peng Z, Yin H, Liu Y, Zhang Y, Li C, Chen G, Wang Q (2024) Multichannel-optical imaging for in vivo evaluating the safety and therapeutic efficacy of stem cells in tumor model in terms of cell tropism, proliferation and NF-κB activity. Biomaterials 307:122510

Hussain S, Mubeen I, Ullah N, Shah SSUD, Khan BA, Zahoor M, Ullah R, Khan FA, Sultan MA (2022) Modern diagnostic imaging technique applications and risk factors in the medical field: a review. Biomed Res Int 2022(1):5164970

Iima M, Kataoka M, Honda M, Le Bihan D (2024) Diffusion-weighted MRI for the assessment of molecular prognostic biomarkers in breast cancer. Korean J Radiol 25:623–633

Illimoottil M, Ginat D (2023) Recent advances in deep learning and medical imaging for head and neck cancer treatment: MRI, CT, and PET scans. Cancers (Basel) 15(13):3267

Imperiale A, Berti V, Burgy M, Cazzato RL, Piccardo A, Treglia G (2024) Molecular imaging and related therapeutic options for medullary thyroid carcinoma: state of the art and future opportunities. Rev Endocr Metab Disord 25(1):187–202

Jin J, Wu X, Yin J, Li M, Shen J, Li J, Zhao Y, Zhao Q, Wu J, Wen Q, Cho CH, Yi T, Xiao Z, Qu L (2019) Identification of genetic mutations in cancer: challenge and opportunity in the new era of targeted therapy. Front Oncol 9:263

Kalamkar S, Mary AG (2023) Multimodal image fusion: a systematic review. Decis Anal J 9: 100327

Kazim M, Yoo E (2024) Recent advances in the development of non-invasive imaging probes for cancer immunotherapy. Angew Chem Int Ed Engl 63(2):e202310694

Kennedy LB, Salama AKS (2020) A review of cancer immunotherapy toxicity. CA Cancer J Clin 70(2):86–104

Kjelle E, Brandsæter IØ, Andersen ER, Hofmann BM (2024) Cost of low-value imaging worldwide: a systematic review. Appl Health Econ Health Policy 22(4):485–501

Krutzek F, Kopka K, Stadlbauer S (2022) Development of radiotracers for imaging of the PD-1/PD-L1 axis. Pharmaceuticals (Basel) 15(6):747

Krzyszczyk P, Acevedo A, Davidoff EJ, Timmins LM, Marrero-Berrios I, Patel M, White C, Lowe C, Sherba JJ, Hartmanshenn C, O'Neill KM, Balter ML, Fritz ZR, Androulakis IP, Schloss RS, Yarmush ML (2018) The growing role of precision and personalized medicine for cancer treatment. Technology (Singap World Sci) 6(3–4):79–100

Kwon M, Jung H, Nam G-H, Kim I-S (2021) The right timing, right combination, right sequence, and right delivery for cancer immunotherapy. J Control Release 331:321–334

LeCun Y, Bengio Y, Hinton G (2015) Deep learning. Nature 521(7553):436–444

Léger M-A, Routy B, Juneau D (2022) FDG PET/CT for evaluation of immunotherapy response in lung cancer patients. Semin Nucl Med 52:707

Li D, Cheng S, Zou S, Zhu D, Zhu T, Wang P, Zhu X (2018) Immuno-PET imaging of 89Zr labeled anti-PD-L1 domain antibody. Mol Pharm 15(4):1674–1681

Li B, Jin J, Guo D, Tao Z, Hu X (2023) Immune checkpoint inhibitors combined with targeted therapy: the recent advances and future potentials. Cancers (Basel) 15(10):2858

Li M, Li T, Wu F, Ren F, Xue S, Li C (2024) Advanced NIR-II fluorescence imaging technology for precise evaluation of nanomedicine delivery in cancer therapy. Chemosensors 12(6):113

Liu M, Anderson R-C, Lan X, Conti PS, Chen K (2020) Recent advances in the development of nanoparticles for multimodality imaging and therapy of cancer. Med Res Rev 40(3):909–930

Liu N, Yan M, Tao Q, Wu J, Chen J, Chen X, Peng C (2023) Inhibition of TCA cycle improves the anti-PD-1 immunotherapy efficacy in melanoma cells via ATF3-mediated PD-L1 expression and glycolysis. J Immunother Cancer 11(9):e007146

Luo Q, Zhang L, Luo C, Jiang M (2019) Emerging strategies in cancer therapy combining chemotherapy with immunotherapy. Cancer Lett 454:191–203

Mahmoudi K, Garvey KL, Bouras A, Cramer G, Stepp H, Jesu Raj JG, Bozec D, Busch TM, Hadjipanayis CG (2019) 5-aminolevulinic acid photodynamic therapy for the treatment of high-grade gliomas. J Neuro-Oncol 141(3):595–607

Mahmud MI, Mamun M, Abdelgawad A (2023) A deep analysis of brain tumor detection from MR images using deep learning networks. Algorithms 16(4):176

Mahoney KM, Rennert PD, Freeman GJ (2015) Combination cancer immunotherapy and new immunomodulatory targets. Nat Rev Drug Discov 14(8):561–584

Maldonado F, Duan F, Raghunath SM, Rajagopalan S, Karwoski RA, Garg K, Greco E, Nath H, Robb RA, Bartholmai BJ, Peikert T (2015) Noninvasive computed tomography-based risk stratification of lung adenocarcinomas in the National Lung Screening Trial. Am J Respir Crit Care Med 192(6):737–744

Mali SB (2024) Bioluminescence in cancer research—applications and challenges. Oral Oncol Rep 9:100127

Mao G-J, Zhang S-Y, Yang T-T, Zhu B, Sun X-Y, Wang Q-Q, Zhang G (2024) A tumor-targeting two-photon fluorescent probe with a far-red to NIR emission for imaging basal hypochlorite in cancer cells and tumor. Talanta 277:126436

Miranda-Gonçalves V, Lameirinhas A, Henrique R, Baltazar F, Jerónimo C (2020) The metabolic landscape of urological cancers: new therapeutic perspectives. Cancer Lett 477:76–87

Moeckel C, Bakhl K, Georgakopoulos-Soares I, Zaravinos A (2023) The efficacy of tumor mutation burden as a biomarker of response to immune checkpoint inhibitors. Int J Mol Sci 24:6710

Mokhtari RB, Homayouni TS, Baluch N, Morgatskaya E, Kumar S, Das B, Yeger HJO (2017) Combination therapy in combating cancer. Oncotarget 8:38022–38043

Najdian A, Beiki D, Abbasi M, Gholamrezanezhad A, Ahmadzadehfar H, Amani AM, Ardestani MS, Assadi M (2024) Exploring innovative strides in radiolabeled nanoparticle progress for multimodality cancer imaging and theranostic applications. Cancer Imaging 24(1):127

Nappi AG, Santo G, Jonghi-Lavarini L, Miceli A, Lazzarato A, La Torre F, Dondi F, Gorica J (2024) Emerging role of [18F]FLT PET/CT in lymphoid malignancies: a review of clinical results. Hematol Rep 16:32–41

Ning J, Li G, Zhao Y, Shi J, Nie Y, Li Y, Zhou Y (2024) Carbon dot-based nanomaterials with enzyme-like activity for fluorescence imaging and potential combination with therapy of cancer. Microchim Acta 191(12):725

Papathanassiou D, Liehn J-C (2008) The growing development of multimodality imaging in oncology. Crit Rev Oncol Hematol 68(1):60–65

Patra M, Eichenberger LS, Fischer G, Holland JP (2019) Photochemical conjugation and one-pot radiolabelling of antibodies for immuno-PET. Angew Chem Int Ed Engl 58(7):1928–1933

Petrella JR, Provenzale JM (2000) MR perfusion imaging of the brain. AJR Am J Roentgenol 175(1):207–219

Purkayastha K, Dhar R, Pethusamy K, Srivastava TP, Shankar A, Rath GS, Karmakar S (2023) The issues and challenges with cancer biomarkers. J Cancer Res Ther 19:20–35

Qiao K, Wei Y, Tao C, Zhu J, Yuan S (2024) Proton therapy for breast cancer: reducing toxicity. Thorac Cancer 15(30):2156–2165

Regmi SK, Sathianathen N, Stout TE, Konety BR (2021) MRI/PET Imaging in elevated PSA and localized prostate cancer: a narrative review. Transl Androl Urol. Jul;10(7):3117-3129. https://doi.org/10.21037/tau-21-374. PMID: 34430415; PMCID: PMC8350235

Rout S, Choudhary R, Thatoi P (2024) Enhanced prognostic assessment of glioblastoma multiforme using machine learning: integrating multimodal imaging and treatment features: a review. Iconic Res Eng J 8. ISSN: 2456-8880

Saeed RF, Awan UA, Saeed S, Mumtaz S, Akhtar N, Aslam S (2023) Targeted therapy and personalized medicine. In: Qazi AS, Tariq K (eds) Therapeutic approaches in cancer treatment. Springer International Publishing, Cham, pp 177–205

Sailer AM, van Zwam WH, Wildberger JE, Grutters JP (2015) Cost-effectiveness modelling in diagnostic imaging: a stepwise approach. Eur Radiol 25(12):3629–3637

Saleh MA, Ali AA, Ahmed K, Sarhan AM (2023) A brief analysis of multimodal medical image fusion techniques. Electronics 12(1):97

Schietinger A, Arina A, Liu RB, Wells S, Huang J, Engels B, Bindokas V, Bartkowiak T, Lee D, Herrmann A, Piston DW, Pittet MJ, Lin PC, Zal T, Schreiber H (2013) Longitudinal confocal microscopy imaging of solid tumor destruction following adoptive T cell transfer. Oncoimmunology 2(11):e26677

Smith AM, Mancini MC, Nie S (2009) Second window for in vivo imaging. Nat Nanotechnol 4(11):710–711
Soliman MA, Guccione J, Reiter AM, Moawad AW, Etchison A, Kamel S, Khatchikian AD, Elsayes KM (2020) Current concepts in multi-modality imaging of solid tumor angiogenesis. Cancers (Basel) 12(11):3239
Subramanian N, Torabi-Parizi P, Gottschalk RA, Germain RN, Dutta B (2015) Network representations of immune system complexity. Wiley Interdiscip Rev Syst Biol Med 7:13–38
Sun X, Huang X, Guo J, Zhu W, Ding Y, Niu G, Wang A, Kiesewetter DO, Wang ZL, Sun S, Chen X (2014) Self-illuminating 64Cu-doped CdSe/ZnS nanocrystals for in vivo tumor imaging. J Am Chem Soc 136(5):1706–1709
Tempany CMC, Jayender J, Kapur T, Bueno R, Golby A, Agar N, Jolesz FA (2015) Multimodal imaging for improved diagnosis and treatment of cancers. Cancer 121(6):817–827
Tubiana M (1996) [Wilhelm Conrad Röntgen and the discovery of X-rays]. Bull Acad Natl Med 180(1):97–108
Vafaei S, Zekiy AO, Khanamir RA, Zaman BA, Ghayourvahdat A, Azimizonuzi H, Zamani M (2022) Combination therapy with immune checkpoint inhibitors (ICIs); a new frontier. Cancer Cell Int 22:2
van de Donk PP, Oosting SF, Knapen DG, van der Wekken AJ, Brouwers AH, Lub-de Hooge MN, de Groot D-JA, de Vries EG (2022) Molecular imaging to support cancer immunotherapy. J Immunother Cancer 10(8):e004949
Vilgelm AE, Johnson DB, Richmond A (2016) Combinatorial approach to cancer immunotherapy: strength in numbers. J Leukoc Biol 100(2):275–290
Villanueva-Meyer JE, Mabray MC, Cha S (2017) Current clinical brain tumor imaging. Neurosurgery 81(3):397–415
Wargo JA, Reuben A, Cooper ZA, Oh KS, Sullivan RJ (2015) Immune effects of chemotherapy, radiation, and targeted therapy and opportunities for combination with immunotherapy. Semin Oncol 42(4):601–616
Xie Y, et al. (2019) Knowledge-based collaborative deep learning for benign-malignant lung nodule classification on chest CT. IEEE Trans Med Imaging 38(4):991–1004. https://doi.org/10.1109/TMI.2018.2876510
Yin R, Zhang Q, Liao S, Lu C, Zhang C, Cai Q, Song G (2024) Medical imaging-based companion diagnostics for solid tumors. EngMedicine 1(1):100009
Yue S, Zhan J, Xu X, Xu J, Bi S, Zhu J-J (2024) A "dual-key-and-lock" DNA nanodevice enables spatially controlled multimodal imaging and combined cancer therapy. Chem Sci 15(29): 11528–11539
Zhang J, Traylor KS, Mountz JM (2020) PET and SPECT imaging of brain tumors. Semin Ultrasound CT MRI 41(6):530–540
Zhang J, Huang D, Saw PE, Song E (2022) Turning cold tumors hot: from molecular mechanisms to clinical applications. Trends Immunol 43(7):523–545
Zhang J, Du B, Wang Y, Cui Y, Wang S, Zhao Y, Li Y, Li X (2024) The role of CD8 PET imaging in guiding cancer immunotherapy. Front Immunol 15:1428541
Zhang H, Zhang M, Xu T, Wang X, Qi J, Wang Y, Liu W, Zhu L, Yuan Z, Si C (2025) Ions and electrons dual transport channels regulated by nanocellulose for mitigating dendrite growth of zinc-ion batteries. Chem Eng J 505:159476
Zhao X, Ma Y, Lei Z (2024a) Advanced optical imaging technology in the near infrared window for cell tracking in vivo. Coord Chem Rev 521:216178
Zhao Y, Hsu JC, Hu S, Cai W (2024b) PET imaging of PD-L1 with a small molecule radiotracer. Eur J Nucl Med Mol Imaging 51(6):1578–1581
Zhou Z, Edil BH, Li M (2023) Combination therapies for cancer: challenges and opportunities. BMC Med 21(1):171

Molecular Imaging and Therapeutic Approaches

5

Rahaman Shaik, Siddharth Krishnan, Rajendra Prasad,
Shweta Meena, Shaheen Mahira, and Nagavendra Kommineni

Abstract

Imaging technologies have been essential for visualizing tumor progression for the last century and even before that. Conventional methods, such as X-ray probing, computed tomography (CT), and magnetic resonance imaging (MRI), have been extensively implemented and studied for the visualization of tumors. This is very useful for measuring the progress of a tumor, which can have applications for diagnosing cancer and measuring treatment efficacy in cancer patients.

Shaheen Mahira and Nagavendra Kommineni contributed equally.

R. Shaik
Department of Pharmacology, Jamia Hamdard, New Delhi, India

S. Krishnan
Department of Bioengineering, University of Pittsburgh, Pittsburgh, PA, USA

R. Prasad
School of Biochemical Engineering, Indian Institute of Technology (BHU) Varanasi, Varanasi, Uttar Pradesh, India
e-mail: rajendra.bce@iitbhu.ac.in

S. Meena
Department of Electronics and Communication Engineering, National Institute of Technology Kurukshetra, Kurukshetra, Haryana, India

S. Mahira (✉)
Department of Obstetrics, Gynecology and Reproductive Sciences, University of Pittsburgh, Pittsburgh, PA, USA

N. Kommineni (✉)
Thermo Fisher Scientific Inc., Cincinnati, OH, USA

R. Prasad, M. Mohan Yallapu (eds.), *Advancements in Cancer Theranostics*,
https://doi.org/10.1007/978-981-95-0938-6_5

Keywords

Solid tumor · Cancer therapy · Nanotheranostics · Molecular imaging · Computed tomography

5.1 Introduction

Imaging technologies have been a vital aspect of early diagnosis and treatment of many solid tumors. However, as diagnosis has improved with better technology, there have been many cases in which imaging resolution is inadequate. In surface-level solid tumors such as melanoma, surgical removal is quite simple due to the mostly noninvasive nature of the procedure and its clear visibility in the epidermal layer of the skin during the early stages of the cancer (Davis et al. 2019). For these reasons, the survival rate of stage 1 melanoma is very high with minimal side effects. In solid tumors of internal organs, such as breast and lung cancers, although imaging the tumor is possible with a CT or MRI scanner, detection may not occur until the tumor has developed significantly. Once a cancer has infiltrated lymph nodes, surgical removal itself will not eliminate the chance of metastasis, and alternative therapeutic options such as adjuvant chemotherapy must be utilized alongside surgery (West and Jin 2015). When using therapeutic treatments in solid tumors, issues tend to arise. As a result of irregular vasculature combined with tumor growth being much faster than angiogenesis, blood flow into the tumor may be insufficient, and hypoxia may arise. Hypoxia can induce treatment resistance in tumor cells. The inadequate vasculature and the higher pressure within tumors are also barriers for treatment entering the tumor (Kabakov and Yakimova 2021).

Molecular imaging is a broad set of modern imaging techniques that aim to visualize the overall cellular processes that occur in live biological tissue. By analyzing the biochemical changes in cells, molecular imaging has the potential to detect the formation of tumors far earlier than can be physically visualized in conventional imaging (Luengo Morato et al. 2021). Even more promisingly, treatment can potentially be assessed earlier by measuring immediate changes in a target downstream effector molecule. The basis of most molecular imaging techniques is "probing," wherein a probe enters the body, typically through intravenous injection into the bloodstream, and interacts with target cell tissue (Chen and Chen 2010). These probe molecules have special properties that allow them to be detected by imaging technologies. For example, for positron-emission tomography (PET), radioactive isotopes such as fluorine-18 and carbon-11 are used as probes because they emit positrons when radioactive decay occurs, which can then be detected by the PET machine (Pees et al. 2023; Alauddin 2012). Existing target molecules that are heavily involved in the cellular processes of tumors can be labeled with radioisotopes, so that accumulations of the radioisotope in tumors can be determined and visualized.

Alongside molecular imaging, theranostics is an emerging technology in cancer treatment that operates similarly. Like molecular imaging, radioactive probes are injected into the body and bind to target molecules. On top of the diagnosis and visualization of tumors, theranostics also uses the radioactive properties of the probe to deliver treatment directly to the tumor cells. Radiation therapy is already frequently used to treat cancer by causing double-strand breaks, which in turn cause cell death. Theranostics uses the radioisotopes intended for visualization to also deliver radiation therapy into the cells, initiating cell death (Burkett et al. 2023). Compared to conventional radiation therapy, the treatment is more localized toward tumor target markers, so less damage would be done to surrounding healthy tissue. One example of a theranostic probing isotope is lutetium-177, a probe that accomplishes both particle emissions for visualization and sufficient radiation to cause fatal DNA damage in tumor cells (Burkett et al. 2023).

In this chapter, various molecular imaging techniques that are currently used in solid tumor detection will be detailed. The functionality of each molecular imaging method and their advantages and disadvantages, as well as current nanotheranostic approaches and recent advances, will be discussed.

5.2 Solid Tumors and Difficulties Encountered in Their Management

Solid tumors are agglutinations of tissue cells growing uncontrollably. They can be classified into benign tumors, which are abnormal growths that remain localized within one region, and malignant tumors, which can infiltrate the bloodstream and cause metastasis in another region of the body. If solid tumors get large enough and infiltrate lymph nodes, conventional therapeutic measures must be taken to combat the growth and potential metastasis of the tumor (West and Jin 2015). When it comes to administering these treatments, however, there are certain obstacles caused by the physiology of solid tumors that prevent the proper delivery of the treatment to the tumor site.

Some tumor cells continue growing for an extended period due to new vasculature being formed to sustain the tumor cells. Vasculature in solid tumors is much more poorly constructed than in healthy tissues, and the uncontrolled nature of the cell growth creates many hypoxic areas, wherein tumor cells are far from vessels (Siemann 2011). Cancer cells undergoing hypoxia can often show resistance to conventional treatment due to the change in their metabolic pathways because of the reduced oxygen supply (Kabakov and Yakimova 2021). Another issue faced in the management of solid tumors is metastasis, which is the infiltration of solid tumor cells into a different area of the body. Hypoxia and the associated necrosis of tumor cells are thought to be a major cause of metastasis (Wang et al. 2015).

Another factor that makes it difficult to manage solid tumors is the heterogeneity of cancer cells. This refers to the existence of many different subpopulations of a cancer cell line with different DNA sequences that can form in a single solid tumor (Gorodetska et al. 2019). This causes solid tumors to have a high genetic diversity, which can pose many problems for cancer treatment. Primarily, this may cause many cells in a solid tumor to develop resistance mutations to therapeutic options. While one treatment may be successful in eradicating most of a solid tumor, there is a chance of resistant tumor cells lingering on and relapse occurring. It is believed that tumor heterogeneity contributes to metastasis because the invasion, intravasation, and extravasation processes are only possible with very specific combinations of genetic mutations (Lawson et al. 2018). Solid tumors also have a higher interstitial pressure compared to healthy tissue, which is in turn due to tumor cell structure having a higher stiffness (Salavati et al. 2022). There are several potential reasons for this disparity, including the extracellular matrix becoming more compact and more erratic blood flow caused by faulty vasculature. Targeted chemotherapy delivery can be difficult to administer to the solid tumor region because the high interstitial pressure can be too strong for passive flow into the tumor to occur (Lunt et al. 2008).

The field of nanotechnology shows promise in getting around these issues. This involves injecting nanoparticles directly into the bloodstream and allowing tumor cells to take up these nanoparticles. While current chemotherapy operates on the same principle, nanoparticles can be designed to selectively be taken in by tumor cells and induce greater damage to them. For example, the carbon nanoparticle graphene oxide (GO) has been tested to deliver chemotherapy to only cells with gastric cancer markers. Because GO has a large surface area and a relatively high binding affinity, tumor-specific ligands can be pre-attached to GO to ensure specific binding and release of chemotherapy into tumor cells, and large amounts of chemotherapeutic drugs can be loaded onto GO (Sharma and Mondal 2020). The active targeting of tumor cells through ligand–receptor chemotaxis can overcome the vasculature insufficiency and interstitial pressure.

5.3 Nanotechnology-Based Molecular Imaging Modalities for Solid Tumors

As previously discussed, nanotechnology can be an impactful tool for more accurately targeting treatment and providing sufficient doses to completely get rid of a solid tumor with minimal damage to surrounding tissue. However, in the case of imaging, nanoparticles are increasingly being researched for their potential in improving existing imaging modalities. Nanoparticles can easily bind to biological tissue, so ligands that bind to tumor membrane receptors of interest can be fitted with nanoparticles and inserted into the body, as discussed in Table 5.1. Doing this increases the selectivity of nanoparticles toward tumor proteins of interest, which improves contrast in the final image (Lan et al. 2023). In this section, common molecular imaging techniques and their nanoparticle-labeled modalities will be discussed.

Table 5.1 Various molecular imaging modalities and their nanoparticle applications

Imaging modality	Principle	Nanosystem	Application	Refs.
PET	Utilizes the positron emission of radioactive compounds to initiate positron–electron collisions which release gamma rays. These gamma rays are detected by the PET scanner	Use of radioactive transition metals, most commonly ^{64}Cu, that provide positron decay for PET as well as the ability to bind organic polymers and ligands for increased targeting and contrast in tumors	– Visualizing cancer/metastasis progression posttreatment – Visualization of brain tissue for neurodegenerative diseases	Kapoor and Kasi (2024), Capriotti et al. (2020)
SPECT	Utilizes radioactive compounds that undergo gamma decay to directly detect gamma rays in the scanner	Use of radioactive transition metals, most commonly involving ^{99m}Tc/^{99}Tc, that undergo gamma decay. It is used in TcO_2 form, which allows for easy chelation to target ligands for improved contrast in tumors	– Diagnosing bone and blood flow-related issues	Crisan et al. (2022), Kane et al. (2024)
Fluorescence imaging	Utilizes fluorophores, which are chemicals that emit a wavelength of light in response to being excited by another wavelength of light. The emitted light wavelength can be detected with a specialized detector	Use of quantum dots, which are extremely small semiconductor metals with fluorescent properties. They have a polymer coat to allow ligands to bind to them to specifically target cells of interest	– Visualizing cell signaling pathways and cell motility – Visualizing organelles of cells in vitro	Ettinger and Wittmann (2014), Zhang et al. (2009)
Photoacoustic imaging	Utilizes the photoacoustic effect of certain materials. When laser pulses hit the material, it releases sound waves which can be detected	Use of gold nanoparticles, which undergo resonance at the electron level when infrared light excites them	– Tumor visualization using exogenous photoacoustic contrast agents – Visualizing vasculature due to photoacoustic properties of hemoglobin	Barbosa and Mendes (2022), Xia et al. (2014), Li and Chen (2015)

(continued)

Table 5.1 (continued)

Imaging modality	Principle	Nanosystem	Application	Refs.
MRI	Checks for radio waves that are released when hydrogen protons shift because of a disturbance in a magnetic field	Use of superparamagnetic iron oxide particles, which can form their own magnetic field when induced by an external magnetic force	– Diagnosing tumors/ metastases and treatment progression – Brain tissue visualization	Berger (2002), Wahajuddin and Arora (2012)
MRS	Analyzes the radio waves from MRI and checks for specific intensities at chemical shift values to determine chemical composition	Use of naturally occurring fluorine, which changes in chemical shift when bound to different organic groups	– Checking for metabolites on tumor surfaces to assess treatment options – Diagnosis of specific neurodegenerative diseases	Tognarelli et al. (2015), Wolf et al. (2000)
Ultrasound	Emits ultrasonic waves into the body and measures the reflection against internal structures	Nanobubbles use the acoustic properties of certain gases surrounded by a lipid layer to provide ultrasound contrast at a nanoscale	– Speedy, noninvasive method for visualizing internal cell tissue in a variety of specialties	Schutt et al. (2003), Pillay et al. (2014)
Optical imaging	Uses light scattering and refraction differences in materials to map out internal structures	Gold nanoparticles have unique Raman scattering properties that make them useful for use in Raman microscopy, a type of optical imaging	– Label-free diagnosis of solid tumors in cancer	Nourizad et al. (2023), Benefits of Nanotechnology for Cancer (n.d.)

5.3.1 Positron Emission Tomography

Positron emission tomography (PET) is an imaging technique that detects positron emission caused by radioactivity. Positrons are released alongside neutrinos in a specific form of radioactive decay known as beta-plus decay, wherein an atomic proton loses its positive charge and becomes a neutron. When these positrons collide with the abundance of electrons around them, they release gamma rays, which are detected by the PET scanner. Prior to a PET, a patient is intravenously supplied with a radioactively labeled molecule, and after some time (depending on the half-life of the radioisotope used), the PET scan can detect accumulation of the radiolabels in different areas of the body (Kapoor and Kasi 2024).

In the field of oncology, the most common radioactively labeled molecule used in clinical settings is ^{18}F-fluorodeoxyglucose (^{18}F-FDG). This is simply an analog of glucose, with one of the hydroxyl groups replaced with F-18, a radioisotope of fluorine. The rate of glucose uptake gives insight into the metabolic rate of cells, and areas with abnormally high glucose uptake rates can be an indication of tumor activity (Almuhaideb et al. 2011). By attaching a radioisotope to glucose, tumors with a high cellular metabolic rate can be clearly visualized by the increased gamma radiation being detected in those areas. In specific tumors, other molecules may be a better option than glucose. In some adrenal cancers, uptake of metomidate is noticeably higher than in normal cells. A ^{11}C-metomidate complex was developed and shows promising results in clinical trials (Minn et al. 2004).

The development of radiolabeled nanoparticles has also been researched. Radioisotopes of metal atoms have been proposed as a nanoparticle label modality for PET. In one study, ^{64}Cu was synthesized and was shown to be a promising radioisotope in testing. The copper ion has a property called chelation, which can allow it to bind organic compounds. For a large variety of cancers, ^{64}Cu was successfully bound to target ligands and antibodies and showed promising signs for potential use in clinical PET settings, as shown in Fig. 5.1 (Capriotti et al. 2020). For example, researchers formed a ^{64}Cu complex with trastuzumab, an inhibitor of the HER2 oncogene in breast cancer, in mice. This was done to monitor the efficacy of the trastuzumab using PET. The study claimed that complex was an effective target for HER2, and that human trials could be on the horizon (Woo et al. 2019).

5.3.2 Single-Photon Emission Computed Tomography

Single photon emission computed tomography (SPECT) is another imaging technique that uses radioactive labeling molecules to measure target effects. While PET relies on the effects of beta-plus radioactive decay, SPECT utilizes a form of radioactive decay known as gamma decay. In gamma decay, an atom releases energy in the form of a gamma ray, as it shifts from an excited state to a stable, relaxed state. These gamma rays are detected by the SPECT scanner, and the location of these gamma rays is visualized. While both PET and SPECT scanning detect gamma rays,

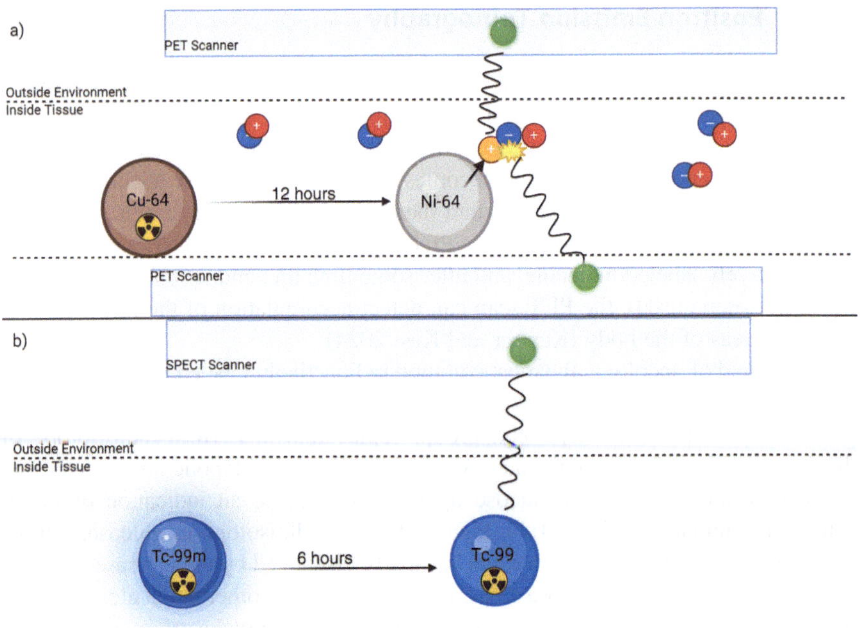

Fig. 5.1 (**a**) Nanoparticles such as Cu-64 are used to achieve increased contrast in PET. Cu-64 undergoes beta decay, wherein a positron (in yellow) is released and the atom changes into the stable Ni-64. The positron hits an electron within the bodily tissue, and this causes the release of two gamma rays in opposite directions, which are detected by the circular PET scanner. (**b**) Tc-99m is a metastable state of Tc-99 that undergoes gamma decay to become Tc-99. This gamma radiation is detected by the SPECT scanner

PET gamma rays are formed through positron–electron interactions, while in SPECT, gamma rays are formed directly through decay (Crisan et al. 2022).

In oncology, SPECT is not frequently used for imaging, as PET technologies are more extensively researched and have better resolution for a typically cheaper price (Hicks and Hofman 2012). The most common isotope used for SPECT is ^{99}Tc, which undergoes gamma decay with a half-life of about 6 h, as addressed in Fig. 5.1 (Kane et al. 2024). It is used in a complex with methyl diphosphonate (MDP), a bisphosphonate that osteoblasts heavily uptake, for detecting metastases in bone tissue (Zhong et al. 2015). ^{99}Tc can also be bound to nanoparticles by reducing it with an acid, most commonly tin (II) chloride. Conventionally, radiopharmaceutical technetium is synthesized in the pertechnetate form (TcO_4^-), which has an oxidation state of +7. By reduction with tin (II) oxide, technetium (IV) dioxide is formed, which has a lower oxidation number, providing more options for binding to organic compounds (Luksic et al. 2015).

5.3.3 Fluorescence Imaging

As the name implies, fluorescence imaging uses fluorescent labels to visualize target molecules. Fluorophores are molecules that emit a certain wavelength of light when activated by light waves hitting them. They have a specific excitation wavelength range, which is the wavelength range of light necessary to cause the fluorophore to emit light, and an emission wavelength range, which is the wavelength of light being emitted by the fluorophore (Ettinger and Wittmann 2014). This effect is exploited in fluorescent microscopy, where molecules of interest are labeled with fluorophores and allowed to circulate through the body (Grimm and Lavis 2022).

In laboratory settings, the excitation and emission wavelengths of fluorophores are not a major concern, but in a clinical setting, it can be more difficult to find fluorophores with an ideal excitation-emission spectrum. Visible light wavelengths are unable to penetrate deep through tissue, so near-infrared fluorophores such as methylene blue and indocyanine green are typically used (Van Manen et al. 2018). One issue with fluorescence microscopy is the photobleaching effect, which is the reduction in fluorescence of a fluorophore for extended periods of excitation. A method in which photobleaching has been minimized is through pulsation patterns of light. As the pulsation frequency of light increases, the fluorophore intensity decay rate decreases significantly (Boudreau et al. 2016).

Recently, nanoparticles known as quantum dots have shown promise in improving fluorescence imaging. Quantum dots are extremely small semiconductors that have strong fluorescent properties. They can be coated with a variety of molecules, depending on what the dots are being used for. For example, this allows them to be easily bound to molecules of interest. In one study, quantum dots were bound to insulin-like growth factor 1 receptor (IGF1R) ligands for fluorescence imaging. IGF1R is a membrane protein that is overexpressed in some breast cancers, and by attaching quantum dots to IGF1 ligands, the exact location of IGF1R accumulations, and thus breast cancer cells, can be found with high accuracy. Furthermore, quantum dots are not significantly affected by photobleaching, which can make it an impactful tool for accurate, long-term imaging of tumors (Zhang et al. 2009).

5.3.4 Photoacoustic Imaging

Some materials have the ability to emit sound waves when light waves excite them, which is known as the photoacoustic effect. This is caused by the increase in kinetic energy in the material from the incidence of light, causing thermal expansion cycles. The material used for this must be able to absorb enough light to allow for this increase in thermal energy. Photoacoustic imaging utilizes this effect by using laser pulses and ultrasonic detectors to map out specific targets (Barbosa and Mendes 2022). Some molecules naturally produced in the body already have this effect. Both melanin and hemoglobin have a high absorbance coefficient over many wavelengths of light, and they have been extensively tested in photoacoustic imaging research

(Barbosa and Mendes 2022). However, one issue that arises is the high amount of these molecules all over the body. Because these will be detected everywhere, the specificity of photoacoustic imaging techniques with internally produced targets is very low. Therefore, clinical photoacoustic imaging most commonly uses exogenous agents, which, much like the imaging techniques previously discussed, are laboratory-produced labeling molecules with the desired photoacoustic effect (Xia et al. 2014). One promising agent was elemental gold in nanoparticle form. Gold has a high absorption coefficient in the near-infrared light range and provides great contrast in its detection. The underdeveloped vasculature typical of solid tumors allows these gold nanoparticles to easily diffuse and accumulate inside tumors, while healthy tissue vasculature is able to prevent the spillage of gold nanoparticles into the tissue (Li and Chen 2015). The combination of the photoacoustic properties of gold and its selective permeability in nanoparticle form makes it a great labeling agent for photoacoustic imaging in tumors.

5.3.5 Magnetic Resonance Imaging

Magnetic resonance imaging (MRI) is a common imaging technique wherein a strong magnetic field is used to measure the alignment of protons in the body. Initially, the protons in the cell tissue align with the magnetic field, and after this occurs, radio wave pulses are emitted toward the area of interest. This causes the protons to temporarily misalign, and when the radio wave pulses are halted, the protons realign again. When charged particles like protons move in a magnetic field, they release energy in the form of radio waves, and an MRI machine detects these radio waves that occur from proton realignment. MRI is very versatile, widely used in clinical practice, and it is particularly effective in measuring proton movement in lipids and water, which are both prevalent in all organs of the body (Berger 2002).

In oncological settings, MRI can be used to check for various features of tumors. Due to the leaky vasculature of tumors, their water content tends to be higher than normal tissue, providing more protons for a brighter MRI image. When gadolinium is used, it also pools in larger amounts within tumor tissue, further enhancing the image. When imaging areas with high proton realignment, such as brain and bone tissue, finding gaps in intensity can indicate a potential metastasis (Fink and Fink 2013). Typically, gadolinium is intravenously injected into the bloodstream prior to an MRI, which improves contrast in the final MRI image. Gadolinium is mainly selected due to its paramagnetic properties, which allow more protons to be aligned by the magnetic field. The time it takes for the realignment of protons after the radiofrequency pulse is also significantly shortened, which makes tissues that absorb gadolinium appear brighter in the image (Ibrahim et al. 2024).

A nanoparticle labeling agent that has been recently developed for use in MRI is the superparamagnetic iron oxide nanoparticle (SPION). Outside of MRIs, SPIONs are used to magnetically transport drugs that are bound to them into tissues of interest. In the case of MRI, iron oxide analogs such as magnetite and maghemite respond strongly to magnetic field changes in MRI and provide great contrast in the final images. SPIONs are coated with polymers that give them the ability to have biological molecules attach to them. When ligands are attached to SPIONS, they have the ability to target specifically cancer cells, and this can provide a more detailed picture of the location and spread of a tumor (Wahajuddin and Arora 2012).

5.3.6 Magnetic Resonance Spectroscopy

The function of magnetic resonance spectroscopy (MRS) is very similar to MRI. In both techniques, magnetic proton alignment is temporarily interrupted by radio frequency signals, and the realignment of the protons can be measured. However, in MRS, instead of focusing only on proton alignment, it also focuses on other charged molecules. This can provide insights into the specific chemical structures and processes that occur within cells at a given time, instead of only on the structure of the cell tissue. The radio waves emitted during magnetic realignment have different waveform patterns for different chemical compounds, because the electron cloud around atomic protons causes an interference pattern (Tognarelli et al. 2015).

In MRS, a quantity known as chemical shift is measured, and different materials have different chemical shift values. Chemical shift is the phase shift in the observed radiofrequency waves in relation to a reference waveform. Common oncological markers that are measured in MRS include choline, creatine, and *N*-acetyl aspartate (NAA), which are all metabolites. The chemical shift relative to tetramethylsilane has been experimentally determined for a large number of metabolite cancer markers. During spectroscopy, peaks observed at the chemical shift value of a target molecule can indicate the amount of that molecule within the tissue imaged, which can be useful for diagnosing and visualizing cancer (Weinberg et al. 2021). MRS can also be conducted with nanolabels such as ^{19}F, which is responsive to magnetic resonance and has a detectable chemical shift. When fluorine is bound to different organic molecules, the chemical shift changes, and this can give insight into the chemical makeup of the compounds that fluorine binds to (Wolf et al. 2000).

5.3.7 Ultrasound

Ultrasound is a widely used clinical imaging technology that uses the echo of high-frequency sound waves to map out the structure of a tissue, as shown in Fig. 5.2. A transducer is placed against the skin, and sound waves are propagated through the skin into the body. When sound waves are reflected, they are detected by the transducer, and the intensities are mapped onto an image. In clinical oncology, ultrasound can be used as a preliminary tool to visualize the structure of an organ, but it is not as detailed or accurate as the previously discussed technologies. In MRI, for example, the properties of tissue are measured at the proton level, which provides far more detail about the intricacies of cellular structure compared to measuring the acoustic properties of a large area. Therefore, very small or emerging tumors may be completely skipped over in ultrasound imaging (Uematsu 2014). In tumors located

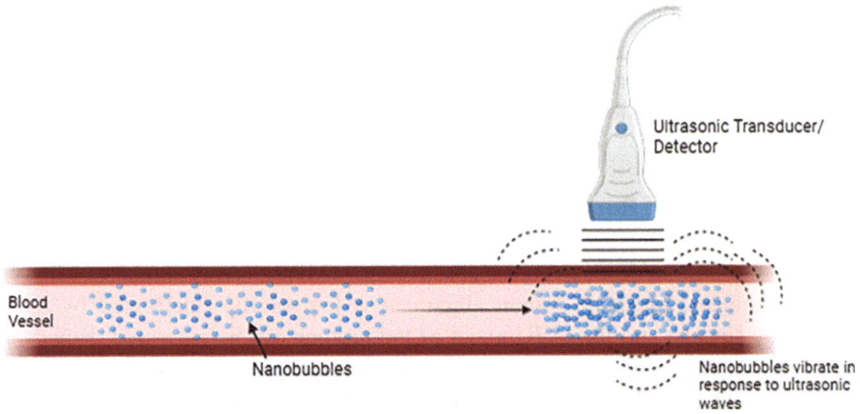

Fig. 5.2 The gases in nanobubbles cause the bubbles to oscillate when ultrasonic signals hit them. These oscillations produce sound waves which can be detected by the ultrasonic transducer

deep within the body, there may be some level of scattering that occurs in the sound waves, which decreases the resolution of the image obtained (Grogan and Mount 2024).

However, ultrasound continues to be used clinically due to its exemplary speed and noninvasive properties. The time it takes from conducting the test to obtaining an image can be as short as 20 min. In suspected tumors, a specialist can conduct an ultrasound to determine whether an irregular mass has been found and to move forward with a more detailed cancer imaging method, such as MRI, CT, or PET scanning. Ultrasound is also used extensively in a clinical setting for extracting internal tumor tissue for biopsies. It provides a rough image of abnormal cell growth and is a useful tool to determine where and how to place a needle for collecting the tumor cells (Guo et al. 2018).

Conventionally, a contrast agent known as a microbubble is a common option if more contrast to ultrasound images is needed. These consist of a lipid bilayer sphere around an insoluble gas that can reflect certain frequencies of sound waves. Microbubbles are extremely useful for visualizing blood movement, but their large size prevents their diffusion into and visualization of tissues (Sridharan et al. 2021).

Recently, smaller versions of microbubbles dubbed "nanobubbles" have been developed to counteract this issue. Nanobubbles have a similar structure to microbubbles, except that the lipids chosen for the shell are much thinner, which allows them to be smaller and more flexible than microbubbles. This allowed them to easily leave the vasculature into the tissues of interest. However, this comes at a tradeoff of reduced stability compared to microbubbles. Seemingly negligible factors such as the type of anesthesia used and slightly incorrect intravenous injection of the nanobubbles can cause nanobubbles to destabilize (Cooley et al. 2023).

5.3.8 Optical

Optical imaging can be loosely categorized into several related technologies that all utilize different optical material properties to form an image. One extensively studied modality of optical imaging in oncology is optical coherence tomography (OCT). OCT utilizes the refractive properties of cells to differentiate between healthy and tumor tissue, and it has been studied for use in many different cancer types. OCT primarily uses UV wavelengths for imaging, due to its increased skin penetration compared to visible light (Wang et al. 2017a). Compared to normal cell structure, solid tumors tend to have an irregular structure, and the variability in the structure between cells is much higher in solid tumors. The change in cell structure can yield a different refractive index, and the amount of light reflected toward the detector varies for tumors and normal cell tissue (Parvin, Ahmed et al. 2021).

Another new optical imaging technique in research is Raman spectroscopy, which utilizes the Raman scattering effect of photons to provide an image of soft cellular tissue. Most light waves do not change in speed and wavelength when they hit a surface, only the direction of their propagation. However, a small portion of the light waves that hit the surface not only change direction but also change their wavelength, and this is known as Raman scattering (Saletnik et al. 2021). Similar to a prism separating the wavelengths of white light into its constituent wavelengths, Raman spectroscopy measures the constituent wavelengths when near-ultraviolet light comes into contact with a refractive surface like cell tissue. The Raman spectra of cancer cells also differ slightly from normal cells in many different cancers, and these differences can be used for diagnostic purposes (Cui et al. 2018).

5.4 Tracers for Molecular Imaging of Disease

Molecular imaging is a powerful technique that plays a crucial role in the early detection, diagnosis, and monitoring of various diseases, as shown in detail in Table 5.2. At the heart of molecular imaging are tracers, consisting of a targeting component that binds to a specific biological target and a signaling component that emits detectable signals, allowing the visualization of targeted processes in the body (Czernin et al. 2017). The choice of tracer depends on the imaging modality being used and the biological process or disease being studied. These tracers can be radioactive, fluorescent, or specific ligands that bind to targets within tissues, aiding in the diagnosis, staging, and monitoring of diseases, including cancer, neurological disorders, and cardiovascular conditions (Jiang 2023).

Some common types of tracers used in molecular imaging include the following.

5.4.1 Radiotracers

Radioactive isotope-labeled molecules used primarily in nuclear medicine techniques like positron emission tomography (PET) and single-photon emission

Table 5.2 Molecular imaging targets with potential probes and tracers for each target

Target	Probes	Nano-based probes	Refs.
Multidrug resistance	99mTc-MIBI, [11C] Verapamil, [18F] Paclitaxel, [18F] Fluoroethyl-L-tyrosine (FET)	Doxorubicin-loaded nanoparticles, gold nanoparticles (AuNPs) conjugated with P-glycoprotein inhibitors, liposomal [18F]Paclitaxel	Wiebe (2004), Becker et al. (2020), Lee et al. (2012)
Glucose metabolism	[18F]FDG (fluorodeoxyglucose), [11C]Acetate	[18F]FDG-loaded liposomes, carbon quantum dots, gold nanoparticles functionalized with glucose	Pulumati et al. (2023), Parihar et al. (2023), Ben-Haim and Ell (2009)
Amino acid metabolism	[11C]Methionine, [18F]Fluciclovine, [18F]FET	[18F]Fluciclovine conjugated liposomes, iron oxide nanoparticles (IONPs) targeting amino acid transporters	Alford et al. (2009), Kastelik-Hryniewiecka et al. (2022)
Tumor vascularization	[64Cu]RGD peptide, [18F]RGD, [68Ga] RGD	Gold nanoshells, [64Cu]-labeled RGD peptide-conjugated nanoparticles, liposomal [68Ga]RGD	Ramzan and Tafti (2023), Rahman et al. (2019), Kumar et al. (2024)
Integrin	[68Ga]RGD, [18F] Galacto-RGD	[68Ga]RGD-conjugated nanoparticles, iron oxide nanoparticles coated with RGD peptides	Novruzov et al. (2022), Kong et al. (2012)
VCAM-1	[89Zr]VCAM-1 antibody, [64Cu] VCAM-1 targeted peptides	Liposomes coated with anti-VCAM-1 antibodies, [64Cu]-labeled VCAM-1 targeting quantum dots	Kuten et al. (2020), Nanni et al. (2007)
VEGF/VEGF	[89Zr]Bevacizumab, [64Cu]VEGF121, [111In]VEGF antibodies, [18F]RGD, [64Cu]RGD	[89Zr]-labeled bevacizumab-loaded nanoparticles, gold nanoparticles functionalized with VEGF antibodies and angiogenesis-targeting peptides, [64Cu]-liposomes targeting VEGFR, [18F]RGD-loaded silica and metal nanoparticles	Salskov et al. (2007), Weis and Cheresh (2011), Cai et al. (2005), Johanssen et al. (2023), Backer and Backer (2012), Pellet-Many et al. (2008), Rylova et al. (2014), Chakravarty et al. (2015)

(continued)

Table 5.2 (continued)

Target	Probes	Nano-based probes	Refs.
Tumor hypoxia	[18F]FMISO (Fluoromisonidazole), [18F]FAZA, [64Cu] ATSM	[18F]FMISO-conjugated nanoparticles, [64Cu] ATSM-loaded polymeric nanoparticles, hypoxia-responsive IONPs	Pecheur et al. (2002), Petrova et al. (2018), Zeng et al. (2016), Zou et al. (2016)
ECM proteins	[99mTc]Collagen-specific tracers, [64Cu] Collagen-binding peptides	Collagen-targeted nanoparticles, iron oxide nanoparticles, [99mTc]-labeled ECM-targeting liposomes	Das et al. (2022), Yu et al. (2016), Carles et al. (2021)
MMPs	[99mTc]MMP inhibitors, [68Ga] MMP-targeted peptides, [18F]MMP inhibitors	[68Ga]-Labeled MMP inhibitor-conjugated nanoparticles, gold nanoparticles targeting MMPs, [99mTc]-MMP targeting liposomes	Wahyudi et al. (2016), Li et al. (2014), Zhou and Lu (2017), Starmans et al. (2013)
Apoptosis	[99mTc]Annexin V, [18F]ML-10, [18F] CP18	[18F]ML-10 encapsulated in liposomes, quantum dots conjugated with Annexin V, [99mTc] Annexin V-loaded nanoparticles	Han and Lu (2017), Zhi et al. (2020), Zhou and Lu (2017), Heath and Bicknell (2009)
Cell tracking-MSCs	[111In]Oxine, [89Zr] Oxine	Iron oxide nanoparticles (SPIONs) labeled MSCs, gold nanoparticles, [89Zr]-labeled MSC-loaded nanoparticles	Breznik et al. (2017), Abdel-Hamid and Abass (2021), Sobhanan et al. (2023)
Somatostain	[68Ga]DOTATATE, [111In]Octreotide,	[68Ga]DOTATATE-functionalized nanoparticles, liposomal [111In] Octreotide,	Liu et al. (2015a), Anani et al. (2016), Zhang et al. (2018)
CD22	Bectumomab	Anti-CD22 antibody-conjugated quantum dots	Mathur et al. (2019)
CEA	Arcitumomab, Satumomab	Anti-CEA antibody-conjugated gold nanoparticles	Yamagata et al. (1998)

(continued)

Table 5.2 (continued)

Target	Probes	Nano-based probes	Refs.
PSMA	Capromab pendetide	PSMA-targeted superparamagnetic iron oxide nanoparticles	Neophytou et al. (2021)
VIP/VIPR-1	[123I]VIP	VIP-conjugated nanoparticles	Feng et al. (2018)
Bombesin, gastrin-releasing peptide/gastrin-releasing peptide receptor	[99mTc]Bombesin	68Ga-NOTA-Aca-BBN gold nanoparticles	Li et al. (2022), Ma et al. (2014), Wang et al. (2024)

VCAM-1 vascular cell adhesion molecule 1, *VEGF* vascular endothelial growth factor, *VEGFR* vascular endothelial growth factor receptor, *ECM* extracellular matrix, *MMPs* matrix metalloproteinases, *MSCs* mesenchymal stem cells, *VIP* vasoactive intestinal peptide, *VIPR-1* vasoactive intestinal peptide receptor 1, *RGD* arginine-glycine-aspartic acid, *FMISO* fluoromisonidazole, *FAZA* fluoroazomycin arabinoside, *ATSM* diacetyl-bis(*N*4-methylthiosemicarbazone), *MIBI* methoxyisobutylisonitrile, *FDG* fluorodeoxyglucose, *FET* fluoroethyl-L-tyrosine, *DOTATATE* DOTA-Tyr3-octreotate, *NOTA* 1,4,7-triazacyclononane-1,4,7-triacetic acid, *Aca* aminocaproic acid, *BBN* bombesin, *PSMA* prostate-specific membrane antigen, *CEA* carcinoembryonic antigen, *SPION* superparamagnetic iron oxide nanoparticle, *IONP* iron oxide nanoparticle, *AuNP* gold nanoparticle

computed tomography (SPECT). Examples include 18F-fluorodeoxyglucose (FDG), 11C-Pittsburgh compound B for amyloid plaques in Alzheimer's disease, and 99mTc-sestamibi for myocardial perfusion imaging (Bao et al. 2018).

5.4.2 Optical Imaging Agents

These tracers emit light, usually in the near-infrared spectrum, when excited. They are used in optical imaging techniques like fluorescence and bioluminescence imaging. For example, indocyanine green for assessing liver function and lymphatic mapping, and 5-aminolevulinic acid for fluorescence-guided surgery in gliomas (Schutt et al. 2003).

5.4.3 MRI Contrast Agents

These tracers alter the magnetic properties of nearby water molecules, enhancing contrast in MRI images. For example, gadolinium-based contrast agents used for tumor imaging and angiography and iron oxide nanoparticles are used for liver imaging and cell tracking (Forte et al. 2019).

5.4.4 Ultrasound Contrast Agents

These are typically gas-filled microbubbles that enhance the acoustic signal in ultrasound imaging, such as perfluorocarbon microbubbles used for vascular imaging and targeted drug delivery (Patamia et al. 2023).

5.4.5 Multimodal Imaging Agents

These tracers combine properties that allow them to be detected by multiple imaging modalities. For example, 4Cu-labeled nanoparticles are used for both PET and MRI (Pillay et al. 2014).

5.4.6 Targeted Molecular Probes

These tracers are designed to bind to specific molecular targets associated with diseases, for example, antibody-based and peptide-based tracers (Khazaei et al. 2023).

The development of novel tracers is an active area of research in molecular imaging, with efforts focused on improving specificity, sensitivity, and safety. For instance, in breast cancer, various tracers have been developed to target key markers like hormone receptors, human epidermal growth factor receptor 2 (HER2), and glucose metabolism. FDG radiotracer is widely used in PET imaging to detect increased glucose uptake in tumor cells, a hallmark of cancer metabolism (Nourizad et al. 2023). HER2-targeted tracers like radiolabeled trastuzumab can highlight HER2-positive lesions, aiding in patient selection for targeted therapies (Benefits of Nanotechnology for Cancer n.d.). These tracers enable real-time visualization of disease processes, offering insights into disease progression, response to therapy, and potential avenues for personalized treatment strategies (Chin et al. 2010).

5.5 Molecular Imaging Strategies for Solid Tumors

5.5.1 Molecular Imaging of Cancer Cell Proliferation and Metabolism

5.5.1.1 Cancer Cell Proliferation Imaging

By using FLT, a nucleoside analog that readily enters cells by carrier-mediated diffusion, analyzing cell proliferation has been possible. The FLT is immobilized when a thymidine kinase (TK) such as TK1 (a surrogate marker), which phosphorylates FLT to fluorothymidine phosphate, cannot undergo further metabolism (Wang et al. 2013a). Therefore, FLT is an ideal tracer for assessing tumor response to any treatment because of its selectivity. Clinically approved and relevant imaging probes have been highlighted in Tables 5.3 and 5.4.

Table 5.3 Timeline of FDA-approved tracers

Tracer name	Type	Primary use	Year approved
18F-FDG	PET	Oncology	1994
18F-Florbetapir (Amyvid)	PET	Alzheimer's disease	2012
18F-Flutemetamol (Vizamyl)	PET	Alzheimer's disease	2013
18-F Florbetaben (Neuraceq)	PET	Alzheimer's disease	2014
68Ga-DOTATATE (Netspot)	PET	Neuroendocrine tumors	2016
18F-FDOPA (Dopamine transporter)	PET	Parkinson's disease	2019
68Ga-PSMA-11	PET	Prostate cancer	2020
18F-PSMA-1007	PET	Prostate cancer	2022
18F-Piflufolastat (Pylarify)	PET	Prostate cancer	2021
64Cu DOTATATE (Detectnet)	PET	Neuroendocrine tumors	2020
18F-Fluciclovine (Axumin)	PET	Prostate cancer	2016
18F NaF	PET	Bone metastases	2011
11C-Choline	PET	Prostate cancer	2012
13N-Ammonia	PET	Myocardial perfusion	2000
82Rb-Chloride	PET	Myocardial perfusion	1989
99mTc-Sestamibi	SPECT	Myocardial perfusion	1990
123I-Ioflupane (DaTscan)	SPECT	Parkinson's disease	2011
18F-FES	PET	Breast cancer	2020
Flurpiridaz (18F)	PET	Coronary artery disease	2024
Flortaucipir (18F)	PET	Alzheimer's disease	2021
Lutetium Lu 177 vipivotide tetraxetan	PET	PSMA-positive mCRPC	2022
(18F)-NaF	PET	Bone imaging	2022

5.5.1.2 Metabolic Imaging

Metabolic imaging depends on the concept that cancer cells absorb separate substrates through different metabolic pathways than normal tissue. Usually, any tiny molecule that contains a radioisotope can be detected, as it moves through the biological system using a variety of detection methods. Positron-emitting compounds are dominant in this class of tracer molecules, mostly because it is suitable to substitute the nonradioactive hydrogen and carbon atoms in any organic molecule by positron-emitting isotopes, including 11C and 18F, respectively (Lin et al. 2018). These positron-emitting analogs of metabolically significant chemical compounds can be identified using PET at picomolar to nanomolar concentrations (Zeng et al. 2015). The highly sensitive nature of this technique permits the introduction of positron-emitting compounds at nonpharmacologic dosages and the imaging of metabolic pathways without affecting molecular, cellular, or subcellular functioning. Several metabolic tracers have been developed to concentrate on the transfer and degradation of amino acids and glucose (Ho Shon and Hogg 2023).

Glucose Metabolism

The FDG-PET technique is used to detect cancer malignancies that accelerate glucose metabolism. The FDG gets phosphorylated by the enzyme hexokinase via

Table 5.4 Molecular tracers currently in clinical trials

Tracer	Modality	Target/function	Condition	NCT number
68Ga-labeled NY108	PET	Anti-PSMA mAbs	Prostate cancer	NCT05623878
Indocyanine green	Near-IR (non-ICG)	Lymph node noncompliance rate	Gastric cancer	NCT05618821
F-18-labeled DX600	PET probe	Standard uptake value (SUV)	Solid tumor	NCT04542863
Ga-68 or F-18-labeled HER2	PET radionuclide	HER2	Breast cancer	NCT04547309
PET radionuclide (Ga-68 or F-18) WL12 peptide	PET/CT	PD-L1	Solid tumors	NCT04304066
68Ga-WL12 PET/CT	PET/CT	PD-L1	Gastrointestinal tumors	NCT04629326
99mTc-labeled FAP	SPECT-CT	Tumor uptake	Head and neck tumor, lung and pancreatic cancer	NCT06438705
[99mTc]Tc-DB8	SPECT	Gastrin-releasing peptide receptors	Prostate and breast cancer	NCT05940298
68Ga-FAPI	PET/CT	Accuracy of FAPI-PET/CT in metastases	Pancreatic, stomach and bile duct neoplasms	NCT05172310
Ga68-FAPI-46	PET/CT	Comparative study	Peritoneal carcinomatosis	NCT06061874
99mTc-P137	SPECT/CT	Lesion scope, size, and metastasis of the tumor	Prostate cancer	NCT05672485
18F-labeled DPA-714	PET	Measure regional brain TSPO levels	Neuroinflammation and cognitive impairment	NCT04364672
68Ga-DOTATATE and 18F-FDG	PET	Tumor texture features as predictors of tumor metabolism and somatostatin receptor expression	Gastroenteropancreatic and neuroendocrine tumors	NCT04804371
68Ga-THP-APN09	PET/CT	PD-L1	Lung cancer melanoma	NCT05156515
68Ga-NOTA-ABSCINT-HER2	PET/CT	Tumor heterogeneity	Metastatic breast cancer	NCT06282588
[18F]-JK-PSMA-7	PET/CT	PSMA expression and tumor grade, tumor stage, and tumor subtype	Soft tissue sarcomas and urothelial cell carcinomas	NCT05522257
64Cu-NOTA-PSMA-PEG-Cy5.5-C′ dot	PET/MRI	Safety, biodistribution/pharmacokinetics	Prostate cancer	NCT04167969

(continued)

Table 5.4 (continued)

Tracer	Modality	Target/function	Condition	NCT number
89Zr-MPDL3280A	PET	Tumor and immune cell PD-L1 expression analysis	Locally advanced or metastatic solid tumors	NCT02453984
68Ga-JH04	PET/CT	Fibroblast activation protein (FAP)	Various types of cancer	NCT06455735
[Al18F]PSMA137	PET/CT	PSMA	PSMA-positive cancer patients	NCT04693169
68Ga-PSMA 68Ga-RM2	PET/MRI	Primary PCa phenotype	Prostate cancer	NCT06484361
68Ga-DOTA-NI-FAPI04	PET/CT	FAP	Various types of cancers	NCT06455761
68Ga-HBED-CC-PSMA	PET/MRI	Comparative study	Prostate cancer	NCT04176497
[18F]FLT	PET/CT	Changes in [18F]FLT uptake in metastatic tumors	Breast neoplasm	NCT02608216
68Ga-Pentixafor	PET/CT	Chemokine receptor CXCR4	Lympho-proliferative diseases	NCT04504526
Ga-68 PSMA ligand	PET/CT	Tumor burden in mCRPC patients	Metastatic prostate cancer	NCT04951817
89Zr-Atezolizumab	PET/CT	PD-L1	Metastatic triple negative breast cancer (MIMIR-mTNBC)	NCT05742269
18F-FAZA	PET	Prostate hypoxia	Prostate cancer	NCT01567800
68 Gallium (68Ga)-DOTATOC	PET/CT	SSTR	Neuroendocrine tumors	NCT03583528
Durvalumab-680LT	Fluorescence molecular endoscopy (FME)	PD-L1	Advanced esophageal cancer	NCT05450484
Ga-68 PSMA11 F-18 PSMA 1007	PET/CT	Demonstrate their utility in detecting prostate cancer	Biochemical recurrence of prostate cancer	NCT03768349

89Zr-Atezolizumab	PET/CT	PD-L1	Large B-cell lymphoma (LBCL)	NCT05404048
[18F]Flutemetamol and FDG	Functional magnetic resonance imaging (fMRI)	Check brain changes	Chemotherapy-induced cognitive impairment	NCT02317783
[18F]FTT	PET/CT	PARP-1	Prostate cancer	NCT03334500
18F-FDOPA 18F-DA 68Ga-DOTATATE	PET	Chromaffin cells	Pheochromocytoma	NCT00004847

PSMA prostate-specific membrane antigen, *PET* positron emission tomography, *CT* computed tomography, *SPECT* single-photon emission computed tomography, *ICG* indocyanine green, *HER2* human epidermal growth factor receptor 2, *PD-L1* programmed death-ligand 1, *FAP* fibroblast activation protein, *FAPI* fibroblast activation protein inhibitor, *TSPO* translocator protein, *FDG* fluorodeoxyglucose, *NOTA* 1,4,7-triazacyclononane-1,4,7-triacetic acid, *MRI* magnetic resonance imaging, *PCa* prostate cancer, *mCRPC* metastatic castration-resistant prostate cancer, *CXCR4* C-X-C chemokine receptor type 4, *DOTATOC* DOTA-Tyr3-octreotide, *SSTR* somatostatin receptor, *FME* fluorescence molecular endoscopy, *LBCL* large B-cell lymphoma, *fMRI* functional magnetic resonance imaging, *PARP-1* poly (ADP-ribose) polymerase 1, *FDOPA* fluoro-L-DOPA, *DA* dopamine, *DOTATATE* DOTA-Tyr3-octreotate, *FLT* fluorothymidine, *FAZA* fluoroazomycin arabinoside, *HBED-CC* *N,N′*-bis[2-hydroxy-5-(carboxyethyl)benzyl]ethylenediamine-*N,N′*-diacetic acid, *THP* tris (hydroxypyridinone), *APN* aminopeptidase N, *FTT* fluorthanatrace

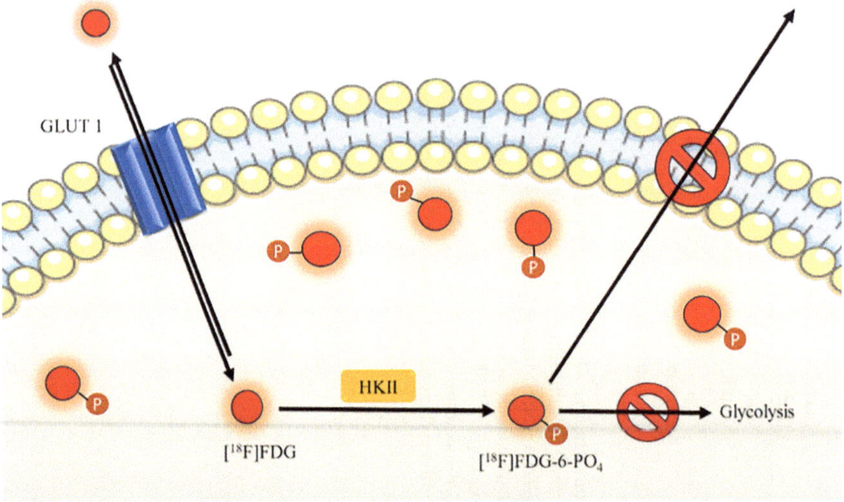

Fig. 5.3 Mechanism of [18F]FDG cellular uptake and phosphorylation through GLUT-1 transporter and hexokinase II in tumor cells

GLUT-1 transporters (Fig. 5.3). As cancer cells need sustained glucose for proliferation even in low insulin conditions, the FDG is shown to be an optimal marker of cancer cells. Radiation therapy or surgery can alter the tumor, thereby assessment and results by structural imaging methods are often ambiguous (Ni et al. 2020). FDG-PET reduces the amount of radiation needed for the treatment of primary tumors while preserving the nonmalignant tissues (Jenkins et al. 2005). The combination of FDG-PET and CT or MRI is being used to understand a more comprehensive view of the volumes of tumor in managing radiotherapy (Jenkins et al. 2003). FDG-PET is sensitive to detect and monitor certain types of cancer as a single modality despite PET/CT (Rajendran et al. 2020). Regardless of its advantages, FDG-PET typically has lower sensitivity for gradually progressing and lower metabolic rate cancers such as bronchioloalveolar cell lung carcinoma, thyroid cancer, prostate cancer, and neuroendocrine cancers (Zhao et al. 2010). Limitations of PET imaging to clearly distinguish structures include the fact that they may not detect small lesions and molecular events, which leads to the miscalculation of radiotracer concentration (Yeo et al. 2014). A further drawback of FDG-PET is linked with metabolic distribution in the kidneys and bowel, which may make it difficult to detect lesions located near those tissues. To overcome this drawback, PET/CT with contrasting agents is administered orally to minimize this problem by enhancing combined structural and functional scans, with better tumor detection and precise identification of FDG uptake (Wu et al. 2018).

Amino Acid Metabolism

The use of methionine and phenylalanine derivatives has been broadly investigated. A latest study reported 91% sensitivity for identification of gliomas and larger tumor volumes by [99mTc] Tc-methionine (half-life of 6 h) (Hu et al. 2012) in contrast to those identified with Gd-enhanced MRI, which might have an influence on radiotherapy management. To address these limitations, 18F-fluorinated amino acid analogs with a longer 109-min half-life have been developed. Several phenylalanine analogs, such as L-DOPA and tyrosine, showed significant capability in diagnosing certain PET applications. Specific analogs of tyrosine, like *O*-(2-18F-fluoroethyl) L-tyrosine (FET) and L-3-18F-α-methyl tyrosine (FMT), have potential for imaging in certain tumors, such as lungs (Kim et al. 2014) and brain (Wang and Maurer 2005). An analog of L-DOPA, 18F-6-fluoro dihydroxyphenylalanine (18F-FDOPA), is accumulated in dopaminergic neurons located in the basal ganglia of the brain. This property is used to visualize Parkinson's disease (Hope et al. 2023), brain tumors (Luo et al. 2023), medullary thyroid cancer, carcinoid malignancies, and melanoma in clinical settings.

5.5.2 Molecular Imaging of Tumor Microenvironment

Tumor microenvironment (TME) shows more consistency across various cancers compared to other cancer-specific biomarkers, which vary among different cancer types. Molecular imaging is the most suitable procedure to diagnose cancer and helps in early detection, accurate diagnosis, and monitoring of treatment response, as shown in Fig. 5.4. Thus, the consistent feature of TME can be used to develop a broad strategy for tumor imaging.

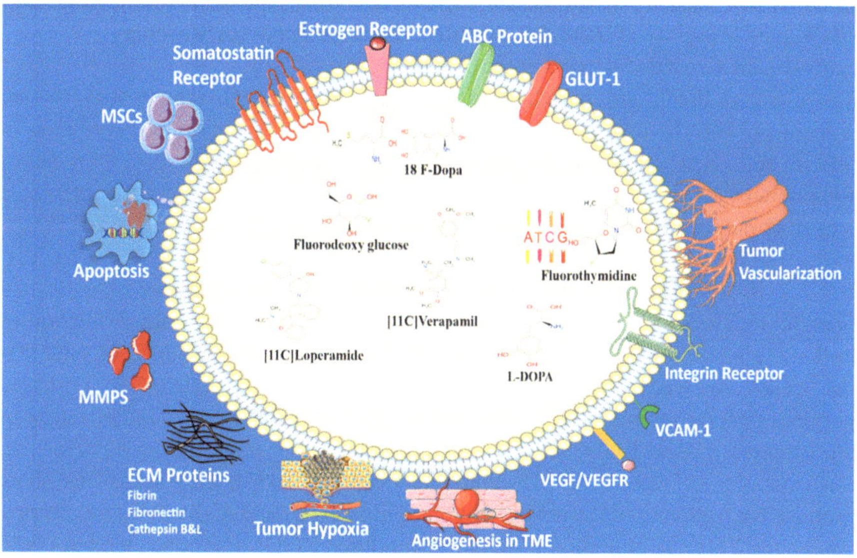

Fig. 5.4 Molecular components and cellular interactions in the TME: a comprehensive overview of key membrane receptors, signaling molecules, and imaging tracers in solid tumors and their roles in tumor development, progression, and visualization through molecular imaging techniques

5.5.2.1 Molecular Imaging of Tumor Vascularization

The blood vessels of tumors show different types of biomolecules that change based on the stages of the tumor. These biomarkers relating to angiogenesis can be detected by using molecular imaging, which helps in the determination and assessment of tumor aggressiveness and guides treatment decisions. It also aims at hindering the blood vessel formation by antiangiogenic therapy and tracking the treatment outcomes.

Molecular Imaging of Cellular Adhesion Molecules

Integrins are the cellular adhesion molecules that are overexpressed in tumor-associated vessels, providing a unique opportunity for targeted imaging and therapeutic interventions (Linden et al. 2006). Different types of imaging agents have been introduced that detect integrins and help in the diagnosis and treatment of cancer. These agents include radioactive tracers in PET or SPECT scans, fluorescent dyes in optical imaging, contrast agents in MRI scans, and microbubbles in ultrasound imaging (Nawaz and Kipreos 2022). Molecules like VCAM-1 also play a similar role to integrins in diagnosing cancer. These molecules are observed at relatively high levels in tumor blood vessels and at the sites of inflammation. Iron particles are easier to detect in small amounts compared to other contrast agents (like Gd) since they show strong T2 susceptibility (Mieog et al. 2022). Iron oxide particles facilitate the early detection of brain metastasis by targeting VCAM-1. Beyond integrins, other crucial molecular targets in tumor angiogenesis include VEGF and its receptors.

Molecular Imaging of Vascular Endothelial Growth Factor/Vascular Endothelial Growth Factor Receptor Expression

Vascular endothelial growth factors (VEGF) play a key role in angiogenesis and are associated with the development of various diseases. The abnormal expression of VEGF and their receptors causes worsening of cancer conditions in patients. Researchers have developed a multimodal imaging agent by using single-chain VEGF to detect VEGFR on tumor blood vessels (Dizdarevic and Peters 2011).

5.5.2.2 Molecular Imaging of Angiogenesis

Specific measurement of activated endothelial cells enables accurate evaluation of tumor angiogenesis, like the detection of $\alpha V\beta 3$ integrins that are found on endothelial cells involved in angiogenesis. Arginine-glycine-aspartate (RGD)-based peptides generally bind to integrins (Mann et al. 2015). Hence, by modification of the structure around RGD, selective binding to specific integrins can be achieved. For example, 18F-labeled RGD compounds can be used in PET scans that show promise for personalized treatment options and antiangiogenic therapies monitoring (Ballinger 2001). A limitation is that these modified RGD compounds bind not only to $\alpha V\beta 3$ integrins on tumor-associated endothelial cells but also to tumor cells. However, measurement of total tumor $\alpha V\beta 3$ integrin levels plays a key role as higher expression of these integrins is linked to greater metastatic potential (Dizdarevic and Peters 2011).

5.5.2.3 Molecular Imaging of Tumor Hypoxia

Hypoxia is a key feature of TME that promotes tumor growth, invasion, and metastasis along with tumor angiogenesis. The hypoxic condition of the tumor initiates increased production of hypoxia-inducible factor-1 (HIF-1) (Alshehri et al. 2020). 18F-labeled nitroimidazole and copper complex compounds are used as hypoxia-specific radiotracers in PET imaging. T1-weighted Gd-based MRI contrast agents with nitroimidazole have been developed for imaging hypoxic tumors (Barua and Mitragotri 2014). MRI probes also detect hypoxia by measuring changes in oxygen tension. Agents that include boron films and dual-emissive nanoparticles like iodide-substituted difluoroboron dibenzoylmethane-poly(lactic acid) (BF2dbm (l)PLA) are also being used in hypoxia imaging (Zhang et al. 2021).

Platinum, ruthenium, palladium, and iridium are used in the development of oxygen-sensitive phosphorescent complexes for imaging tumor hypoxia by using optical techniques. Nitroreductase (NTR) shows hyperactivity in hypoxic tumors and can be used as an indicator for hypoxia in the TME. NTR activity is detected by fluorescent probes through quenching strategies (Zhang et al. 2022). Bioreduction-activated fluorescent probes can be developed by using small molecules like 2-nitroimidazole and indolequinone, which are reduced in the hypoxic TME. Non-invasive detection of tumor hypoxia plays a key role in approaching treatments like intensity-modulated radiation therapy and new options like photodynamic therapy (PDT) (Pramod Kumar et al. 2020). 18F-fluoromisonidazole PET imaging can show early changes and detect the efficiency of the treatment approach, so this technique is considered valuable in patients undergoing chemotherapy for advanced head and neck cancers (Verkhovskii et al. 2023).

5.5.2.4 Molecular Imaging of Extracellular Matrix Proteins and Proteases

Molecular Imaging of Extracellular Matrix Proteins

Cancer progression initiates specific and significant remodeling of the tumor extracellular matrix (ECM) that supports cancer initiation, growth, invasion, and metastasis. In TME, excess amounts of collagen get deposited and undergo structural changes of subtypes (I, II, III, and IV). Therefore, these collagen abnormalities can be used as targets for cancer molecular imaging. A peptide called collagen mimetic peptide (CMP) mimics natural collagen and shows a strong binding to collagen. This specific feature of CMPs can be utilized in cancer imaging by targeting collagen (Elumalai et al. 2024). Several imaging CMP agents have been developed for the early detection and management of cancer. Some of them include an engineered fluorophore-labeled M13 bacteriophage modified with Gly-Pro-Pro repeats (7GPP) (CMP) in optical imaging that diagnoses A549 lung adenocarcinoma cells by detecting abnormal collagen. A probe called nitrobenzyl caged CMP undergoes a triple helix fold when triggered by light and binds to denatured collagen I. The folded caged CMP (IR-Ahx-(GPO)9) conjugates with near-infrared fluorescent (NIRF) dye through an aminohexanoic acid (Ahx) spacer that allows specific binding to abnormal collagen I (Li and Huang 2010). The testing of this probe in vivo in mice with PC-3 prostate tumor xenografts revealed effective accumulation in tumors, whereas a nontargeting control (IR-Ahx-SG9P909) showed minimum tumor accumulation (Lee et al. 2022).

Fibrin plays a crucial role in cancer imaging since it is produced in abnormal quantities in TME during cancer invasion and metastasis, along with angiogenesis. Autoradiography can be used to observe tumor-attached fibrin-binding peptides and antibodies that are being developed for the detection of targeted tumor cells (Warszyńska et al. 2023). Another ECM protein, fibronectin, also plays a key role as it gets upregulated in TME, especially at metastatic sites and initiates the accumulation of tumor-promoting cells, growth factors, and cytokines (Liew et al. 2020). Significantly, small peptides like the CREKA peptide (Cys-Arg-Glu-Lys-Ala) and CLT1 (Cysteine-Glycine-Leucine-Isoleucine-Isoleucine-Glutamine-Lysine-Asparagine-Glutamic acid-Cysteine) bind to fibronectin-associated tumor complexes for imaging. A superparamagnetic iron oxide nanoparticle system was developed to detect the efficacy of these peptides that bind to ECM biomarkers (Lei et al. 2017). Generally, the TME expresses huge amounts of fibronectin-associated complexes, and thus they have become promising molecular targets for MRI. Recently, CREKA-Tris (Gd-DOTA)3 (a pentapeptide CREKA-targeted Gd-based MRI contrast agent) that offers strong contrast enhancement in tumors was developed (Cheng et al. 2020). Extra Domain B fibronectin (EDB-FN), an oncofetal isoform, is the most promising fibronectin biomarker that was found in the TME. This makes EDB-FN the most valuable biomarker for cancer detection and diagnosis. Glycoproteins like tenascin-C are some other important ECM components that are overexpressed during tumor growth, invasion, and metastasis, including angiogenesis. Tenascin-C can be detected by using radiolabeled specific monoclonal antibodies and nucleic acid aptamers (Graván et al. 2023).

Molecular Imaging of Tumor-Associated Proteases

During cancer progression, the TME expresses abnormal levels of proteases like MMPs and cathepsins (Bukhari et al. 2021). Thus, molecular imaging of these proteases helps in the assessment of tumor malignancy and aids in diagnosis and treatment monitoring (Tang et al. 2023). Based on Förster Resonance Energy Transfer (FRET), quantum dots (QD) can be used as sensors in which a quencher dye is attached to a peptide on the QD surface. The cleavage of peptide by proteases causes the release of quenchers that leads to the reduction of FRET efficiency and thus allows the QD to emit a signal (Beloqui et al. 2016). Probes can also be developed by using the Bioluminescence Resonance Energy Transfer technique for monitoring protease activity. Tumor to background fluorescence ratios were improved by nanotechnology, which increases the sensitivity of protease imaging (Pathak et al. 2023). Tumor cells also express other groups of proteases called cathepsins, which can be detected by NIRF probes (Kumar 2021), and protease-sensitive MRI probes have been introduced recently. In MRI, poly-L-glutamate is used as a probe that develops Chemical Exchange Saturation Transfer and detects glutamate moieties after cathepsins cause breakdown of lysosomal proteins (Yadav et al. 2009).

5.5.2.5 Molecular Imaging of pH Changes

The acidic nature of TME acts as an imaging tool for the detection and study of tumor growth, metabolism, changes in TME, and assessment of treatment approach (Patil et al. 2022). Fluorescence probes and MRI contrast agents have been developed as pH-activatable imaging agents for imaging acidic TME (Paliwal et al. 2020). Ultra pH-sensitive (UPS) nanoprobes respond to pH changes between 4 and 7.4. The activation of probes by fluorescence causes a 50-fold increase of UPS nanoprobes. The hydrophobic cores of nanoprobes and attached fluorophores undergo aggregation densely at high pH levels, which causes fluorescence self-quenching. A decrease in pH causes ionizable blocks of nanoprobes to become protonated and hydrophilic. This leads to the breakdown of nanoprobes and disaggregation of fluorophores and thus turns on the fluorescence (Karamchedu et al. 2020). pH-sensitive MRI contrast agents were also designed to achieve precise off-on contrast control at pathological pH levels (Naz et al. 2019). Other researchers introduced nanosized MRI contrast agents, PEG-GMnCap nanoparticles, which amplify the MR signal in an acidic environment. These nanoparticles contain a calcium phosphate core that encapsulates Mn^{2+} and a PEG shell. Rapid hydrolysis of PEGMnCap nanoparticles takes place at acidic conditions and initiates the release of Mn^{2+} that causes stronger MR signals by enhancing reflexivity with protein binding (Zafar et al. 2020).

5.6 Molecular Imaging of Cell Death

Molecular imaging of apoptosis allows researchers to analyze programmed cell death in living organisms without invasive procedures. This process helps to identify how productive the treatment is without any changes in glucose metabolism and cell death, resulting in a clear output of therapy (Bozzuto et al. 2024). Annexin V binds to phosphatidylserine, a molecule which translocates in the cell membrane's outer layer surface at the time of programmed cell death, and this binding can be marked by radioisotopes, either 18F or 99mTc, or 124I (Wadhwa et al. 2022). This method enhances the ability to observe programmed cell death in biological organisms. Super paramagnetic iron-based contrast agents, in addition to other paramagnetic contrast agents combined with targeted carrier nanoparticles, are under development in animal models to specify molecules or biologic processes for detection using MRI (Sun et al. 2008).

5.7 Molecular Imaging of Cell Tracking

Cell tracking can be achieved by engineering cells to continuously produce a reporter protein, which can be detected by using optical imaging or PET (Wahajuddin and Arora 2012). For example, bioluminescent prostate cancer and breast cancer models were developed to examine the therapeutic effects of cytotoxic therapy using optical imaging. Additionally, researchers have established luciferase and GFP in genetically modified cells during gene therapy to observe the efficiency of transfection by using optical imaging or PET (Shokrollahi 2013; Talluri and Malla 2020). In recent times, researchers have utilized MRI to track cells in animal models by implementing paramagnetic contrast agents or superparamagnetic iron oxide nanoparticles delivery (Sun et al. 2008).

5.7.1 Molecular Imaging of Mesenchymal Stromal Cell Tracking

Mesenchymal stem cells (MSCs) release therapeutic compounds directly to tumor sites, resulting in immune responses against cancer cells (Tran et al. 2017). Marked cells can be detected through imaging methods like optically charged coupled devices, PET, MRI, SPECT, fluorescence imaging, and bioluminescence imaging (Wagstaff et al. 2012). Distinct marking agents like QD, fluorophores, radionuclides, and superparamagnetic iron oxide particles (SPIO) are used to tag MSCs to visualize their transfer (Obaidat et al. 2019).

QD-labeled MSCs have been effectively integrated in co-culture systems, exhibiting their ability to interact with other cells, thereby allowing researchers to analyze behavior, differentiation, and treatment efficacy of MSCs (Cui et al. 2023). In a past study, researchers labeled MSCs with the NIR fluorophore IR-786 and delivered them to the coronary artery in a swine model of myocardial infarction, which was effectively observed at 90-min postinjection, allowing for real-time tracking of MSC behavior. Radionuclides used for labeling MSCs are Technetium-99m, Indium-111, fluorine-18 (18F; half-life 109 min), and copper-64 (Tran et al. 2017). Zirconium-89 is growing in popularity due to its half-life, which allows extended monitoring of MSCs by using PET imaging. This property allows researchers to observe the behavior and distribution of MSCs over a prolonged period, increasing the understanding of their therapeutic effects in various applications (He et al. 2023).

5.8 Molecular Imaging of Cancer-Specific Receptors

5.8.1 Somatostatin Receptors

Neuroendocrine tumors contain higher amounts of somatostatin receptors (SSTRs). Whole body imaging with radiolabeled versions of octreotide is used to image these tumors by targeting subtype 2 of SSTRs (SSTR2). 111In-pentetreotide is mostly used for SPECT imaging. In addition to this, SSTR-targeted tracers like 68Ga-DOTATOC and Gluc-Lys (18F-fluoro propionyl-TOCA) were developed for PET imaging (Fu et al. 2023).

5.8.2 Estrogen Receptors

Estrogen receptor (ER) targeting tracers can be used to assess ER expression using 18F-fluorinated estrogen analogs and 16α-18F-fluoroestradiol-17β (FES) (Zhang et al. 2019). In monitoring breast cancer therapy using tamoxifen, FES-PET shows a faster response than that of FDG-PET at 7–10 days after treatment initiation. FES-PET increases overall treatment response from 23% to 34% in patients with HER2eu-negative breast cancer (Dong et al. 2020).

5.9 Molecular Imaging of Disease Recurrence and Drug Resistance

Accurately monitoring and diagnosing recurrent disease has become a major challenge in preclinical and clinical oncology. It is critically essential to recognize when tumor recurrence or drug resistance has occurred. Radiologists can use CT to determine suspicious mass-like structures, but the anatomic detail and functional information are not available. Intrinsic and acquired drug resistance are the key factors leading to cancer treatment failure. Multidrug resistance and specific ABC transporters can be scanned using radiopharmaceuticals that are either MDR substrates or inhibitors (Martinez Pancorbo et al. 2019).

Molecular optical imaging aids in identifying microscopic proliferation of tumor cells in biopsy samples. One such technique uses a contrast dye linked to folate, which is essential for metabolic activities involved in DNA and RNA synthesis, epigenetic processes, cellular proliferation, and survival (Wang et al. 2013b). Intraoperative optical imaging does not require ionizing radiation, permits deeper tissue visualization, and it is simple to analyze in real time during surgery. Hence, this technique is secure for both patients and surgeons in carrying out the procedure (Arcos Rosero et al. 2024).

5.10 Nanotheranostics for Solid Tumors

Nanotheranostics is the latest research in personalized cancer medicine. The single regimen was allowed for accurate cancer diagnosis and treatment by nanotheranostics through various noninvasive diagnostic approaches. The treatment effectiveness of theranostic nanomedicine mainly depends on prolonged blood circulation, intratumoral penetration, and tumor cell internalization of the nanoparticles (NPs) (Shi et al. 2021). Rapid elimination of circulating NPs in the reticuloendothelial system (RES) during systemic delivery is a major obstacle in nanomedicine (Illes et al. 2018). Hence, it is essential to minimize the factors that reduce the blood circulation half-life and biodistribution of NPs while designing them for theranostic applications. The factors, including NP composition, size, surface charge, stability, and surface modifications, are influenced by NP-RES interactions.

5.10.1 Polymer-Based Nanotheranostics

Polymer-based nanoparticles (NPs) have shown promise in cancer treatment by enhancing drug delivery, improving circulation time, and targeting tumors effectively (Tomczak et al. 2013). Polyethylene glycol (PEG) coating creates an inert surface, further extending NP circulation time and passive tumor targeting (Yang et al. 2018). In the TME, NPs can trigger specific responses, such as oxygen production to relieve tumor hypoxia and boost photodynamic therapy (PDT) efficacy (Rahman et al. 2015). Combining PDT with immunotherapies, like PD-L1 (programmed death ligand 1) antibodies, has been shown to significantly increase immune response and tumor infiltration by cytotoxic T-cells (Oliveira et al. 2022). Additionally, pH-responsive charge conversion was used to improve mitochondrial targeting, thus improving the efficiency of PDT (Liu et al. 2015b). Another interesting strategy is a pH-sensitive "ligand-reversible-shielding micelle" to improve tumor accumulation and the effectiveness of PDT (Wang et al. 2017b). A TME pH-responsive "Cluster Bomb" was developed, which consists of a drug (DOX) with loaded mesoporous silica NPs (MSN) and capped with tumor-homing or tumor-penetrating peptide tLyP-1-modified tungsten disulfide quantum dots (WS2-HP), via pH-sensitive benzoic-imine bonds for the targeted tumor therapy (Fuchs et al. 2013). Size-responsive NPs have also been explored as the NPs smaller than 10 nm are rapidly removed through renal excretion, resulting in decreased tumor accumulation. The large NPs accumulate in tumors using the EPR effect and then shrink or degrade into smaller particles in the acidic TME, enhancing tumor penetration and therapeutic outcomes (Kumari et al. 2020). For instance, catalase-loaded NPs help decompose tumor-rich H_2O_2, augmenting PDT efficacy (Shinn et al. 2022). These advancements of polymer-based nanotheranostics provide strategies for targeted cancer therapy, combining therapeutic delivery and diagnostic imaging for better clinical outcomes.

5.10.2 Lipid-Based Nanotheranostics

Lipid nanoparticles (LNPs) are better absorbed by the cancer cells, which is an important factor for developing a theranostic system. They are well tolerated without significant immune response, metabolizing, and excretion over time, which could minimize toxicity and potentially surpass physiological limitations. Additionally, these nanocarriers can be modified to enhance penetration in the target cells, leading to better therapeutic outcomes. Based on the target site in vivo, the size of LNPs may vary (Sun and Gu 2016). Some lipid molecules alter in response to temperature, particularly known as thermo-responsive LNPs (TLNs), which "self-assemble" or organize themselves by detecting the changes in temperature, which facilitates the permeability of drugs across the blood-brain barrier, targeting glioblastoma cells (Mo et al. 2014). Among these LNPs, other types have gained attention in recent years, such as solid lipid carriers (SLNs) composed of solid lipids, nanostructured lipid carriers (NLCs) which combine solid and liquid lipids, lipid nanocapsules

(LNCs) which are similar to SLNs, liposomes, and core-shell lipid nanoparticles (CLNs) where the drug is surrounded by a shell of lipids (Nsairat et al. 2022). SLNs and NLCs provided new possibilities as oral drug carriers, which led scientists to further research on lipid-based nanocarriers (Lee and Im 2019). Additionally, gadolinium (Gd), a lipid-based theranostic system, has been developed, and a dendrimer-based lipid nanotheranostic platform designed for targeted drug delivery of mRNA and for cancer imaging in vivo (Liu et al. 2022).

The NLCs demonstrated several diagnostic applications by forming structures that encapsulate diagnostic agents such as iodixanol, which is tagged with a radioactive isotope 64Cu for PET and CT imaging for more comprehensive targeting of lung cancer cells (Zhu et al. 2021). When the NLC is attached to the radioactive isotope, it achieves high radiolabeling efficacy and maintains structure and function as intended. Hence, well-defined CT imaging and identification of minute changes in PET imaging are achieved, which provides comprehensive and precise detection of lung cancer in early stages. On the other hand, SLNs are used as carriers for oral drugs for angiogenesis, which are called angiogenesis inhibitors. These SLNs are modified on their surface and encapsulated with curcumin, which has demonstrated potential inhibition of angiogenesis (Liu et al. 2023). In addition to this, SLNs have demonstrated the ability to enhance the effectiveness of various anticancer agents such as 5-fluorouracil for colorectal cancer (Li et al. 2019), Nutlin-3a and Asiatic acid for treating glioblastoma (Mo et al. 2014), epigallocatechin gallate, docetaxel, letrozole, and variabilin are used for breast cancer (Aebisher et al. 2024) and morin hydrate for cervical cancer treatment (Aroso et al. 2021). Moreover, to ensure extended and concurrent release of doxorubicin (DOX) and paclitaxel (PTX), the lipid-calcium-phosphate has shown antitumor activity in human lung cancer A549 cells (Ailioaie et al. 2022). Similarly, LNCs are incorporated in the treatment of breast cancer by concurrently delivering docetaxel (DTX) and thymoquinone (THQ) to increase the effectiveness of the treatment against drug-resistant breast cancer cells (Ganz et al. 2005). Furthermore, LNCs can be used as carriers of various chemotherapeutic agents to improve immunotherapy of cancers like glioblastoma, colorectal cancer, and breast cancer (Lee et al. 2017).

5.10.3 Magnetotheranostics

Magnetic nanoparticles (MNPs) are a specialized type of nanomaterials composed of metals such as iron, nickel, cobalt, and other metals and their oxides. Their structural analysis is determined by techniques like X-ray diffraction (XRD), scanning electron microscopy (SEM), and transmission electron microscopy (TEM) (Jiang et al. 2023). Few examples of MNPs are silica-coated magnetic nanoparticles (enhance biocompatibility), lipid-coated magnetic nanoparticles (enhance stability), polymer-coated magnetic nanoparticles (for controlled drug release), vesicle-type magnetic nanoparticles (for a specific type of drug delivery), and superparamagnetic iron oxide nanoparticles (SPIONs), which have a diameter of less than 50 nm; they are superparamagnetic in nature (Luo et al. 2016).

MNPs have shown potential diagnostic results using magnetic resonance imaging (MRI). Since silver nanoparticles are metal, they presumably enhance the CT contrast effects. Also, MNP nanoprobes are being explored for their ability to provide real-time imaging, which is crucial for early detection of cancer cells like HepG2 cells (Yang et al. 2018). Additionally, MNPs, particularly superparamagnetic iron oxide nanoparticles (SPIONs), which are easily metabolized, safely administered, and are used not only for imaging but also as therapeutic agents/carriers to target specific cells or tissues (Wang and Yang 2024). The application of the MNP extends to both in vitro and in vivo, helping detect and monitor disease.

NMR-based diagnostics exploit MNPs as proximity sensors (contrast agents). Compared with other contrast agents, superparamagnetic contrast agents have the advantage of producing enhanced proton relaxation in MRI, and therefore, less doses are needed to inject into the human body (Ghosh et al. 2023). SPIONs have the ability to generate T2/negative contrast and are useful in detecting cancers (Han and Choi 2021). IONPs can bind to specific cancer biomarkers or overexpress surface moieties, for example, MUC-1, PSMA, EGFR, and HER2 on cancer cells through the modification of molecular targeting ligands produced covalently or noncovalently (Zeng et al. 2022). For example, receptors overexpressed in tumor cells can bind to the surface of IONPs, so that IONPs can be accumulated and retained in tumor tissues, and T2 signal contrast can be generated to detect the tumors. Several researchers have been working on developing unique MNPs with specific structural properties, for example, gold-plated iron oxide containing platinum-based anticancer drugs like cisplatin, effective against human ovarian cancer cell line A2780 (Duan et al. 2023). Magnetic hyperthermia, a treatment where magnetic nanoparticles (MNPs), particularly Fe_3O_4 nanoparticles or iron oxide, are used to generate heat when exposed to an external magnetic field to destroy cancer cells (Kadkhoda et al. 2022). It is most commonly used alongside regional hyperthermia (RHT).

Likewise, the extensive use of the optical characteristics of nanoparticles for diagnosis and treatment is often called theranostics, involving photothermal therapy (PTT), photodynamic therapy (PDT), and photoacoustic imaging (PAI).

5.10.4 Plasmonic Theranostics

Plasmonic materials are effective photothermal agents that have been demonstrated to have tissue compatibility and excellent uptake of drugs in NIR bio-window. The key factor involved in therapeutic efficiency in PTT is its ability to deliver heat effectively, which relies on light-to-heat conversion efficiency and light extinction capacity. Metals are ideal as plasmonic materials as they have a large number of conduction electrons. When the metal nanoparticle size is smaller than the wavelength of the incident light, the plasmon becomes confined to the particle surface, which is called local surface plasmon resonance (LSPR). This confinement enhances the particle's ability to absorb and convert light into heat, hence making it more efficient for PTT. Photothermal imaging leverages local surface plasmon resonance (LSPR) nanoparticles, which have the ability to absorb light and convert into heat, making thermal imaging of cancerous tissues more visible. Photoacoustic imaging, on the other hand, similar to excitation of laser nanoparticles, rapid heating from the laser causes the tissue to undergo thermoelastic expansion, generating ultrasonic waves. These create detailed images of tissues and other abnormalities.

Plasmonic nanoparticles such as gold and silver are highly effective in enhancing the sensitivity of Surface-Enhanced Raman Scattering (SERS) for extremely sensitive detection of biological markers like circulating tumor DNA (ctDNA) and miRNAs [209]. These nanoparticles are modified with DNA or RNA probes, and these SERS nanotags, when bound to tissues, amplify Raman-based signals, allowing for the detection of even single-base mutations. For example, a SERS-based assay using a hairpin-shaped DNA-RNA-DNA probe was developed to detect circulating tumor DNA (ctDNA) in lung cancer patients. When the probe binds with the target ctDNA, it triggers the activity of an enzyme called RNase HII, which cuts the RNA segment in the hairpin structure. This action further enhances the Raman signal, enabling extremely sensitive detection of the cancer biomarkers.

The development of versatile plasmonic platforms improved both therapeutic and diagnostic abilities in cancer treatment. A nanoplatform composed of lanthanide ions in gold nanorods (AuNRs) was developed. Also, Prussian blue modified $CaMoO_4$ nanoparticles, conjugated with HER2 antibodies, collectively showed improvement in contrast imaging of cancer cells, focusing on the capability for diagnostic and treatment of cancer

Plasmonic theranostics entail core-shell nanoparticle systems, for example, Au-WO_3 core-shell, which enhances the imaging of human cancer cells; see Tables 5.5 and 5.6. Also, plasmonic nanoparticles (NPs) are used for observing variations in cellular pH levels. For example, a gold (Au) nanoparticle is attached to the cell membrane to observe pH variations around the cell. This demonstrated the applications of plasmonic nanoparticles in real-time monitoring of intracellular activities in response to different treatments.

5.10.5 Hybrid Nanotheranostics

Nanoparticles can be further subdivided into two types: organic and inorganic nanoparticles. Organic nanoparticles consist of polymers and lipid groups, and their biodegradability and ease of production make them a common method of nanoparticle transport. Inorganic nanoparticles are typically metallic particles such as copper or gold, and they have a variety of optical and thermal properties that make them useful for cellular transport, as well as having increased stability (Alshammari, Lashin et al. 2023).

Combining organic and inorganic features has the potential to reap the benefits of both types of nanoparticles. An example of hybrid nanoparticles that have been studied is lipid-coated gold nanoparticles. Gold nanoparticles are normally used as inorganic nanoparticles for use in photoacoustic imaging and phototherapy, and thus

Table 5.5 Summary of studies on plasmonic nanomaterials-based phototheranostics

Sample	Theranostic functionalities	Mechanism
AuNRs/IR-E1050	Fluorescence cancer imaging	Local field enhancement
Au/UCNP MIM	Upconversion luminescence cell imaging	Local field enhancement
CuPc@HG@BN	SERS-based microRNA monitoring	Local field enhancement
Au-nanofève	SERS-based cancer cell imaging	Local field enhancement
P-GERT	SERS-based cancer cell and lymph node imaging	Local field enhancement
Au@GO NP-NACs	SERS-based cancer imaging and PTT	Local field enhancement and energy dissipation
$CaMoO_4$: Eu@SiO_2@GNR	SERS-based cancer imaging and PTT	Local field enhancement and energy dissipation
Au-SiO_2-WO_3	SERS-based cancer imaging	Local field enhancement
AuNBPs motor/AuNP	SERS-based microRNA detection and PTT	Local field enhancement and energy dissipation
AuNPs	SERS-based intracellular pH measurement	Local field enhancement
AgNPs	SERS-based cell membrane imaging	Local field enhancement
PpIX-conjugated Au NPs	SERS-based cancer cell imaging and PDT	Local field enhancement
Plasmonic GVs-Ce6	PTT, PAI, PDT	Local field enhancement and energy dissipation
AuNSs	Circulating cancer sensing	Hot carrier generation
Au-Ag-GOx HTNs	PDT	Hot carrier generation
NPD@M	PDT	Hot carrier generation
Z-scheme SnS 1.68-WO 2.41 nanocatalyst	Hydrogen therapy, hole therapy	Hot carrier generation
Pd@Au-PEG	Catalytic oxygen production	Hot carrier generation
CF-bPEA@siRNA NRs	PDT, fluorescence imaging, and PTT	Hot carrier generation, local field enhancement, and energy dissipation
Cu_3BiS_3 NC	PTT	Energy dissipation
CuS superstructures	PTT	Energy dissipation
Au-Cu_9S_5 NPs	PTT and CT	Energy dissipation
AuNPs-condensed CP	Thermal delivery	Energy dissipation
Anti-EGFR-CuO/ Cu_2O TNCs	PTT, PDT, fluorescence imaging, PAI, and PTI	Hot carrier generation, local field enhancement, and energy dissipation
Au@Au-Ag dot-in-cubic nanoframes	PAI, SERS-based cancer cell imaging	Local field enhancement and energy dissipation
AuNCs@SiO_2	PAI and PTT	Local field enhancement and energy dissipation
Au NEs-PEG	PAI and fluorescence imaging	Energy dissipation

Table 5.6 Nanoparticles and their theranostic capabilities

Name	Theranostic type	Diagnostic implementation	Therapeutic implementation	Refs.
Gold nanoparticles	Plasmonic	Gold has very good photoacoustic properties, making it a great contrast agent for photoacoustic imaging	Gold nanoparticles are used in phototherapy because of their ability to convert light into thermal energy. Increased heat can activate certain chemotherapeutic drugs that are applied separately	Li and Chen (2015), Wang et al. (2013b), Arcos Rosero et al. (2024)
Quantum dots	Polymeric and hybrid	Quantum dots have excellent fluorescent properties for fluorescence imaging	Quantum dots can be coated with polymers that can bind with chemotherapeutic drugs	Zhang et al. (2009), Tomczak et al. (2013)
GO	Graphene	Oxygen functional groups can bind to imaging contrast nanoparticles, such as copper and technetium	Oxygen functional groups can form bonds with many chemotherapeutic agents	Yang et al. (2018), Rahman et al. (2015), Oliveira et al. (2022)
Ferritin	Biomimetic	Ferritin naturally holds iron, so iron nanoparticles for MRI can also be easily loaded	Ferritin can hold therapeutic drugs inside its hollow structure as well	Zhu et al. (2021)

have valuable theranostic properties. Adding a lipid coating allows for the loading and release of hydrophobic drugs, as well as the binding of cell membrane proteins (Arcos Rosero et al. 2024).

Metallic nanoparticles can also be coated with polymers on their surface. One of the most common polymers used is polyethylene glycol (PEG), which helps inhibit recognition by the immune system. In most cases, PEG is normally used to bind to liposomes or other organic nanoparticles (Shi et al. 2021). However, PEG can also be modified to be bound to metal nanoparticles, an example being gold. In one study, PEG was thiolated to form strong Au-S covalent bonds with the gold nanoparticle (Rahme, Chen et al. 2013). Magnetite-based nanoparticles, used as contrasting agents for MRI, have also been PEGylated experimentally using a modified version of PEG compounded with acrylic acid (Illes et al. 2018).

Semiconductor nanoparticles, such as quantum dots, are also polymerized to optimize transport in the bloodstream. Quantum dots are a nanoparticle contrast agent for fluorescent imaging, and the polymerization of quantum dots can help improve its stability and binding to other organic functional groups. Similar to gold nanoparticles, a thiolated anchor molecule was used to connect the zinc-based outer core of the quantum dot with organic polymer functional groups such as PEG (Tomczak et al. 2013).

5.10.6 Graphene Nanotheranostics

Graphene is a form of carbon that is composed of a sheet of hexagonal lattice. Each carbon atom binds to three other carbon atoms, leaving one of its valence electrons free to interact with other molecules (Yang et al. 2018). The main graphene analog used for targeted cellular delivery is graphene oxide (GO). GO differs from conventional graphene due to the oxygen-based reactive groups interspersed within the sheet. This increases the hydrophilicity of the graphene sheet, which allows it to transport better within aqueous environments such as blood plasma. Furthermore, the various oxygen groups add areas of hydrogen bonding and weak ionic bonding alongside the covalent bonds formed by carbon, which increases the variety of drugs and imaging labels that can be transported (Sharma and Mondal 2020). As a result of its large surface area and ability to easily bind with other organic molecules, graphene has been studied extensively as a nanoparticle able to transport a large number of target molecules to cells. In the context of diagnosis, a single sheet of graphene can hold many nanolabels for use in imaging (Rahman et al. 2015). Gold nanoparticles, which are heavily used in photoacoustic molecular imaging, normally have a neutral charge. However, they often have a positive charge in nanoparticle form caused by the use of positively charged surfactants in particle synthesis (Guo, Armstrong et al. 2015). This makes them able to bind to GO due to their weak electrostatic interactions with the negatively charged oxygen groups.

For the therapeutic aspect, chemotherapeutic drugs can also be loaded onto a GO sheet. For example, doxorubicin (DOX) has been studied extensively for paired use in GO for a variety of cancers because of its ability to form strong pi-pi ionic stacking bonds. In human prostate, breast, cervical, and liver cancer cells, the DOX-GO loading mechanism showed promising results, with decent loading efficiencies and a significant increase in drug release in slightly acidic conditions (Oliveira et al. 2022). Ligands can also be attached to GO to induce specific targeting of cancer cells. For example, transferrin has been studied as a targeting mechanism, due to the overexpression of transferrin receptors on solid tumor membranes as a result of increased dependence on iron ions for proliferation (Daniels, Bernabeu et al. 2012). In one study, transferrin was combined with dihydroartemisinin (DHA), a chemotherapeutic agent, on GO to initiate apoptosis in tumor cells specifically. The study concluded that significant cell uptake of DHA occurred with high specificity in tumor cells (Liu et al. 2015b).

All in all, the ability of graphene, especially GO, to bind and load a large amount and variety of molecules. Various imaging contrast agents, chemotherapeutic drugs, and targeting molecules can be concurrently attached to a large sheet of graphene, which makes it a promising option for theranostics.

5.10.7 Bio-Responsive Soft Theranostics

Bio-responsivity refers to theranostics that are activated in response to an abnormal state in the body. Tumors differ from normal cell tissue in many qualities, such as acidity, blood oxygen levels, and glucose concentrations, and treatments as well as contrast agents can be designed to only be released in these conditions (Wang et al. 2017b). One method of bioresponsive theranostic technology is the use of enzyme-cleaved nanoparticles, as given in Fig. 5.5. For example, the enzyme hepsin, which is involved in the metabolism of glucose, is overexpressed in prostate cancer. One study created a spherical polymer-peptide coating that was able to hold "payload" molecules, which in the context of theranostics can be contrasting agents and chemotherapeutic drugs. Some sections of the spherical coating contained hepsin-cleavable sequences, which enabled the contents within the nanoparticle to be released in areas with high amounts of hepsin (Fuchs et al. 2013).

One major differentiating feature of tumors is the hypoxic conditions within tumor tissue. Tumor vasculature is poorly constructed, and oxygen does not properly diffuse into tumors. Hypoxia-responsive nanoparticle systems also use enzyme-based cleavage to release its payload. As an example, hypoxic tumors have high amounts of nitroreductase because of the enzyme's ability to function independently of oxygen. Polymer-based nanoparticle membranes that contain sections of nitroaromatic compounds were developed for the contents of the nanoparticle to be released in areas with nitroreductase such as hypoxic tumors (Kumari et al. 2020).

Similarly, nanoparticle and related theranostics containers can be opened by slight changes in acidity, as highlighted in Table 5.6. The pH of tumors is slightly lower than normal cell tissue, and bioresponsive nanoparticles have been developed to take advantage of this disparity. Acid-responsive nanoparticle membranes are specifically spherical polymers, which are cleaved in slightly acidic environments, causing a transition of the membrane from hydrophobic to hydrophilic. This causes the influx of water into the nanoparticle and ultimately, lysis of the membrane, releasing its contents into the tumor microenvironment (Shinn et al. 2022). Once again, these nanoparticles can be loaded with contrast agents and chemotherapeutic drugs to act as a selective theranostic application for tumor cells.

Because of the lack of extracellular ATP and the ubiquity of intracellular ATP, ATP-responsive nanoparticles are very useful in ensuring intracellular specificity of nanoparticle payload (Sun and Gu 2016). When paired with tumor-specific ligands, ATP-responsive treatment can be extremely effective in inducing apoptosis in cancer cells. In one study, ATP-specific aptamers, which are nucleic acids that bind to other nucleic acids to ATP, were bound to modified strands of DNA. The cystine and guanine bases were bound to doxorubicin (DOX), a widely used chemotherapeutic drug. When ATP binds to the aptamer, the DNA unfurls, causing DOX to be released within the cell, inducing apoptosis (Mo et al. 2014).

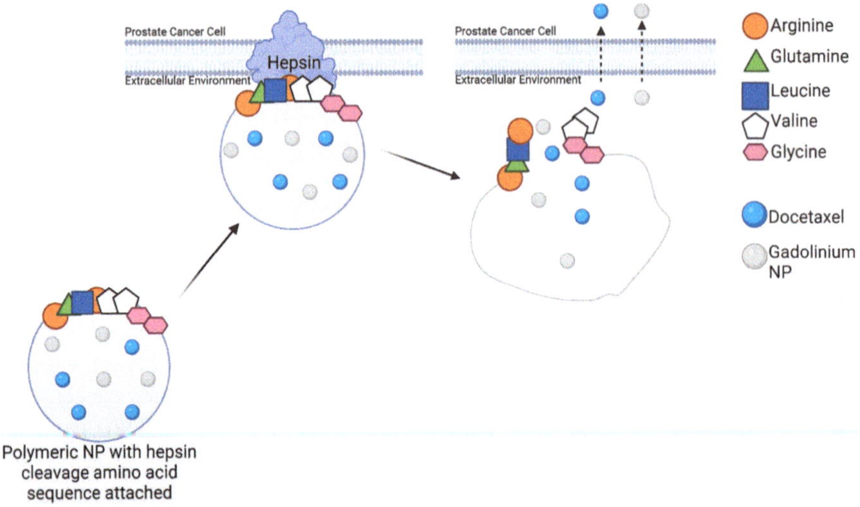

Fig. 5.5 A potential theranostic approach to prostate cancer using an enzyme-cleavable nanoparticle. Prostate cancer cell membranes have highly upregulated levels of hepsin, and adding a hepsin-cleavable amino acid sequence of RGLRVVGG allows the nanoparticle to be opened only by hepsin. Common prostate cancer chemotherapeutic drugs, such as docetaxel, can be loaded and targeted toward prostate cancer tissue, as well as imaging contrast NPs such as gadolinium for MRI

5.11 Biomimetic Theranostics

One method of designing nanoparticles for cellular transport is by utilizing and improving natural cell transport mechanisms. One of the earliest nanoparticles developed was the liposome, which is simply a spherical phospholipid bilayer membrane. This is essentially an artificial version of a vesicle, which can be loaded with hydrophilic molecules inside the liposome and hydrophobic molecules in the inner layer of the membrane (Nsairat et al. 2022). In several studies, DOX was loaded alongside imaging contrast agents such as indocyanine green for fluorescence imaging and gadolinium-bound polymers for MRI (Lee and Im 2019). Liposomes can be opened into target cells using endocytosis. Naturally, endocytosis occurs when vesicles come in contact with the cell membrane, and the vesicle opens its contents into the cell. Liposomes use this same feature to release imaging contrast agents and chemotherapeutic treatments into target cells (Liu et al. 2022).

Ferritin is another natural protein that has favorable properties for use in nanoparticles. Ferritin stores large amounts of iron by forming a spherical container with its 24 subunits. Another property of ferritin that makes it useful for molecular delivery is its self-assembly. At different pH values, the subunits of ferritin can either be fully assembled or independent, making it have bioresponsive properties for theranostic molecule release (Zhu et al. 2021).

Another important criterion for effective nanoparticle transport is biocompatibility. Unmodified liposomes and other organic nanoparticles are cleared out of the bloodstream within a few hours due to being discovered and removed by the immune system. Many studies have explored the coating of nanoparticles with cell membrane fragments to evade immune detection of the nanoparticles to improve their half-life within the body (Liu et al. 2023). For example, erythrocyte membranes have been shown to be effective in polymer-based and metal nanoparticles, increasing the half-life of the nanoparticle within the bloodstream (Li et al. 2019).

5.12 Concluding Remarks

Nanotechnology-based modalities are becoming the emerging strategies for the detection, diagnosis, and treatment of cancer. In this review, we gave insights into various nanotechnology-based imaging and diagnostic approaches to better detect and treat cancer cells. Due to tremendous advancements in the field of nanotheranostics, designing more effective cancer diagnostics and therapeutics is possible, allowing us to understand cancer biology in a better way. We also outlined nanotheranostics with emphasis on various imaging techniques and therapeutic modalities that are at different stages of clinical phases, with specific applications. The future of these imaging techniques is very promising as they allow noninvasive body imaging with high sensitivity for detecting cancer-targeted therapy, which mainly depends on the use of the right imaging technique and the molecular tracers. Nanotechnology-based diagnostic imaging provides a unique opportunity to build better detection strategies that may lead to significant progress in personalized medicine. Most of these platforms are flexible and tunable for a wide variety of applications for precise imaging of biological targets. In spite of the bright future, several challenges have to be addressed to support the clinical integration of these technologies. There is a need of an in-depth understanding of nanoparticle-tumor interactions, toxicity assessment, scaled-up synthesis, and regulatory protocols for nanotheranostics to enhance their clinical use.

Acknowledgments Rajendra Prasad acknowledges the funding support from Anusandhan National Research Foundation (ANRF/IRG/2024/000620/LS).

Author Contribution R.K., S.K., and S.M. have done thorough literature and primary editing, and R.P., S.M., and N.K have done final review and editing.

References

Abdel-Hamid NM, Abass SA (2021) Matrix metalloproteinase contribution in management of cancer proliferation, metastasis and drug targeting. Mol Biol Rep 48(9):6525–6538. https://doi.org/10.1007/s11033-021-06635-z

Aebisher D, Rogóż K, Myśliwiec A, Dynarowicz K, Wiench R, Cieślar G, Kawczyk-Krupka A, Bartusik-Aebisher D (2024) The use of photodynamic therapy in medical practice. Front Oncol 14. https://doi.org/10.3389/fonc.2024.1373263

Ailioaie LM, Ailioaie C, Litscher G (2022) Light as a cure in COVID-19: a challenge for medicine. Photonics 9(10):686. https://doi.org/10.3390/photonics9100686

Alauddin MM (2012) Positron emission tomography (PET) imaging with (18)F-based radiotracers. Am J Nucl Med Mol Imaging 2(1):55–76

Alford R, Ogawa M, Choyke PL, Kobayashi H (2009) Molecular probes for the in vivo imaging of cancer. Mol Biosyst 5(11):1279. https://doi.org/10.1039/b911307j

Almuhaideb A, Papathanasiou N, Bomanji J (2011) 18F-FDG PET/CT imaging in oncology. Ann Saudi Med 31(1):3–13

Alshehri S, Imam SS, Rizwanullah M, Akhter S, Mahdi W, Kazi M, Ahmad J (2020) Progress of cancer nanotechnology as diagnostics, therapeutics, and theranostics nanomedicine: preclinical promise and translational challenges. Pharmaceutics 13(1):24

Alshammari BH, Lashin MMA, Mahmood MA, et al. (2023) Organic and inorganic nanomaterials: fabrication, properties and applications. RSC Adv 13(20):13735-13785. Published 2023 May 5. https://doi.org/10.1039/d3ra01421e

Anani T, Panizzi P, David AE (2016) Nanoparticle-based probes to enable noninvasive imaging of proteolytic activity for cancer diagnosis. Nanomedicine 11(15):2007–2022. https://doi.org/10.2217/nnm-2016-0027

Arcos Rosero WA et al (2024) Review of advances in coating and functionalization of gold nanoparticles: from theory to biomedical application. Pharmaceutics 16(2):255

Aroso RT, Schaberle FA, Arnaut LG, Pereira MM (2021) Photodynamic disinfection and its role in controlling infectious diseases. Photochem Photobiol Sci 20(11):1497–1545. https://doi.org/10.1007/s43630-021-00102-1

Backer MV, Backer JM (2012) Imaging key biomarkers of tumor angiogenesis. Theranostics 2(5):502

Ballinger JR (2001) 99mTc-tetrofosmin for functional imaging of P-glycoprotein modulation in vivo. J Clin Pharmacol 41(S7):39S–47S

Bao Y, Sherwood JA, Sun Z (2018) Magnetic iron oxide nanoparticles asT1contrast agents for magnetic resonance imaging. J Mater Chem C 6(6):1280–1290. https://doi.org/10.1039/c7tc05854c

Barbosa RCS, Mendes PM (2022) A comprehensive review on photoacoustic-based devices for biomedical applications. Sensors (Basel) 22(23):9541

Barua S, Mitragotri S (2014) Challenges associated with penetration of nanoparticles across cell and tissue barriers: a review of current status and future prospects. Nano Today 9(2):223–243

Becker J, Schwarzenböck SM, Krause BJ (2020) FDG PET hybrid imaging. In: Recent results in cancer research. Springer International Publishing, pp 625–667. https://doi.org/10.1007/978-3-030-42618-7_19

Beloqui A, Solinís MÁ, Rodríguez-Gascón A, Almeida AJ, Préat V (2016) Nanostructured lipid carriers: promising drug delivery systems for future clinics. Nanomedicine 12(1):143–161. https://doi.org/10.1016/j.nano.2015.09.004

Benefits of Nanotechnology for Cancer (n.d.) Cancer.Gov. https://www.cancer.gov/nano/cancer-nanotechnology/benefits. Accessed 28 Oct 2024

Ben-Haim S, Ell P (2009) 18F-FDG PET and PET/CT in the evaluation of cancer treatment response. J Nucl Med 50(1):88–99. https://doi.org/10.2967/jnumed.108.054205

Berger A (2002) Magnetic resonance imaging. BMJ 324(7328):35

Boudreau C et al (2016) Excitation light dose engineering to reduce photo-bleaching and photo-toxicity. Sci Rep 6:30892

Bozzuto G, Calcabrini A, Colone M, Condello M, Dupuis ML, Pellegrini E, Stringaro A (2024) Phytocompounds and nanoformulations for anticancer therapy: a review. Molecules 29(16):3784. https://doi.org/10.3390/molecules29163784

Breznik B, Motaln H, Lah Turnšek T (2017) Proteases and cytokines as mediators of interactions between cancer and stromal cells in tumours. Biol Chem 398(7):709–719. https://doi.org/10.1515/hsz-2016-0283

Bukhari SZ, Zeth K, Iftikhar M, Rehman M, Usman Munir M, Khan WS, Ihsan A (2021) Supramolecular lipid nanoparticles as delivery carriers for non-invasive cancer theranostics. Curr Res Pharmacol Drug Discov 2:100067. https://doi.org/10.1016/j.crphar.2021.100067

Burkett BJ et al (2023) A review of theranostics: perspectives on emerging approaches and clinical advancements. Radiol Imaging Cancer 5(4):e220157

Cai W, Sam Gambhir S, Chen X (2005) Multimodality tumor imaging targeting integrin αvβ3. BioTechniques 39(Suppl 6). https://doi.org/10.2144/000112091

Capriotti G et al (2020) Copper-64 labeled nanoparticles for positron emission tomography imaging: a review of the recent literature. Q J Nucl Med Mol Imaging 64(4):346–355

Carles M, Fechter T, Grosu AL, Sörensen A, Thomann B, Stoian RG, Wiedenmann N, Rühle A, Zamboglou C, Ruf J, Martí-Bonmatí L, Baltas D, Mix M, Nicolay NH (2021) 18F-FMISO-PET hypoxia monitoring for head-and-neck cancer patients: radiomics analyses predict the outcome of chemo-radiotherapy. Cancer 13(14):3449. https://doi.org/10.3390/cancers13143449

Chakravarty R, Chakraborty S, Dash A (2015) Molecular imaging of breast cancer: role of RGD peptides. Mini Rev Med Chem 15(13):1073–1094

Chen K, Chen X (2010) Design and development of molecular imaging probes. Curr Top Med Chem 10(12):1227–1236

Cheng X, He L, Xu J, Fang Q, Yang L, Xue Y, Wang X, Tang R (2020) Oxygen-producing catalase-based prodrug nanoparticles overcoming resistance in hypoxia-mediated chemo-photodynamic therapy. Acta Biomater 112:234–249. https://doi.org/10.1016/j.actbio.2020.05.035

Chin PTK, Buckle T, Aguirre de Miguel A, Meskers SCJ, Janssen RAJ, van Leeuwen FWB (2010) Dual-emissive quantum dots for multispectral intraoperative fluorescence imaging. Biomaterials 31(26):6823–6832. https://doi.org/10.1016/j.biomaterials.2010.05.030

Cooley MB et al (2023) Real-time imaging of nanobubble ultrasound contrast agent flow, extravasation, and diffusion through an extracellular matrix using a microfluidic model. Lab Chip 23(15):3453–3466

Crisan G et al (2022) Radiopharmaceuticals for PET and SPECT imaging: a literature review over the last decade. Int J Mol Sci 23(9):5023

Cui S, Zhang S, Yue S (2018) Raman spectroscopy and imaging for cancer diagnosis. J Healthc Eng 2018:8619342

Cui X, Ruan Q, Zhuo X, Xia X, Hu J, Fu R, Li Y, Wang J, Xu H (2023) Photothermal nanomaterials: a powerful light-to-heat converter. Chem Rev 123(11):6891–6952

Czernin J, Israel O, Herrmann K, Barrio M, Nathanson D, Allen-Auerbach M (2017) Clinical applications of PET/CT and SPECT/CT imaging. In: Imaging in medical diagnosis and therapy. CRC Press, pp 439–472. https://doi.org/10.1201/9781315374383-2

Daniels TR, Bernabeu E, Rodríguez JA, Patel S, Kozman M, Chiappetta DA, Holler E, Ljubimova JY, Helguera G, Penichet ML (2012) The transferrin receptor and the targeted delivery of therapeutic agents against cancer. Biochimica et biophysica acta 1820(3):291–317. https://doi.org/10.1016/j.bbagen.2011.07.016

Das S, Indurthi HK, Asati P, Sharma DK (2022) Small molecule fluorescent probes for sensing and bioimaging of nitroreductase. ChemistrySelect 7(1). https://doi.org/10.1002/slct.202102895

Davis LE, Shalin SC, Tackett AJ (2019) Current state of melanoma diagnosis and treatment. Cancer Biol Ther 20(11):1366–1379

Dizdarevic S, Peters AM (2011) Imaging of multidrug resistance in cancer. Cancer Imaging 11(1):1

Dong Z, Xue X, Liang H, Guan J, Chang L (2020) DNA nanomachines for identifying cancer biomarkers in body fluids and cells. Anal Chem 93(4):1855–1865

Duan S, Hu Y, Zhao Y, Tang K, Zhang Z, Liu Z, Wang Y, Guo H, Miao Y, Du H, Yang D (2023) Nanomaterials for photothermal cancer therapy. RSC Adv 13(21):14443–14460

Elumalai K, Srinivasan S, Shanmugam A (2024) Review of the efficacy of nanoparticle-based drug delivery systems for cancer treatment. Biomed Technol 5:109–122. https://doi.org/10.1016/j.bmt.2023.09.001

Ettinger A, Wittmann T (2014) Fluorescence live cell imaging. Methods Cell Biol 123:77–94

Feng L, Dong Z, Tao D, Zhang Y, Liu Z (2018) The acidic tumor microenvironment: a target for smart cancer nano-theranostics. Natl Sci Rev 5(2):269–286

Fink KR, Fink JR (2013) Imaging of brain metastases. Surg Neurol Int 4(Suppl 4):S209–S219

Forte E, Fiorenza D, Torino E, Costagliola di Polidoro A, Cavaliere C, Netti PA, Salvatore M, Aiello M (2019) Radiolabeled PET/MRI nanoparticles for tumor imaging. J Clin Med 9(1):89. https://doi.org/10.3390/jcm9010089

Fu L, Lin CT, Karimi-Maleh H, Chen F, Zhao S (2023) Plasmonic nanoparticle-enhanced optical techniques for cancer biomarker sensing. Biosensors 13(11):977

Fuchs AV et al (2013) Enzyme cleavable nanoparticles from peptide based triblock copolymers. Nanoscale 5(11):4829–4839

Ganz RA, Viveiros J, Ahmad A, Ahmadi A, Khalil A, Tolkoff MJ, Nishioka NS, Hamblin MR (2005) Helicobacter pylori in patients can be killed by visible light. Lasers Surg Med 36(4): 260–265. https://doi.org/10.1002/lsm.20161

Ghosh S, Chitgupi U, Sunar U, Lovell JF (2023) Chemophototherapeutic ablation of doxorubicin-resistant human ovarian tumor cells. Photochem Photobiol 99(2):844–849

Gorodetska I, Kozeretska I, Dubrovska A (2019) BRCA genes: the role in genome stability, cancer stemness and therapy resistance. J Cancer 10(9):2109–2127

Graván P, Aguilera-Garrido A, Marchal JA, Navarro-Marchal SA, Galisteo-González F (2023) Lipid-core nanoparticles: classification, preparation methods, routes of administration and recent advances in cancer treatment. Adv Colloid Interf Sci 314:102871. https://doi.org/10.1016/j.cis.2023.102871

Grimm JB, Lavis LD (2022) Caveat fluorophore: an insiders' guide to small-molecule fluorescent labels. Nat Methods 19(2):149–158

Grogan SP, Mount CA (2024) Ultrasound physics and instrumentation. In: StatPearls. StatPearls Publishing, Treasure Island, FL

Guo R et al (2018) Ultrasound imaging technologies for breast cancer detection and management: a review. Ultrasound Med Biol 44(1):37–70

Guo, Jianfeng Armstrong, Mark O'Driscoll, Caitriona Holmes, Justin Rahme, Kamil (2015) Positively Charged, Surfactant-Free Gold Nanoparticles for Nucleic Acid Delivery. RSC Adv. 5. https://doi.org/10.1039/C4RA16294C

Han HS, Choi KY (2021) Advances in nanomaterial-mediated photothermal cancer therapies: toward clinical applications. Biomedicine 9(3):305

Han Z, Lu Z-R (2017) Targeting fibronectin for cancer imaging and therapy. J Mater Chem B 5(4): 639–654. https://doi.org/10.1039/c6tb02008a

He Z, Zhang CY, Lei Y, Song G, Yao Y (2023) Plasmonic nanomaterials: a versatile phototheranostic platform of cancers. Mater Today 62:168–189

Heath VL, Bicknell R (2009) Anticancer strategies involving the vasculature. Nat Rev Clin Oncol 6(7):395–404. https://doi.org/10.1038/nrclinonc.2009.52

Hicks RJ, Hofman MS (2012) Is there still a role for SPECT-CT in oncology in the PET-CT era? Nat Rev Clin Oncol 9(12):712–720

Ho Shon I, Hogg PJ (2023) Imaging of cell death in malignancy: targeting pathways or phenotypes? Nucl Med Biol 124–125:108380. https://doi.org/10.1016/j.nucmedbio.2023.108380

Hope TA, Allen-Auerbach M, Bodei L, Calais J, Dahlbom M, Dunnwald LK, Graham MM, Jacene HA, Heath CL, Mittra ES, Wright CL (2023) SNMMI procedure standard/EANM practice guideline for SSTR PET: imaging neuroendocrine tumors. J Nucl Med 64(2):204–210

Hu S, Lu P, Zhang L, Li F, Chen Z, Wu N, Meng H, Lin J, Feng H (2012) In vivo magnetic resonance imaging tracking of SPIO-labeled human umbilical cord mesenchymal stem cells. J Cell Biochem 113(3):1005–1012. https://doi.org/10.1002/jcb.23432

Ibrahim MA, Hazhirkarzar B, Dublin AB (2024) Gadolinium magnetic resonance imaging. In: StatPearls. StatPearls Publishing, Treasure Island, FL

Illes E et al (2018) PEGylation of superparamagnetic iron oxide nanoparticles with self-organizing polyacrylate-PEG brushes for contrast enhancement in MRI diagnosis. Nanomaterials (Basel) 8(10):776

Jenkins DE, Oei Y, Hornig YS, Yu S-F, Dusich J, Purchio T, Contag PR (2003) Bioluminescent imaging (BLI) to improve and refine traditional murine models of tumor growth and metastasis. Clin Exp Metastasis 20(8):733–744. https://doi.org/10.1023/b:clin.0000006815.49932.98

Jenkins DE, Hornig YS, Oei Y, Dusich J, Purchio T (2005) Bioluminescent human breast cancer cell lines that permit rapid and sensitive in vivodetection of mammary tumors and multiple metastases in immune deficient mice. Breast Cancer Res 7(4). https://doi.org/10.1186/bcr1026

Jiang S (2023) Fluorescence imaging-guided surgery. In: Biomedical photonic technologies, pp 137–163. https://doi.org/10.1002/9783527823550.ch5

Jiang H, Zuo J, Li B, Chen R, Luo K, Xiang X, Lu S, Huang C, Liu L, Tang J, Gao F (2023) Drug-induced oxidative stress in cancer treatments: angel or devil? Redox Biol 63:102754. https://doi.org/10.1016/j.redox.2023.102754

Johanssen VA, Zarghami N, Sibson NR (2023) Magnetic resonance imaging of neuroinflammation. In: Progress in inflammation research. Springer International Publishing, pp 91–117. https://doi.org/10.1007/978-3-031-23661-7_3

Kabakov AE, Yakimova AO (2021) Hypoxia-induced cancer cell responses driving radioresistance of hypoxic tumors: approaches to targeting and radiosensitizing. Cancers (Basel) 13(5):1102

Kadkhoda J, Tarighatnia A, Barar J, Aghanejad A, Davaran S (2022) Recent advances and trends in nanoparticles based photothermal and photodynamic therapy. Photodiagn Photodyn Ther 37: 102697

Kane SM et al (2024) Technetium-99m. In: StatPearls. StatPearls Publishing, Treasure Island, FL

Kapoor M, Kasi A (2024) PET scanning. In: StatPearls. StatPearls Publishing, Treasure Island, FL

Karamchedu S, Tunki L, Kulhari H, Pooja D (2020) Morin hydrate loaded solid lipid nanoparticles: characterization, stability, anticancer activity, and bioavailability. Chem Phys Lipids 233: 104988. https://doi.org/10.1016/j.chemphyslip.2020.104988

Kastelik-Hryniewiecka A, Jewula P, Bakalorz K, Kramer-Marek G, Kuźnik N (2022) Targeted PET/MRI imaging super probes: a critical review of opportunities and challenges. Int J Nanomedicine 16:8465–8483. https://doi.org/10.2147/ijn.s336299

Khazaei M, Hosseini MS, Haghighi AM, Misaghi M (2023) Nanosensors and their applications in early diagnosis of cancer. Sens Biosens Res 41:100569. https://doi.org/10.1016/j.sbsr.2023.100569

Kim JE, Kalimuthu S, Ahn B-C (2014) In vivo cell tracking with bioluminescence imaging. Nucl Med Mol Imaging 49(1):3–10. https://doi.org/10.1007/s13139-014-0309-x

Kong F-L, Yang J, D. (2012) Amino acid transporter-targeted radiotracers for molecular imaging in oncology. Curr Med Chem 19(20):3271–3281. https://doi.org/10.2174/092986712801215946

Kumar V (2021) Diagnostic and therapeutic applications of smart nanocomposite dendrimers. Front Biosci 26(3):518–536. https://doi.org/10.2741/4905

Kumar P, Kumar A, Nagaraj C, Sadashiva N, Saini J, Mangalore S, Rajan A, Sitani K, Beniwal M, Santosh V, Basavaraja H, Hazari PP, Mishra AK (2024) Evaluating the diagnostic efficacy of 99mtc-methionine single-photon emission computed tomography–computed tomography: a head-to-head comparison with 11c-methionine positron emission tomography–magnetic resonance imaging in glioma patients. Cancer Biother Radiopharm 39(5):349–357. https://doi.org/10.1089/cbr.2023.0147

Kumari R, Sunil D, Ningthoujam RS (2020) Hypoxia-responsive nanoparticle based drug delivery systems in cancer therapy: an up-to-date review. J Control Release 319:135–156

Kuten J, Linevitz A, Lerman H, Freedman N, Kestenbaum M, Shiner T, Giladi N, Even-Sapir E (2020) [18F] FDOPA PET may confirm the clinical diagnosis of Parkinson's disease by imaging the nigro-striatal pathway and the sympathetic cardiac innervation: proof-of-concept study. J Integr Neurosci 19(3):489. https://doi.org/10.31083/j.jin.2020.03.196

Lan H et al (2023) The role of nanoparticles and nanomaterials in cancer diagnosis and treatment: a comprehensive review. Am J Cancer Res 13(12):5751–5784

Lawson DA et al (2018) Tumour heterogeneity and metastasis at single-cell resolution. Nat Cell Biol 20(12):1349–1360

Lee W, Im HJ (2019) Theranostics based on liposome: looking back and forward. Nucl Med Mol Imaging 53(4):242–246

Lee P, Kupelian P, Czernin J, Ghosh P (2012) Current concepts in F18 FDG PET/CT-based radiation therapy planning for lung cancer. Front Oncol 2. https://doi.org/10.3389/fonc.2012.00071

Lee HH, Choi M-G, Hasan T (2017) Application of photodynamic therapy in gastrointestinal disorders: an outdated or re-emerging technique? Korean J Intern Med 32(1):1–10. https://doi.org/10.3904/kjim.2016.200

Lee D, Kwon S, Jang S, Park E, Lee Y, Koo H (2022) Overcoming the obstacles of current photodynamic therapy in tumors using nanoparticles. Bioactive Mater 8:20–34. https://doi.org/10.1016/j.bioactmat.2021.06.019

Lei Q, Wang S-B, Hu J-J, Lin Y-X, Zhu C-H, Rong L, Zhang X-Z (2017) Stimuli-responsive "cluster bomb" for programmed tumor therapy. ACS Nano 11(7):7201–7214. https://doi.org/10.1021/acsnano.7b03088

Li W, Chen X (2015) Gold nanoparticles for photoacoustic imaging. Nanomedicine (Lond) 10(2):299–320

Li S-D, Huang L (2010) Stealth nanoparticles: high density but sheddable PEG is a key for tumor targeting. J Control Release 145(3):178–181. https://doi.org/10.1016/j.jconrel.2010.03.016

Li Y, Foss CA, Pomper MG, Yu SM (2014) Imaging denatured collagen strands in vivo and ex vivo via photo-triggered hybridization of caged collagen mimetic peptides. J Vis Exp 83. https://doi.org/10.3791/51052

Li H et al (2019) Size dependency of circulation and biodistribution of biomimetic nanoparticles: red blood cell membrane-coated nanoparticles. Cells 8(8):881

Li R, Gao R, Zhao Y, Zhang F, Wang X, Li B, Wang L, Ma L, Du J (2022) pH-responsive graphene oxide loaded with targeted peptide and anticancer drug for OSCC therapy. Front Oncol 12:930920

Liew SS, Qin X, Zhou J, Li L, Huang W, Yao SQ (2020) Smart design of nanomaterials for mitochondria-targeted nanotherapeutics. Angew Chem Int Ed 60(5):2232–2256. https://doi.org/10.1002/anie.201915826

Lin G, Chen S, Mi P (2018) Nanoparticles targeting and remodeling tumor microenvironment for cancer theranostics. J Biomed Nanotechnol 14(7):1189–1207. https://doi.org/10.1166/jbn.2018.2546

Linden HM, Stekhova SA, Link JM, Gralow JR, Livingston RB, Ellis GK, Petra PH, Peterson LM, Schubert EK, Dunnwald LK, Krohn KA (2006) Quantitative fluoroestradiol positron emission tomography imaging predicts response to endocrine treatment in breast cancer. J Clin Oncol 24(18):2793–2799

Liu Q, Wang J, Boyd BJ (2015a) Peptide-based biosensors. Talanta 136:114–127. https://doi.org/10.1016/j.talanta.2014.12.020

Liu L et al (2015b) Dihydroartemisinin and transferrin dual-dressed nano-graphene oxide for a pH-triggered chemotherapy. Biomaterials 62:35–46

Liu P, Chen G, Zhang J (2022) A review of liposomes as a drug delivery system: current status of approved products, regulatory environments, and future perspectives. Molecules 27(4):1372

Liu H et al (2023) Cell membrane-coated nanoparticles: a novel multifunctional biomimetic drug delivery system. Drug Deliv Transl Res 13(3):716–737

Luengo Morato Y et al (2021) Recent advances in multimodal molecular imaging of cancer mediated by hybrid magnetic nanoparticles. Polymers (Basel) 13(17):2989

Luksic R, et al (2015) Incorporating technetium in minerals and other solids: A review, Journal of Nuclear Materials, Volume 466, 2015, Pages 526–538, ISSN 0022-3115, https://doi.org/10.1016/j.jnucmat.2015.08.052

Lunt SJ et al (2008) Interstitial fluid pressure, vascularity and metastasis in ectopic, orthotopic and spontaneous tumours. BMC Cancer 8:2

Luo D, Carter KA, Miranda D, Lovell JF (2016) Chemophototherapy: an emerging treatment option for solid tumors. Adv Sci 4(1). https://doi.org/10.1002/advs.201600106

Luo F, Wangerin KA, Buckley C, van de Ven S, Long E (2023) Molecular imaging in oncology and immuno-oncology. Biomed J Sci Techn Res 48(1):39014–39021

Ma X, Wang Y, Zhao T, Li Y, Su L-C, Wang Z, Huang G, Sumer BD, Gao J (2014) Ultra-pH-sensitive nanoprobe library with broad pH tunability and fluorescence emissions. J Am Chem Soc 136(31):11085–11092. https://doi.org/10.1021/ja5053158

Mann A, Semenenko I, Meir M, Eyal S (2015) Molecular imaging of membrane transporters' activity in cancer: a picture is worth a thousand tubes. AAPS J 17:788–801

Martinez Pancorbo P, Thummavichai K, Clark L, Tabish TA, Mansfield J, Gardner B, Chang H, Stone N, Zhu Y (2019) Novel au–sio2–wo3 core–shell composite nanoparticles for surface-enhanced Raman spectroscopy with potential application in cancer cell imaging. Adv Funct Mater 29(46). https://doi.org/10.1002/adfm.201903549

Mathur S, Turnbull A, Akaev I, Stevens C, Agrawal N, Chopra M, Mincher D (2019) Design of a new peptide substrate probe of the putative biomarker legumain with potential application in prostate cancer diagnosis ex vivo. Int J Pept Res Ther 26(4):1965–1980. https://doi.org/10.1007/s10989-019-09994-1

Mieog JS, Achterberg FB, Zlitni A, Hutteman M, Burggraaf J, Swijnenburg RJ, Gioux S, Vahrmeijer AL (2022) Fundamentals and developments in fluorescence-guided cancer surgery. Nat Rev Clin Oncol 19(1):9–22

Minn H et al (2004) Imaging of adrenal incidentalomas with PET using (11)C-metomidate and (18) F-FDG. J Nucl Med 45(6):972–979

Mo R et al (2014) ATP-triggered anticancer drug delivery. Nat Commun 5:3364

Nanni C, Fanti S, Rubello D (2007) 18F-DOPA PET and PET/CT. J Nucl Med 48(10):1577–1579. https://doi.org/10.2967/jnumed.107.041947

Nawaz FZ, Kipreos ET (2022) Emerging roles for folate receptor FOLR1 in signaling and cancer. Trends Endocrinol Metab 33(3):159–174

Naz S, Shamoon M, Wang R, Zhang L, Zhou J, Chen J (2019) Advances in therapeutic implications of inorganic drug delivery nano-platforms for cancer. Int J Mol Sci 20(4):965. https://doi.org/10.3390/ijms20040965

Neophytou CM, Panagi M, Stylianopoulos T, Papageorgis P (2021) The role of tumor microenvironment in cancer metastasis: molecular mechanisms and therapeutic opportunities. Cancer 13(9):2053. https://doi.org/10.3390/cancers13092053

Ni J-S, Li Y, Yue W, Liu B, Li K (2020) Nanoparticle-based cell trackers for biomedical applications. Theranostics 10(4):1923–1947. https://doi.org/10.7150/thno.39915

Nourizad A, Golmohammadi S, Aghanejad A, Tohidkia MR (2023) Recent trends in aptamer-based nanobiosensors for detection of vascular endothelial growth factors (VEGFs) biomarker: a review. Environ Res 236:116726. https://doi.org/10.1016/j.envres.2023.116726

Novruzov E, Mori Y, Antke C, Dabir M, Schmitt D, Kratochwil C, Koerber SA, Haberkorn U, Giesel FL (2022) A role of non-fdg tracers in lung cancer? Semin Nucl Med 52(6):720–733. https://doi.org/10.1053/j.semnuclmed.2022.05.004

Nsairat H et al (2022) Liposomes: structure, composition, types, and clinical applications. Heliyon 8(5):e09394

Obaidat IM, Narayanaswamy V, Alaabed S, Sambasivam S, Muralee Gopi CVV (2019) Principles of magnetic hyperthermia: a focus on using multifunctional hybrid magnetic nanoparticles. Magnetochemistry 5(4):67. https://doi.org/10.3390/magnetochemistry5040067

Oliveira AML et al (2022) Graphene oxide thin films with drug delivery function. Nanomaterials (Basel) 12(7):1149

Paliwal R, Paliwal SR, Kenwat R, Kurmi BD, Sahu MK (2020) Solid lipid nanoparticles: a review on recent perspectives and patents. Expert Opin Ther Pat 30(3):179–194. https://doi.org/10.1080/13543776.2020.1720649

Parihar AS, Dehdashti F, Wahl RL (2023) FDG pet/ct–based response assessment in malignancies. Radiographics 43(4). https://doi.org/10.1148/rg.220122
Parvin, T, Ahmed, K, Alatwi, AM et al. (2021) Differential optical absorption spectroscopy-based refractive index sensor for cancer cell detection. Opt Rev 28, 134–143. https://doi.org/10.1007/s10043-021-00644-w
Patamia V, Zagni C, Brullo I, Saccullo E, Coco A, Floresta G, Rescifina A (2023) Computer-assisted design of peptide-based radiotracers. Int J Mol Sci 24(7):6856. https://doi.org/10.3390/ijms24076856
Pathak A, Naqvi S, Jain K (2023) Functionalized dendrimers: promising nanocarriers for theranostic applications. In: Multifunctional and targeted theranostic nanomedicines. Springer Nature, Singapore, pp 163–185. https://doi.org/10.1007/978-981-99-0538-6_8
Patil P, Killedar S, More H, Vambhurkar G (2022) Development and characterization of 5-fluorouracil solid lipid nanoparticles for treatment of colorectal cancer. J Pharm Innov 17(4):1268–1281. https://doi.org/10.1007/s12247-021-09605-x
Pecheur I, Peyruchaud O, Serre CM, Guglielmi J, Voland C, Bourre F, Margue C, Cohen-Solal M, Buffet A, Kieffer N, Clézardin P (2002) Integrin αvβ3 expression confers on tumor cells a greater propensity to metastasize to bone. FASEB J 16(10):1266–1268
Pees A et al (2023) Recent developments in carbon-11 chemistry and applications for first-in-human PET studies. Molecules 28(3):931
Pellet-Many C, Frankel P, Jia H, Zachary I (2008) Neuropilins: structure, function and role in disease. Biochem J 411(2):211–226
Petrova V, Annicchiarico-Petruzzelli M, Melino G, Amelio I (2018) The hypoxic tumour microenvironment. Oncogenesis 7(1):10
Pillay V, Frank D, Tyagi C, Tomar LK, Choonara YE, du Toit LC, Kumar P, Penny C (2014) Overview of the role of nanotechnological innovations in the detection and treatment of solid tumors. Int J Nanomedicine 589. https://doi.org/10.2147/ijn.s50941
Pramod Kumar EK, Um W, Park JH (2020) Recent developments in pathological pH-responsive polymeric nanobiosensors for cancer theranostics. Front Bioeng Biotechnol 8. https://doi.org/10.3389/fbioe.2020.601586
Pulumati A, Pulumati A, Dwarakanath BS, Verma A, Papineni RVL (2023) Technological advancements in cancer diagnostics: improvements and limitations. Cancer Rep 6(2). https://doi.org/10.1002/cnr2.1764
Rahman M et al (2015) Emerging advances in cancer nanotheranostics with graphene nanocomposites: opportunities and challenges. Nanomedicine (Lond) 10(15):2405–2422
Rahman WT, Wale DJ, Viglianti BL, Townsend DM, Manganaro MS, Gross MD, Wong KK, Rubello D (2019) The impact of infection and inflammation in oncologic 18F-FDG PET/CT imaging. Biomed Pharmacother 117:109168. https://doi.org/10.1016/j.biopha.2019.109168
Rahme K, Chen L, Hobbs RG, Morris MA, O'Driscoll C, Holmes JD. (2013) PEGylated gold nanoparticles: polymer quantification as a function of PEG lengths and nanoparticle dimensions. RSC Adv 3(17):6085–6094. https://doi.org/10.1039/C3RA22739A
Rajendran RL, Jogalekar MP, Gangadaran P, Ahn B-C (2020) Noninvasive in vivo cell tracking using molecular imaging: a useful tool for developing mesenchymal stem cell-based cancer treatment. World J Stem Cells 12(12):1492–1510. https://doi.org/10.4252/wjsc.v12.i12.1492
Ramzan A, Tafti D (2023) Nuclear medicine PET/CT gastrointestinal assessment, protocols, and interpretation. In: StatPearls. StatPearls Publishing, Treasure Island, FL
Rylova SN, Barnucz E, Fani M, Braun F, Werner M, Lassmann S, Maecke HR, Weber WA (2014) Does imaging αvβ3 integrin expression with PET detect changes in angiogenesis during bevacizumab therapy? J Nucl Med 55(11):1878–1884
Salavati H et al (2022) Interstitial fluid pressure as an emerging biomarker in solid tumors. Biochim Biophys Acta Rev Cancer 1877(5):188792
Saletnik A, Saletnik B, Puchalski C (2021) Overview of popular techniques of Raman spectroscopy and their potential in the study of plant tissues. Molecules 26(6):1537

Salskov A, Tammisetti VS, Grierson J, Vesselle H (2007) FLT: measuring tumor cell proliferation in vivo with positron emission tomography and 3′-deoxy-3′-[18f]fluorothymidine. Semin Nucl Med 37(6):429–439. https://doi.org/10.1053/j.semnuclmed.2007.08.001

Schutt EG, Klein DH, Mattrey RM, Riess JG (2003) Injectable microbubbles as contrast agents for diagnostic ultrasound imaging: the key role of perfluorochemicals. Angew Chem Int Ed 42(28): 3218–3235. https://doi.org/10.1002/anie.200200550

Sharma H, Mondal S (2020) Functionalized graphene oxide for chemotherapeutic drug delivery and cancer treatment: a promising material in nanomedicine. Int J Mol Sci 21(17):6280

Shi L et al (2021) Effects of polyethylene glycol on the surface of nanoparticles for targeted drug delivery. Nanoscale 13(24):10748–10764

Shinn J et al (2022) Smart pH-responsive nanomedicines for disease therapy. J Pharm Investig 52(4):427–441

Shokrollahi H (2013) Contrast agents for MRI. Mater Sci Eng C 33(8):4485–4497. https://doi.org/10.1016/j.msec.2013.07.012

Siemann DW (2011) The unique characteristics of tumor vasculature and preclinical evidence for its selective disruption by tumor-vascular disrupting agents. Cancer Treat Rev 37(1):63–74

Sobhanan J, Anas A, Biju V (2023) Nanomaterials for fluorescence and multimodal bioimaging. Chem Rec 23(3). https://doi.org/10.1002/tcr.202200253

Sridharan A et al (2021) Ultrasound contrast agents: microbubbles made simple for the pediatric radiologist. Pediatr Radiol 51(12):2117–2127

Starmans LWE, van Duijnhoven SMJ, Rossin R, Aime S, Daemen MJAP, Nicolay K, Grüll H (2013) SPECT imaging of fibrin using fibrin-binding peptides. Contrast Media Mol Imaging 8(3):229–237. https://doi.org/10.1002/cmmi.1521

Sun W, Gu Z (2016) ATP-responsive drug delivery systems. Expert Opin Drug Deliv 13(3): 311–314

Sun C, Lee JSH, Zhang M (2008) Magnetic nanoparticles in MR imaging and drug delivery. Adv Drug Deliv Rev 60(11):1252–1265. https://doi.org/10.1016/j.addr.2008.03.018

Talluri S, Malla RR (2020) Superparamagnetic iron oxide nanoparticles (spions) for diagnosis and treatment of breast, ovarian and cervical cancers. Curr Drug Metab 20(12):942–945. https://doi.org/10.2174/1389200220666191016124958

Tang C-H, Chen H-L, Dong J-R (2023) Solid lipid nanoparticles (slns) and nanostructured lipid carriers (nlcs) as food-grade nanovehicles for hydrophobic nutraceuticals or bioactives. Appl Sci 13(3):1726. https://doi.org/10.3390/app13031726

Tognarelli JM et al (2015) Magnetic resonance spectroscopy: principles and techniques: lessons for clinicians. J Clin Exp Hepatol 5(4):320–328

Tomczak N, Liu R, Vancso JG (2013) Polymer-coated quantum dots. Nanoscale 5(24): 12018–12032

Tran S, DeGiovanni P, Piel B, Rai P (2017) Cancer nanomedicine: a review of recent success in drug delivery. Clin Transl Med 6(1). https://doi.org/10.1186/s40169-017-0175-0

Uematsu T (2014) Ultrasonographic Findings of Missed Breast Cancer: Pitfalls and Pearls. Breast Cancer 21:10–19. https://doi.org/10.1007/s12282-013-0498-7

Van Manen L et al (2018) A practical guide for the use of indocyanine green and methylene blue in fluorescence-guided abdominal surgery. J Surg Oncol 118(2):283–300

Verkhovskii RA, Ivanov AN, Lengert EV, Tulyakova KA, Shilyagina NY, Ermakov AV (2023) Current principles, challenges, and new metrics in pH-responsive drug delivery systems for systemic cancer therapy. Pharmaceutics 15(5):1566

Wadhwa P, Sharma S, Sahu S, Sharma A, Kumar D (2022) A review of nanoparticles characterization techniques. Curr Nanomater 7(3):202–214. https://doi.org/10.2174/2405461507666220405113715

Wagstaff AJ, Brown SD, Holden MR, Craig GE, Plumb JA, Brown RE, Schreiter N, Chrzanowski W, Wheate NJ (2012) Cisplatin drug delivery using gold-coated iron oxide nanoparticles for enhanced tumour targeting with external magnetic fields. Inorg Chim Acta 393:328–333. https://doi.org/10.1016/j.ica.2012.05.012

Wahajuddin, Arora S (2012) Superparamagnetic iron oxide nanoparticles: magnetic nanoplatforms as drug carriers. Int J Nanomedicine 7:3445–3471. https://doi.org/10.2147/ijn.s30320

Wahyudi H, Reynolds AA, Li Y, Owen SC, Yu SM (2016) Targeting collagen for diagnostic imaging and therapeutic delivery. J Control Release 240:323–331. https://doi.org/10.1016/j.jconrel.2016.01.007

Wang J, Maurer L (2005) Positron emission tomography: applications in drug discovery and drug development. Curr Top Med Chem 5(11):1053–1075

Wang Z, Yang L (2024) Natural-product-based, carrier-free, noncovalent nanoparticles for tumor chemo-photodynamic combination therapy. Pharmacol Res 203:107150. https://doi.org/10.1016/j.phrs.2024.107150

Wang Y, Zhou K, Huang G, Hensley C, Huang X, Ma X, Zhao T, Sumer BD, DeBerardinis RJ, Gao J (2013a) A nanoparticle-based strategy for the imaging of a broad range of tumours by nonlinear amplification of microenvironment signals. Nat Mater 13(2):204–212. https://doi.org/10.1038/nmat3819

Wang L, Jiang X, Ji Y, Bai R, Zhao Y, Wu X, Chen C (2013b) Surface chemistry of gold nanorods: origin of cell membrane damage and cytotoxicity. Nanoscale 5(18):8384. https://doi.org/10.1039/c3nr01626a

Wang RA, Lu YY, Fan DM (2015) Reasons for cancer metastasis: a holistic perspective. Mol Clin Oncol 3(6):1199–1202

Wang J, Xu Y, Boppart SA (2017a) Review of optical coherence tomography in oncology. J Biomed Opt 22(12):1–23

Wang J et al (2017b) Emerging advances in nanotheranostics with intelligent bioresponsive systems. Theranostics 7(16):3915–3919

Wang X, Ding Q, Groleau RR, Wu L, Mao Y, Che F, Kotova O, Scanlan EM, Lewis SE, Li P, Tang B, James TD, Gunnlaugsson T (2024) Fluorescent probes for disease diagnosis. Chem Rev 124(11):7106–7164. https://doi.org/10.1021/acs.chemrev.3c00776

Warszyńska M, Repetowski P, Dąbrowski JM (2023) Photodynamic therapy combined with immunotherapy: recent advances and future research directions. Coord Chem Rev 495:215350. https://doi.org/10.1016/j.ccr.2023.215350

Weinberg BD et al (2021) Clinical applications of magnetic resonance spectroscopy in brain tumors: from diagnosis to treatment. Radiol Clin North Am 59(3):349–362

Weis SM, Cheresh DA (2011) Tumor angiogenesis: molecular pathways and therapeutic targets. Nat Med 17(11):1359–1370

West HJ, Jin JO (2015) Adjuvant therapy. JAMA Oncol 1(5):698

Wiebe LI (2004) PET radiopharmaceuticals for metabolic imaging in oncology. Int Congr Ser 1264:53–76. https://doi.org/10.1016/j.ics.2003.12.102

Wolf W, Presant CA, Waluch V (2000) 19F-MRS studies of fluorinated drugs in humans. Adv Drug Deliv Rev 41(1):55–74

Woo SK et al (2019) Development of (64)Cu-NOTA-Trastuzumab for HER2 targeting: a radiopharmaceutical with improved pharmacokinetics for human studies. J Nucl Med 60(1):26–33

Wu T-J, Chiu H-Y, Yu J, Cautela MP, Sarmento B, das Neves J, Catala C, Pazos-Perez N, Guerrini L, Alvarez-Puebla RA, Vranješ-Đurić S, Ignjatović NL (2018) Nanotechnologies for early diagnosis, in situ disease monitoring, and prevention. In: Nanotechnologies in preventive and regenerative medicine. Elsevier, pp 1–92. https://doi.org/10.1016/b978-0-323-48063-5.00001-0

Xia J, Yao J, Wang LV (2014) Photoacoustic tomography: principles and advances. Electromagn Waves (Camb) 147:1–22

Yadav VR, Suresh S, Devi K, Yadav S (2009) Novel formulation of solid lipid microparticles of curcumin for anti-angiogenic and anti-inflammatory activity for optimization of therapy of inflammatory bowel disease. J Pharm Pharmacol 61(3):311–321. https://doi.org/10.1211/jpp/61.03.0005

Yamagata M, Hasuda K, Stamato T, Tannock I (1998) The contribution of lactic acid to acidification of tumours: studies of variant cells lacking lactate dehydrogenase. Br J Cancer 77(11): 1726–1731. https://doi.org/10.1038/bjc.1998.289

Yang G et al (2018) Structure of graphene and its disorders: a review. Sci Technol Adv Mater 19(1): 613–648

Yeo DC, Wiraja C, (Sakis) Mantalaris A, Xu C (2014) Nanosensors for regenerative medicine. J Biomed Nanotechnol 10(10):2722–2746. https://doi.org/10.1166/jbn.2014.1992

Yu J, Zhang Y, Hu X, Wright G, Gu Z (2016 Jun) Hypoxia-sensitive materials for biomedical applications. Ann Biomed Eng 44:1931–1945

Zafar S, Akhter S, Ahmad I, Hafeez Z, Alam Rizvi MM, Jain GK, Ahmad FJ (2020) Improved chemotherapeutic efficacy against resistant human breast cancer cells with co-delivery of docetaxel and thymoquinone by chitosan grafted lipid nanocapsules: formulation optimization, in vitro and in vivo studies. Colloids Surf B Biointerfaces 186:110603. https://doi.org/10.1016/j.colsurfb.2019.110603

Zeng W, Wang X, Xu P, Liu G, Eden HS, Chen X (2015) Molecular imaging of apoptosis: from micro to macro. Theranostics 5(6):559–582. https://doi.org/10.7150/thno.11548

Zeng Y, Ma J, Zhang S, Wu D (2016) Imaging agents in targeting tumor hypoxia. Curr Med Chem 23(17):1775–1800. https://doi.org/10.2174/0929867323666160321122023

Zeng W, Li Z, Chen H, Zeng X, Mei L (2022) An optimal portfolio of photothermal combined immunotherapy. Cell Rep Phys Sci 3(6):100898

Zhang H et al (2009) Detection and downregulation of type I IGF receptor expression by antibody-conjugated quantum dots in breast cancer cells. Breast Cancer Res Treat 114(2):277–285

Zhang C, Wu W, Li R, Qiu W, Zhuang Z, Cheng S, Zhang X (2018) Peptide-based multifunctional nanomaterials for tumor imaging and therapy. Adv Funct Mater 28(50). https://doi.org/10.1002/adfm.201804492

Zhang J, Dong Y, Zhu W, Xie D, Zhao Y, Yang D, Li M (2019) Ultrasensitive detection of circulating tumor DNA of lung cancer via an enzymatically amplified SERS-based frequency shift assay. ACS Appl Mater Interfaces 11(20):18145–18152

Zhang M, Gao S, Yang D, Fang Y, Lin X, Jin X, Liu Y, Liu X, Su K, Shi K (2021) Influencing factors and strategies of enhancing nanoparticles into tumors in vivo. Acta Pharm Sin B 11(8): 2265–2285. https://doi.org/10.1016/j.apsb.2021.03.033

Zhang P, Chen D, Li L, Sun K (2022) Charge reversal nano-systems for tumor therapy. J Nanobiotechnol 20(1). https://doi.org/10.1186/s12951-021-01221-8

Zhao C, Tian M, Zhang H (2010) In vivo stem cell imaging. Open Nucl Med J 2(1):171–177. https://doi.org/10.2174/1876388x01002010171

Zhi D, Yang T, Yang J, Fu S, Zhang S (2020) Targeting strategies for superparamagnetic iron oxide nanoparticles in cancer therapy. Acta Biomater 102:13–34. https://doi.org/10.1016/j.actbio.2019.11.027

Zhong ZA et al (2015) (99m)TC-Methylene diphosphonate uptake at injury site correlates with osteoblast differentiation and mineralization during bone healing in mice. Bone Res 3:15013

Zhou Z, Lu Z-R (2017) Molecular imaging of the tumor microenvironment. Adv Drug Deliv Rev 113:24–48. https://doi.org/10.1016/j.addr.2016.07.012

Zhu Y et al (2023) Ferritin-based nanomedicine for disease treatment. Med Rev (2021) 3(1):49–74

Zou X, Pan T, Chen L, Tian Y, Zhang W (2016) Luminescence materials for pH and oxygen sensing in microbial cells—structures, optical properties, and biological applications. Crit Rev Biotechnol 37(6):723–738. https://doi.org/10.1080/07388551.2016.1223011

Choice of Nanomedicine in Cancer Imaging and Therapeutics

6

Krishna Jadhav, Eknath Kole, Jitendra Naik, Rahul Kumar Verma, Amol Gholap, Satyawati Rawat, Shweta Meena, and Satish Rojekar

Abstract

This chapter comprehensively explores how nanotechnology is reshaping the landscape of cancer imaging and therapeutics. Nanomedicine offers a promising avenue for personalized cancer treatment by integrating diagnostic and therapeutic approaches. The discussion delves into the distinctive characteristics, applications, and challenges associated with various nanoparticles, counting organic (liposomes, micelles) and inorganic (gold, iron oxide, silica, quantum dots, carbon nanotubes), and within the field of oncology. Emphasis is placed on how these nanomaterials facilitate advanced cancer treatments by enhancing drug delivery systems, enabling precise photothermal therapy, and advancing imaging

K. Jadhav · R. K. Verma
Pharmaceutical Nanotechnology Lab, Institute of Nano Science and Technology, Mohali, Punjab, India

E. Kole · J. Naik
Department of Pharmaceutical Technology, University Institute of Chemical Technology, KBC North Maharashtra University, Jalgaon, Maharashtra, India

A. Gholap
Department of Pharmaceutics, St. John Institute of Pharmacy and Research, Palghar, Maharashtra, India

S. Rawat
Department of Botany, Kurukshetra University, Kurukshetra, Haryana, India

S. Meena
Department of Electronics and Communication Engineering, National Institute of Technology Kurukshetra, Kurukshetra, Haryana, India

S. Rojekar (✉)
Department of Pharmacological Sciences, Icahn School of Medicine at Mount Sinai, New York, NY, USA
e-mail: satish.rojekar@mssm.edu

R. Prasad, M. Mohan Yallapu (eds.), *Advancements in Cancer Theranostics*,
https://doi.org/10.1007/978-981-95-0938-6_6

techniques such as computed tomography (CT), magnetic resonance imaging (MRI), and fluorescence imaging. This chapter underscores the potential of nanoparticles to elevate therapeutic efficacy, mitigate systemic toxicity, and offer controlled and targeted drug release, thus opening new avenues for effective cancer treatment. It also addresses critical issues related to the biocompatibility and safety of nanomaterials, highlighting the necessity for ongoing research to optimize their clinical applications. The chapter concludes by considering the future of nanotechnology in predictive oncology and tailored medicine, envisioning a transformative impact on early detection, individualized therapies, and overall cancer care.

Keywords

Bioimaging · Cancer nanomedicine · Nanotheranostics · Translational · Phototherapy · Solid tumor

6.1 Introduction

As the leading cause of death worldwide, cancer demands early detection and the implementation of effective treatments to combat its progression. The field of nanomedicine has fundamentally transformed cancer diagnosis and therapy, leveraging the unique properties of nanoparticles to achieve unprecedented precision and efficacy. These nanomaterials, typically ranging between 10 to 100 nm in size, show exceptional physical, chemical, and optical characteristics that set them apart from conventional agents used in oncology (Wang et al. 2024; Oehler et al. 2024). Their nanoscale dimensions allow them to navigate complex biological environments, permeating tissues and organs more efficiently than traditional molecules, as presented in Fig. 6.1. This ability enables nanoparticles to selectively target diseased areas, such as cancerous tumors while minimizing off-target effects in healthy tissues (Cheng et al. 2023). In addition to this precise targeting capability, nanoparticles can be conjugated with therapeutic agents, imaging probes, or both, offering a dual role in cancer diagnostics and therapy (Al-Thani et al. 2024). Due to their small size, comparable to or smaller than DNA, nanoparticles can penetrate biological barriers, including the fenestrations in the epithelium, resulting in enhanced delivery of therapeutic payloads and improved outcomes in cancer care (Karahmet Sher et al. 2024).

Integrating nanoparticles into medical applications has led to a paradigm shift, particularly in melanoma imaging, where traditional methods often lack specificity and sensitivity. The ability of nanoparticles to enhance the contrast in several imaging modes, such as magnetic resonance imaging (MRI), computed tomography (CT), positron emission tomography (PET), and ultrasound, is a significant advancement. These nanomaterials can be engineered with chemical and surface modifications to optimize their distribution, biodurability, and overall performance in imaging tasks (Navya et al. 2019; Toy et al. 2014). For instance, nanoparticles

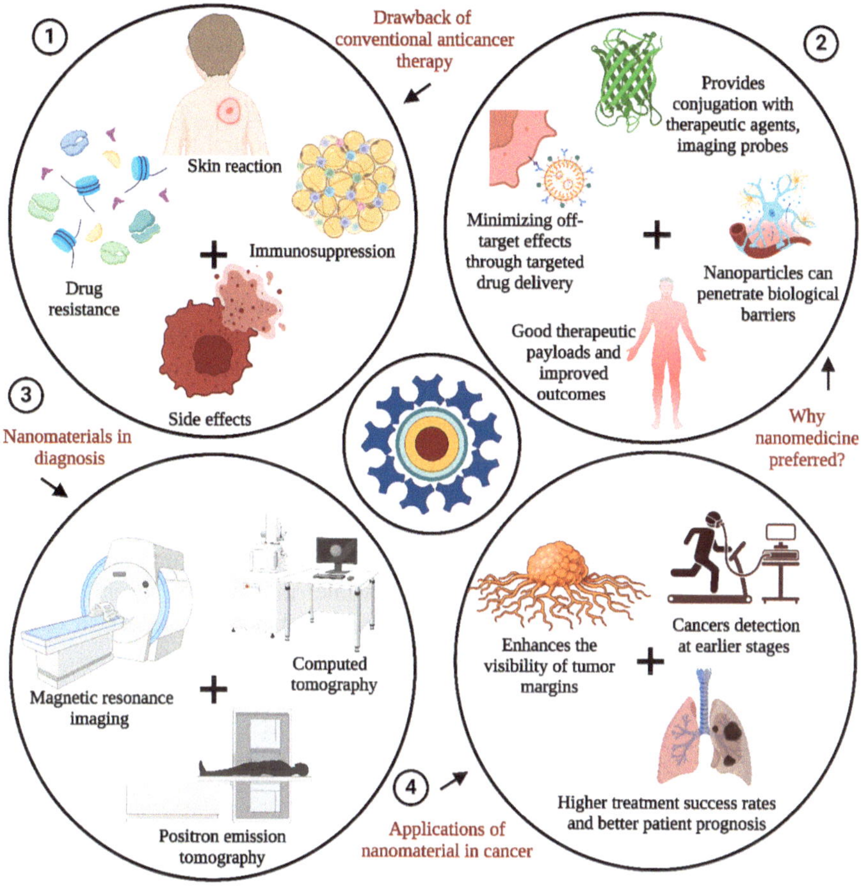

Fig. 6.1 Summary of nanomaterials used in diagnosis and therapeutics included (1) Drawbacks of conventional anticancer therapy. (2) Why nanomedicines are required in cancer. (3) Implementation of nanomaterials in the diagnosis. (4) Applications of nanomaterials in cancer. (Created by using Boiorender.com)

such as gold, iron oxide, or quantum dots offer superior optical properties, enabling more precise, high-resolution images of tumor regions. Magnetic and metallic nanoparticles can be incorporated into imaging agents to significantly enhance the visibility of tumor margins, allowing clinicians to detect cancers at earlier stages with more precision than ever before. This improvement in early cancer detection is crucial, as it directly correlates with higher treatment success rates and better patient prognosis (Siddique and Chow 2020; Padmanabhan et al. 2016).

Beyond their utility in imaging, nanoparticles have become instrumental in cancer therapeutics, offering solutions to some of the most significant challenges in oncology, such as drug resistance, non-specific toxicity, and the inability of many drugs to reach deep-seated or protected tumor sites (Yao et al. 2020). Nanoparticles can encapsulate chemotherapeutic agents, protecting them from premature degradation in the bloodstream and enabling controlled, sustained release at the tumor site (Rojekar et al. 2024). This targeted delivery system reduces the systemic side effects commonly associated with chemotherapy and increases the therapeutic efficacy of the drugs (Vaghasiya et al. 2021; Kole et al. 2023, 2024; Jadhav et al. 2023, 2024; Jakhar et al. 2023). Nanoparticles are also central to advancing novel therapeutic approaches, such as gene, photothermal, photodynamic, and immunotherapy (Kadkhoda et al. 2022). In photothermal therapy, for example, gold nanoparticles are designed to absorb light at specific wavelengths and convert it into heat, which selectively destroys tumor cells. Similarly, in photodynamic therapy, nanoparticles deliver light-sensitive agents that, upon activation, produce reactive oxygen species (ROS) capable of inducing cancer cell death (Duan et al. 2023; Vines et al. 2019). The field of theranostic nanomedicine, which combines diagnostic and therapeutic functionalities within a single nanoparticle platform, represents one of the most exciting frontiers in cancer treatment. These multifunctional nanoparticles permit simultaneous drug delivery monitoring, enabling clinicians to assess therapeutic efficacy as treatment progresses (Al-Thani et al. 2024). Theranostic systems also facilitate the co-delivery of multiple therapeutic agents, which may include a combination of chemotherapy, gene therapy, and immune modulators, maximizing treatment impact while minimizing adverse effects (Fernandez-Fernandez et al. 2011). The statistics of nanomedicine use in cancer imaging and therapeutics from 2014 to 2024 are presented in Fig. 6.2.

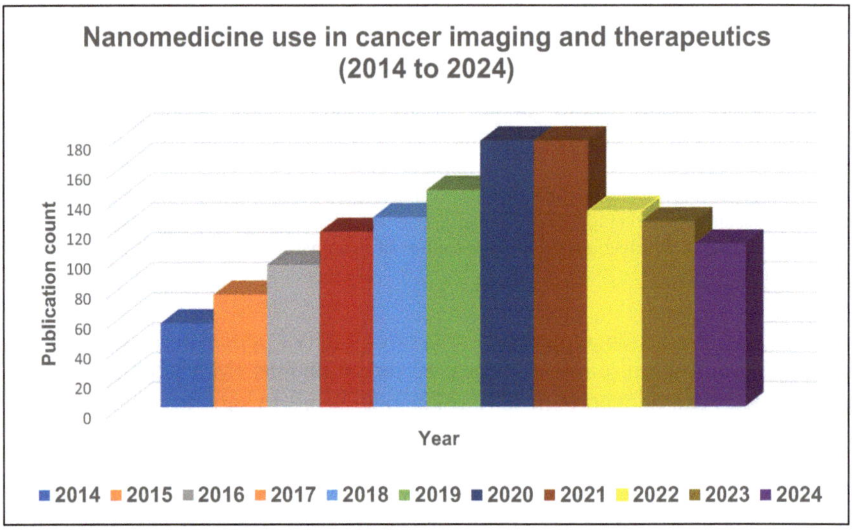

Fig. 6.2 The statistics for nanomedicine use in cancer imaging and therapeutics from 2014 to 2024. (Data obtained from https://pubmed.ncbi.nlm.nih.gov/?term=Nanomedicine+in+Cancer+Imaging+and+Therapeutics&filter=pubt.review&filter=datesearch.y_10)

This chapter provides an in-depth exploration of nanomedicine evolving role in cancer imaging and therapeutics, discussing the wide range of nanomaterials utilized, their properties, and their applications in the clinical setting. It will delve into the advancements in nanoparticle-based imaging, the breakthroughs in nanoparticle-facilitated drug delivery, and the novel therapeutic approaches emerging from nanomedicine. Furthermore, the chapter will examine future directions in nanomedicine, including the integration of AI in nanomedicine, and explore the advancement of next-generation nanoparticles intended for improved safety, efficiency, and patient-specific applications. As the field continues to grow, nanomedicine holds the potential to revolutionize cancer care, improving patient outcomes and offering hope for more effective, personalized cancer treatment strategies.

6.2 Fundamentals of Nanomedicine

Nanomedicine is an innovative field that focuses on the development of nanosystems, or nanocarriers, which typically range in size from a few nanometers to several hundred nanometers. These nanosystems are engineered to carry therapeutic agents or diagnostic probes, offering unprecedented opportunities in cancer imaging and therapy (Seynhaeve et al. 2020). Encapsulation within nanosystems enhances therapeutic efficacy through several mechanisms. Firstly, it reduces the volume of drug distribution, concentrating the drug at the tumor site. Secondly, it extends the circulation time of the drug in the bloodstream, allowing for prolonged retention at the target site, as presented in Fig. 6.3 (Priester and ten Hagen 2023; Bertrand et al. 2014; Anselmo and Mitragotri 2016). Thirdly, nanocarriers enable the co-encapsulation of multiple agents, which can work synergistically to improve treatment outcomes. Finally, they can interact with the unique conditions of the tumor microenvironment, like altered pH, enzyme activity, or hypoxia, to enhance targeting and release behavior (Shi et al. 2016; Mura et al. 2013). The encapsulation of therapeutic agents within nanocarriers also significantly influences the pharmacokinetics and biodistribution of the drugs (Ernsting et al. 2013). By protecting the drug from premature degradation in the systemic circulation, nanocarriers increase the likelihood that the therapeutic agent will reach the tumor intact. This improved stability and the ability to deliver the drug to the specific site precisely maximize the therapeutic outcome while diminishing off-target toxicity. Notably, the structural stability of the nanocarrier itself is a critical consideration, as a stable system ensures reliable delivery, particularly in combination therapies where multiple agents must be delivered to the tumor site concurrently (ud Din et al. 2017; Perrigue et al. 2021).

Nanocarriers also support complex treatment strategies, such as the delivery of multiple drugs that can be activated by specific external triggers. For example, some nanosystems are designed to release their payload in response to near-infrared (NIR) radiation through photothermal therapy while simultaneously facilitating photodynamic therapy by generating ROS (Gholap et al. 2024a, b, c, d). This dual functionality can enhance the therapeutic effect by combining localized heat with increased

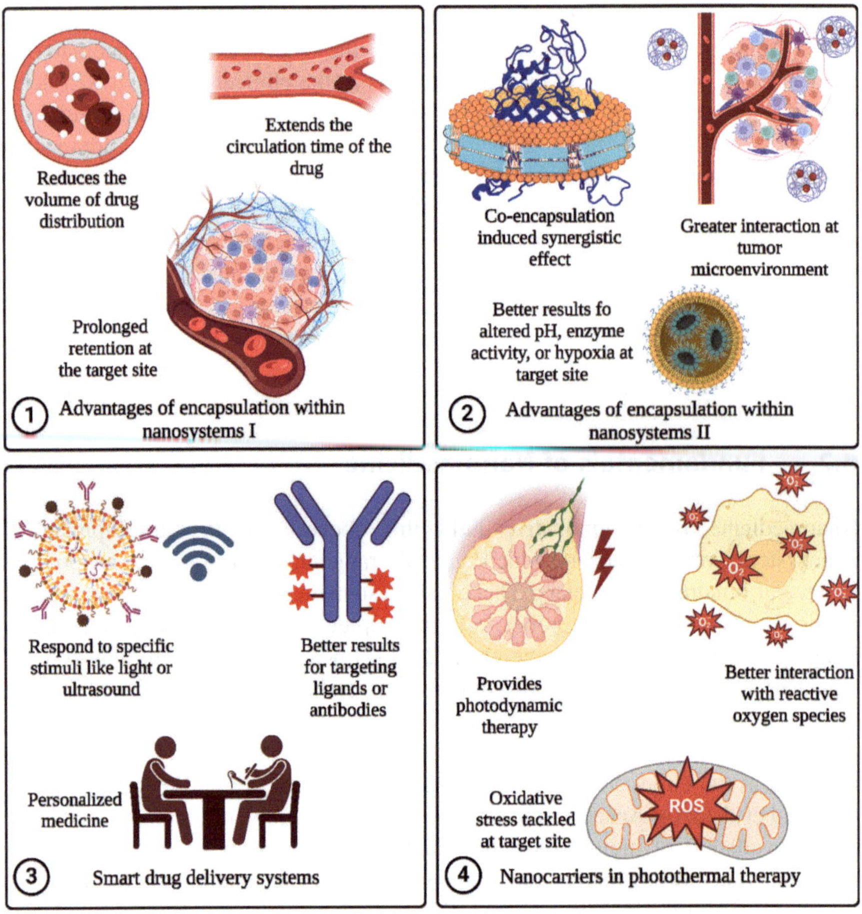

Fig. 6.3 The summary for applications of nanocarriers in cancer involves. (1) Advantages of encapsulation within the nanosystem part I, focusing on prolonged retention at the target site and reduced volume of distribution. (2) Advantages of encapsulation within the nanosystem part II include greater interaction at the target site with better results in the presence of altered pH, enzyme activity and hypoxia. (3) Smart drug delivery platforms engineered to respond to exogenous stimuli, such as light irradiation or ultrasound exposure, enabling spatiotemporally controlled drug release. (4) Nanocarrier-based modalities for photothermal therapy, often integrated with photodynamic approaches, to synergistically generate and modulate ROS. (Created by using Boiorender.com)

chemotherapeutic activity and oxidative stress at the tumor site. The ability to monitor the delivery of these systems using imaging techniques and to activate them precisely when needed further underscores the potential of nanomedicine for highly targeted and effective cancer treatment (Lucky et al. 2015; Cheng et al. 2014; Tong et al. 2012; Wang et al. 2016). Smart drug delivery systems (SDDS) represent the next generation of nanomedicine. These advanced systems are designed to respond to specific stimuli, either endogenous or exogenous triggers such as light or ultrasound. SDDS can release their therapeutic payload precisely, improving the precision and timing of drug delivery. They can also be modified to incorporate targeting ligands or antibodies on their surface, further enhancing their ability to home in on tumor cells while thrifty, healthy tissue. Furthermore, SDDS allows for multifunctional approaches, combining therapeutic and diagnostic abilities in a single stage (Priester and ten Hagen 2023; Rana et al. 2023). Refining these systems into innovative drug delivery platforms will open new doors for personalized, efficient, and minimally invasive cancer therapies as research continues.

6.3 Imaging Moieties in Cancer Imaging

Imaging moieties are vital in research and clinical settings, facilitating the present imagining and quantification of drug delivery. Such external substances temporarily enhance the visibility of morphologies, internal structures, and biological functions (Gholap et al. 2023a, b, c, 2024c, d; Mesut et al. 2024). For improved delivery and monitoring, a differentiation is made between nanomaterials that possess inherent imaging capabilities and external imaging agents, such as probes and labels (Sharma et al. 2024; Wallyn et al. 2019). These moieties include various contrast agents, radionuclides, and fluorescent markers that can be incorporated into nanomedicine platforms to improve imaging outcomes. Imaging moieties can also be encapsulated within a nanosystem individually or in combination. While some agents are designed solely for imaging, others are multifunctional, contributing to triggered drug release through mechanisms such as photothermal or photosensitizing effects (Sarbadhikary et al. 2021; Jung et al. 2018).

Contrast agents (CAs) are pivotal in enhancing imaging by improving the differentiation between healthy and malignant tissues. Solid tumors, with their abnormal vascular networks characterized by increased permeability and poor lymphatic drainage, often show higher accumulation and prolonged retention of CAs compared to normal tissues (Seynhaeve et al. 2020). This property is harnessed in image-guided drug delivery to monitor nanosystem location, retention, and drug delivery efficiency (Ponce et al. 2007). CAs are categorized based on imaging modality: optical agents, such as dyes, proteins, and autofluorescent substances, and metal-based agents used for MRI or CT imaging. Optical contrast agents are typically bright, photostable, and biodegradable, with fluorophores emitting across a broad spectrum from ultraviolet to infrared (Kirui and Ferrari 2015). These dyes offer temporary, non-specific contrast for real-time in vivo monitoring. Some require metabolic conversion to emit fluorescence, while others, like carboxyfluorescein (CF), are designed to avoid cellular uptake to prevent fluorescence changes from intracellular interactions (Joshi and Wang 2018; ten Hagen et al. 2021).

NIR fluorophores, operating within the 700–3000 nm range, provide enhanced optical properties such as deeper tissue penetration, higher spatial resolution, and reduced background noise (Luo et al. 2011; Potara et al. 2021). Common NIR dyes include indocyanine green (ICG), Cy5.5, and IR780 (Jiao et al. 2020). Although ICG is FDA-approved and used widely due to its low toxicity and minimal autofluorescence, it has a short half-life and poor photostability (Joshi and Wang 2018; Zeng et al. 2021). In contrast, IR780 offers greater fluorescence intensity, better stability, and improved tumor targeting (Potara et al. 2021; Wang and Niu 2021). Fluorescent proteins, categorized as endogenous or exogenous, also contribute to imaging. Endogenous sources include transgenic models with fluorescent reporters, fluorescent cell implantation, or targeted gene delivery methods (Kirui and Ferrari 2015). Bioluminescence, providing a better signal-to-noise ratio, enables deep tissue imaging without excitation. Exogenous fluorescent agents like labeled dextrans are used to study tumor blood flow (Mezzanotte et al. 2017; Seynhaeve and ten Hagen 2018). Challenges such as background autofluorescence, especially in the visible light spectrum, are addressed through strategies like shifting to NIR wavelengths or employing autofluorescent drugs such as doxorubicin (DOX) and topotecan, which serve dual roles as therapeutic agents and fluorescent probes (Li et al. 2017).

In addition to optical contrast agents, other types of contrast agents are used in imaging. MRI contrast agents, such as gadolinium-based chelates, are classified into T1-weighted or T2-weighted agents based on their effect on longitudinal (T1) or transverse (T2) relaxation times. T1-weighted agents, like Magnevist®, enhance positive contrast for brighter images, while T2-weighted agents produce darker images through negative contrast. Despite the widespread use of gadolinium agents, they can have side effects such as nephrotoxicity, prompting the development of safer alternatives (Wahsner et al. 2019; Hu 2020; Anani et al. 2021). CT contrast agents enhance imaging based on X-ray attenuation. Heavy elements such as iodine, bismuth, and gold are commonly used, with iodine-based agents being prevalent but limited by short circulation times and low specificity (Zhang et al. 2018; Meng et al. 2021). Bismuth and gold offer improved CT contrast due to higher atomic numbers and densities, though their circulation times are still relatively short. These elements also possess photothermal properties, making them suitable for multimodal imaging applications (Ma et al. 2018; Aslan et al. 2020).

Radionuclides are utilized in nuclear imaging techniques such as single photon emission computed tomography (SPECT) and positron emission tomography (PET). SPECT uses gamma-emitting radioisotopes like technetium-99m and indium-111, while PET relies on positron-emitting radionuclides such as fluorine-18 and copper-64 (Man et al. 2019; Polyak and Ross 2018). Radionuclides offer advantages like boundless diffusion depth for whole-body imaging, but challenges include the potential detachment of radionuclides from radiolabeled compounds, which can lead to undesired tissue uptake and interference with imaging (Man et al. 2019; Chakravarty et al. 2014). Encapsulation of radionuclides in nanosystems can enhance sensitivity and contrast and allow for combined therapeutic and diagnostic functions (Chakravarty et al. 2014).

6.4 Types of Nanomaterials Used in Cancer

Recent advancements in nanotechnology have significantly enhanced clinical diagnostics by improving sensitivity and enabling early cancer detection (Gholap et al. 2023c, d, e). Engineered nanoparticles demonstrate significant potential in cancer diagnosis and therapy, including metallic, polymeric, metal oxides, quantum dots, graphene, fullerenes, liposomes, carbon nanotubes, and dendrimers (Zhang et al. 2019a; Jaishree and Gupta 2012). To further enhance their detection capabilities, nanoparticles can be functionalized with targeting moieties, including antibodies, aptamers, carbohydrates, peptides, and other small molecules that specifically bind to target molecules at the desired site (Pavitra et al. 2021). For example, cancer-targeting antibodies were conjugated with polyethylene glycol (PEG) to detect breast adenocarcinoma cells. The antibody-PEG complex was then anchored to the surface of nanoparticles through a sulfur-containing group located at the PEG linker's terminal end (Alrushaid et al. 2023).

6.4.1 Liposomes

Liposomes are self-assembling, spherical vesicles composed of phospholipids. Due to their amphiphilic nature, these lipids create a structure with a hydrophilic aqueous core encased by a hydrophobic bilayer, as presented in Fig. 6.4 (Vaghasiya et al. 2019).

Both traditional and stealth liposomes have been investigated for use in image-guided drug delivery. Fluorescence labeling is a prevalent adjustment, achieved by incorporating fluorescent lipids like Lissamine Rhodamine B (Rho-PE) into the lipid film during preparation, which enables the membrane to emit light. Beyond these inherent properties, imaging agents can be incorporated into the liposome's core or bilayer. Lipophilic tracers like DiD and DiO are embedded in the lipid bilayer, while hydrophilic fluorophores such as calcein and CF reside in the aqueous core (Priester and ten Hagen 2023). For instance, Doxil® has been labeled with the lipophilic dye DiO, allowing for visualization of liposomal distribution. Using intravital microscopy, Seynhaeve and colleagues experimentally observed a substantial rise in Doxil®-DiO penetration into tumor tissues following vascular manipulation with low-dose TNFα, resulting in a 7.1-fold upsurge in signal intensity within the tumor interstitium (Seynhaeve et al. 2007).

In addition to phospholipid composition, functionalizing liposomes plays a critical role in determining their timing, location, and duration of action. One technique for precisely evaluating tumor distribution and uptake involves using dual-labeled liposomes. These liposomes contain two imaging agents, an optical probe and a contrast agent, allowing for multimodal imaging within a single organism. A commonly used pairing is near-infrared fluorescence (NIRF) and MRI (Anani et al. 2021). These methods offer high spatial resolution but differ in penetration depth and the type of information they provide. Using two imaging modalities together allows for real-time, detailed visualization of tumors and

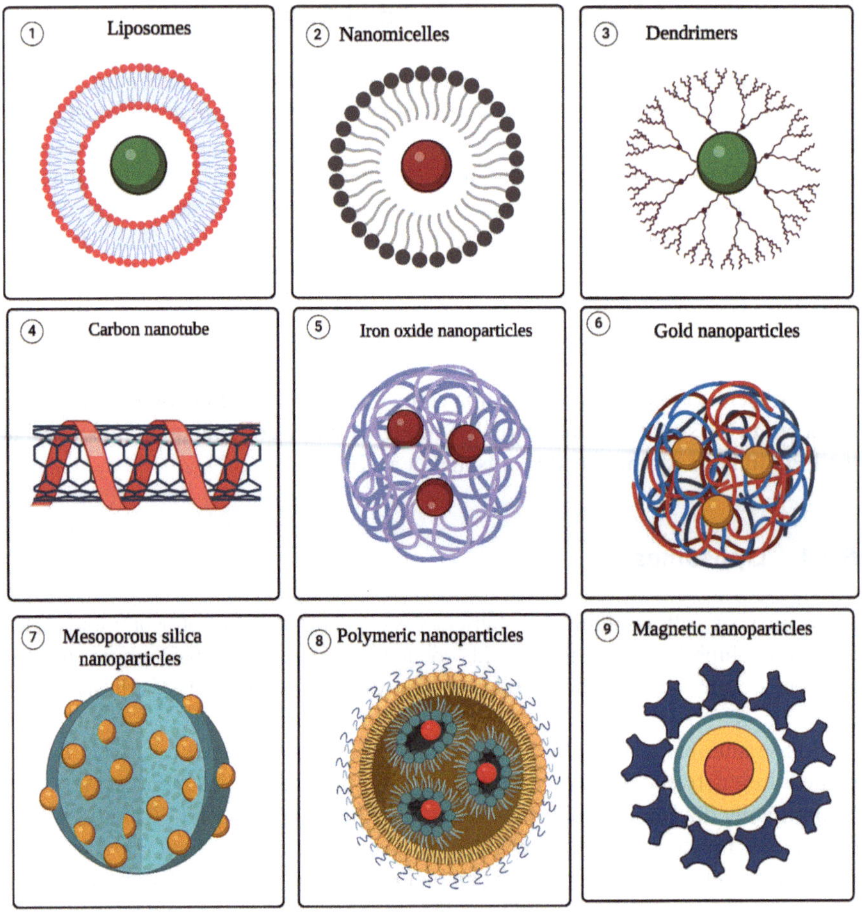

Fig. 6.4 Summary of nanocarriers used in cancer diagnosis and treatment, including liposomes, nanomicelle, dendrimers, carbon nanotubes, iron oxide nanoparticles, gold nanoparticles, mesoporous silica nanoparticles, polymeric nanoparticles, and magnetic nanoparticles. (Created by using Boiorender.com)

liposomes, with side-by-side comparisons verifying the results. Moreover, more controlled drug delivery is possible using MRI-guided high-intensity focused ultrasound, which enables precise monitoring of tumoral temperatures for improved drug release (Priester and ten Hagen 2023; Li et al. 2016).

Liposomes encapsulating gadolinium-based contrast agents have facilitated high-resolution imaging of the microvasculature. This improvement is due to the prolonged blood pool residence time, leading to clinically significant contrast enhancement even at reduced dosages (Bennett et al. 2014; Ayyagari et al. 2006). Magnetic resonance (MR) contrast compounds are often used in conjunction with whole-body imaging in investigations to map biodistribution and evaluate therapeutic efficacy. For example, Xiao et al. developed liposomes co-loaded with sorafenib (SF) and gadolinium, allowing exact monitoring of the liposomes' distribution via MR imaging across several time points. Compared to Magnevist®, these liposomes exhibited superior signal enhancement and extended imaging duration. However, a critical safety issue associated with gadolinium liposomes is toxicity. If gadolinium chelates remain in the bloodstream beyond their stable period, decomplexation can occur, leading to Gd^{3+} deposition in tissues (Xiao et al. 2016).

Radiolabeled liposomes present another option for high-resolution imaging. These liposomes may be labeled by incorporating radiolabeled drugs, labeling the liposome surface or bilayer itself, or using both methods (Man et al. 2019; Aranda-Lara et al. 2020). Using several imaging modalities, dual-labeling approaches allow drug uptake quantification and tumor surveillance. For example, Lamichhane et al. developed dual-tracer imaging with ^{111}In-labeled PEGylated liposomes encapsulating an ^{18}F-labeled carboplatin derivative. Although similar biodistribution patterns were observed in vivo, PET-based imaging allowed better visualization of drug accumulation over time compared to SPECT imaging of the liposomes (Lamichhane et al. 2017). Nuclear imaging offers high sensitivity and resolution but carries risks due to ionizing radiation and potential tracer scattering from liposomal degradation (van der Geest et al. 2016).

The development of stimuli-responsive liposomes for drug delivery has evolved significantly, beginning with the creation of thermosensitive liposomes. Modern designs employ various triggers, including pH, light, and heat, to release drug payloads (Yatvin et al. 1978). Real-time monitoring of the delivery system can be seen in the work by Ponce et al., who continuously captured MR images during a 75-min treatment session. They observed induced drug release within the tumor using manganese-loaded lysolipid-based thermosensitive liposomes; nevertheless, this approach presupposes co-localization of manganese and the drug after release (Ponce et al. 2007).

6.4.2 Micelles

Unlike many other nanocarrier systems reviewed, micelles stand out due to their simplicity in design and structure. Functional modifications can be made to the polymers forming the micelles or the core, which can house multiple therapeutic or imaging agents. For example, whole-body fluorescence imaging confirmed that these micelles efficiently accumulated in solid tumors, with peak intratumoral levels observed 24 h after injection. Photoacoustic imaging further supported this, where micelles encapsulating IR780 and DOX showed a 1.5-fold higher signal than free theranostic agents. Upon NIR irradiation (808 nm, 1 W/cm^2, 5 min), the micelles confirmed their presence in the tumor, visualized through thermal imaging that revealed rapid heat generation. This photothermal effect was absent in untreated controls. Animals receiving micellar IR780/DOX and NIR irradiation showed complete tumor inhibition with no recurrence for the 30-day study, unlike the groups treated with either IR780 or DOX alone, which failed to eradicate the tumor (Yang et al. 2018).

Kim et al. utilized a high-definition 3D multiparametric photoacoustic computed tomography (PACT) system to investigate primary and metastatic tumors in living mice. By employing micelle-loaded IR780 dye, the study demonstrated the capability of micelles to enhance imaging contrast and facilitate the non-invasive monitoring of tumor characteristics, including angiogenesis and hypoxia. The results indicated that the micelle-loaded dye provided clear visualization of tumor structures and allowed for real-time pharmacokinetics tracking. This was validated through both in vivo and ex vivo fluorescence imaging, confirming the effectiveness of micelles in improving the resolution and specificity of tumor imaging (Kim et al. 2024).

Cherukla et al. developed micelle-based nanoplatforms featuring a thermosensitive lipid coating to enhance tumor targeting. This lipid layer operates through two fundamental mechanisms: first, it experiences charge alteration of the zwitterionic lipid DPPC in response to the acidic tumor microenvironment; second, it experiences lipid deshielding when exposed to NIR irradiation. These actions expose the underlying micellar core, incorporating polyethyleneimine (PEI) for improved internalization and lithocholic acid to boost drug penetration. This core encapsulates ICG and PTX. Fluorescence imaging demonstrated peak tumor accumulation of these LIL-PTX nanoparticles 24 h after administration, with concurrent clearance from other major organs such as the liver and spleen. At this optimal point, laser irradiation (808 nm, 2 W/cm^2, 10 min) was applied, resulting in a marked increase in tumor temperature, as shown by thermal imaging. In alignment with therapeutic studies, the LIL-PTX nanoparticles, when paired with laser treatment, led to ample tumor removal without return over the 14-day study, whereas those without laser irradiation exhibited no tumor-suppressing effects (Cherukula et al. 2019).

In a related approach, Zhang et al. introduced a hierarchical pomegranate-like nanostructure composed of multiple micelles. The inner micelles were based on H30-PCL, encapsulating IR780 and camptothecin (CPT), and coated with positively charged helical PPI chains, further functionalized with alkynyl-modified guanidinium groups to improve tissue penetration. These micelles were enmeshed within negatively charged PEI (–COOH/FA) chains, forming the outer nanostructure. This design relies on the acidic tumor microenvironment to trigger a charge reversal, leading to the micelles' release from the larger structure. In vivo photoacoustic (PA) and ultrasound (US) imaging showed a steady increase in tumor accumulation of the CPT/IR780@H30-PCL-PPI(L−)/PEI(–COOH/FA) nanostructure, reaching a peak 24 h post-injection. Subsequent NIR irradiation enabled effective chemo-photothermal therapy. Efficacy studies demonstrated that tumor suppression relied on both heat-triggered drug release and photothermal effects, as animals lacking either component of the treatment showed incomplete tumor inhibition and poor survival outcomes (Zhang et al. 2019b).

6.4.3 Dendrimers

As nanosystems for solid tumor treatment, dendrimers are highly regarded due to their generational architecture. Research indicates that high-generation, non-modified dendrimers can enhance localized drug delivery in various cancers (Crintea et al. 2023). However, the incorporation of multiple agents, either encapsulated in the core or conjugated into branching units, expands the potential of these nanosystems (Bober et al. 2022). For example, Cao et al. created a G5 PAMAM dendrimer with an imaging ligand on the surface that contained paclitaxel (PTX) within its hydrophobic core. This ligand improved tumor selectivity and made fluorescence imaging possible. It was based on an RGD peptide and an NIR dye. These imaging ligands enhance visualization because biological autofluorescence can obscure ordinary PAMAM dendrimers (Cao et al. 2018). Phthalocyanines (Pc) represent another therapeutic strategy. Known for their excellent NIR optical properties, Pc molecules tend to aggregate due to their hydrophobicity, leading to self-quenching. However, encapsulating Pc within 60 nm, dendrimers successfully enabled fluorescence imaging and photodynamic therapy. Though not a complete image-guided drug delivery example, this strategy demonstrates the versatility of dendrimers in combining multiple functions (Taratula et al. 2013).

For MRI, dendrimers are typically used as carriers for contrast agents (Lee et al. 2005). Gadolinium (Gd)-functionalized dendrimers demonstrate increased sensitivity in tumors due to their hyperbranched structure, which enables higher Gd^{3+} loading and better retention (Bober et al. 2022; McMahon and Bulte 2018). Modifications like PEGylation can minimize toxicity, as shown in studies where PEGylated PAMAM dendrimers exhibited negligible Gd leakage and strong MR contrast effects (Mekuria et al. 2017). Zhou et al. developed a dendritic contrast agent (DCA) with zwitterionic molecules using a β-cyclodextrin (β-CD) core surrounded by biodegradable polyesters. This high-generation (G4) dendrimer demonstrated extended circulation in the body and achieved a 3.5-fold increase in the R1 value when tested in non-tumor-bearing animals, compared to Magnevist®. In healthy hepatic tissues, the signal remained low, while blood vessels exhibited high signal intensity, suggesting that the zwitterionic molecules helped reduce DCA absorption by hepatocytes and Kupffer cells. In tumor-bearing animals, T1-weighted imaging displayed a marked increase in signal intensity, highlighting metastatic tumors within 30 min of administration (Zhou et al. 2017).

In efforts to develop safer alternatives to Gd-based agents, organic radical contrast agents (ORCAs) have emerged. Radical dendrimers, such as those functionalized with PROXYL radical units, enhance MRI while overcoming hydrophobicity issues through modifications like adding amino acid linkers (Zhang et al. 2020). For example, PPH-based radical dendrimers with tyrosine linkers have shown strong intratumoral contrast in glioblastoma models at much lower doses compared to Gd-based controls (Zhang et al. 2022). A more advanced dendrimer type, core-shell to dendrimers (CTSDs), has improved imaging application performance. For example, Song et al. considered a CSTD with a G5 PAMAM core and a G3 dendrimer shell functionalized with gadolinium and other ligands. These complexes exhibited enhanced MR signals and higher signal-to-noise ratios due to amplified enhanced permeability and retention (EPR) effects (Song et al. 2021). Further development of multifunctional dendrimers, such as those incorporating gold nanoparticles (AuNPs) and targeting peptides like RGD, may hold great promise for future image-guided drug delivery applications (Liu et al. 2021).

6.4.4 Carbon Nanotubes

Carbon nanotubes (CNTs) have emerged as promising tools in cancer imaging and therapy due to their unique features, such as strong intrinsic fluorescence and large surface area, which enhance their potential for theranostic applications. CNTs, especially single-walled carbon nanotubes (SWNTs), are often functionalized with tumor-targeting ligands like RGD peptides to improve specificity for cancer cells (Liu et al. 2007). For instance, SWNTs conjugated with PEG chains of different lengths and RGD peptides have demonstrated enhanced tumor accumulation and reduced liver uptake, especially when the PEG chains are longer. This improved biodistribution was confirmed through microPET imaging, which visualized SWNTs' affinity for tumors with varying $\alpha_v\beta_3$ expression, highlighting their potential for targeted imaging and therapy. The slow elimination of ^{64}Cu-labeled SWNTs indicated prolonged tumor retention, albeit raising concerns about potential long-term radioactivity. To mitigate this, non-radioactive NIRF imaging has been used, where SWNTs with PEG and RGD ligands displayed clear signals of tumor vasculature with minimal autofluorescence and deep tissue penetration (Welsher et al. 2009).

Dynamic contrast-enhanced imaging with SWNTs allows for actual one-to-one care of blood flow and tumor circulation, further improving imaging accuracy. SWNT-based systems also show potential in photoacoustic imaging (PA), as De la Zerda et al. demonstrated an eightfold increase in photoacoustic signals in $\alpha_v\beta_3$-positive tumors following SWNT administration. This suggests that SWNTs enhance imaging contrast and therapeutic outcomes when combined with photoacoustic techniques (De la Zerda et al. 2008). Similarly, multi-walled carbon nanotubes (MWNTs) have been explored for their ability to deliver multiple therapeutic agents and improve cancer treatment efficacy. For instance, dual-functionalized MWNTs equipped with targeting ligands like iRGD and candesartan (CaD) demonstrated enhanced tumor specificity and co-delivery efficiency. Upon entering the tumor microenvironment, the MWNTs release therapeutic agents through stimuli-responsive mechanisms, leading to significant tumor inhibition. Imaging studies revealed that MWNTs functionalized with both iRGD and CaD exhibited greater intratumoral fluorescence and reduced liver accumulation, demonstrating the advantages of dual-targeting strategies (Su et al. 2017).

To achieve more effective cancer eradication, external facilitators such as laser irradiation are often required in conjunction with CNT-based therapies. MWNTs functionalized with folic acid (FA) and loaded with ICG have been used for photodynamic and photothermal therapy. These systems enhance tumor targeting and enable localized treatment through light-triggered mechanisms. The tumor-homing capabilities of these MWNT complexes were validated by whole-animal fluorescence and photoacoustic imaging. During laser irradiation, these complexes markedly raised intratumoral temperatures, which resulted in total tumor ablation in animal models. This demonstrates that when combined with external stimuli like light, CNT-based systems can offer effective imaging and therapeutic solutions for cancer treatment (Hu et al. 2018).

6.4.5 Iron Oxide Nanoparticles

Superparamagnetic iron oxide nanoparticles (SPIONs) are gaining significant attention in cancer imaging and therapy due to their superior magnetic resonance (MR) contrast capabilities, primarily attributed to the superparamagnetic nature of their iron oxide core. The size and surface coating of SPIONs influence their imaging properties, with smaller SPIONs (<40 nm) typically exhibiting T1 contrast, while larger SPIONs are more suited for T2-weighted imaging (Alipour et al. 2018; Nelson et al. 2020). One notable example of conventional SPION design comes from Wang et al., who developed a multi-layered SPION system with a Fe_3O_4 core coated with polypyrrole (PPy) and further functionalized with DOX and PEG. The system formed nanoclusters about 50 nm in size, exhibiting T2 contrast properties. This SPION complex was delivered intratumorally and combined with mild NIR laser irradiation, which induced localized heating and facilitated nanoparticle internalization and controlled DOX release. This treatment approach led to an 88% inhibition of tumor growth, as demonstrated by T2-weighted MR imaging, which displayed significant darkening in the tumor area post-treatment. The treatment was further optimized by increasing the laser power, which achieved complete tumor eradication in some cases without significant side effects (Wang et al. 2013).

SPIONs are also integrated into hybrid nanosystems to enhance their therapeutic and imaging performance. For instance, Zheng et al. designed a hybrid liposomal system incorporating SPIONs in the aqueous core and paclitaxel (PTX) within the lipid bilayer, combined with a pH-responsive targeting peptide. This system, larger than 150 nm, demonstrated prolonged circulation time and tumor accumulation, as evidenced by strong fluorescent signals in the tumor microenvironment. MR imaging further confirmed SPION accumulation at the tumor site, with a steady lessening in signal strength over a period, indicative of effective targeting. Although the initial tumor growth inhibition reached 90%, further tumor progression was noted after 25 days, indicating the need for additional therapeutic strategies (Zheng et al. 2018).

Externally guided SPION delivery systems have shown remarkable efficiency in tumor targeting. Wang et al. designed a system using biotin-functionalized exosomes loaded with DOX, coated with polydopamine (PDA)-decorated SPIONs. These SPIONs were directed to tumor sites by applying an external magnetic field, resulting in enhanced accumulation and strong fluorescence signals. Combined with NIR laser irradiation, this approach led to a dramatic increase in intratumoral temperature and a 97.57% reduction in tumor size after 31 days of treatment. However, larger SPION sizes (above 50 nm) can lead to undesirable accumulation in major organs, limiting their long-term safety (Wang et al. 2021). To address issues related to organ accumulation, ultrasmall SPIONs (<5 nm) are being explored for their potential to enhance renal clearance and improve imaging properties. For instance, Ma et al. developed ultrasmall, zwitterionic SPIONs around 3 nm, demonstrating strong T1-weighted MR imaging properties. These SPIONs showed rapid tumor accumulation within 45 min post-injection and maintained strong signal intensities throughout a 3-h imaging period. Such designs hold promise not only for imaging but also for targeted drug delivery applications in cancer therapy (Ma et al. 2017).

An innovative approach merging SPIONs with stem cells has been explored for cell-based tumor therapy. Thin et al. nurtured luciferase-transfected adipocyte-derived stem cells (ADSCs) with SPIONs characterized by radionuclide ^{111}In (ADSC-^{111}In-SPION). This gave the cells tri-modal imaging properties, combining MR, optical, and nuclear imaging. In vivo studies bare that intracardiac administration of these ADSC-^{111}In-SPION complexes resulted in higher tumor uptake compared to intravenous administration, showcasing the potential of SPIONs in enhancing the precision of stem cell-based therapies (Zaw Thin et al. 2020). These diverse applications of SPIONs underscore their growing role in cancer imaging and therapy, particularly when combined with innovative targeting strategies and multi-modal imaging techniques.

6.4.6 Gold Nanoparticles (AuNPs)

AuNPs have been explored in various geometries, such as nanoshells, nanorods, nanocages, and nanostars, for their imaging and therapeutic potential (Bansal et al. 2020). These nanosystems are naturally capable of enhancing CT, PA, and thermal imaging and are also effective in tumor cell internalization. Once inside the tumor cells, they facilitate drug release and induce apoptosis, often utilizing photothermal effects (Priester and ten Hagen 2023). Gold nanoparticles are recognized for their strong, tunable optical absorption due to surface plasmon resonance, making them ideal photoacoustic contrast agents. For instance, Liu et al. utilized photoacoustic and fluorescent imaging to observe the tumor-targeting ability of gold nanoprism systems, which were PEGylated and functionalized with PD-L1 targeting ligands and Ce6 photosensitizers. Their study showed increased photoacoustic signal intensity in the tumor region within 1 h post-injection, which continued to rise until 24 h. At this point, NIR irradiation triggered a temperature rise to 43.5 °C, compared to 36.3 °C in the saline control. Even though the temperature differences between targeted and non-targeted nanoparticles were minimal, the targeted system exhibited more significant tumor growth inhibition over time (Liu et al. 2020). Similarly, Xu et al. explored porous gold nanoshells (PGNSs) functionalized with PEG and trastuzumab (HER) to achieve multimodal imaging (CT, PA, and PT) and drug delivery of DM1. Enhanced tumor contrast and sustained PA brightness were observed for up to 24 h post-administration. NIR laser exposure triggered an intratumoral temperature of 45 °C in the mPEG/HER-PGNS group compared to

38 °C in the saline control, resulting in 83.9% tumor growth inhibition over 20 days (Xu et al. 2021).

AuNPs have been integrated into hybrid nanosystems for enhanced cancer therapy. Li et al. developed a self-assembling nanostructure combining ultrasmall AuNPs with trivalent gadolinium ions (Gd^{3+}), forming a PEGylated system modified for matrix metalloproteinase-2 (MMP-2) targeting. This system encapsulated the IR820 dye for dual-modal imaging. T1-weighted MRI and fluorescence imaging revealed that tumor accumulation peaked 12 h post-injection, after which NIR laser irradiation was applied. This leads to efficient tumor cell ablation and substantially reduced tumor volume (Li et al. 2021). Similarly, Yang et al. designed MMP-responsive AuNPs to promote extracellular assembly within the tumor microenvironment. These PEG-coated nanoparticles, loaded with DOX, aggregated in the tumor microenvironment through interactions of DNA strands exposed following MMP cleavage. Imaging confirmed peak tumor accumulation at 24 h, and subsequent NIR laser irradiation led to rapid localized heating, which resulted in uniform tumor temperature elevation. This approach demonstrated a synergistic antitumor effect, significantly inhibiting tumor growth (Yang et al. 2019a). Although both nanosystems were engineered for MMP-targeted tumor accumulation, they demonstrated variations in spatiotemporal behavior and therapeutic efficacy. Tumor accumulation peaked at different time points, leading to differences in survival outcomes. These results underscore that even similarly designed nanosystems can yield diverse therapeutic effects due to their structure and behavioral differences within the in vivo environment.

Furthermore, Yang et al. developed a multifunctional nanoplatform of Bi_2Se_3/Au@PLGA-PEG-DOX (BS/Au@PP-DOX) designed for synergistic photothermal, photocatalytic, and chemotherapy, all simultaneously triggered by NIR light. This platform also exhibits dual-modal infrared thermal and CT imaging capabilities, enhancing its potential in anti-tumor applications. The BS/Au heterostructures significantly improved photothermal conversion efficiency (37.85%) compared to single BS (30.29%). Additionally, the presence of in situ-grown AuNPs on the BS improved the photocatalytic properties by promoting photo-induced electron-hole separation, which enhances redox activity and the generation of ROS. Upon exposure to an 808 nm laser, the BS/Au@PP-DOX system induced significant apoptosis or necrosis in cancer cells, leading to substantial tumor suppression. Concurrently, the platform demonstrated remarkable IRT/CT imaging capabilities, further facilitating cancer diagnosis (Yang et al. 2021).

6.4.7 Mesoporous Silica Nanoparticles

Mesoporous silica nanoparticles (MSNs) are available in various designs, such as well-ordered, hollow, or MSNs. However, hollow mesoporous organosilica nanoparticles (HMONs) suffer from significant drug leakage, particularly with hydrophobic drugs, due to their small, superficial pores and weak adsorption capacity. To mitigate this, functionalization strategies must be implemented to enhance drug retention and prevent premature release (Vaghasiya et al. 2020). Yang et al. developed small-sized hollow mesoporous organosilica nanoparticles (HMPDINs@TP-SN38) functionalized with PEG2000-PDI-silane. Perylene diimide (PDI), a thermosensitive polymer, was utilized for fluorescent and photoacoustic imaging, as well as photothermal therapy. Co-hydrolysis with PEG2000-PDI-silane led to polymer growth inside the hollow core, enhancing the drug loading capacity and preventing the undesired release of the anticancer drug SN38. Upon injection, photoacoustic and fluorescent imaging confirmed that the nanoparticles accumulated in tumors through peak intensity observed 24 h post-injection. Subsequent NIR laser irradiation induced temperatures as high as 55 °C, leading to tumor ablation and prolonged survival in treated animals (Yang et al. 2019b). Another approach to enhancing drug retention in MSNs was explored by Wu et al., who designed HMONs functionalized with ultrasmall $Cu_{2-x}Se$ particles for the treatment of glioblastoma multiforme (GBM). These particles served as photoacoustic contrast agents, reducing pore size to prevent drug leakage. The system was used to deliver DOX, and disulfide linkers facilitated controlled degradation in response to the redox environment within tumors. Focused ultrasound (FUS) further enhanced the nanoparticle penetration across the blood-brain barrier (BBB), significantly increasing tumor-specific accumulation and therapeutic efficacy, as evidenced by a 91.1% inhibition of tumor growth and extended survival over 60 days (Wu et al. 2018).

Given the critical role of the tumor microenvironment in tumor progression and drug resistance, there is growing interest in developing theranostic nanosystems responsive to tumor microenvironment factors, such as ROS. Rao et al. introduced ROS-responsive hollow MSNs (R-MSNs) encapsulating DOX within mesopores sealed by Gd-DOTA complexes. These complexes remain stable during circulation but are cleaved by singlet oxygen ($^1O_2$) within the tumor microenvironment, triggering drug release. The surface of the R-MSNs was functionalized with PEG-conjugated Ce6 to enhance ROS production and accumulation in tumors. Upon NIR laser irradiation, the system showed significant tumor inhibition and minimal systemic toxicity, demonstrating the potential for precise, ROS-triggered chemotherapy delivery with sustained intratumoral retention and therapeutic activity (Vijayakameswara Rao et al. 2018).

Feng et al. developed rattle-type MSNs featuring a Fe_3O_4-Au core for magnetic resonance (MR) and micro-computed tomography (μCT) imaging, surrounded by a SiO_2 outer shell. This nanosystem was loaded with the photosensitizer methylene blue and surface-modified with a peptide (PDPPA-1) that could be cleaved by matrix metalloproteinase-2 (MMP-2), enabling immune checkpoint blockade (ICB) by inhibiting PD-L1 expression. After intravenous administration, μCT and MR imaging showed peak tumor accumulation at 8 h, with the nanoparticles remaining localized for up to 48 h. The synergistic effects of photodynamic therapy and ICB were investigated using a bilateral tumor model, where intermittent NIR irradiation significantly reduced primary tumor volume by 98.49%. Interestingly, the distal tumors also exhibited inhibited growth, suggesting that the combination of photodynamic therapy and ICB triggered apoptosis in targeted cells and stimulated an enhanced anti-tumor immune response in untreated regions (Feng et al. 2021).

6.5 Nanoparticle-Enhanced Advanced Diagnostic Techniques

6.5.1 Magnetic Resonance Imaging (MRI)

The application of nanoparticles in MRI is revolutionary in medical diagnostics, enhancing imaging capabilities via amended contrast and targeted imaging. Nanoparticles developed from iron oxide, gadolinium, and manganese offer inimitable magnetic properties that amend relaxation time while taming intensity and contrast. The compatibility and degradability of nanoparticles in the body make them ideal for MRI by allowing them to be tailored for imaging, which reduces background noise and increases diagnostic accuracy. Fabrication of theragnostic nanoparticles further extends the capability by combining images with therapeutic functions, enabling real-time monitoring and personalized treatment. Magnetic resonance spectroscopy and magnetic resonance angiography approaches are advantageous because these innovative nanoparticles offer detailed insight into tissue biochemistry and blood vessel visualization. Magnetic resonance angiography is especially useful for evaluating tumor angiogenesis and vascular properties, while it faces challenges in differentiating tumor grades and benign versus malignant tumors. Continuous research and technological progressions are anticipated to address these limitations, refining the precision and efficacy of MRI-based diagnostics and treatments and, ultimately, showing patient outcomes (Al-Thani et al. 2024). The agents often utilize magnetic materials such as paramagnetic, ferromagnetic, and superparamagnetic iron oxide nanoparticles (SPIONs). For example, ultrasmall superparamagnetic iron oxide nanoparticles (USPIONs), like ferumoxytol, can shorten relaxation times and target liver metastases effectively. Gold nanoparticles (AuNPs) are preferred for their low toxicity and ability to generate heat when exposed to near-infrared radiation, making them useful for photothermal therapy.

In comparison, while commonly used, gadolinium-based agents have a short circulation time due to their low molecular weight. Macromolecular agents like Gd-DTPA-polylysine provide extended imaging due to slower blood clearance. Recent innovations include dual-mode contrast agents combining materials like bismuth and iron oxide to enhance MRI and CT imaging. Although bismuth slightly reduces MRI contrast, these composite nanoparticles perform well in both imaging techniques. Additionally, MRI is increasingly integrated with therapeutic approaches such as chemotherapy and thermal treatments to create hybrid diagnostic and treatment solutions. Advances like P2-bMRI offer cost-effective and customizable tumor diagnosis and prognosis. Multimodal imaging methods, including various MRI techniques, are combined with therapeutic strategies and responsive nanomaterials to improve diagnostic precision and treatment outcomes (Gupta et al. 2024).

6.5.2 Computed Tomography (CT)

CT scans, which use X-rays to create detailed internal images, are crucial for evaluating the size, location, and spread of cancer. Although CT provides high resolution and 3D imaging, it faces challenges distinguishing between tissues with similar physical properties. To address this, contrast agents such as tungsten and iodine compounds and polymer-based nanoparticles with iodine and PEG have been developed to improve imaging quality and cancer detection. Combining CT with imaging modalities, including MRI, PET, and molecular imaging, enhances diagnostic precision. For example, using 2-deoxy-D-glucose-labeled gold nanoparticles and F18 fluorodeoxyglucose as contrast agents has improved cancer detection, with PET/CT combinations offering greater accuracy compared to CT alone. Contrast-enhanced CT also plays a significant role in diagnosing renal cell carcinomas by differentiating between malignant and benign neoplasms, thus aiding in more accurate diagnosis and prognosis (Gupta et al. 2024). Nanoparticles, such as those containing gold or iodine, enhance CT imaging by increasing contrast and sensitivity, allowing for better visualization of tumors and vascular abnormalities. This advancement aids in personalized cancer treatment by improving specificity and enabling precise monitoring of therapy effectiveness. CT scans are critical for diagnosing conditions like gastrointestinal stromal tumors and for guiding targeted treatments such as radiofrequency ablation. However, challenges remain, including difficulties distinguishing between benign and malignant tumors, variations in contrast enhancement, and concerns about radiation exposure. Addressing these issues is crucial for optimizing CT imaging's effectiveness and ensuring better patient outcomes (Al-Thani et al. 2024).

6.5.3 Positron Emission Tomography (PET)

PET scans, which use radiolabeled tracers to target cancer-specific molecules, are highly effective for early cancer detection and evaluating metabolic activity. PET's advantages include excellent sensitivity and deep tissue penetration. Combining PET with MRI or CT, known as dual imaging, further refines molecular and functional information. For example, FDG PET/CT is valuable for assessing liver cancer and monitoring treatment, though diffusion-weighted MRI may offer better diagnostic performance in some cases. FDG PET/CT also helps predict patient outcomes in head and neck cancers by analyzing tumor-infiltrating lymphocytes. In prostate cancer, PSMA PET/CT has demonstrated high sensitivity for detecting pelvic lymphadenopathy and is more cost-effective than traditional imaging methods. Advances like Ga PET and cerium oxide nanoparticles with Zr-89 are showing promise for detecting cancers in the breast and urinary tract, improving diagnostic accuracy and treatment effectiveness (Gupta et al. 2024).

6.5.4 Optical Imaging

Optical imaging (OI) utilizes light reflectance and absorbance to produce images, covering fluorescence and luminescence techniques. Fluorescence imaging, commonly used in preclinical studies, offers high sensitivity, repeated dosing capability, and avoids radioactivity. However, it faces limitations with visible light due to poor tissue penetration (Yuan et al. 2009; Schouw et al. 2021). Near-infrared (NIR) light (700–3000 nm) improves visualization by providing deeper penetration and reduced background fluorescence, though it is still restricted to a few millimeters of tissue depth. Optimal NIR wavelengths (650–950 nm) are crucial, as light below 650 nm penetrates poorly, and light above 950 nm suffers from increased water absorption. Fluorescence-guided drug delivery at the (sub)cellular level employs various fluorescent agents, including dyes, proteins, and autofluorescent substances. These agents, like carboxyfluorescein and 5-aminolevulinic acid, offer high sensitivity for tracking drug release and cellular activities. Fluorescent dyes emit light across a broad spectrum and are useful for real-time imaging, but their effectiveness is limited by tissue penetration. NIR dyes, however, provide better performance with deeper tissue penetration and reduced background interference. Cyanine-based dyes, such as ICG and IR780, are commonly used in diagnostics, though ICG has issues like a short half-life and poor stability. IR780, on the other hand, offers better fluorescence and targeting. Additionally, fluorescent proteins and bioluminescence techniques are used in research for deep tissue imaging and tumor visualization, although light penetration remains a constraint (Priester and ten Hagen 2023; Walter et al. 2020; Pierce et al. 2008). Photoacoustic imaging (PAI) improves contrast using natural chromophores like hemoglobin and melanin or synthetic alternatives such as polydopamine. While natural chromophores are advantageous for their biocompatibility, synthetic agents like organic dyes (e.g., indocyanine green and methylene blue) enhance imaging performance. Incorporating these agents into nanosystems can improve contrast and specificity, leading to more precise imaging and diagnostics. For effective image-guided drug delivery and surgery, it is important to address background autofluorescence by using NIR wavelengths or adjusting tissue conditions to improve imaging accuracy and effectiveness (Priester and ten Hagen 2023; Liu et al. 2023; Etrych et al. 2016).

6.5.5 Single Photon Emission Computed Tomography (SPECT)

SPECT is a non-invasive imaging technique where a radioactive tracer is vaccinated into a patient's bloodstream. The tracer collects in targeted tissues and emits gamma rays captured by a specific camera. By rotating the camera around the patient, multiple 2D images are obtained from different angles and then processed to create a 3D image of the tissue. SPECT is widely used in oncology to identify cancerous cells, assess tumor size and location, and evaluate treatment efficacy, particularly for cancers like bone and thyroid. It also helps in guiding biopsies and surgical procedures. The procedure is generally well-tolerated, with minimal side effects from the tracers. However, SPECT has some limitations, including lower resolution and less detailed anatomical information than imaging methods like MRI or CT, which may reduce its accuracy in sensing small tumors or lesions. Moreover, the use of radioactive tracers involves a risk of radiation acquaintance, necessitating careful safety protocols for both patients and medical staff. Despite these drawbacks, SPECT remains a valuable tool in cancer diagnosis and treatment planning, offering critical insights into the concentration and distribution of cancerous cells (Gupta et al. 2024; Yandrapalli and Puckett 2024).

6.5.6 Photoacoustic Tomography (PAT)

Photoacoustic tomography (PAT) is an advanced imaging method that combines optical and ultrasound methods to imagine tissue structures and bodily functions. PAT directs a laser beam at biological tissues, causing them to emit ultrasonic waves. These waves are captured by an ultrasound transducer, which then helps create images of the tissue. PAT is especially useful in cancer detection and monitoring, as it can target molecules like hemoglobin, melanin, and lipids, which strongly absorb light at specific wavelengths. This allows PAT to identify cancerous tissues and monitor how tumors respond to treatment by tracking changes in the generated signals (Xia et al. 2014; Li et al. 2018; Li and Wang 2021). Unlike CT and MRI, PAT also provides functional information about tissue oxygenation, molecular distribution, and metabolic activity. Nanoparticles enhance PAT's capabilities, particularly gold nanorods and silver, palladium, and platinum nanoparticles. Gold nanorods, for example, are effective due to their strong photoacoustic signal when exposed to near-infrared light. Silver nanoparticles, which gradually dissolve in biological systems, offer better contrast and sensitivity for cancer detection than gold nanoparticles. Palladium and platinum nanoparticles further improve PAT by being functionalized with targeting molecules, increasing the specificity of cancer detection. These nanoparticles also have high photothermal conversion efficiency and stability, making them valuable in non-invasive cancer treatments like photothermal therapy and imaging (Gupta et al. 2024; Mandal 2024; Maccora et al. 2019).

6.5.7 Raman Spectroscopy

Raman Spectroscopy is a vibrational spectroscopic technique that leverages nanoparticles, such as silver and gold, to enhance diagnostic capabilities by amplifying Raman signals. This amplification allows for the detection of molecular fingerprints associated with diseases like cancer, offering detailed insights into the biochemical makeup of tissues. By functionalizing nanoparticles for specific targets, Raman Spectroscopy can more accurately detect particular chemicals or cellular structures, making it a valuable tool in diagnosing cancer and infectious diseases by identifying unique Raman signatures (Al-Thani et al. 2024; Miao et al. 2024).

However, Raman Spectroscopy faces challenges such as inconsistencies in signal intensity due to nanoparticle manufacturing variations and the equipment's high cost and complexity. Future efforts should focus on standardizing nanoparticle production and creating more affordable, user-friendly Raman Spectroscopy systems to increase its clinical application. Incorporating nanoparticles into various imaging methods, including MRI, CT, and photoacoustic imaging, has significantly improved diagnostic precision, contrast, and sensitivity. This integration of nanoparticles with imaging techniques offers great probability for forward-modified medicine and accurate diagnostics (Al-Thani et al. 2024).

6.6 Dual-Functional Nanoparticles for Theranostic

Dual-functional nanoparticles represent a significant advancement in cancer theranostics by combining therapeutic and diagnostic capabilities in a single platform (Jani et al. 2020). These nanoparticles are engineered to deliver targeted therapy while simultaneously providing real-time imaging of tumor cells. These nanoparticles enhance treatment precision and monitoring by integrating therapeutic agents, such as chemotherapeutic drugs or gene-silencing materials, and imaging agents, such as fluorescent dyes or MRI contrast agents (Kandasamy and Maity 2021). For instance, nanoparticles loaded with chemotherapy drugs can selectively release their payload within tumor cells. At the same time, embedded imaging agents allow for real-time tracking of treatment efficacy and tumor progression. This dual functionality improves targeting accuracy and reduces off-target effects, leading to more effective and personalized cancer treatment (Priester and ten Hagen 2023; Shrestha et al. 2021). Moreover, using dual-functional nanoparticles in cancer therapy offers significant advantages in monitoring and evaluating treatment responses. The imaging component enables continuous observation of tumor behavior and drug distribution, providing critical insights into how well the therapy works. This capability allows for timely adjustments to treatment strategies based on real-time data, potentially improving patient outcomes. Additionally, the ability to imagine the circulation and localization of therapeutic agents helps in optimizing dosages and minimizing adverse effects (Al-Thani et al. 2024; Chehelgerdi et al. 2023; Li et al. 2022). Overall, dual-functional nanoparticles represent a capable method of enhancing cancer therapies' effectiveness and advancing personalized medicine.

6.7 Conclusion

In conclusion, exploring nanotechnology in cancer imaging and therapeutics reveals a transformative potential for the field of oncology. Nanoparticles have emerged as pivotal tools, offering significant advancements in drug delivery, imaging precision, and therapeutic strategies. Their ability to precisely target cancer cells and minimize collateral damage underscores a major shift toward more effective and personalized cancer treatments. Innovations such as targeted drug delivery, advanced imaging techniques, and novel approaches like photothermal and photodynamic therapies highlight the profound impact of nanomedicine on improving patient outcomes. Nevertheless, several challenges persist, including biocompatibility, long-term toxicity, and the seamless integration of nanomedicines with existing treatment modalities. Addressing these challenges requires ongoing research and development to enhance the safety and efficacy of nanomaterials. The future of cancer treatment is poised to benefit greatly from advancements in predictive oncology, where nanotechnology plays a crucial role in early detection and precision therapy. Integrating diagnostics and therapeutics through nanotheranostics represents a promising avenue for achieving more effective, less invasive, and highly individualized cancer care. As research continues to evolve, nanomedicine is expected to drive significant progress in the fight against cancer, paving the way for innovative and personalized therapeutic strategies.

Author Contribution K.J., E.K., J.N., R.K.V., A.G., and S.R. have prepared the outline and the first draft of the chapter. S.M. has done primary editing. S.R. has finally reviewed and edited the chapter.

References

Alipour A, Soran-Erdem Z, Utkur M, Sharma VK, Algin O, Saritas EU, Demir HV (2018) A new class of cubic SPIONs as a dual-mode T1 and T2 contrast agent for MRI. Magn Reson Imaging 49:16–24. https://doi.org/10.1016/j.mri.2017.09.013

Alrushaid N, Khan FA, Al-Suhaimi EA, Elaissari A (2023) Nanotechnology in cancer diagnosis and treatment. Pharmaceutics 15:1025. https://doi.org/10.3390/pharmaceutics15031025

Al-Thani AN, Jan AG, Abbas M, Geetha M, Sadasivuni KK (2024) Nanoparticles in cancer theragnostic and drug delivery: a comprehensive review. Life Sci 352:122899. https://doi.org/10.1016/j.lfs.2024.122899

Anani T, Rahmati S, Sultana N, David AE (2021) MRI-traceable theranostic nanoparticles for targeted cancer treatment. Theranostics 11:579–601. https://doi.org/10.7150/thno.48811

Anselmo AC, Mitragotri S (2016) Nanoparticles in the clinic. Bioeng Transl Med 1:10–29. https://doi.org/10.1002/btm2.10003

Aranda-Lara L, Morales-Avila E, Luna-Gutiérrez MA, Olivé-Alvarez E, Isaac-Olivé K (2020) Radiolabeled liposomes and lipoproteins as lipidic nanoparticles for imaging and therapy. Chem Phys Lipids 230:104934. https://doi.org/10.1016/j.chemphyslip.2020.104934

Aslan N, Ceylan B, Koç MM, Findik F (2020) Metallic nanoparticles as X-ray computed tomography (CT) contrast agents: a review. J Mol Struct 1219:128599. https://doi.org/10.1016/j.molstruc.2020.128599

Ayyagari AL, Zhang X, Ghaghada KB, Annapragada A, Hu X, Bellamkonda RV (2006) Long-circulating liposomal contrast agents for magnetic resonance imaging. Magn Reson Med 55:1023–1029. https://doi.org/10.1002/mrm.20846

Bansal SA, Kumar V, Karimi J, Singh AP, Kumar S (2020) Role of gold nanoparticles in advanced biomedical applications. Nanoscale Adv 2:3764–3787. https://doi.org/10.1039/D0NA00472C

Bennett KM, Jo J, Cabral H, Bakalova R, Aoki I (2014) MR imaging techniques for nano-pathophysiology and theranostics. Adv Drug Deliv Rev 74:75–94. https://doi.org/10.1016/j.addr.2014.04.007

Bertrand N, Wu J, Xu X, Kamaly N, Farokhzad OC (2014) Cancer nanotechnology: the impact of passive and active targeting in the era of modern cancer biology. Adv Drug Deliv Rev 66:2–25. https://doi.org/10.1016/j.addr.2013.11.009

Bober Z, Bartusik-Aebisher D, Aebisher D (2022) Application of dendrimers in anticancer diagnostics and therapy. Molecules 27:3237. https://doi.org/10.3390/molecules27103237

Cao J, Ge R, Zhang M, Xia J, Han S, Lu W, Liang Y, Zhang T, Sun Y (2018) A triple modality BSA-coated dendritic nanoplatform for NIR imaging, enhanced tumor penetration and anticancer therapy. Nanoscale 10:9021–9037. https://doi.org/10.1039/C7NR09552J

Chakravarty R, Hong H, Cai W (2014) Positron emission tomography image-guided drug delivery: current status and future perspectives. Mol Pharm 11:3777–3797. https://doi.org/10.1021/mp500173s

Chehelgerdi M, Chehelgerdi M, Allela OQB, Pecho RDC, Jayasankar N, Rao DP, Thamaraikani T, Vasanthan M, Viktor P, Lakshmaiya N, Saadh MJ, Amajd A, Abo-Zaid MA, Castillo-Acobo RY, Ismail AH, Amin AH, Akhavan-Sigari R (2023) Progressing nanotechnology to improve targeted cancer treatment: overcoming hurdles in its clinical implementation. Mol Cancer 22:169. https://doi.org/10.1186/s12943-023-01865-0

Cheng L, Wang C, Feng L, Yang K, Liu Z (2014) Functional nanomaterials for phototherapies of cancer. Chem Rev 114:10869–10939. https://doi.org/10.1021/cr400532z

Cheng X, Xie Q, Sun Y (2023) Advances in nanomaterial-based targeted drug delivery systems. Front Bioeng Biotechnol 11. https://doi.org/10.3389/fbioe.2023.1177151

Cherukula K, Uthaman S, Park I-K (2019) "Navigate-dock-activate" anti-tumor strategy: tumor micromilieu charge-switchable, hierarchically activated nanoplatform with ultrarapid tumor-tropic accumulation for trackable photothermal/chemotherapy. Theranostics 9:2505–2525. https://doi.org/10.7150/thno.33280

Crintea A, Motofelea AC, Şovrea AS, Constantin A-M, Crivii C-B, Carpa R, Duţu AG (2023) Dendrimers: advancements and potential applications in cancer diagnosis and treatment—an overview. Pharmaceutics 15:1406. https://doi.org/10.3390/pharmaceutics15051406

De la Zerda A, Zavaleta C, Keren S, Vaithilingam S, Bodapati S, Liu Z, Levi J, Smith BR, Ma T-J, Oralkan O, Cheng Z, Chen X, Dai H, Khuri-Yakub BT, Gambhir SS (2008) Carbon nanotubes as photoacoustic molecular imaging agents in living mice. Nat Nanotechnol 3:557–562. https://doi.org/10.1038/nnano.2008.231

Duan S, Hu Y, Zhao Y, Tang K, Zhang Z, Liu Z, Wang Y, Guo H, Miao Y, Du H, Yang D, Li S, Zhang J (2023) Nanomaterials for photothermal cancer therapy. RSC Adv 13:14443–14460. https://doi.org/10.1039/D3RA02620E

Ernsting MJ, Murakami M, Roy A, Li S-D (2013) Factors controlling the pharmacokinetics, biodistribution and intratumoral penetration of nanoparticles. J Control Release 172:782–794. https://doi.org/10.1016/j.jconrel.2013.09.013

Etrych T, Lucas H, Janoušková O, Chytil P, Mueller T, Mäder K (2016) Fluorescence optical imaging in anticancer drug delivery. J Control Release 226:168–181. https://doi.org/10.1016/j.jconrel.2016.02.022

Feng Y, Xie X, Zhang H, Su Q, Yang G, Wei X, Li N, Li T, Qin X, Li S, Wu C, Zheng C, Zhu J, You F, Wang G, Yang H, Liu Y (2021) Multistage-responsive nanovehicle to improve tumor penetration for dual-modality imaging-guided photodynamic-immunotherapy. Biomaterials 275:120990. https://doi.org/10.1016/j.biomaterials.2021.120990

Fernandez-Fernandez A, Manchanda R, McGoron AJ (2011) Theranostic applications of nanomaterials in cancer: drug delivery, image-guided therapy, and multifunctional platforms. Appl Biochem Biotechnol 165:1628–1651. https://doi.org/10.1007/s12010-011-9383-z

Gholap AD, Hatvate NT, Khuspe PR, Mandhare TA, Kashid P, Gaikwad VD (2023a) Proteomics in oncology: retrospect and prospects. In: Novel technologies in biosystems, biomedical & drug delivery. Springer Nature, Singapore, pp 243–269. https://doi.org/10.1007/978-981-99-5281-6_10

Gholap AD, Sawant SD, Sayyad SF, Hatvate NT, Chavan M, Rojekar S, Faiyazuddin M (2023b) Food biopackaging for human benefits: status and perspectives. In: Biobased packaging materials. Springer Nature, Singapore, pp 1–35. https://doi.org/10.1007/978-981-99-6050-7_1

Gholap AD, Gupta JS, Kamandar PA, Banchhod GV, Hatvate NT (2023c) Antibody-drug conjugates for cancer therapy: an up-to-date review on the chemistry and pharmacology. In: Bioconjugated materials part 2—applications in drug delivery, vaccine formulations and important conjugates for cancer therapy. Elsevier, pp 105–190. https://doi.org/10.1016/bs.coac.2023.02.007

Gholap AD, Said RP, Pawar RD, Ambore GS, Hatvate NT (2023d) Importance of carbohydrate-drug conjugates in vaccine development: a detailed review. In: Bioconjugated materials part 2—applications in drug delivery, vaccine formulations and Important conjugates for cancer therapy. Elsevier, pp 191–256. https://doi.org/10.1016/bs.coac.2023.03.002

Gholap AD, Bhowmik DD, Deshmukh AY, Hatvate NT (2023e) Quintessential impact of dendrimer bioconjugates in targeted drug delivery. In: Bioconjugated materials part 2—applications in drug delivery, vaccine formulations and Important conjugates for cancer therapy. Elsevier, pp 257–302. https://doi.org/10.1016/bs.coac.2023.02.009

Gholap AD, Pardeshi SR, Hatvate NT, Dhorkule N, Sayyad SF, Faiyazuddin M, Khalid M (2024a) Environmental implications and nanotechnological advances in octocrylene-enriched sunscreen formulations: a comprehensive review. Chemosphere 358:142235. https://doi.org/10.1016/j.chemosphere.2024.142235

Gholap AD, Uddin MJ, Faiyazuddin M, Omri A, Gowri S, Khalid M (2024b) Advances in artificial intelligence for drug delivery and development: a comprehensive review. Comput Biol Med 178:108702. https://doi.org/10.1016/j.compbiomed.2024.108702

Gholap AD, Hatvate NT, Khuspe PR, Mandhare TA, Kashid P, Vora LK (2024c) Decoding of nanoneutraceuticals and their delivery. In: Sustainable green nanotechnology. CRC Press, New York, pp 284–309. https://doi.org/10.1201/9781003389408-16

Gholap AD, Gupta J, Kamandar P, Bhowmik DD, Rojekar S, Faiyazuddin M, Hatvate NT, Mohanto S, Ahmed MG, Subramaniyan V, Kumarasamy V (2024d) Harnessing nanovaccines for effective immunization—a special concern on COVID-19: facts, fidelity, and future prospective. ACS Biomater Sci Eng 10:271–297. https://doi.org/10.1021/acsbiomaterials.3c01247

Gupta D, Roy P, Sharma R, Kasana R, Rathore P, Gupta TK (2024) Recent nanotheranostic approaches in cancer research. Clin Exp Med 24:8. https://doi.org/10.1007/s10238-023-01262-3

Hu H (2020) Recent advances of bioresponsive nano-sized contrast agents for ultra-high-field magnetic resonance imaging. Front Chem 8. https://doi.org/10.3389/fchem.2020.00203

Hu Y, Wang R, Zhou Y, Yu N, Chen Z, Gao D, Shi X, Shen M (2018) Targeted dual-mode imaging and phototherapy of tumors using ICG-loaded multifunctional MWCNTs as a versatile platform. J Mater Chem B 6:6122–6132. https://doi.org/10.1039/C8TB01870G

Jadhav K, Jhilta A, Singh R, Ray E, Sharma N, Shukla R, Singh AK, Verma RK (2023) Clofazimine nanoclusters show high efficacy in experimental TB with amelioration in paradoxical lung inflammation. Biomater Adv 154:213594. https://doi.org/10.1016/j.bioadv.2023.213594

Jadhav K, Jhilta A, Singh R, Ray E, Kumar V, Yadav AB, Singh AK, Verma RK (2024) Effective cerebral tuberculosis treatment via nose-to-brain transport of anti-TB drugs using mucoadhesive nano-aggregates. Nanoscale. https://doi.org/10.1039/D4NR02621G

Jaishree V, Gupta PD (2012) Nanotechnology: a revolution in cancer diagnosis. Indian J Clin Biochem 27:214–220. https://doi.org/10.1007/s12291-012-0221-z

Jakhar DK, Vishwakarma VK, Singh R, Jadhav K, Shah S, Arora T, Verma RK, Yadav HN (2023) Fat fighting liraglutide based nano-formulation to reverse obesity: design, development and animal trials. Int J Pharm 634:122585. https://doi.org/10.1016/j.ijpharm.2023.122585

Jani P, Subramanian S, Korde A, Rathod L, Sawant KK (2020) Theranostic nanocarriers in cancer: dual capabilities on a single platform. In: Functional bionanomaterials. Springer, pp 293–312. https://doi.org/10.1007/978-3-030-41464-1_13

Jiao J, Zhang J, Yang F, Song W, Han D, Wen W, Qin W (2020) Quicker, deeper and stronger imaging: a review of tumor-targeted, near-infrared fluorescent dyes for fluorescence guided surgery in the preclinical and clinical stages. Eur J Pharm Biopharm 152:123–143. https://doi.org/10.1016/j.ejpb.2020.05.002

Joshi BP, Wang TD (2018) Targeted optical imaging agents in cancer: focus on clinical applications. Contrast Media Mol Imaging 2018:1–19. https://doi.org/10.1155/2018/2015237

Jung HS, Verwilst P, Sharma A, Shin J, Sessler JL, Kim JS (2018) Organic molecule-based photothermal agents: an expanding photothermal therapy universe. Chem Soc Rev 47:2280–2297. https://doi.org/10.1039/C7CS00522A

Kadkhoda J, Tarighatnia A, Barar J, Aghanejad A, Davaran S (2022) Recent advances and trends in nanoparticles based photothermal and photodynamic therapy. Photodiagn Photodyn Ther 37: 102697. https://doi.org/10.1016/j.pdpdt.2021.102697

Kandasamy G, Maity D (2021) Multifunctional theranostic nanoparticles for biomedical cancer treatments—a comprehensive review. Mater Sci Eng C 127:112199. https://doi.org/10.1016/j.msec.2021.112199

Karahmet Sher E, Alebić M, Marković Boras M, Boškailo E, Karahmet Farhat E, Karahmet A, Pavlović B, Sher F, Lekić L (2024) Nanotechnology in medicine revolutionizing drug delivery for cancer and viral infection treatments. Int J Pharm 660:124345. https://doi.org/10.1016/j.ijpharm.2024.124345

Kim J, Lee J, Choi S, Lee H, Yang J, Jeon H, Sung M, Kim WJ, Kim C (2024) 3D multiparametric photoacoustic computed tomography of primary and metastatic tumors in living mice. ACS Nano 18:18176–18190. https://doi.org/10.1021/acsnano.3c12551

Kirui D, Ferrari M (2015) Intravital microscopy imaging approaches for image-guided drug delivery systems. Curr Drug Targets 16:528–541. https://doi.org/10.2174/1389450116666150330114030

Kole E, Jadhav K, Shirsath N, Dudhe P, Verma RK, Chatterjee A, Naik J (2023) Nanotherapeutics for pulmonary drug delivery: an emerging approach to overcome respiratory diseases. J Drug Deliv Sci Technol 81:104261. https://doi.org/10.1016/j.jddst.2023.104261

Kole E, Jadhav K, Singh R, Mandpe S, Abhang A, Verma RK, Naik J (2024) Recent developments in tyrosine kinase inhibitor-based nanotherapeutics for EGFR-resistant non-small cell lung cancer. Curr Drug Deliv 21. https://doi.org/10.2174/0115672018278617231207051907

Lamichhane N, Dewkar G, Sundaresan G, Mahon R, Zweit J (2017) [18F]-Fluorinated carboplatin and [111In]-liposome for image-guided drug delivery. Int J Mol Sci 18:1079. https://doi.org/10.3390/ijms18051079

Lee CC, MacKay JA, Fréchet JMJ, Szoka FC (2005) Designing dendrimers for biological applications. Nat Biotechnol 23:1517–1526. https://doi.org/10.1038/nbt1171

Li L, Wang LV (2021) Recent advances in photoacoustic tomography. BME Front. https://doi.org/10.34133/2021/9823268

Li X, Zhang X-N, Li X-D, Chang J, Li X, Zhang X-N, Li X-D, Chang J (2016) Multimodality imaging in nanomedicine and nanotheranostics. Cancer Biol Med 13:339–348. https://doi.org/10.20892/j.issn.2095-3941.2016.0055

Li C, Qian M, Wang S, Jiang H, Du Y, Wang J, Lu W, Murthy N, Huang R (2017) Aptavalve-gated mesoporous carbon nanospheres image cellular mucin and provide on-demand targeted drug delivery. Theranostics 7:3319–3325. https://doi.org/10.7150/thno.18692

Li M, Tang Y, Yao J (2018) Photoacoustic tomography of blood oxygenation: a mini review. Photoacoustics 10:65–73. https://doi.org/10.1016/j.pacs.2018.05.001

Li B, Sun L, Li T, Zhang Y, Niu X, Xie M, You Z (2021) Ultra-small gold nanoparticles self-assembled by gadolinium ions for enhanced photothermal/photodynamic liver cancer therapy. J Mater Chem B 9:1138–1150. https://doi.org/10.1039/D0TB02410D

Li C, Li Y, Li G, Wu S (2022) Functional nanoparticles for enhanced cancer therapy. Pharmaceutics 14:1682. https://doi.org/10.3390/pharmaceutics14081682

Liu Z, Cai W, He L, Nakayama N, Chen K, Sun X, Chen X, Dai H (2007) In vivo biodistribution and highly efficient tumour targeting of carbon nanotubes in mice. Nat Nanotechnol 2:47–52. https://doi.org/10.1038/nnano.2006.170

Liu B, Qiao G, Han Y, Shen E, Alfranca G, Tan H, Wang L, Pan S, Ma L, Xiong W, Liu Y, Cui D (2020) Targeted theranostics of lung cancer: PD-L1-guided delivery of gold nanoprisms with chlorin e6 for enhanced imaging and photothermal/photodynamic therapy. Acta Biomater 117: 361–373. https://doi.org/10.1016/j.actbio.2020.09.040

Liu R, Guo H, Ouyang Z, Fan Y, Cao X, Xia J, Shi X, Guo R (2021) Multifunctional core–shell tecto dendrimers incorporated with gold nanoparticles for targeted dual mode CT/MR imaging of tumors. ACS Appl Bio Mater 4:1803–1812. https://doi.org/10.1021/acsabm.0c01525

Liu R, Xu Y, Zhang N, Qu S, Zeng W, Li R, Dai Z (2023) Nanotechnology for enhancing medical imaging. In: Nanomedicine. Springer, pp 99–156. https://doi.org/10.1007/978-981-16-8984-0_8

Lucky SS, Soo KC, Zhang Y (2015) Nanoparticles in photodynamic therapy. Chem Rev 115:1990–2042. https://doi.org/10.1021/cr5004198

Luo S, Zhang E, Su Y, Cheng T, Shi C (2011) A review of NIR dyes in cancer targeting and imaging. Biomaterials 32:7127–7138. https://doi.org/10.1016/j.biomaterials.2011.06.024

Ma D, Chen J, Luo Y, Wang H, Shi X (2017) Zwitterion-coated ultrasmall iron oxide nanoparticles for enhanced T1-weighted magnetic resonance imaging applications. J Mater Chem B 5:7267–7273. https://doi.org/10.1039/C7TB01588G

Ma G, Liu X, Deng G, Yuan H, Wang Q, Lu J (2018) A novel theranostic agent based on porous bismuth nanosphere for CT imaging-guided combined chemo-photothermal therapy and radiotherapy. J Mater Chem B 6:6788–6795. https://doi.org/10.1039/C8TB02189A

Maccora D, Dini V, Battocchio C, Fratoddi I, Cartoni A, Rotili D, Castagnola M, Faccini R, Bruno I, Scotognella T, Giordano A, Venditti I (2019) Gold nanoparticles and nanorods in nuclear medicine: a mini review. Appl Sci 9:3232. https://doi.org/10.3390/app9163232

Man F, Gawne PJ, de Rosales RTM (2019) Nuclear imaging of liposomal drug delivery systems: a critical review of radiolabelling methods and applications in nanomedicine. Adv Drug Deliv Rev 143:134–160. https://doi.org/10.1016/j.addr.2019.05.012

Mandal AK (2024) Platinum nanoparticles as delivery system in combating various diseases. J Drug Deliv Ther 14. https://doi.org/10.22270/jddt.v14i7.6644

McMahon MT, Bulte JWM (2018) Two decades of dendrimers as versatile MRI agents: a tale with and without metals. WIREs Nanomed Nanobiotechnol 10. https://doi.org/10.1002/wnan.1496

Mekuria SL, Debele TA, Tsai H-C (2017) Encapsulation of gadolinium oxide nanoparticle (Gd2O3) contrasting agents in PAMAM dendrimer templates for enhanced magnetic resonance imaging in vivo. ACS Appl Mater Interfaces 9:6782–6795. https://doi.org/10.1021/acsami.6b14075

Meng X, Wu Y, Bu W (2021) Functional CT contrast nanoagents for the tumor microenvironment. Adv Healthc Mater 10. https://doi.org/10.1002/adhm.202000912

Mesut B, Al-Mohaya M, Gholap AD, Yeşilkaya E, Das U, Akhtar MS, Sah R, Khan S, Moin A, Faiyazuddin M (2024) Demystifying the potential of lipid-based nanocarriers in targeting brain malignancies. Naunyn Schmiedebergs Arch Pharmacol. https://doi.org/10.1007/s00210-024-03212-6

Mezzanotte L, van't Root M, Karatas H, Goun EA, Löwik CWGM (2017) In vivo molecular bioluminescence imaging: new tools and applications. Trends Biotechnol 35:640–652. https://doi.org/10.1016/j.tibtech.2017.03.012

Miao Y, Wu L, Qiang J, Qi J, Li Y, Li R, Kong X, Zhang Q (2024) The application of Raman spectroscopy for the diagnosis and monitoring of lung tumors. Front Bioeng Biotechnol 12. https://doi.org/10.3389/fbioe.2024.1385552

Mura S, Nicolas J, Couvreur P (2013) Stimuli-responsive nanocarriers for drug delivery. Nat Mater 12:991–1003. https://doi.org/10.1038/nmat3776

Navya PN, Kaphle A, Srinivas SP, Bhargava SK, Rotello VM, Daima HK (2019) Current trends and challenges in cancer management and therapy using designer nanomaterials. Nano Converg 6. https://doi.org/10.1186/s40580-019-0193-2

Nelson N, Port J, Pandey M (2020) Use of superparamagnetic iron oxide nanoparticles (SPIONs) via multiple imaging modalities and modifications to reduce cytotoxicity: an educational review. J Nanotheranostics 1:105–135. https://doi.org/10.3390/jnt1010008

Oehler JB, Rajapaksha W, Albrecht H (2024) Emerging applications of nanoparticles in the diagnosis and treatment of breast cancer. J Pers Med 14:723. https://doi.org/10.3390/jpm14070723

Padmanabhan P, Kumar A, Kumar S, Chaudhary RK, Gulyás B (2016) Nanoparticles in practice for molecular-imaging applications: an overview. Acta Biomater 41:1–16. https://doi.org/10.1016/j.actbio.2016.06.003

Pavitra E, Dariya B, Srivani G, Kang S-M, Alam A, Sudhir P-R, Kamal MA, Raju GSR, Han Y-K, Lakkakula BVKS, Nagaraju GP, Huh YS (2021) Engineered nanoparticles for imaging and drug delivery in colorectal cancer. Semin Cancer Biol 69:293–306. https://doi.org/10.1016/j.semcancer.2019.06.017

Perrigue PM, Murray RA, Mielcarek A, Henschke A, Moya SE (2021) Degradation of drug delivery nanocarriers and payload release: a review of physical methods for tracing nanocarrier biological fate. Pharmaceutics 13:770. https://doi.org/10.3390/pharmaceutics13060770

Pierce MC, Javier DJ, Richards-Kortum R (2008) Optical contrast agents and imaging systems for detection and diagnosis of cancer. Int J Cancer 123:1979–1990. https://doi.org/10.1002/ijc.23858

Polyak A, Ross TL (2018) Nanoparticles for SPECT and PET imaging: towards personalized medicine and theranostics. Curr Med Chem 25:4328–4353. https://doi.org/10.2174/0929867324666170830095553

Ponce AM, Viglianti BL, Yu D, Yarmolenko PS, Michelich CR, Woo J, Bally MB, Dewhirst MW (2007) Magnetic resonance imaging of temperature-sensitive liposome release: drug dose painting and antitumor effects. JNCI J Natl Cancer Inst 99:53–63. https://doi.org/10.1093/jnci/djk005

Potara M, Nagy-Simon T, Focsan M, Licarete E, Soritau O, Vulpoi A, Astilean S (2021) Folate-targeted Pluronic-chitosan nanocapsules loaded with IR780 for near-infrared fluorescence imaging and photothermal-photodynamic therapy of ovarian cancer. Colloids Surf B Biointerfaces 203:111755. https://doi.org/10.1016/j.colsurfb.2021.111755

Priester MI, ten Hagen TLM (2023) Image-guided drug delivery in nanosystem-based cancer therapies. Adv Drug Deliv Rev 192:114621. https://doi.org/10.1016/j.addr.2022.114621

Rana A, Adhikary M, Singh PK, Das BC, Bhatnagar S (2023) "Smart" drug delivery: a window to future of translational medicine. Front Chem 10. https://doi.org/10.3389/fchem.2022.1095598

Rojekar S, Gholap AD, Togre N, Bhoj P, Haeck C, Hatvate N, Singh N, Vitore J, Dhoble S, Kashid S, Patravale V (2024) Current status of mannose receptor-targeted drug delivery for improved anti-HIV therapy. J Control Release 372:494–521. https://doi.org/10.1016/j.jconrel.2024.06.002

Sarbadhikary P, George BP, Abrahamse H (2021) Recent advances in photosensitizers as multi-functional theranostic agents for imaging-guided photodynamic therapy of cancer. Theranostics 11:9054–9088. https://doi.org/10.7150/thno.62479

Schouw HM, Huisman LA, Janssen YF, Slart RHJA, Borra RJH, Willemsen ATM, Brouwers AH, van Dijl JM, Dierckx RA, van Dam GM, Szymanski W, Boersma HH, Kruijff S (2021) Targeted optical fluorescence imaging: a meta-narrative review and future perspectives. Eur J Nucl Med Mol Imaging 48:4272–4292. https://doi.org/10.1007/s00259-021-05504-y

Seynhaeve ALB, ten Hagen TLM (2018) Intravital microscopy of tumor-associated vasculature using advanced dorsal skinfold window chambers on transgenic fluorescent mice. J Vis Exp. https://doi.org/10.3791/55115

Seynhaeve ALB, Hoving S, Schipper D, Vermeulen CE, aan de Wiel-Ambagtsheer G, van Tiel ST, Eggermont AMM, ten Hagen TLM (2007) Tumor necrosis factor α mediates homogeneous distribution of liposomes in murine melanoma that contributes to a better tumor response. Cancer Res 67:9455–9462. https://doi.org/10.1158/0008-5472.CAN-07-1599

Seynhaeve ALB, Amin M, Haemmerich D, van Rhoon GC, ten Hagen TLM (2020) Hyperthermia and smart drug delivery systems for solid tumor therapy. Adv Drug Deliv Rev 163–164:125–144. https://doi.org/10.1016/j.addr.2020.02.004

Sharma N, Mohammad W, Le Guével X, Shanavas A (2024) Gold nanoclusters as high resolution NIR-II theranostic agents. Chem Biomed Imaging 2:462–480. https://doi.org/10.1021/cbmi.4c00021

Shi J, Kantoff PW, Wooster R, Farokhzad OC (2016) Cancer nanomedicine: progress, challenges and opportunities. Nat Rev Cancer 17:20–37. https://doi.org/10.1038/nrc.2016.108

Shrestha B, Wang L, Brey EM, Uribe GR, Tang L (2021) Smart nanoparticles for chemo-based combinational therapy. Pharmaceutics 13:853. https://doi.org/10.3390/pharmaceutics13060853

Siddique S, Chow JCL (2020) Application of nanomaterials in biomedical imaging and cancer therapy. Nanomaterials 10:1700. https://doi.org/10.3390/nano10091700

Song C, Ouyang Z, Guo H, Qu J, Gao Y, Xia J, Shen M, Shi X (2021) Core–shell tecto dendrimers enable enhanced tumor MR imaging through an amplified EPR effect. Biomacromolecules 22:2181–2188. https://doi.org/10.1021/acs.biomac.1c00262

Su Y, Hu Y, Wang Y, Xu X, Yuan Y, Li Y, Wang Z, Chen K, Zhang F, Ding X, Li M, Zhou J, Liu Y, Wang W (2017) A precision-guided MWNT mediated reawakening the sunk synergy in RAS for anti-angiogenesis lung cancer therapy. Biomaterials 139:75–90. https://doi.org/10.1016/j.biomaterials.2017.05.046

Taratula O, Schumann C, Naleway MA, Pang AJ, Chon KJ, Taratula O (2013) A multifunctional theranostic platform based on phthalocyanine-loaded dendrimer for image-guided drug delivery and photodynamic therapy. Mol Pharm 10:3946–3958. https://doi.org/10.1021/mp400397t

ten Hagen TLM, Dreher MR, Zalba S, Seynhaeve ALB, Amin M, Li L, Haemmerich D (2021) Drug transport kinetics of intravascular triggered drug delivery systems. Commun Biol 4:920. https://doi.org/10.1038/s42003-021-02428-z

Tong R, Hemmati HD, Langer R, Kohane DS (2012) Photoswitchable nanoparticles for triggered tissue penetration and drug delivery. J Am Chem Soc 134:8848–8855. https://doi.org/10.1021/ja211888a

Toy R, Bauer L, Hoimes C, Ghaghada KB, Karathanasis E (2014) Targeted nanotechnology for cancer imaging. Adv Drug Deliv Rev 76:79–97. https://doi.org/10.1016/j.addr.2014.08.002

ud Din F, Aman W, Ullah I, Qureshi OS, Mustapha O, Shafique S, Zeb A (2017) Effective use of nanocarriers as drug delivery systems for the treatment of selected tumors. Int J Nanomedicine 12:7291–7309. https://doi.org/10.2147/IJN.S146315

Vaghasiya K, Sharma A, Kumar K, Ray E, Adlakha S, Katare OP, Hota SK, Verma RK (2019) Heparin-encapsulated metered-dose topical “nano-spray gel” liposomal formulation ensures rapid on-site management of frostbite injury by inflammatory cytokines scavenging. ACS Biomater Sci Eng 5:6617–6631. https://doi.org/10.1021/acsbiomaterials.9b01486

Vaghasiya K, Ray E, Sharma A, Katare OP, Verma RK (2020) Matrix metalloproteinase-responsive mesoporous silica nanoparticles cloaked with cleavable protein for “self-actuating” on-demand controlled drug delivery for cancer therapy. ACS Appl Bio Mater 3:4987–4999. https://doi.org/10.1021/acsabm.0c00497

Vaghasiya K, Ray E, Singh R, Jadhav K, Sharma A, Khan R, Katare OP, Verma RK (2021) Efficient, enzyme responsive and tumor receptor targeting gelatin nanoparticles decorated with concanavalin-A for site-specific and controlled drug delivery for cancer therapy. Mater Sci Eng C 123:112027. https://doi.org/10.1016/j.msec.2021.112027

van der Geest T, Laverman P, Metselaar JM, Storm G, Boerman OC (2016) Radionuclide imaging of liposomal drug delivery. Expert Opin Drug Deliv 13:1231–1242. https://doi.org/10.1080/17425247.2016.1205584

Vijayakameswara Rao N, Han HS, Lee H, Nguyen VQ, Jeon S, Jung D-W, Lee J, Yi G-R, Park JH (2018) ROS-responsive mesoporous silica nanoparticles for MR imaging-guided

photodynamically maneuvered chemotherapy. Nanoscale 10:9616–9627. https://doi.org/10.1039/C8NR00888D
Vines JB, Yoon J-H, Ryu N-E, Lim D-J, Park H (2019) Gold nanoparticles for photothermal cancer therapy. Front Chem 7. https://doi.org/10.3389/fchem.2019.00167
Wahsner J, Gale EM, Rodríguez-Rodríguez A, Caravan P (2019) Chemistry of MRI contrast agents: current challenges and new frontiers. Chem Rev 119:957–1057. https://doi.org/10.1021/acs.chemrev.8b00363
Wallyn J, Anton N, Akram S, Vandamme TF (2019) Biomedical imaging: principles, technologies, clinical aspects, contrast agents, limitations and future trends in nanomedicines. Pharm Res 36:78. https://doi.org/10.1007/s11095-019-2608-5
Walter A, Paul-Gilloteaux P, Plochberger B, Sefc L, Verkade P, Mannheim JG, Slezak P, Unterhuber A, Marchetti-Deschmann M, Ogris M, Bühler K, Fixler D, Geyer SH, Weninger WJ, Glösmann M, Handschuh S, Wanek T (2020) Correlated multimodal imaging in life sciences: expanding the biomedical horizon. Front Phys 8. https://doi.org/10.3389/fphy.2020.00047
Wang L, Niu C (2021) IR780-based nanomaterials for cancer imaging and therapy. J Mater Chem B 9:4079–4097. https://doi.org/10.1039/D1TB00407G
Wang C, Xu H, Liang C, Liu Y, Li Z, Yang G, Cheng L, Li Y, Liu Z (2013) Iron oxide @ polypyrrole nanoparticles as a multifunctional drug carrier for remotely controlled cancer therapy with synergistic antitumor effect. ACS Nano 7:6782–6795. https://doi.org/10.1021/nn4017179
Wang S, Huang P, Chen X (2016) Stimuli-responsive programmed specific targeting in nanomedicine. ACS Nano 10:2991–2994. https://doi.org/10.1021/acsnano.6b00870
Wang J, Chen P, Dong Y, Xie H, Wang Y, Soto F, Ma P, Feng X, Du W, Liu B-F (2021) Designer exosomes enabling tumor targeted efficient chemo/gene/photothermal therapy. Biomaterials 276:121056. https://doi.org/10.1016/j.biomaterials.2021.121056
Wang B, Hu S, Teng Y, Chen J, Wang H, Xu Y, Wang K, Xu J, Cheng Y, Gao X (2024) Current advance of nanotechnology in diagnosis and treatment for malignant tumors. Signal Transduct Target Ther 9:200. https://doi.org/10.1038/s41392-024-01889-y
Welsher K, Liu Z, Sherlock SP, Robinson JT, Chen Z, Daranciang D, Dai H (2009) A route to brightly fluorescent carbon nanotubes for near-infrared imaging in mice. Nat Nanotechnol 4:773–780. https://doi.org/10.1038/nnano.2009.294
Wu M, Chen W, Chen Y, Zhang H, Liu C, Deng Z, Sheng Z, Chen J, Liu X, Yan F, Zheng H (2018) Focused ultrasound-augmented delivery of biodegradable multifunctional nanoplatforms for imaging-guided brain tumor treatment. Adv Sci 5. https://doi.org/10.1002/advs.201700474
Xia J, Yao J, Wang LHV (2014) Photoacoustic tomography: principles and advances (invited review). Prog Electromagn Res 147:1–22. https://doi.org/10.2528/PIER14032303
Xiao Y, Liu Y, Yang S, Zhang B, Wang T, Jiang D, Zhang J, Yu D, Zhang N (2016) Sorafenib and gadolinium co-loaded liposomes for drug delivery and MRI-guided HCC treatment. Colloids Surf B Biointerfaces 141:83–92. https://doi.org/10.1016/j.colsurfb.2016.01.016
Xu P, Wang R, Yang W, Liu Y, He D, Ye Z, Chen D, Ding Y, Tu J, Shen Y (2021) A DM1-doped porous gold nanoshell system for NIR accelerated redox-responsive release and triple modal imaging guided photothermal synergistic chemotherapy. J Nanobiotechnol 19:77. https://doi.org/10.1186/s12951-021-00824-5
Yandrapalli S, Puckett Y (2024) SPECT imaging
Yang Z, Cheng R, Zhao C, Sun N, Luo H, Chen Y, Liu Z, Li X, Liu J, Tian Z (2018) Thermo- and pH-dual responsive polymeric micelles with upper critical solution temperature behavior for photoacoustic imaging-guided synergistic chemo-photothermal therapy against subcutaneous and metastatic breast tumors. Theranostics 8:4097–4115. https://doi.org/10.7150/thno.26195
Yang K, Liu Y, Wang Y, Ren Q, Guo H, Matson JB, Chen X, Nie Z (2019a) Enzyme-induced in vivo assembly of gold nanoparticles for imaging-guided synergistic chemo-photothermal therapy of tumor. Biomaterials 223:119460. https://doi.org/10.1016/j.biomaterials.2019.119460

Yang Z, Fan W, Zou J, Tang W, Li L, He L, Shen Z, Wang Z, Jacobson O, Aronova MA, Rong P, Song J, Wang W, Chen X (2019b) Precision cancer theranostic platform by in situ polymerization in perylene diimide-hybridized hollow mesoporous organosilica nanoparticles. J Am Chem Soc 141:14687–14698. https://doi.org/10.1021/jacs.9b06086

Yang C, Chang M, Yuan M, Jiang F, Ding B, Zhao Y, Dang P, Cheng Z, Al Kheraif AA, Ma P, Lin J (2021) NIR-triggered multi-mode antitumor therapy based on Bi2Se3/Au heterostructure with enhanced efficacy. Small 17. https://doi.org/10.1002/smll.202100961

Yao Y, Zhou Y, Liu L, Xu Y, Chen Q, Wang Y, Wu S, Deng Y, Zhang J, Shao A (2020) Nanoparticle-based drug delivery in cancer therapy and its role in overcoming drug resistance. Front Mol Biosci 7. https://doi.org/10.3389/fmolb.2020.00193

Yatvin MB, Weinstein JN, Dennis WH, Blumenthal R (1978) Design of liposomes for enhanced local release of drugs by hyperthermia. Science 202:1290–1293. https://doi.org/10.1126/science.364652

Yuan B, Burgess SA, Iranmahboob A, Bouchard MB, Lehrer N, Bordier C, Hillman EMC (2009) A system for high-resolution depth-resolved optical imaging of fluorescence and absorption contrast. Rev Sci Instrum 80. https://doi.org/10.1063/1.3117204

Zaw Thin M, Allan H, Bofinger R, Kostelec TD, Guillaume S, Connell JJ, Patrick PS, Hailes HC, Tabor AB, Lythgoe MF, Stuckey DJ, Kalber TL (2020) Multi-modal imaging probe for assessing the efficiency of stem cell delivery to orthotopic breast tumours. Nanoscale 12:16570–16585. https://doi.org/10.1039/D0NR03237A

Zeng W-N, Yu Q-P, Wang D, Liu J-L, Yang Q-J, Zhou Z-K, Zeng Y-P (2021) Mitochondria-targeting graphene oxide nanocomposites for fluorescence imaging-guided synergistic phototherapy of drug-resistant osteosarcoma. J Nanobiotechnol 19:79. https://doi.org/10.1186/s12951-021-00831-6

Zhang D-Y, Zheng Y, Zhang H, Yang G-G, Tan C-P, He L, Ji L-N, Mao Z-W (2018) Folate receptor-targeted theranostic IrS x nanoparticles for multimodal imaging-guided combined chemo-photothermal therapy. Nanoscale 10:22252–22262. https://doi.org/10.1039/C8NR08095J

Zhang Y, Li M, Gao X, Chen Y, Liu T (2019a) Nanotechnology in cancer diagnosis: progress, challenges and opportunities. J Hematol Oncol 12:137. https://doi.org/10.1186/s13045-019-0833-3

Zhang J, Qiao Z, Liu H-Y, Song J, Yin J (2019b) Positively charged helical chain-modified stimuli-responsive nanoassembly capable of targeted drug delivery and photoacoustic imaging-guided chemo-photothermal synergistic therapy. Biomater Sci 7:2050–2060. https://doi.org/10.1039/C9BM00055K

Zhang S, Lloveras V, Pulido D, Liko F, Pinto LF, Albericio F, Royo M, Vidal-Gancedo J (2020) Radical dendrimers based on biocompatible oligoethylene glycol dendrimers as contrast agents for MRI. Pharmaceutics 12:772. https://doi.org/10.3390/pharmaceutics12080772

Zhang S, Lloveras V, Lope-Piedrafita S, Calero-Pérez P, Wu S, Candiota AP, Vidal-Gancedo J (2022) Metal-free radical dendrimers as MRI contrast agents for glioblastoma diagnosis: ex vivo and in vivo approaches. Biomacromolecules 23:2767–2777. https://doi.org/10.1021/acs.biomac.2c00088

Zheng X-C, Ren W, Zhang S, Zhong T, Duan X, Yin Y-F, Xu M-Q, Hao Y-L, Li Z-T, Li H, Liu M, Li Z-Y, Zhang X (2018) The theranostic efficiency of tumor-specific, pH-responsive, peptide-modified, liposome-containing paclitaxel and superparamagnetic iron oxide nanoparticles. Int J Nanomedicine 13:1495–1504. https://doi.org/10.2147/IJN.S157082

Zhou X, Ye M, Han Y, Tang J, Qian Y, Hu H, Shen Y (2017) Enhancing MRI of liver metastases with a zwitterionized biodegradable dendritic contrast agent. Biomater Sci 5:1588–1595. https://doi.org/10.1039/C7BM00126F

Clinical Validations of Localized Cancer Nanomedicine

7

Hitesh Harsukhbhai Chandpa, Shovan Naskar,
Prakash Ashok Kumbhar, Abhijeet Singh, and Jairam Meena

Abstract

A rapidly advancing area of cancer treatment, "nanomedicine" uses materials at the nanoscale to target cancer cells precisely while limiting harm to healthy tissues and lowering systemic toxicity. Localized cancer therapies leverage nanotechnology by utilizing carriers such as nanoparticles, liposomes, and dendrimers, to enhance drug solubility, bioavailability, and targeted delivery to the tumor site. The enhanced permeability and retention (EPR) effect facilitates the retention of these nanomedicines at cancer sites, while surface modifications and specific targeting ligands increase therapeutic efficacy and personalization. Theranostic agents enable simultaneous therapy and diagnosis, offering real-time monitoring and optimized treatment. Despite the promising potential, challenges such as regulatory approvals, production scalability, and tumor microenvironment (TME) complexity hinder clinical translation. Regulatory frameworks in regions like the United States, Europe, and Japan are evolving to address these issues. Continuous research and technological advancements are essential for overcoming the current limitations and achieving clinical success in cancer nanomedicine.

Keywords

Solid cancer · Bioimaging · Biodistribution · Cancer theranostics · Enhanced permeability and retention effect · Translational

H. H. Chandpa · S. Naskar · P. A. Kumbhar · A. Singh · J. Meena (✉)
Immuno Engineering and Therapeutics Laboratory, Department of Pharmaceutical Engineering and Technology, Indian Institute of Technology (Banaras Hindu University) Varanasi, Varanasi, India
e-mail: jairam.phe@itbhu.ac.in

R. Prasad, M. Mohan Yallapu (eds.), *Advancements in Cancer Theranostics*,
https://doi.org/10.1007/978-981-95-0938-6_7

7.1 Introduction

Currently, cancer is a major global health burden, and it needs advanced therapeutic approaches to overcome its inherent complexity and heterogeneity. Nanomedicine, one of the applications of nanotechnology, utilizes nanoscale materials, normally ranging from 1 to 100 nanometers (nm) for the prevention of severe diseases like cancer (Gavas et al. 2021). A specialized subset—localized cancer nanomedicine—focuses on delivering therapeutic agents to the specific sites, for example, the tumor microenvironment (TME). Due to the less undesirable effects and systemic toxicity, this approach is considered as superior to the conventional methods (Peer et al. 2020). Nanomedicines, which include nanoparticles (NPs), liposomes, and dendrimers, have shown promise to deal with these issues via targeted drug delivery, improved solubility, and lower toxicity. The clinical impact of nanomedicines on therapeutic outcomes in cancer therapy is mainly attributed to distinct properties, like nanoscale size, greater surface area, and surface modification ability (Petros and DeSimone 2010). Because of these characteristics, nanomedicines can enhance drug solubility and bioavailability, which is crucial for particularly hydrophobic drugs. One of the primary mechanisms is enhanced permeability and retention (EPR), which NPs follow to transport therapeutic agents to a specified location in the body. Due to the leaky vasculature of tumor tissues, it allows NPs to accumulate preferentially at tumor site with minimum interaction of toxic drugs to the nearby normal cells (Maeda et al. 2013).

Nanomedicines can be modified using specific targeting ligands, like antibodies, peptides, or any small molecules to enhance their specificity for tumor cells (Fan et al. 2023). These targeted medicines can decrease adverse effects and increase treatment efficacy. Moreover, nanocarriers can also be formulated to deliver therapeutics in a controlled way for prolonged delivery. This approach allows the combination therapy with multiple therapeutic agents being delivered together to target multiple aspects of tumor progression and metastasis (Barenholz 2001). The potential of nanomedicines to integrate diagnostic and therapeutic functions has led to the advancement of theranostic agents. These agents offer both therapeutic and diagnostic properties simultaneously and enable us to monitor the tumor progression as well as therapeutic efficacy (Wang et al. 2024a). For instance, the incorporation of imaging agents such as quantum dots and magnetic nanoparticles enables the use of non-invasive imaging modalities, including MRI, CT scans, and fluorescence imaging. This enables clinicians to assess the in vivo distribution and accumulation of nanomedicines, allowing them to optimize treatment plans accordingly (Rashidi et al. 2024).

Although nanomedicines have a lot of potential, translating preclinical findings into therapeutic uses presents a lot of challenges. Securing regulatory approval stands as one of the primary challenges, which demands significant safety and efficacy data prior to the entry of nanomedicines into clinical trials following entry into the market (Đorđević et al. 2022). The large-scale production of nanomedicines is still a difficult and expensive process, along with maintaining their long-term stability (Muthu and Wilson 2012). The constant and effective delivery of nanomedicines to every tumor location gets further complicated by the heterogeneity and complexity of the TME (Wang et al. 2024b). Therefore, a deep understanding of how nanomedicines interact with both the TME and the immune system is essential for their clinical success. While the EPR effect presents a promising method for targeted drug delivery to tumors, its universal application is hindered by variations among different cancer types and even changes in the same TME over time (Wu 2021). Additionally, factors such as elevated interstitial fluid pressure within tumors can impede the penetration of nanoparticles (NPs), diminishing their therapeutic effectiveness (Zhang et al. 2021). Therefore, to achieve reliable and reproducible treatment outcomes, further optimization of NP formulations, drug release mechanisms, and targeting techniques is essential for the clinical implementation of cancer nanomedicine.

7.2 Significance of Localized Cancer Therapy

Localized cancer therapy refers to the approaches that transport the therapeutic molecules to the tumor-specific location with less harm to neighboring healthy tissues. As a result, the chances of systemic toxicity of the agent are lower, and this improves the treatment outcomes. Because conventional therapies like chemotherapy and radiation therapy lack targeted distribution, they often spread non-specifically throughout the body, leading to serious side effects and diminishing patients' quality of life (Bae and Park 2011). Here, nanomedicines can overcome these challenges. The therapeutic NPs can readily move across the biological barriers. They use passive and active targeting to reach tumor cells and deliver medications in a prolonged manner. Additionally, to become responsive to certain stimuli-for example, differences of pH or temperature in TME-these nanoparticles are engineered in very specific ways that improve their accuracy and efficacy (Hong et al. 2009).

7.3 Global Regulatory Landscape on Nanomaterials and Their Limitations

Safety assessments are essential for product development; early and continuous safety assessments decide whether the product is ready for the market or not. Nanotechnology helps to improve material properties, reduce waste, and lower environmental emissions. However, with the benefits, new challenges are emerging. To tackle those challenges, global regulation and thorough investigation of side effects and toxicity studies are required. Sharing research results and facilities helps to create stronger regulations for nanomedicines and nanotechnology products (Hodge et al. 2009). In this topic, regulations on nanomedicines by different regulatory bodies from different regions globally and their limitations have been discussed.

7.3.1 The United States

In the United States, the Food and Drug Administration (FDA) is the regulatory body of utmost importance working on regulations for nanomedicines. The FDA's objective is to create review standards for nanotechnology products and improve knowledge of nanomaterials (Hariyadi et al. 2020). It regulates nanotechnology products, such as nanomedicines, using existing laws and guidelines. They have given guidance for NPs in food, cosmetics, and animal feed, but not specifically for nanomedicine. However, in 2018, industries received guidelines on Liposome Drug Products that covered human pharmacokinetics, bioavailability, manufacturing and controls, chemistry, and labeling requirements. These guidelines cover the registration and approval of nanomedicines that contain nanomaterials (FDA 2018). The FDA evaluates each nanotechnology individually, rather than broadly declaring it safe or harmful (Ridge 2018).

The FDA participates in the National Nanotechnology Initiative (NNI), a collaborative effort that coordinates federal research and development (R&D) activities in nanoscale science, engineering, and technology. Objectives of the NNI include sustaining cutting-edge R&D in the field of nanotechnology, transforming innovative technologies into marketable products to spur economic growth, creating educational resources and developing a skilled workforce, and promoting the responsible advancement of nanotechnology. Research is carried out by FDA centers to support their regulatory goals for items involving nanotechnology. Through the NNI, the FDA and other agencies can define their research needs (FDA 2024a). The FDA Nanotechnology Task Force was created in August 2006, and it works on regulatory approaches to safely develop nanotechnology products and fills knowledge gaps. They support research by funding and training FDA staff as well as by participating in global nanotechnology standard development (FDA 2024b).

The FDA collaborates with countries like Canada and India to build capacity and participates in international standardization efforts for nanomaterial characterization. Although progress has been made, the number of approved nanotechnology products remains low. Despite positive findings in cancer research, more efforts are required to bring these products to market. Submissions regarding nanomaterial-based products that the FDA is receiving are higher than ever, with many in clinical testing and some approved for clinical use (Tyner 2019).

7.3.2 Europe

Requests for nanomedicine approval have increased in the European Union and its member states during the past 20 years. This presents a challenge for agencies, as there is a need for a balance between regulating and encouraging innovation in this fast-evolving area. Overall, the European Union (EU) has taken a positive and proactive perspective toward nanotechnology and nanomedicines, acknowledging the possible advantages these developments may bring to the economy and society (Vencken and Greene 2018).

On June 7, 2005, the Commission Communication released "Nanoscience and Nanotechnologies: An action plan for Europe 2005–2009." This strategic plan outlined a detailed series of coordinated actions intended to ensure safe, integrated, and responsible development and utilization of nanoscience and nanotechnologies throughout Europe (Bochon 2011). The EU legislation on nanomedicine addresses specifics such as particle size, chemical and physical characteristics, toxicological profile, safety data, exposure conditions, and identification. These comprehensive guidelines provide a robust framework for managing nanomedicine effectively (Rauscher et al. 2017). The EU has advanced in defining "nanomaterial" by establishing task forces and consortiums that produce reports on food, environment, and health. Government-funded initiatives such as the European Nanomedicine Characterization Laboratory (EU-NCL) and the REFINE project are contributing to progress in this domain. Unlike the United Kingdom, the European Medicines Union (EMU) has issued preliminary guidelines for the preparation of nanomedicine; however, these are still in the public consultation phase and have not yet been formalized into regulatory guidance. The EU-NCL collaborates with regulatory agencies to influence decisions regarding the safety and regulation of nanomedicine (Foulkes et al. 2020).

To share regulatory experience and address scientific challenges in nanomedicine evaluation, the European Medicines Agency (EMA) published its first reflection paper on nanotechnology-based medicinal products in 2006 and launched a dedicated webpage. In response to growing industry interest, the European Nanomedicines Expert Group was established in 2009, bringing together academics and regulatory specialists to collaborate with global agencies like the FDA and Japan's Pharmaceutical and Medical Devices Agency (PMDA). Reinforced in 2011 with experts in manufacturing, safety, and pharmacokinetics, the group focused on key challenges such as characterization, scale-up, pharmacokinetics, and long-term safety. Their role includes drafting technical documents to guide sponsors and researchers on regulatory expectations for nanomedicine approval in the EU (Pita et al. 2016).

The European Nano-Safety Cluster has produced a strategic document that outlines the key priorities for nano-safety research from 2015 to 2025. This document is a collaborative effort and includes several research projects that have received funding from various EU Member States. The strategic vision within the document focuses on a comprehensive understanding of the features of engineered nanomaterials (ENMs), the mechanisms by which they may pose hazards, the potential exposure to these materials, and the assessment and management of related risks. This approach ensures that the research addresses all critical aspects of nano-safety, promoting the safe use of nanotechnologies. This document outlines several milestones to be achieved between 2015 and 2025. These include materials classification, measurement principles, bio-nano interactions, and the safety assessments of engineered nanomedicines. These goals aim to ensure a comprehensive understanding and management of nano-safety (Savolainen 2013).

7.3.3 The United Kingdom

The Medicines and Healthcare products Regulatory Authority (MHRA) is in charge of medicine regulations throughout the United Kingdom, yet nanotechnology-based products have no tailored regulations (Ali et al. 2023). Current regulatory frameworks for chemicals, cosmetics, and medicines are broad enough to cover nanotechnologies in the United Kingdom. Given this, an additional regulatory body or framework is deemed unnecessary. These existing frameworks already address the major areas impacted by nanomaterials. For now, they are considered sufficient to manage nanotechnology-related products and processes (Enderby and Dowling 2004). Most UK laws on nanomedicine come from European law and UK health and safety laws. Researchers are motivated to work with the MHRA Innovation Office for guidance on the development of new nanomedicines (Bremer-Hoffmann et al. 2018).

7.3.4 Canada

In Canada, several government agencies are involved in regulating the approval, marketing, and evaluation of nanomedicine products. Key authorities include Health Canada, Environment and Climate Change Canada, Agriculture and Agri-Food Canada, and the Canadian Food Inspection Agency. As the principal body responsible for public health, Health Canada defines nanomaterials as manufactured products, substances, or systems that measure between 1 and 100 nm in at least one spatial dimension, or those that are outside this range but exhibit nanoscale phenomena (Cullis 2019). In case of approval of nanotechnology products, Health Canada depends on the existing regulatory framework and follows the guidelines established by the OECD (2017). Canada collaborates closely with the United States concerning nanomedicine and has adopted a stance similar to that of the USFDA. Remarkably, over the past decade, there have been no major updates to the nanomedicine guidelines in Canada.

7.3.5 Australia

In Australia, many regulatory bodies oversee nanotechnology and the use of nanomaterials across various market sectors such as pharmaceuticals, agriculture, and more. Among these, the Therapeutic Goods Administration (TGA) deals with NPs in therapeutic use and medical devices with reference to the *Therapeutic Goods Act 1989*. According to a review published by Monash University, this is enough for the regulation of the nanoparticles because the TGA considers the nanomedicine as complementary medicine and claims that, it does not pose any significant threat than conventional medicines. (Harford et al. 2007; Priestly et al. 2007). Even though the industrial use of nanomaterials has been under the umbrella of the Australian

Industrial Chemicals Introduction Scheme (AICIS) since 2020, specific regulatory mandates for nanomaterials have not been specified by these bodies (Ludlow et al. 2007).

7.3.6 Japan

In Japan, the oversight of pharmaceuticals falls under the jurisdiction of several government entities, including the Ministry of Health, Labour and Welfare (MHLW), the Pharmaceutical and Medical Devices Agency (PMDA), the National Institute of Health Sciences (NIHS), and the Agency for Medical Research and Development (AMED). The regulation of nanotechnology in Japan is primarily governed by existing legislation, such as the *Chemical Substances Control Act* and the *Industrial Safety and Health Act* (Nasu and Faunce 2013). Notably, there is currently no dedicated legal framework specifically addressing the safety concerns unique to nanomaterials. To address this, the Japanese government has assigned the Ministry of Economy, Trade and Industry (METI) the responsibility of gathering data on nanomaterials, while the Environment Ministry (ME) is tasked with evaluating their potential environmental impacts (Ali et al. 2023; Allan et al. 2021).

7.3.7 Russia

Russia has an established regulatory framework for the regulation of nanomedicine in the form of federal law with amendments on 6 December 2011, applied to all pharmaceuticals and medical purpose products, irrespective of use or non-use of nanotechnology (OECD 2013). Federal service for Surveillance on Consumer Rights Protection (Rospotrebnadzor), along with the Ministry of Health and Social Development, and also RUSNANO, an essential private entity in advising the federation on the topics of nanotechnology (Glushkova et al. 2016).

7.3.8 India

In India, efforts are underway to establish a comprehensive regulatory framework for nanomedicine, aiming to align with international standards. This framework emphasizes a multi-tiered governance system, incorporating four key principles: regulatory scope, policy framework, governance structures, and the lifecycle of nanomedicine (Bhatia and Chugh 2017). Collaborative efforts spearheaded by the Department of Biotechnology (DBT) have resulted in the development of "Guidelines for Evaluation of Nanopharmaceuticals" in India. These guidelines encompass a broad range of nanopharmaceuticals, including both formulations and active pharmaceutical ingredients (API) that have been modified at the nanoscale, regardless of whether they are new entities or modifications of previously approved drugs. These modifications encompass alterations in size, properties, or behavior resulting from the application of nanotechnology. The guidelines focus on nanopharmaceuticals intended for human use in the treatment, diagnosis, mitigation, cure, or prevention of diseases and disorders (DBT 2019).

7.4 Clinical Impact of Nanomedicines in Cancer Management

Nanomedicine represents a frontier in oncology, leveraging nanoparticles to deliver precise, targeted treatment modalities that can significantly improve clinical outcomes. These nanoparticles offer several benefits over traditional cancer treatments because of their unique physio-chemical properties. These particles have been used for different types of formulations, following approval for clinical use in the treatment and diagnosis of cancer. The first nanoformulation to ever be approved for the treatment of cancer was Doxil®, a PEG-coated liposomal doxorubicin (PL-DOX) in 1995 by USFDA. This formulation was not only able to deliver more drug to the cancer and produce equivalent effect at the same dose, but also was able to reduce the side effect like cardiotoxicity (Barenholz 2012). This approval started a race and until the end of twentieth century, 7 nanoformulations were approved by USFDA for the chemotherapy of cancer and out of those 7 formulations, 5 were liposomal nanoformulations (Wang et al. 2024a). By the year 2000, more than 20 nanoformulations had been approved by different regulatory agencies throughout the globe. These approvals include the formulations made up of different materials, including proteins, polymers, lipids, crystals, and metals (Rodríguez et al. 2022; Wicki et al. 2015). Table 7.1 contains the list of approved nanoformulations for the treatment and diagnosis of cancer (Wang et al. 2024a; Rodríguez et al. 2022; Wicki et al. 2015).

As many as there are approved nanoformulations for treating and diagnosing cancer, countless more are presently navigating different clinical trial phases worldwide. For evaluation of pharmacokinetics parameters, safety, and efficacy, clinical trials are trivial for any new drug or new formulation or for new indications of old drug or formulation. After the approval of PL-DOX in 1995 for ovarian cancer, the same formulations has been approved for Kaposi's sarcoma, breast cancer, and multiple myeloma (Lee et al. 2023). The PL-DOX is still being evaluated for the combinations with different drugs and formulations as well as for different indications. There are more than 10 clinical trials, listed on the website of the National Cancer Institute, that are testing PL-DOX (National Cancer Institute 2025a). Similarly, in 2005, the FDA approved an albumin-bound nanoformulation of paclitaxel specifically to treat metastatic breast cancer (MBC), which is now being tested for other indications and in combination with other drugs in more than 50 clinical trials (National Cancer Institute 2025b). This list of nanoformulations in clinical trials can be extended at length, but it is briefly summarized in Table 7.2. The list in Table 7.2 has been prepared by using keywords cancer and nanoparticles to shortlist the search at https://clinicaltrials.gov/ and important novel formulations that are under clinical trials were selected.

Table 7.1 Globally approved nanotheranostics for diagnosis and management of cancer

Approval country (year)	Drug	Nanoparticle	Product name	Indications	Type of therapy
The United States (1995)	Doxorubicin	PEG-coated liposome	Doxil	Kaposi's sarcoma, breast cancer, ovarian cancer, multiple myeloma	Chemotherapy
The United States (1996)	Doxorubicin	PEG-coated liposome	Caelyx	Metastatic breast, Kaposi's sarcoma, multiple myeloma	Chemotherapy
The United States (1996)	Daunorubicin	Liposomes	DaunoXome	Kaposi's sarcoma	Chemotherapy
The United States (1996)	Doxorubicin	Liposomes	Myocet	Metastatic breast cancer	Chemotherapy
The United States (1999)	Cytarabine	Liposomes	DepoCyt	Lymphomatous meningitis	Chemotherapy
The United States (2002)	Leuprolide	Polymeric micelles	Eligard	Prostate cancer	Chemotherapy
China (2003)	Paclitaxel	Liposomes	Lipusu	Ovarian cancer, metastatic gastric cancer	Chemotherapy
The United States (2005)	Paclitaxel	Albumin-bound nanoparticles	Abraxane	Breast and pancreatic cancer, non-small-cell lung cancer	Chemotherapy
The United States (1994, 2006)	Pegaspargase	Polymer protein conjugate	Oncaspar	Acute lymphoblastic leukemia	Chemotherapy
South Korea (2007)	Paclitaxel	PEG-PLA polymeric micelle	Genexol-PM	Breast, lung, ovarian cancer	Chemotherapy
Europe (2009)	Mifamurtide MTP-PE	Liposome	Mepact	Osteosarcoma	Chemotherapy
Europe (2010)	–	Iron-oxide NP	NanoTherm	Recurrent glioblastoma	Hyperthermia
The United States (2011)	5-Aminolevulinic acid	Liposomes	Ameluz	Nodular basal cell carcinoma	Chemotherapy
The United States (2012)	Vincristine	Liposome	Marqibo	Acute lymphoblastic leukemia	Chemotherapy

(continued)

Table 7.1 (continued)

Approval country (year)	Drug	Nanoparticle	Product name	Indications	Type of therapy
The United States (2013)	Ferric oxide	Iron-oxide NP	NanoTherm	Glioblastoma, prostate cancer, pancreatic cancer	Hyperthermia
China (2013)	Doxorubicin hydrochloride	Liposome	Lipodox	Kaposi's sarcoma, ovarian cancer, multiple myeloma, metastatic pancreatic cancer	Chemotherapy
The United States (2013)	Emtansine	Protein	Kadcyla	Breast cancer	Chemotherapy
India (2014)	Paclitaxel	Polymer/lipid NP	ONCO-PICN	Metastatic breast cancer	Chemotherapy
The United States (2015)	Paclitaxel	Polymeric nanoparticle	Paclical	Pancreatic cancer, metastatic breast cancer	Chemotherapy
The United States (2015)	Irinotecan	PEGylated liposomes	Onivyde	Metastatic pancreatic cancer	Chemotherapy
Korea (2016)	Paclitaxel	Lipid nanoparticle	DHP107	Advanced gastric cancer	Chemotherapy
The United States (2017)	Cytarabine/ daunorubicin (5: 1)	Liposomes	VYXEOS	Acute myeloid leukemia	Chemotherapy
The United States (2018)	Cisplatin	Liposomes	Lipoplatin™, Nanoplatin™	Pancreatic cancer, lung cancer	Chemotherapy
The United States (2019)	Paclitaxel	Albumin-bound nanoparticle	Pazenir	Metastatic breast cancer, metastatic pancreatic cancer, non-small cell lung cancer	Chemotherapy
Europe (2019)	–	Hafnium oxide NP	Hensify	Locally advanced soft tissue sarcoma	Radiotherapy enhancer

Table 7.2 Cancer nanotheranostics in different phases of clinical trials

Drug name	Nanoformulation	Product name	Indication(s)	Phase	Trial ID	Type of therapy
ThrCer6 and NY-ESO-1	PLGA nanoparticle	PRECIOUS-01	NY-ESO-1 positive cancers	1	NCT04751786	Immunotherapy
–	Gold metal shell with silica core nanoparticle	AuroLase	Prostate carcinoma	NA	NCT04240639	Radiotherapy
–	Carbon nanoparticle-loaded iron	CNSI-Fe(II)	Advanced solid tumors	1	NCT06048367	Chemotherapy
–	Polysiloxane and gadolinium chelates nanoparticle	NANOPRO	Recurrent cancer	2	NCT04784221	Radiation by proton therapy
Paclitaxel	Albumin-stabilized nanoparticle	Nab-paclitaxel	Metastatic breast cancer	2	NCT00609791	Chemotherapy
–	Carbon nanoparticles	–	Papillary thyroid cancer	3	NCT06791005	Imaging
Akt-1 antisense oligonucleotide	Lipid nanoparticle	WGI-0301	Advanced solid tumors	1	NCT05267899	Chemotherapy
–	Iron-oxide nanoparticles	NanoEcho Particle-1	Rectal cancer	2	NCT06693375	Imaging
Camptothecin	Cyclodextrin based nanoparticle	EP0057	Small cell lung cancer	1 and 2	NCT02769962	Chemotherapy
–	Gadolinium based nanoparticle	AGuIX	Pancreatic cancer and lung tumors	1 and 2	NCT04789486	Radiotherapy
–	Hafnium oxide nanoparticles	NBTXR3	Advanced solid tumors	1 and 2	NCT05039632	Radiotherapy
Paclitaxel	Albumin-stabilized nanoparticle	Nab-paclitaxel	Locally advanced or metastatic breast cancer	2	NCT01463072	Chemotherapy
Doxorubicin	Liposome	ThermoDox	Liver cancer	1	NCT02181075	Chemotherapy
Paclitaxel	Liposome	LEP-ETU	Neoplasm	1	NCT00080418	Chemotherapy
Quercetin	PLGA-PEG NPs	Nano-QUT	Oral cancer	2	NCT05456022	Chemotherapy

(continued)

Table 7.2 (continued)

Drug name	Nanoformulation	Product name	Indication(s)	Phase	Trial ID	Type of therapy
Paclitaxel	Micellar nanoparticle	NK 105	Breast cancer	3	NCT01644890	Chemotherapy
Docetaxel	Polymeric micelles	Docetaxel-PM	Esophagus squamous cell carcinoma, metastatic cancer	2	NCT03585673	Chemotherapy
Docetaxel	Polymeric micelles	BIND-014	KRAS positive non-small cell lung cancer, squamous cell non-small cell lung cancer	2	NCT02283320	Chemotherapy
Cetuximab	Polymeric micelles	NC-6004	Head and neck neoplasms	1	NCT02817113	Chemotherapy
Oxaliplatin	Liposome	MBP-426	Advanced or metastatic solid tumors	1	NCT00355888	Chemotherapy
Topotecan	Liposomes	FF-10850	Advanced solid tumors	1	NCT04047251	Chemotherapy
Doxorubicin	PEGylated liposomes	ABT-888	Ovarian cancer	2	NCT01113957	Chemotherapy
Vinorelbine tartrate	Liposomes	LipoVNB	Advanced malignancy	1 and 2	NCT02925000	Chemotherapy
KRAS G12D siRNA	Exosomes	iExosomes	Metastatic pancreas cancer	1	NCT03608631	Gene therapy
Tumor antigen	Exosome	CSET 1437	Non-small cell lung cancer	2	NCT01159288	Immunotherapy
Paclitaxel	Cationic liposomes	EndoTAG-1	Breast cancer	2	NCT01537536	Chemotherapy
Irinotecan hydrochloride	Liposome	LY01610	Recurrent small-cell lung cancer	3	NCT06128837	Chemotherapy

ThrCer6 threitolceramide-6, *NY-ESO-1* New York Esophageal Squamous Cell Carcinoma-1, *KRAS* Kirsten rat sarcoma viral oncogene homologue

The number of approved nanoformulations that have undergone clinical trials for cancer theranostics shows that the use of nanoparticles has provided many benefits over conventional drugs. As it has been shown with PL-DOX, where it was able to reduce the cardiotoxicity, produce the equivalent effect, and improve the circulation time in comparison to conventional doxorubicin (Franco et al. 2018). In a clinical trial phase III study, DaunoXome® (liposomal daunorubicin) showed equivalent efficacy, mean survival time, and treatment failure time when compared to standard doxorubicin, bleomycin, and vincristine (ABV) therapy. It was proven to be a safe and effective therapy for the AIDS-related Kaposi's sarcoma (AIDS-KS), which was approved by the FDA in 1996 (Gill et al. 1996). Similarly, in a clinical trial phase III study of PL-DOX against ABV for the therapy of AIDS-KS, the overall response produced by PL-DOX (45.9%) was significantly higher than that produced by combination therapy (24.8%) (Northfelt et al. 1998). This led to the approval of Doxil®, for the treatment of AIDS-KS in 2004 (Udhrain et al. 2007). Liposomal formulations of doxorubicin, daunorubicin, and cytarabine were approved and are still being tested for different indications with or without the combination of chemotherapeutic or immunotherapeutic drugs (Eckardt et al. 1994; Giri et al. 2023; Glantz et al. 1999; Moeremans et al. 2004).

Abraxane® was the first nanoformulation to get approval with protein-bound nanoparticles (Lu et al. 2021). This nanoformulation, consisting of paclitaxel bound to albumin, received regulatory approval for the treatment of MBC following the successful completion of phase III clinical trials. This trial demonstrated that the response rate of Abraxane® in metastatic breast cancer (MBC) patients was 33%, significantly surpassing the control arm's response rate of 19%. The control arm received polyethylated castor oil-based standard paclitaxel. This formulation also shows reduced incidence of side effects like grade 4 neutropenia (9% for Abraxane® and 22% for the control arm) and febrile neutropenia ($<$2% for both arms). Incidence of grade 3 sensory neuropathy was a little higher (10% in Abraxane® and 2% in control arm), but easily manageable (Gradishar et al. 2005). Abraxane® combined with cisplatin (AP) was evaluated against the gemcitabine and cisplatin combination (GP) in cases of metastatic triple-negative breast cancer. In this report, the AP group showed an 81.1% response rate, while the GP group was at 56.3%, which was significantly higher and also reduced the incidence rate of thrombocytopenia (3.9% in the AP group and 29.4% in the GP group). 18.8% of patients with neuropathy and 3.9% of patients with serious side effects were reported, which was higher in the AP group compared to the GP group, where none of the patients experienced neuropathy and 2.4% patients were presented with serious side effects (Hu et al. 2020). In another study with Abraxane® in combination with gemcitabine versus gemcitabine monotherapy for pancreatic cancer, similar results were obtained. The combination was able to significantly increase the overall survival (8.5 months), progression-free survival (5.5 months) as well as response rate (23%), but incidence of side effects like fatigue (17%), neuropathy (17%) and neutropenia (38%) were higher in comparison to gemcitabine monotherapy group (Hoff et al. 2013).

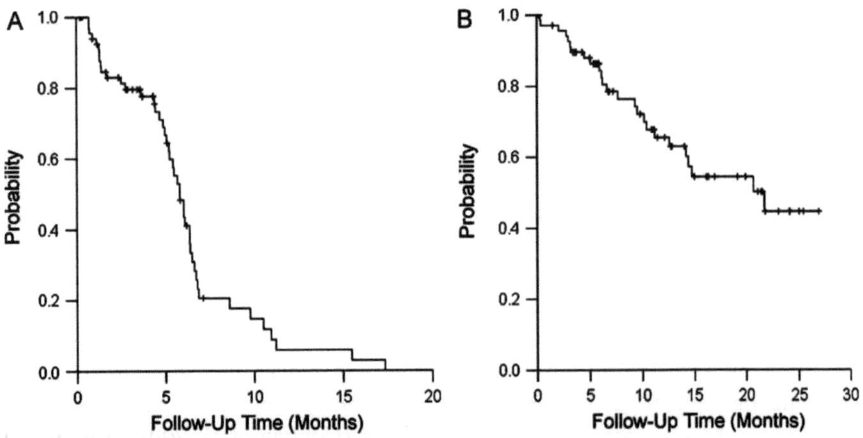

Fig. 7.1 Kaplan-Meier survival curve of advanced NSCLC patients after therapy with a combination of Genexol-PM® and cisplatin. Here, plot (**a**) shows progression-free survival and plot (**b**) shows overall survival of patients. (This figure has been reprinted from Annals of Oncology, Vol. 18(12), D.-W. Kim, S.-Y. Kim, H.-K. Kim, S.-W. Kim, S.W. Shin, J.S. Kim, K. Park, M.Y. Lee, D.S. Heo, Multicenter phase II trial of Genexol-PM, a novel Cremophor-free, polymeric micelle formulation of paclitaxel, with cisplatin in patients with advanced non-small-cell lung cancer, Pages 2009-2014, Copyright 2007, with permission from Elsevier)

The approval and studies of different types of nanoformulations are growing continuously for cancer theranostics. Researchers are working on different types of biocompatible materials like polymers, metals, lipids, etc. The first polymeric nanoformulation that was approved for cancer treatment is Genexol-PM in South Korea (Werner et al. 2013). This formulation, despite allowing the higher doses of paclitaxel, was able to show significant efficacy and relatively lesser incidences of side effects in advanced non-small cell lung cancer (NSCLC) patients. Overall response rate of this combination was 37.7% and median overall survival time was 21.7 months (Fig. 7.1), while median progression time was 5.8 months (Kim et al. 2007). Following the approval of Genexol-PM®, further clinical investigations were conducted. These studies included a phase III trial evaluating Genexol-PM® in MBC patients that did not overexpress the HER2 protein. Additionally, clinical trials explored the efficacy of Genexol-PM® in combination with carboplatin for the treatment of advanced epithelial ovarian cancer and other gynecological malignancies. In all these studies, the obtained results were similar in the sense that at higher doses of the drug, the efficacy was superior without additional side effects than in control groups (Kim et al. 2012; Park et al. 2017; Nam et al. 2023).

After liposomes, protein-bound nanoparticles, and polymeric nanoparticles, iron-oxide nanoparticles (IONPs) based formulation got approval for hyperthermia therapy of cancer. This therapy was evaluated with a lower dose of radiation therapy in patients with recurrent glioblastoma, where the injection of IONPs was given into the tumor, and with an alternative magnetic field, the temperature was increased. This temperature change was monitored by the thermometry catheter (Fig. 7.2).

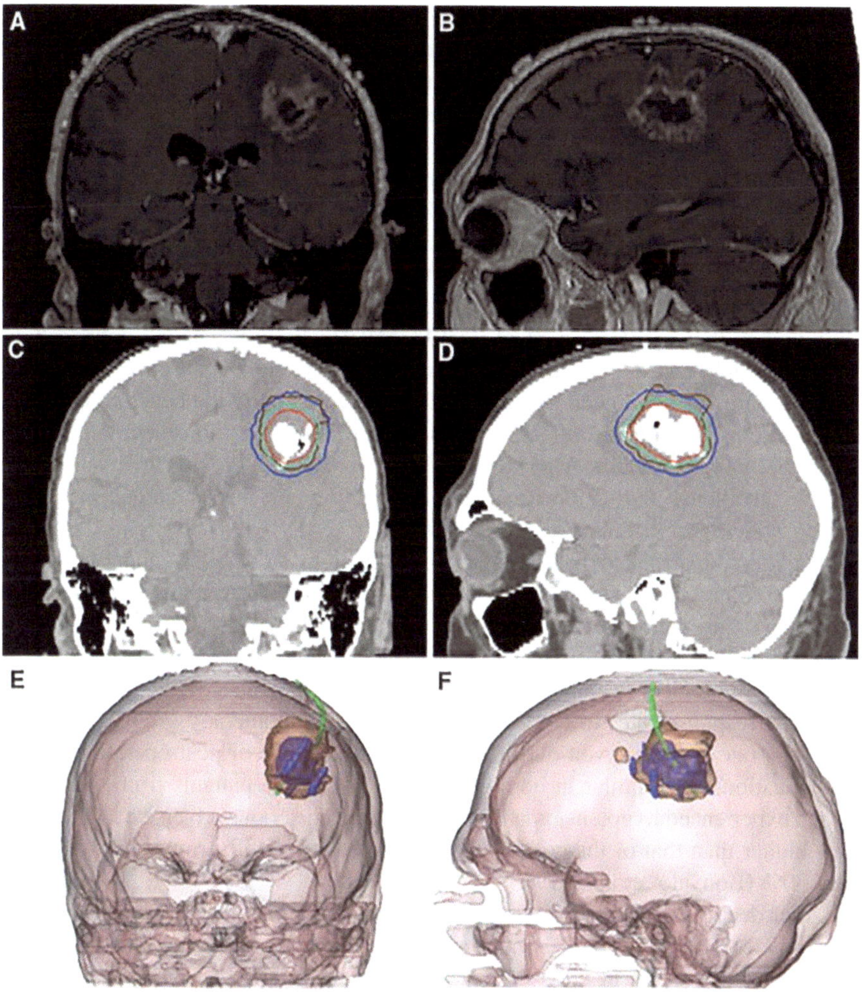

Fig. 7.2 Thermotherapy of patients with recurrent glioblastoma in combination with a lower dose of radiotherapy (**a**) sagittal and (**b**) coronal slice of brain MRI showing the brain tumor before the injection of IONPs. (**c**) Sagittal and (**d**) coronal slices of brain CT scan after the injection of IONPs, which can be seen as a black color spot inside the tumor, red line shows the calculated temperature of 50 °C, blue line shows 40 °C, and brown line shows the area of the tumor. (**e**) and (**f**) are the 3D-construct by fusing MRI and CT scan showing tumor (Brown color), IONPs (Blue color), and thermometry catheter (green color). (This figure is reprinted from the Journal of Neuro-Oncology Vol. 103, Maier-Hauff, K., Ulrich, F., Nestler, D., Niehoff, H., Wust, P., Thiesen, B., Orawa, H., Budach, V. and Jordan, A., Efficacy and safety of intratumoral thermotherapy using magnetic iron-oxide nanoparticles combined with external beam radiotherapy on patients with recurrent glioblastoma multiforme, Pages 317–324, copyright 2010 under the CC BY license)

Patients undergoing this combination therapy following the diagnosis of their first tumor recurrence had experienced a median overall survival time of 13.4 months, while those diagnosed with their first tumor had a median overall survival time of 23.2 months. Additionally, the side effects caused by this formulation were mild and any serious complications were not observed (Maier-Hauff et al. 2011).

In another study where the thermotherapy of recurrent glioblastoma with radiotherapy was evaluated, it was shown that the IONPs were able to produce an inflammatory reaction in the tumor area, which led to the induction of antitumor immune reaction to stabilize the long-term recurrence of glioblastoma. Not only that, but from that study, two patients had treatment response for over 23 months without any further therapy (Grauer et al. 2019). MRI techniques can be enhanced by utilizing IONPs as diagnostic agents for cancer detection. In a clinical study of patients with head and neck cancer, IONPs were able to provide better contrast and reliable result for the identification of metastatic as well as non-metastatic lymph nodes before the resection. MRIs using IONPs were able to provide 96% sensitivity, 78.9% specificity, and 85.7% accuracy for detecting non-metastatic and metastatic lymph nodes (Curvo-Semedo et al. 2006).

Another metal called hafnium has come to light as a radio-enhancer for the therapy of cancer in combination with radiotherapy. The formulation composed of hafnium oxide nanoparticles (HONPs) was approved by the European Medical Agency in 2019. In the clinical trial, HONPs in combination with radiotherapy showed pathological response in double number of patients, which was 14 out of 87 (16% response rate) in comparison to the only radiotherapy group, where 8% response rate was observed out of 89 patients. Incidence of post-operative wound complications was similar in both groups at 9%, but 7% patients from the HONP group experienced hypotension and 39% experienced serious adverse events, which was higher than that of the control group, where the rate of serious adverse events was 30% (Bonvalot et al. 2019).

After decades of research and approval of so many nanoformulations for cancer theranostics, it seems like it is not enough, as the expected outcome is not being achieved by the current state of research. So far, results have been in favor of nanoformulations in terms of efficacy, stability, and tolerability, but the side effects that are being observed in different studies, although manageable, cannot be ignored, and this sought for more research and technological developments.

7.5 Therapeutic Limitations of Cancer Nanomedicine

Although there are numerous advantages of nanomedicine, several limitations hinder its clinical translation. The efficacy in clinical trials is limited by the complex TME, variations in drug responses, and challenges with large-scale reproducibility (Wilhelm et al. 2016). Clinical trials are ongoing to evaluate novel nanocarrier-based therapies, but many of them have failed because of insufficient therapeutic effects, unintended toxicities, and regulatory challenges (Fan et al. 2023). Addressing these limitations is crucial to improve nanomedicine design for effective cancer treatment. Only a few nanomedicines get regulatory approval despite extensive research, mostly because of issues with their pharmacokinetics and pharmacodynamics (Park et al. 2021). These two main types of limitations are described below.

7.5.1 Pharmacokinetic Limitations

Pharmacokinetics (PK) plays a crucial role in determining the therapeutic success of nanomedicines. The ADME (absorption, distribution, metabolism, and excretion) profile of nanoparticles is affected by several factors, including size, surface charge, and composition (Yuan et al. 2023). Nanomedicines, especially designed to utilize the EPR effect, frequently exhibit variable PK profiles due to the variations in tumor vasculature and the diverse nature of tumor tissues. The length of time that NPs circulate throughout the body and their concentration at the tumor location can both be affected by their size and surface properties (Allen and Cullis 2013). Also, NPs may be cleared by the reticuloendothelial system (RES), especially the liver and spleen, limiting their bioavailability and therapeutic potential (Liu et al. 2017). Unpredictable variations in tumor perfusion and permeability can lead to low efficacy of NPs in certain tumors. Despite the EPR effect, not all tumors exhibit the same level of NP uptake, and some tumors may be resistant to NP-based therapies due to high interstitial fluid pressure, dense extracellular matrix, or lack of suitable receptors (Torchilin 2011). These factors can lead to inconsistent drug delivery and lower therapeutic efficacy.

7.5.2 Pharmacodynamic Limitations

Several variables can affect the pharmacodynamics of nanomedicines, which refers to the interactions between the drug and its target. The local TME, which alters the drug's effective release from the NP carrier, may not always facilitate the intended release. For example, the acidic pH of the TME can trigger the release of drugs from pH-sensitive NPs, and variations in pH across different tumor sites can lead to uneven drug release (Oberdörster et al. 2005). Likewise, enzymes or other biomolecules in the tumor can degrade NPs, altering their pharmacodynamics and lowering therapeutic effects (Li et al. 2020). Also, because of the varied character of TME, cancer cells frequently do not absorb NPs efficiently, which can lead to sub-optimal drug delivery with less therapeutic effect (Mitchell et al. 2021). This can happen in the case of cancer cells becoming resistant to drugs because of the overexpression of the efflux pumps, which leads to diminished effect of the drug over time (Housman et al. 2014).

7.6 Conclusion

Cancer is the leading health issue worldwide, and while traditional therapies have made strides in treatment and diagnosis, they often come with limitations such as non-specificity and significant side effects. By enhancing targeting and lowering toxicity, nanomedicines present a prospective substitute that could completely transform the treatment of cancer. However, their clinical use is still in its infancy and has its own set of difficulties, necessitating further work to gain a balance between efficacy and safety.

Author Contribution H.H.C., S.N., P.A.K., and A.S. have done the literature survey and prepared the chapter. J.M. has done the final editing and review of this chapter.

References

Ali F, Neha K, Parveen S (2023) Current regulatory landscape of nanomaterials and nanomedicines: a global perspective. J Drug Deliv Sci Technol 80:104118

Allan J, Belz S, Hoeveler A, Hugas M, Okuda H, Patri A et al (2021) Regulatory landscape of nanotechnology and nanoplastics from a global perspective. Regul Toxicol Pharmacol 122: 104885

Allen TM, Cullis PR (2013) Liposomal drug delivery systems: from concept to clinical applications. Adv Drug Deliv Rev 65(1):36–48

Bae YH, Park K (2011) Targeted drug delivery to tumors: myths, reality and possibility. Elsevier, pp 198–205

Barenholz Y (2001) Liposome application: problems and prospects. Curr Opin Colloid Interface Sci 6(1):66–77

Barenholz Y (2012) Doxil®—the first FDA-approved nano-drug: lessons learned. J Control Release 160(2):117–134

Bhatia P, Chugh A (2017) A multilevel governance framework for regulation of nanomedicine in India. Nanotechnol Rev 6(4):373–382

Bochon A (2011) Evolution of the European commission recommendation for a code of conduct for responsible nanosciences and nanotechnology research. Nanotechnol Law Bus 8:117

Bonvalot S, Rutkowski PL, Thariat J, Carrère S, Ducassou A, Sunyach M-P et al (2019) NBTXR3, a first-in-class radioenhancer hafnium oxide nanoparticle, plus radiotherapy versus radiotherapy alone in patients with locally advanced soft-tissue sarcoma (Act.In.Sarc): a multicentre, phase 2–3, randomised, controlled trial. Lancet Oncol 20(8):1148–1159

Bremer-Hoffmann S, Halamoda-Kenzaoui B, Borgos SE (2018) Identification of regulatory needs for nanomedicines. J Interdiscip Nanomed 3(1):4–15

Cullis P (2019) Nanomedicines innovation network (NMIN): Canada's national centre of excellence in nanomedicines. Global summit on regulatory science

Curvo-Semedo L, Diniz M, Migueis J, Julião MJ, Martins P, Pinto A et al (2006) USPIO-enhanced magnetic resonance imaging for nodal staging in patients with head and neck cancer. J Magn Reson Imaging 24(1):123–131

DBT (2019) Guidelines for evaluation of nanopharmaceuticals in India. Department of Biotechnology, Government of India. http://dbtindia.gov.in/sites/default/files/uploadfiles/Guidelines%20For%20Evaluation%20of%20Nanopharmaceuticals%20in%20India.pdf

Đorđević S, Gonzalez MM, Conejos-Sánchez I, Carreira B, Pozzi S, Acúrcio RC et al (2022) Current hurdles to the translation of nanomedicines from bench to the clinic. Drug Deliv Transl Res 12:1–26

Eckardt JR, Campbell E, Burris HA, Weiss GR, Rodriguez GI, Fields SM et al (1994) A phase II trial of DaunoXome, liposome-encapsulated daunorubicin, in patients with metastatic adenocarcinoma of the colon. Am J Clin Oncol 17(6):498–501
Enderby J, Dowling A (2004) Nanoscience and nanotechnologies: opportunities and uncertainties. The Royal Society & The Royal Academy of Engineering Report, London, p 7
Fan D, Cao Y, Cao M, Wang Y, Cao Y, Gong T (2023) Nanomedicine in cancer therapy. Signal Transduct Target Ther 8(1):293
FDA (2018) Liposome drug products: chemistry, manufacturing, and controls; human pharmacokinetics and bioavailability; and labeling documentation. https://www.fda.gov/regulatory-information/search-fda-guidance-documents/liposome-drug-products-chemistry-manufacturing-and-controls-human-pharmacokinetics-and
FDA (2024a) Nanotechnology programs at FDA. https://www.fda.gov/science-research/science-and-research-special-topics/nanotechnology-programs-fda
FDA (2024b) Nanotechnology task force. https://www.fda.gov/science-research/nanotechnology-programs-fda/nanotechnology-task-force
Foulkes R, Man E, Thind J, Yeung S, Joy A, Hoskins C (2020) The regulation of nanomaterials and nanomedicines for clinical application: current and future perspectives. Biomater Sci 8(17): 4653–4664
Franco YL, Vaidya TR, Ait-Oudhia S (2018) Anticancer and cardio-protective effects of liposomal doxorubicin in the treatment of breast cancer. Breast Cancer (Dove Med Press) 10:131–141
Gavas S, Quazi S, Karpiński TM (2021) Nanoparticles for cancer therapy: current progress and challenges. Nanoscale Res Lett 16(1):173
Gill PS, Wernz J, Scadden DT, Cohen P, Mukwaya GM, von Roenn JH et al (1996) Randomized phase III trial of liposomal daunorubicin versus doxorubicin, bleomycin, and vincristine in AIDS-related Kaposi's sarcoma. J Clin Oncol 14(8):2353–2364
Giri PM, Banerjee A, Layek B (2023) A recent review on cancer nanomedicine. Cancers (Basel) 15(8):2256
Glantz MJ, Jaeckle KA, Chamberlain MC, Phuphanich S, Recht L, Swinnen LJ et al (1999) A randomized controlled trial comparing intrathecal sustained-release cytarabine (DepoCyt) to intrathecal methotrexate in patients with neoplastic meningitis from solid tumors. Clin Cancer Res 5(11):3394–3402
Glushkova A, Refsnes MA, Låg M, Khlebnikova N (2016) Risk assessment and regulations of nanomaterials in Norway and Russia
Gradishar WJ, Tjulandin S, Davidson N, Shaw H, Desai N, Bhar P et al (2005) Phase III trial of nanoparticle albumin-bound paclitaxel compared with polyethylated castor oil-based paclitaxel in women with breast cancer. J Clin Oncol 23(31):7794–7803
Grauer O, Jaber M, Hess K, Weckesser M, Schwindt W, Maring S et al (2019) Combined intracavitary thermotherapy with iron oxide nanoparticles and radiotherapy as local treatment modality in recurrent glioblastoma patients. J Neuro-Oncol 141(1):83–94
Harford AJ, Edwards JW, Priestly BG, Wright PFA (2007) Current OHS best practices for the Australian nanotechnology industry: a position paper by the NanoSafe Australia Network. J Occup Health Saf Aust NZ 23:315–331
Hariyadi DM, Athiyah U, Pathak YV (2020) Nanomedicine: risk, safety, regulation, and public health. In: Das MK, Pathak YV (eds) Nano medicine and nano safety: recent trends and clinical evidences. Springer, Singapore, pp 561–578
Hodge GA, Bowman D, Ludlow K (2009) New global frontiers in regulation: the age of nanotechnology. Edward Elgar Publishing
Hoff DDV, Ervin T, Arena FP, Chiorean EG, Infante J, Moore M et al (2013) Increased survival in pancreatic cancer with nab-paclitaxel plus gemcitabine. N Engl J Med 369(18):1691–1703
Hong M, Zhu S, Jiang Y, Tang G, Pei Y (2009) Efficient tumor targeting of hydroxycamptothecin loaded PEGylated niosomes modified with transferrin. J Control Release 133(2):96–102
Housman G, Byler S, Heerboth S, Lapinska K, Longacre M, Snyder N et al (2014) Drug resistance in cancer: an overview. Cancer 6(3):1769–1792

Hu X, Wang B, Zhang J, Wang Z, Sun T, Wang S et al (2020) 282MO Abraxane plus cisplatin compared with gemcitabine plus cisplatin as first-line treatment in patients with metastatic triple-negative breast cancer (GAP): a multicenter, randomized, open-label, phase III trial. Ann Oncol 31:S354

Kim DW, Kim SY, Kim HK, Kim SW, Shin SW, Kim JS et al (2007) Multicenter phase II trial of Genexol-PM, a novel Cremophor-free, polymeric micelle formulation of paclitaxel, with cisplatin in patients with advanced non-small-cell lung cancer. Ann Oncol 18(12):2009–2014

Kim Y, Lee S, Kim Y, Kim B, Kang S (2012) Multi-institutional phase I clinical trial of Cremophor-free, polymeric micelle formulation of paclitaxel (Genexol-PM) with carboplatin for patients with advanced epithelial ovarian cancer. Gynecol Oncol 125:S32

Lee J, Choi MK, Song IS (2023) Recent advances in doxorubicin formulation to enhance pharmacokinetics and tumor targeting. Pharmaceuticals (Basel) 16(6):802

Li M, Zhao G, Su W-K, Shuai Q (2020) Enzyme-responsive nanoparticles for anti-tumor drug delivery. Front Chem 8:647

Liu L, Ye Q, Lu M, Chen S-T, Tseng H-W, Lo Y-C et al (2017) A new approach to deliver anti-cancer nanodrugs with reduced off-target toxicities and improved efficiency by temporarily blunting the reticuloendothelial system with intralipid. Sci Rep 7(1):16106

Lu H, Zha S, Zhang W, Wang Q, Jiang D, Xu X et al (2021) A systematic review and meta-analysis of nab-paclitaxel mono-chemotherapy for metastatic breast cancer. BMC Cancer 21(1):830

Ludlow K, Bowman D, Hodge G (2007) A review of possible impacts of nanotechnology on Australia's regulatory framework. Report to Commonwealth Dept of Industry, Tourism and Resources

Maeda H, Nakamura H, Fang J (2013) The EPR effect for macromolecular drug delivery to solid tumors: Improvement of tumor uptake, lowering of systemic toxicity, and distinct tumor imaging in vivo. Adv Drug Deliv Rev 65(1):71–79

Maier-Hauff K, Ulrich F, Nestler D, Niehoff H, Wust P, Thiesen B et al (2011) Efficacy and safety of intratumoral thermotherapy using magnetic iron-oxide nanoparticles combined with external beam radiotherapy on patients with recurrent glioblastoma multiforme. J Neuro-Oncol 103(2): 317–324

Mitchell MJ, Billingsley MM, Haley RM, Wechsler ME, Peppas NA, Langer R (2021) Engineering precision nanoparticles for drug delivery. Nat Rev Drug Discov 20(2):101–124

Moeremans K, Annemans L, Morris J (2004) Liposomal cytarabine (DepoCyte™Injection) is cost-effective compared with standard cytarabine for the intrathecal treatment of patients with Lymphomatous meningitis. Blood 104(11):267

Muthu MS, Wilson B (2012) Challenges posed by the scale-up of nanomedicines. Taylor & Francis, pp 307–309

Nam SH, Lee S-W, Lee Y-J, Kim YM (2023) Safety and tolerability of weekly Genexol-PM, a Cremophor-free polymeric micelle formulation of paclitaxel, with carboplatin in gynecologic cancer: a phase I STUDY. Cancer Res Treat 55(4):1346–1354

Nasu H, Faunce T (2013) Nanotechnology in Japan: a route to energy security after Fukushima? Bull At Sci 69(5):68–74

National Cancer Institute (2025a) Clinical trials using pegylated liposomal doxorubicin hydrochloride. National Institutes of Health

National Cancer Institute (2025b) Clinical trials using nab-paclitaxel. National Institutes of Health

Northfelt DW, Dezube BJ, Thommes JA, Miller BJ, Fischl MA, Friedman-Kien A et al (1998) Pegylated-liposomal doxorubicin versus doxorubicin, bleomycin, and vincristine in the treatment of AIDS-related Kaposi's sarcoma: results of a randomized phase III clinical trial. J Clin Oncol 16(7):2445–2451

Oberdörster G, Oberdörster E, Oberdörster J (2005) Nanotoxicology: an emerging discipline evolving from studies of ultrafine particles. Environ Health Perspect 113(7):823–839

OECD (2013) Regulatory frameworks for nanotechnology in foods and medical products: summary results of a survey activity. OECD Publishing

OECD (2017) Test No. 318: Dispersion stability of nanomaterials in simulated environmental media, OECD guidelines for the testing of chemicals, section 3. OECD Publishing, Paris

Park IH, Sohn JH, Kim SB, Lee KS, Chung JS, Lee SH et al (2017) An open-label, randomized, parallel, phase III trial evaluating the efficacy and safety of polymeric micelle-formulated paclitaxel compared to conventional Cremophor EL-based paclitaxel for recurrent or metastatic HER2-negative breast cancer. Cancer Res Treat 49(3):569–577

Park J, Lee Y-K, Park I-K, Hwang SR (2021) Current limitations and recent progress in nanomedicine for clinically available photodynamic therapy. Biomedicine 9(1):85

Peer D, Karp JM, Hong S, Farokhzad OC, Margalit R, Langer R (2020) Nanocarriers as an emerging platform for cancer therapy. In: Nano-enabled medical applications. Jenny Stanford Publishing, pp 61–91

Petros RA, DeSimone JM (2010) Strategies in the design of nanoparticles for therapeutic applications. Nat Rev Drug Discov 9(8):615–627

Pita R, Ehmann F, Papaluca M (2016) Nanomedicines in the EU—regulatory overview. AAPS J 18: 1576–1582

Priestly BG, Harford AJ, Sim MR (2007) Nanotechnology: a promising new technology—but how safe? Med J Aust 186(4):187–188

Rashidi N, Davidson M, Apostolopoulos V, Nurgali K (2024) Nanoparticles in cancer diagnosis and treatment: Progress, challenges, and opportunities. J Drug Deliv Sci Technol 95:105599

Rauscher H, Rasmussen K, Sokull-Klüttgen B (2017) Regulatory aspects of nanomaterials in the EU. Chem Ing Tech 89(3):224–231

Ridge S (2018) A regulatory framework for nanotechnology. Homeland Security Affairs

Rodríguez F, Caruana P, De la Fuente N, Español P, Gámez M, Balart J et al (2022) Nano-based approved pharmaceuticals for cancer treatment: present and future challenges. Biomolecules 12(6):784

Savolainen K (2013) Nanosafety in Europe 2015-2025: towards safe and sustainable nanomaterials and nanotechnology innovations. Finnish Institute of Occupational Health

Torchilin V (2011) Tumor delivery of macromolecular drugs based on the EPR effect. Adv Drug Deliv Rev 63(3):131–135

Tyner K (2019) Regulatory research supporting the development of drug products containing nanomaterials: a US-FDA perspective. Private Communication at Global Summit on Regulatory Science 2019 on Nanotechnology and Nanoplastics

Udhrain A, Skubitz KM, Northfelt DW (2007) Pegylated liposomal doxorubicin in the treatment of AIDS-related Kaposi's sarcoma. Int J Nanomedicine 2(3):345–352

Vencken SF, Greene CM (2018) A review of the regulatory framework for nanomedicines in the European Union. In: Inorganic frameworks as smart nanomedicines. William Andrew, pp 641–679

Wang B, Hu S, Teng Y, Chen J, Wang H, Xu Y et al (2024a) Current advance of nanotechnology in diagnosis and treatment for malignant tumors. Signal Transduct Target Ther 9(1):200

Wang G, Zhang M, Lai W, Gao Y, Liao S, Ning Q et al (2024b) Tumor microenvironment responsive RNA drug delivery systems: intelligent platforms for sophisticated release. Mol Pharm 21(9):4217–4237

Werner ME, Cummings ND, Sethi M, Wang EC, Sukumar R, Moore DT et al (2013) Preclinical evaluation of Genexol-PM, a nanoparticle formulation of paclitaxel, as a novel radiosensitizer for the treatment of non-small cell lung cancer. Int J Radiat Oncol Biol Phys 86(3):463–468

Wicki A, Witzigmann D, Balasubramanian V, Huwyler J (2015) Nanomedicine in cancer therapy: challenges, opportunities, and clinical applications. J Control Release 200:138–157

Wilhelm S, Tavares AJ, Dai Q, Ohta S, Audet J, Dvorak HF et al (2016) Analysis of nanoparticle delivery to tumours. Nat Rev Mater 1(5):1–12

Wu J (2021) The enhanced permeability and retention (EPR) effect: the significance of the concept and methods to enhance its application. J Pers Med 11(8):771

Yuan L, Chen Q, Riviere JE, Lin Z (2023) Pharmacokinetics and tumor delivery of nanoparticles. J Drug Deliv Sci Technol 83:104404

Zhang M, Gao S, Yang D, Fang Y, Lin X, Jin X et al (2021) Influencing factors and strategies of enhancing nanoparticles into tumors in vivo. Acta Pharm Sin B 11(8):2265–2285

Near-Infrared Light-Mediated Photodynamic Cancer Therapy

8

Hemant and Nancy Sharma

Abstract

Photodynamic treatment (PDT), a new approach, uses photosensitizers that release reactive oxygen species (ROS) when they are triggered by near-infrared (NIR) light. This minimally invasive technique reduces off-target toxicity while allowing for the targeted destruction of cancerous tissues. Deep-tissue penetration, tumor selectivity, and photosensitizer stability have all improved with recent developments in nanotechnology. Furthermore, PDT-induced immunogenic cell death (ICD) improves systemic anticancer immunity and may work in concert with chemotherapy and immunotherapy. AI-driven treatment planning and advancements in light delivery, such as activation mediated by X-rays and ultrasounds, are increasing the therapeutic uses of photodynamic therapy. With the aim of improving treatment efficacy, NIR-mediated photodynamic therapy (PDT) combines photophysics, nanomedicine, and immunotherapy.

Keywords

Photodynamic therapy · Photosensitizer · Illumination · Tumor · Reactive oxygen species

Hemant
Institute of Nano Science and Technology, Sector-81, Mohali, Punjab, India

Academy of Scientific and Innovative Research (AcSIR), Ghaziabad, India

N. Sharma (✉)
Institut Parisien de Chimie Moléculaire (IPCM), CNRS, Sorbonne Université, Paris, France

R. Prasad, M. Mohan Yallapu (eds.), *Advancements in Cancer Theranostics*,
https://doi.org/10.1007/978-981-95-0938-6_8

8.1 Introduction

Photodynamic therapy (PDT) is a cutting-edge, minimally invasive therapeutic technique that combines physics, biology, and chemistry concepts to target and kill diseased cells specifically (Zheng et al. 2022; Pratavieira et al. 2022; Zhu and Finlay 2008; Gunaydin et al. 2021a; Hu et al. 2022). PDT has become well known as a very successful therapy method for a number of illnesses, such as microbial infections (Mackay 2022; Kim et al. 2023; Akhtar et al. 2024), dermatological diseases (Xu et al. 2023; Dey et al. 2022), and cancers (Macchi et al. 2021; Karges 2022; Chen et al. 2021a; Nasseri et al. 2022; Zhao et al. 2024; Li et al. 2021). The basic process of photodynamic therapy involves giving a photosensitizing medication that preferentially builds up in ill tissues and then exposure to light with a specific wavelength. The photosensitizer changes into an excited state when light is activated, producing harmful reactive oxygen species (ROS) such as hydroxyl radicals and singlet oxygen (O_2). Through oxidative damage to cellular constituents, these ROS eventually cause cell death through necrosis, autophagy, or apoptosis (Vafadar et al. 2024; Zhang et al. 2023; Esmaeili et al. 2022).

Because the cytotoxic effects of PDT are limited to the illuminated area (Xie et al. 2021), it has the important advantage of spatial selectivity, which lowers systemic toxicity while maintaining healthy surrounding tissues. Compared to traditional treatment techniques like chemotherapy and radiotherapy, which frequently cause severe collateral damage, this tailored approach offers major advantages. PDT is a recommended choice for individuals with recurring or refractory illnesses because of its non-invasive nature, which also permits repeated treatments with few side effects.

Numerous studies on photosensitizer pharmacokinetics (Tabrizi et al. 2024; Li et al. 2022a; Luo et al. 2021), light delivery devices (Chen et al. 2022; Cui et al. 2023), and molecular mechanisms underpinning PDT-induced cytotoxicity have fueled the development of PDT as a therapeutic intervention. In order to ensure deeper tissue penetration and increased efficacy, modern photosensitizers have been designed to show improved tumor selectivity, extended retention time, and superior absorption qualities within the therapeutic window (600–800 nm) (Chen et al. 2021b; Li et al. 2022b). At the same time, developments in light-emitting technology, such lasers and light-emitting diodes (LEDs), have made it possible to precisely adjust treatment parameters, which has improved therapy results even more. Additionally, PDT's antibacterial qualities have been used to fight fungal infections, biofilm-associated illnesses, and bacterial infections that are resistant to many drugs, providing a good substitute for traditional antimicrobial treatments (Sikder et al. 2021).

8.2 Mechanism of PDT

Photodynamic Therapy (PDT) operates through a highly specific and localized mechanism that involves three essential elements: (a) photosensitizer, (b) light source, and (c) molecular oxygen. The success of PDT is dependent on the selective uptake of the photosensitizer by target cells, the appropriate wavelength of light for activation, and the presence of sufficient oxygen to generate cytotoxic reactive oxygen species (ROS). Upon administration, the photosensitizer preferentially accumulates in diseased tissues due to its unique affinity for hyperproliferative cells, abnormal vasculature, or altered metabolic pathways within tumors or pathological sites. Once localized, the photosensitizer remains in a non-toxic ground state until activated by an external light source. The light used in PDT typically falls within the therapeutic window of 600–800 nm, allowing for optimal tissue penetration while minimizing absorption by endogenous chromophores such as hemoglobin (Jiang et al. 2019) and melanin (Udrea et al. 2023). Upon illumination, the photosensitizer transitions from its ground state to an excited singlet state, which can then undergo intersystem crossing to a longer-lived triplet state. This highly reactive triplet-state photosensitizer interacts with molecular oxygen in the surrounding environment through two primary photochemical pathways. In type I reactions, the photosensitizer transfers an electron to cellular biomolecules, producing free radicals and superoxide anions that induce oxidative damage. In Type II reactions, energy transfer from the excited photosensitizer directly generates singlet oxygen ($^1O_2$), a highly reactive ROS responsible for inducing cellular damage (Teng et al. 2022) (Fig. 8.1).

The ROS generated during PDT interacts with vital cellular components such as lipids and nucleic acids, including phospholipids in cell membranes, proteins, and amino acid residues in enzymes, leading to oxidative stress and structural damage. This results in various forms of cell death, including apoptosis (programmed cell death), necrosis (unregulated cell lysis), and autophagy (self-digestion of cellular components). Apoptosis is the predominant mechanism in PDT and is initiated through mitochondrial dysfunction, activation of caspase pathways, and disruption of intracellular signaling networks (Zhuang et al. 2023). In contrast, necrosis occurs

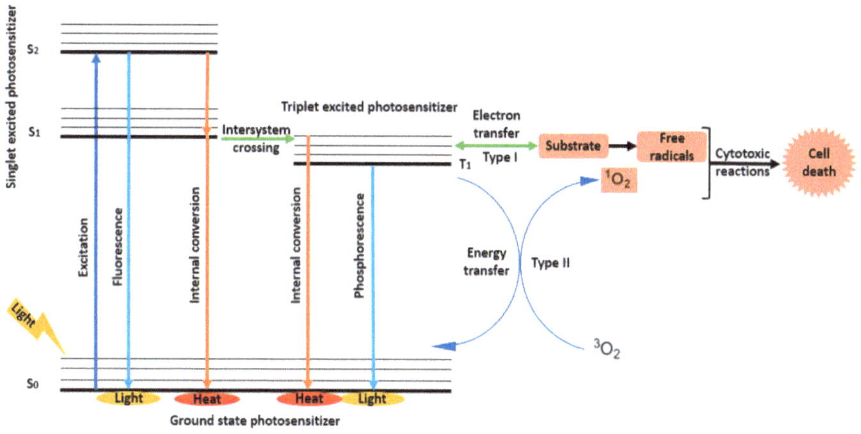

Fig. 8.1 Schematic illustration showing the mechanism of PDT. Adapted with permission from reference Niculescu and Grumezescu (2021). Copyright Creative Commons Licenses CC by 4.0, MDPI

when oxidative damage overwhelms cellular repair mechanisms, leading to membrane rupture and inflammatory responses. The effectiveness of PDT depends on several factors, including photosensitizer pharmacokinetics, tissue oxygenation, light dose, and irradiation timing (Zhao et al. 2024). Advances in nanotechnology and targeted delivery strategies have improved photosensitizer selectivity and bioavailability, enhanced therapeutic efficacy while minimized off-target effects. Overall, the unique mechanism of PDT enables precise and localized treatment, making it a promising modality for oncology, dermatology, and antimicrobial applications

8.3 Materials-Based Photodynamic Therapy

The transition from heliotherapy to modern photodynamic therapy (PDT) marked significant scientific advancements. Early twentieth-century discoveries showed that certain dyes could render bacteria light-sensitive, revealing the potential of combining light with chemical treatments. First-generation photosensitizers (PSs) like Hematoporphyrin derivatives (HpD) and Photofrin demonstrated photo-induced cytotoxicity for cancer treatment but had drawbacks such as poor tumor selectivity, prolonged skin photosensitivity, and low water solubility (Gunaydin et al. 2021a, b).

To overcome these limitations, second-generation PSs, including porphyrazines, phthalocyanines, and chlorins, were developed with improved tumor selectivity, enhanced singlet oxygen production, and better absorption in the therapeutic window (600–800 nm). Some, like 5-aminolevulinic acid (5-ALA), act as prodrugs, converting into active photosensitizers within malignant cells for greater specificity (Abrahamse and Hamblin 2016). Recent advancements have led to third-generation PSs, which integrate nanotechnology and targeted drug delivery, improving selectivity, biocompatibility, and photodynamic efficiency by coupling PSs with proteins, nanoparticles, or antibodies.

8.3.1 Photosensitizer-Based PDT

Photosensitizers (PS), in conjunction with reactive oxygen species and a suitable light source, are a key part of the photodynamic treatment (PDT) mechanism. PS are materials that absorb particular light wavelengths and cause photophysical or photochemical reactions.

Recently, Xu and his co-workers have designed a structurally improved photosensitizer CySG-2 based on IR-780 dye. Depending on the enzymatic identification of MRSA's β-lactamase TEM-1, CySG-2 has demonstrated the ability to enhance the clearance of internalized MRSA by promoting the autophagy pathway, which acts synergistically with its capability to generate reactive oxygen species (ROS) under near-infrared (NIR) laser irradiation during photodynamic therapy (PDT) (Xu et al. 2022).

Gourlot and his team successfully achieved antimicrobial photodynamic therapy (a-PDT) within the optical therapeutic window by engineering a photosensitizer with

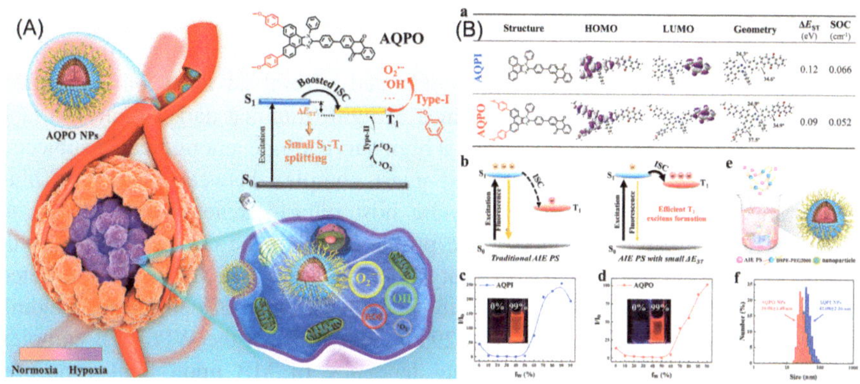

Fig. 8.2 (**A**) Schematic representation of in vitro and in vivo strategies for hypoxia-resilient cancer therapy. (**B**) (**a**) Molecular structures, frontier molecular orbital distributions, optimized geometries, and intramolecular torsional angles of AQPI and AQPO, as computed using DFT. (**b**) Schematic depiction of the triplet exciton generation pathways in conventional AIE PSs versus those in AIE PSs engineered with a minimized singlet–triplet energy gap upon photoexcitation. (**c**, **d**) Plots showing the relative PL intensity (I/I_0) of AQPI and AQPO, respectively, in THF/water solvent mixtures, reflecting their aggregation behavior. (**e**) Schematic illustration of the nanoprecipitation method used for the fabrication of nanoparticles (NPs). (**f**) Dynamic light scattering analysis depicting the size distribution profiles of AQPI and AQPO nanoparticles dispersed in deionized (DI) water. Reprinted (adapted) with permission from reference Xiao et al. (2022). Copyright {2022} American Chemical Society

an extended π-conjugation system. The peptide linkage is a viable candidate for PDT because it imparts amphiphilic property without affecting singlet oxygen production. Upon irradiation at 720 nm for 15 min, the system exhibited potent photobactericidal activity against both Gram-positive Staphylococcus aureus and Gram-negative Escherichia coli. Even at a low photosensitizer concentration of 1 μM, a significant reduction in bacterial viability was observed, corresponding to a 3 $\log_{10}$ decrease in colony-forming units (CFUs) (Gourlot et al. 2022).

Two AIE-active phenanthro[9,10-d]imidazole derivatives (AQPO and AQPI) with minimal ΔEST values were developed and synthesized by Xiao et al. as highly effective PSs. They also increased the radical production efficiency of AQPO by adding electron-rich anisole for hypoxia-overcoming PDT. The authors demonstrated that the derivatives achieved a threefold increase in ROS generation (Fig. 8.2) (Xiao et al. 2022).

By adding various chain lengths of tethered morpholine groups to the structure, Mai et al. enhanced a number of the photophysical characteristics of heptamethine cyanine IR780, an NIR absorbing photosensitizer. The photostability of chemically modified IR780 derivatives was found to be significantly enhanced compared to the unmodified IR780 molecule. This improvement was attributed to structural modifications involving the direct substitution of the meso-chlorine atom with a nitrogen atom, which disrupted the pull–push π-conjugation system (Mai et al. 2020). The proposed strategy significantly enhances the therapeutic potential of IR780 by improving its photostability, enabling dual-targeting capabilities, extending near-infrared (NIR) fluorescence imaging duration, and optimizing its performance in NIR-mediated photodynamic therapy (PDT). Furthermore, both in vitro and in vivo evaluations demonstrated several advantageous attributes, including low cytotoxicity in the absence of light (minimal dark toxicity), a lack of observable adverse side effects, and a high survival rate among treated subjects.

Shangguan et al. developed a host–guest complex between methylene blue (MB) and the water-soluble macrocyclic host pillar[6]arene (WP6), aiming to enhance the photodynamic therapeutic profile of MB. The complexation significantly mitigated the inherent dark toxicity associated with free MB, as demonstrated through cytotoxicity assays. In vitro studies using a range of cell lines, including MCF-7, HeLa, HepG2, HL7702, and 293 T, revealed that the WP6–MB complex effectively reduced the photobleaching tendency of MB upon light exposure, prolonging the time for ROS formation (Shangguan et al. 2017).

Desgranges et al. synthesized amphiphilic derivatives of protoporphyrin IX that exhibit improved properties, including enhanced water solubility and potent photodynamic activity. In a separate study, the Faustino group developed modified derivatives of TMPyP, which demonstrated superior singlet oxygen generation compared to the parent compound. These derivatives also exhibited pronounced phototoxic effects by inducing autophagy-mediated cell death in cancer cells, highlighting their potential as effective agents for photodynamic (Fig. 8.3) (Desgranges et al. 2022).

8.3.2 Nanoparticle-Based PDT

Nanoparticles enhance NIR-mediated PDT by improving PS delivery, stability, and activation efficiency. Various nanoparticles have been developed to optimize PDT outcomes.

P. Jia et al. have developed IR808 dye-sensitized, glutathione-clad Au–Bi bimetallic nanoparticles (Au–Bi-GSH@IR808) for enhanced cancer therapy. These nanoparticles exhibit strong near-infrared (808 nm) absorption, high photothermal conversion efficiency (34.2%), and effective reactive oxygen species (ROS) generation, enabling combined photothermal and photodynamic therapy. In vitro and in vivo experiments demonstrated significant tumor cell ablation. Additionally, the high atomic number of bismuth provides excellent contrast for computed tomography imaging, suggesting potential clinical applications for precise cancer diagnosis and treatment (Jia et al. 2021).

Mangalath et al. developed BODIPY-graphene quantum dot nanoconjugates (GQD-BDPA) that exhibit enhanced photophysical properties, including high triplet excited state populations and superior singlet oxygen quantum yields, alongside excellent water solubility. The GQD-BDPA conjugates achieved a singlet oxygen quantum yield of 90%, exceeding the 85% yield recorded for the free BDPA molecule. Additionally, a redshift in the absorption maximum from 530 nm to 550 nm was observed, coupled with an increase in triplet-state lifetime from 0.49 μs (free BDPA) to 0.55 μs (GQD-BDPA), indicating improved photodynamic

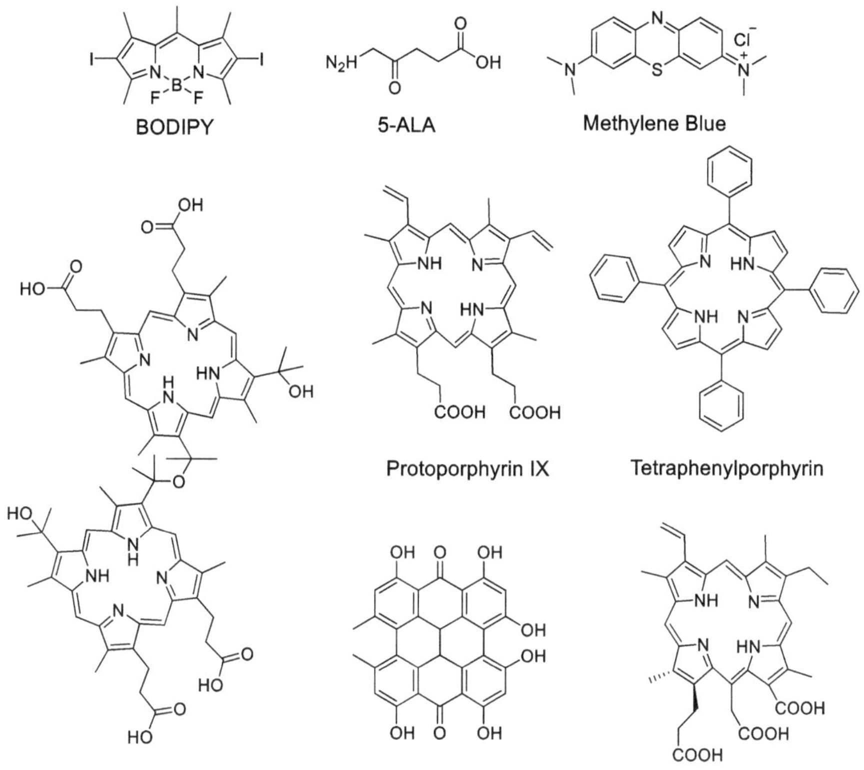

Fig. 8.3 Molecular structures of some classical photosensitizers

efficiency. Detailed structural and spectroscopic analyses confirmed the successful conjugation of BDPA to graphene quantum dots. These enhancements in photophysical performance position GQD-BDPA as a promising candidate for advanced photodynamic therapy (PDT) applications (Mangalath et al. 2021).

Neelakandan et al. present a supramolecular nanocomposite comprising BODIPY, tryptophan, and gold nanoparticles. This heavy-atom-free photosensitizer demonstrates high photostability and efficient singlet oxygen generation. The incorporation of gold nanoparticles induces an external heavy atom effect, leading to enhanced singlet oxygen production (46%). Nanocomposites significantly increased ROS levels in C6 glioma cells, leading to strong phototoxicity (~25% cell viability) under light but minimal toxicity (~80% viability) in the dark. The presence of BODIPY reduced dark cytotoxicity compared to tryptophan-capped AuNPs, likely due to lower ligand toxicity and nanoparticle size differences (Kumar et al. 2019).

8.3.3 Liposome-Based and Polymeric Nanoparticles

Lipid-based and polymeric nanoparticles play a crucial role in enhancing the solubility, bioavailability, and stability of photosensitizers (PSs), thereby improving their selective accumulation in tumor tissues. Nanotechnology-based delivery systems, including liposomes, polymeric micelles, and various nanoparticles, have been extensively utilized to enhance the tumor-specific localization of PSs while minimizing off-target effects, thereby reducing systemic toxicity (Truong et al. 2024). Moreover, the integration of photodynamic therapy (PDT) with immunotherapy has emerged as a promising strategy to trigger systemic antitumor immune responses. PDT-induced immunogenic cell death (ICD) leads to the release of damage-associated molecular patterns (DAMPs), which can activate antigen-presenting cells (APCs) and stimulate cytotoxic T lymphocytes, thereby enhancing immune-mediated tumor eradication (Birmpilis et al. 2022). Recent studies highlight the development of multifunctional nanosystems that combine PDT with immune checkpoint inhibitors, cytokine delivery, or tumor vaccines to enhance therapeutic outcomes. Such advancements in nanotechnology and combinatorial treatment strategies offer a promising avenue for improving PDT efficacy in cancer therapy. Table 8.1 presents different types of nanotechnology-based delivery systems commonly utilized for various photodynamic therapeutic agents.

The nanoparticle-based photodynamic therapy has been used in recent years for treating various cancers as described in the next section.

Table 8.1 Nanotechnology-based delivery systems are commonly used for nanotheranostics

Materials	Characteristics	Applications	References
Liposomes	Amphiphilic	Combining hydrophilic and hydrophobic medications using nanocarriers	Tian et al. (2022), Cheng et al. (2021)
Polymers	Modifiable and biocompatible	Drug intelligent response, functional coating	Yi et al. (2021), Elian et al. (2024)
Gold nanostructures	SPR effect and higher singlet oxygen generation efficiency	PA imaging, SERS, radiotherapy, fluorescence imaging	Vinita et al. (2023), Santhakumar et al. (2023)
Quantum dots	Superior stability and non-radiative decay	Delivery platform, PDT	Ding et al. (2021), Murali et al. (2022)
Graphene	Higher surface area and strong absorption of NIR light	NIR fluorescence probes, PA imaging, carriers	Lee and Kim (2021), Tamtaji et al. (2021)

8.4 Current Clinical Applications

Despite their frequent effectiveness, traditional cancer treatments including chemotherapy, radiation, and surgery have several drawbacks. Because of treatment resistance, metastasis, and patient recurrence, some tumors do not react well to these conventional approaches, making cancer therapy more difficult and complex. By making use of the special qualities of nanoparticles, PDT can target tumors more precisely, increase PS's durability and effectiveness, greatly boost therapeutic results, and lessen adverse effects on healthy tissues. We will now go over how nanoparticle-based PDT is used for the most common types of cancer.

Gastric Cancer: Due in great part to improvements in lifestyle and healthcare, the prevalence of stomach cancer has steadily decreased in recent decades. However, new research has shown that younger people are increasingly affected by stomach cancer. The global incidence of stomach cancer exceeded 968,000 new cases in 2022, leading to about 660,000 deaths, highlighting the disease's ongoing importance as a major global health concern.

IRCB@M, a new nanoparticle developed by Li et al., improves PDT via two different processes. For homologous targeting, these nanoparticles were wrapped in cancer cell membranes. They self-assembled to incorporate the photosensitizer IR780 and the glutaminase inhibitor CB-839. By inhibiting aerobic respiration and glutamine metabolism, NADPH and glutathione levels dropped, reducing hypoxia and enhancing the PDT impact that IR780 mediates. These nanoparticles significantly improved the anticancer effect on gastric cancer in both in vitro and in vivo tests, providing a new way to improve PDT and other ROS-dependent treatments (Li et al. 2024).

Situ et al. developed ACSN/Fe3O4@MSNs-iRGD, a targeted drug delivery platform that combines the active targeting peptide iRGD with new ACSN to increase PDT's effectiveness. The drug's favorable active targeting potential toward stomach cancer cells was revealed by experimental results, which facilitated effective mitochondrial penetration. By drastically reducing mitochondrial aerobic respiration, the PS produced by ACSN preserved oxygen while producing a significant amount of ROS. Additionally, by providing outstanding T2 dark contrast, the Fe_3O_4 core-shell structure made it possible to use magnetic resonance imaging for real-time tumor monitoring and its inhibitory effect on gastric cancer MGC-803 cells. With the ability to overcome the limitations of traditional cancer medicines, this pharmacological platform offers a thorough and effective approach to tumor identification and therapy by combining PDT and MRI imaging functions (Situ et al. 2024).

Breast Cancer: With almost 2.3 million new cases, or 11.6% of all cancer cases, female breast cancer is the second most frequent malignancy worldwide, according to recent data. Furthermore, with 666,000 deaths—or 6.9% of all cancer deaths—it ranks as the fourth most common cause of cancer-related deaths globally.

Montaseri et al. engineered spherical Janus nanoparticles (JNPs) composed of gold and polyacrylic acid (Au/PAA), designed for multifunctional photodynamic therapy (PDT). One hemisphere of the JNPs was loaded with 5-aminolevulinic acid (5-ALA), serving as the photosensitizer precursor, while the opposite hemisphere was functionalized with folic acid (FA) and thiol-terminated polyethylene glycol amine (SH-PEG-NH_2) to ensure enhanced biocompatibility and selective tumor targeting via folate receptor-mediated endocytosis. In vitro photodynamic assays demonstrated pronounced cytotoxic effects against MCF-7 breast cancer cells, indicating the potential of FA-Au/PAA-ALA JNPs as a targeted nanoplatform for PDT-based tumor ablation therapy (Montaseri and Abrahamse 2024).

Wu et al. designed a multifunctional biomimetic nanoplatform named 4T1Mem@PGA-Ce6/Ola (MPCO) for the co-delivery of chlorin e6 (Ce6) and Olaparib (Ola), aiming to overcome the limitations of photodynamic therapy (PDT) associated with tumor cell repair mechanisms and immunosuppressive tumor microenvironmnents (Wu et al. 2023). Under laser irradiation, our nanoplatform produced notable synergistic anticancer effects by effectively targeting tumors and internalizing cells through cell membrane camouflage. Furthermore, MPCO was shown to activate the cGAS–STING signaling pathway, leading to the production of pro-inflammatory cytokines and the initiation of a systemic antitumor immune response through the induction of immunogenic cell death and the maturation of dendritic cells. This research highlights the therapeutic potential of multifunctional nanoplatforms that synergistically combine photodynamic therapy, chemotherapy, and immunotherapy, offering a comprehensive approach for effective cancer treatment.

Lung Cancer: PDT has been used to treat lung cancer patients with central airway obstructive lesions. To improve respiratory function and reduce airway congestion, PDT targets specific tumor areas inside the bronchial tree. Recent studies have concentrated on improving light delivery methods and investigating the potential benefits of integrating PDT with other treatment modalities, including immune checkpoint inhibitors and targeted medicines.

Güleryüz et al. investigated the synergistic therapeutic effects of co-delivering IR780 and curcumin within mesoporous silica nanoparticles (MSNs) for targeted treatment of lung cancer cells. The engineered nanocarrier system, denoted as Cur&IR780@MSN, integrates near-infrared (NIR)-triggered photodynamic therapy with the chemotherapeutic properties of curcumin. Compared to the individual treatments, this dual-modal approach significantly enhanced anticancer efficacy, resulting in up to a 70% reduction in lung cancer cell viability, demonstrating the potential of combinatorial nanoplatforms for improved therapeutic outcomes (Güleryüz et al. 2024).

Liao et al. used a straightforward and environmentally friendly self-assembly approach to create new carrier-free NPs between theophylline derivative compounds D17 and Ce6, which had an average hydrated particle size of 85 nm and a surface charge of −33 mV. In order to provide anticancer effects, this nanomedicine may be efficiently absorbed by NSCLC tumor cells and shows good stability in vitro. Thus, in preclinical research, it might be a promising candidate for chemo-photodynamic synergistic antitumor therapy, opening a new path for the creation of combination antitumor therapy (Liao et al. 2024). (Fig. 8.4).

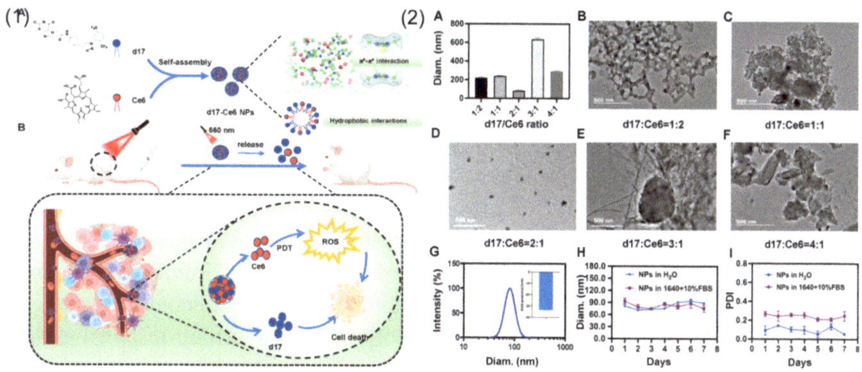

Fig. 8.4 (1) Schematic illustration of the d17–Ce6 NPs; (**a**) The d17–Ce6 NPs self-assembled through π–π stacking and hydrogen bond interactions between d17 and Ce6; (**b**) d17–Ce6 NPs were targeted to tumor tissues by PDT and chemotherapy. (2) Characterization of d17–Ce6 NPs. (A) Particle size. (**b**–**f**) TEM images. (**g**) Hydrophilic particle size and zeta (**h**) variation in hydrophilic particle size of d17–Ce6 (2:1) NPs in H_2O and 1640 with 10% FBS. (**i**) Differences in PDI of d17–Ce6 (2:1) NPs in H_2O and 1640 with 10% FBS. Reprinted (adapted) with permission from reference Liao et al. (2024). Copyright {2024} American Chemical Society

8.5 Future Directions

8.5.1 Development of Advanced Photosensitizers

Next-generation photosensitizers are being designed to enhance PDT's therapeutic potential by improving light absorption, singlet oxygen generation, and biocompatibility while minimizing toxicity and side effects. Strategies for optimizing photosensitizers include:

- **Red-shifted absorption:** Developing photosensitizers with longer-wavelength absorption (near-infrared, 700–800 nm) to allow deeper tissue penetration.
- **Targeted delivery:** Conjugating photosensitizers with nanoparticles, antibodies, or peptides to enhance tumor specificity and reduce off-target effects.
- **Biodegradability:** Engineering photosensitizers that are rapidly cleared from the body to minimize long-term toxicity.

8.5.2 Integration of PDT with Emerging Technologies

The combination of PDT with cutting-edge biomedical technologies is expected to broaden its therapeutic scope and address existing limitations. Key integration strategies include:

- **Photopharmacology**: The development of light-activated drugs that offer spatial and temporal control over therapeutic effects, reducing systemic toxicity and enhancing tumor selectivity.
- **Gene Editing and PDT**: CRISPR-Cas9 and other gene-editing technologies can be utilized to enhance PDT responsiveness by modifying genes involved in oxidative stress resistance, apoptosis, or hypoxia adaptation in tumors.
- **Synergistic Therapies**: PDT can be combined with immunotherapy, chemotherapy, and radiotherapy to improve overall treatment efficacy. PDT-induced immunogenic cell death (ICD) can enhance the efficacy of immune checkpoint inhibitors, making it a promising combination approach in oncology.
- **Smart Drug Delivery Systems**: The incorporation of stimuli-responsive drug carriers, such as pH-sensitive or enzyme-activated nanoparticles, can improve the controlled release of photosensitizers and co-delivered therapeutic agents, optimizing PDT effectiveness.

8.6 Challenges and Considerations

Despite its promising therapeutic potential, photodynamic therapy (PDT) faces several challenges that hinder its widespread clinical application. One major limitation is the restricted tissue penetration of light, particularly in deep-seated tumors, as most photosensitizers absorb in the visible range (400–700 nm), limiting PDT efficacy to superficial or endoscopically accessible lesions. The variability in photosensitizer uptake and distribution across different tumor types and patients further complicates treatment outcomes, necessitating the development of more tumor-selective and bioavailable photosensitizers. Additionally, precise light delivery systems are essential to ensure uniform illumination and optimal dosimetry, especially for irregularly shaped or deeply located tumors. Overcoming these challenges requires advancements in nanotechnology, targeted drug delivery, and innovative light sources such as upconversion nanoparticles and two-photon excitation strategies. Further research into combining PDT with other therapeutic modalities, such as immunotherapy and gene editing, may enhance its efficacy and broaden its clinical applications.

8.7 Conclusion

Photodynamic therapy (PDT) has emerged as a powerful and minimally invasive treatment modality, evolving from ancient light-based healing practices into a precise, targeted therapeutic approach. Its unique mechanism, combining photosensitizers, light activation, and oxygen-dependent cytotoxicity, has made it a promising option for treating cancers, infections, and other diseases. Despite challenges such as limited tissue penetration and variability in treatment response, continuous advancements in nanotechnology, photosensitizer design, and combination therapies are expanding PDT's clinical applications. With the integration of personalized medicine, artificial intelligence, and emerging biomedical technologies, PDT holds immense potential to revolutionize modern healthcare. As research continues to refine its efficacy and accessibility, PDT is set to become a cornerstone of non-invasive, targeted therapy, offering new hope for patients worldwide.

Acknowledgments Hemant (201920-191620089158) thanks the University Grants Commission and INST Mohali, respectively, for research fellowships and facilities.

Conflict of Interest The authors declare no conflict of interest.

References

Abrahamse H, Hamblin MR (2016) New photosensitizers for photodynamic therapy. Biochem J 473(4):347–364. https://doi.org/10.1042/BJ20150942

Akhtar F, Misba L, Khan AU (2024) The dual role of photodynamic therapy to treat cancer and microbial infection. Drug Discov Today 29(8):104099. https://doi.org/10.1016/j.drudis.2024.104099

Birmpilis AI, Paschalis A, Mourkakis A, Christodoulou P, Kostopoulos IV, Antimissari E, Terzoudi G, Georgakilas AG, Armpilia C, Papageorgis P, Kastritis E, Terpos E, Dimopoulos MA, Kalbacher H, Livaniou E, Christodoulou M-I, Tsitsilonis OE (2022) Immunogenic cell death, DAMPs and prothymosin α as a putative anticancer immune response biomarker. Cells 11(9):1415. https://doi.org/10.3390/cells11091415

Chen D, Long Z, Zhong C, Chen L, Dang Y, Hu J-J, Lou X, Xia F (2021a) Highly efficient near-infrared photosensitizers with aggregation-induced emission characteristics: rational molecular design and photodynamic cancer cell ablation. ACS Appl Bio Mater 4(6):5231–5239. https://doi.org/10.1021/acsabm.1c00398

Chen H, Shi T, Wang Y, Liu Z, Liu F, Zhang H, Wang X, Miao Z, Liu B, Wan M, Mao C, Wei J (2021b) Deep penetration of nanolevel drugs and micrometer-level T cells promoted by nanomotors for cancer immunochemotherapy. J Am Chem Soc 143(31):12025–12037. https://doi.org/10.1021/jacs.1c03071

Chen H, Sun T, Zeng W, Zeng X, Mei L, Jiang C, Zhao Y (2022) NIR-light-intensified hypoxic microenvironment for cascaded supra-prodrug activation and synergistic chemo/photodynamic cancer therapy. ACS Mater Lett 4(1):111–119. https://doi.org/10.1021/acsmaterialslett.1c00677

Cheng X, Gao J, Ding Y, Lu Y, Wei Q, Cui D, Fan J, Li X, Zhu E, Lu Y, Wu Q, Li L, Huang W (2021) Multi-functional liposome: a powerful theranostic nano-platform enhancing photodynamic therapy. Adv Sci 8(16):2100876. https://doi.org/10.1002/advs.202100876

Cui X, Li X, Peng C, Qiu Y, Shi Y, Liu Y, Fei J-F (2023) Beyond external light: on-spot light generation or light delivery for highly penetrated photodynamic therapy. ACS Nano 17(21):20776–20803. https://doi.org/10.1021/acsnano.3c05619

Desgranges S, Juzenas P, Vasovic V, Gederaas OA, Lindgren M, Warloe T, Peng Q, Contino-Pépin C (2022) Amphiphilic protoporphyrin IX derivatives as new photosensitizing agents for the improvement of photodynamic therapy. Biomedicine 10(2):423. https://doi.org/10.3390/biomedicines10020423

Dey A, Singhvi G, Puri A, Kesharwani P, Dubey SK (2022) An insight into photodynamic therapy towards treating major dermatological conditions. J Drug Deliv Sci Technol 76:103751. https://doi.org/10.1016/j.jddst.2022.103751

Ding S, Liu Z, Huang C, Zeng N, Jiang W, Li Q (2021) Novel engineered bacterium/black phosphorus quantum dot hybrid system for hypoxic tumor targeting and efficient photodynamic therapy. ACS Appl Mater Interfaces 13(8):10564–10573. https://doi.org/10.1021/acsami.0c20254

Elian C, Méallet R, Versace D-L (2024) Photoactive dye-loaded polymer materials: a new cutting edge for antibacterial photodynamic therapy. Adv Funct Mater 34(44):2407228. https://doi.org/10.1002/adfm.202407228

Esmaeili Y, Yarjanli Z, Pakniya F, Bidram E, Łos MJ, Eshraghi M, Klionsky DJ, Ghavami S, Zarrabi A (2022) Targeting autophagy, oxidative stress, and ER stress for neurodegenerative disease treatment. J Controlled Rel 345:147–175. https://doi.org/10.1016/j.jconrel.2022.03.001

Gourlot C, Gosset A, Glattard E, Aisenbrey C, Rangasamy S, Rabineau M, Ouk T-S, Sol V, Lavalle P, Gourlaouen C, Ventura B, Bechinger B, Heitz V (2022) Antibacterial photodynamic therapy in the near-infrared region with a targeting antimicrobial peptide connected to a π-extended porphyrin. ACS Infect Dis 8(8):1509–1520. https://doi.org/10.1021/acsinfecdis.2c00131

Güleryüz B, Işık A, Gülsoy M (2024) Synergistic effect of mesoporous silica nanocarrier-assisted photodynamic therapy and anticancer agent activity on lung cancer cells. Lasers Med Sci 39(1):91. https://doi.org/10.1007/s10103-023-03969-x

Gunaydin G, Gedik ME, Ayan S (2021a) Photodynamic therapy current limitations and novel approaches. Front Chem 9:9. https://doi.org/10.3389/fchem.2021.691697

Gunaydin G, Gedik ME, Ayan S (2021b) Photodynamic therapy for the treatment and diagnosis of cancer—a review of the current clinical status. Front Chem 9:9. https://doi.org/10.3389/fchem.2021.686303

Hu X, Zhang H, Wang Y, Shiu B-C, Lin J-H, Zhang S, Lou C-W, Li T-T (2022) Synergistic antibacterial strategy based on photodynamic therapy: progress and perspectives. Chem Eng J 450:138129. https://doi.org/10.1016/j.cej.2022.138129

Jia P, Ji H, Liu S, Zhang R, He F, Zhong L, Yang P (2021) Integration of IR-808 and thiol-capped au–bi bimetallic nanoparticles for NIR light mediated photothermal/photodynamic therapy and imaging. J Mater Chem B 9(1):101–111. https://doi.org/10.1039/D0TB02378G

Jiang L, Bai H, Liu L, Lv F, Ren X, Wang S (2019) Luminescent, oxygen-supplying, hemoglobin-linked conjugated polymer nanoparticles for photodynamic therapy. Angew Chem Int Ed 58(31):10660–10665. https://doi.org/10.1002/anie.201905884

Karges J (2022) Clinical development of metal complexes as photosensitizers for photodynamic therapy of cancer. Angew Chem Int Ed 61(5):e202112236. https://doi.org/10.1002/anie.202112236

Kim H, Lee YR, Jeong H, Lee J, Wu X, Li H, Yoon J (2023) Photodynamic and photothermal therapies for bacterial infection treatment. Smart Mol 1(1):e20220010. https://doi.org/10.1002/smo.20220010

Kumar PPP, Yadav P, Shanavas A, Thurakkal S, Joseph J, Neelakandan PP (2019) A three-component supramolecular nanocomposite as a heavy-atom-free photosensitizer. Chem Commun 55(39):5623–5626. https://doi.org/10.1039/C9CC02480H

Lee S, Kim SY (2021) Gold nanorod/reduced graphene oxide composite nanocarriers for near-infrared-induced cancer therapy and photoacoustic imaging. ACS Appl Nano Mater 4(11):11849–11860. https://doi.org/10.1021/acsanm.1c02419

Li K, Zhang Y, Hussain A, Weng Y, Huang Y (2021) Progress of photodynamic and RNAi combination therapy in cancer treatment. ACS Biomater Sci Eng 7(9):4420–4429. https://doi.org/10.1021/acsbiomaterials.1c00765

Li C, Gao Y, Huang R, Fang L, Sun Y, Yang Y, Gou S, Zhao J (2022a) An effective supramolecular approach to boost the photodynamic therapy efficacy of a near-infrared activating perylene diimide-based photosensitizer. ACS Mater Lett 4(4):657–664. https://doi.org/10.1021/acsmaterialslett.1c00818

Li S, Ma Z, Zhang K, Zhang W, Song Z, Wang W, Yu X, Han H (2022b) A two-pronged strategy for enhanced deep-tumor penetration and nir-ii multimodal imaging-monitored photothermal therapy. ACS Appl Mater Interfaces 14(37):41684–41694. https://doi.org/10.1021/acsami.2c08930

Li Z, Li X, Lu Y, Zhu X, Zheng W, Chen K, Liu S, Wu J, Guan W (2024) Improved photodynamic therapy based on glutaminase blockage via tumor membrane coated CB-839/IR-780 nanoparticles. Small 20(10):2305174. https://doi.org/10.1002/smll.202305174

Liao Y, Ye J, Xie L, Mao L, Wei J, Liu C, Liang J, Zhou J, Yuan M, Qin Z (2024) Carrier-free nanodrug for the chemical-photodynamic synergistic treatment of lung cancer. ACS Appl Nano Mater 7(5):5628–5636. https://doi.org/10.1021/acsanm.4c00659

Luo T, Nash GT, Xu Z, Jiang X, Liu J, Lin W (2021) Nanoscale metal–organic framework confines zinc-phthalocyanine photosensitizers for enhanced photodynamic therapy. J Am Chem Soc 143(34):13519–13524. https://doi.org/10.1021/jacs.1c07379

Macchi S, Zubair M, Hill R, Alwan N, Khan Y, Ali N, Guisbiers G, Berry B, Siraj N (2021) Improved photophysical properties of ionic material-based combination chemo/PDT nanomedicine. ACS Appl Bio Mater. 4(10):7708–7718. https://doi.org/10.1021/acsabm.1c00961

Mackay AM (2022) The evolution of clinical guidelines for antimicrobial photodynamic therapy of skin. Photochem Photobiol Sci 21(3):385–395. https://doi.org/10.1007/s43630-021-00169-w

Mai Y, Qu X, Ding S, Lv J, Li X, Gao P, Liu Y, Yuan Z (2020) Improved IR780 derivatives bearing morpholine group as tumor-targeted therapeutic agent for near-infrared fluorescence imaging and photodynamic therapy. Dyes Pigments 177:107979. https://doi.org/10.1016/j.dyepig.2019.107979

Mangalath S, Saneesh Babu PS, Nair RR, Manu PM, Krishna S, Nair SA, Joseph J (2021) Graphene quantum dots decorated with boron dipyrromethene dye derivatives for photodynamic therapy. ACS Appl Nano Mater 4(4):4162–4171. https://doi.org/10.1021/acsanm.1c00486

Montaseri H, Abrahamse H (2024) Targeted photodynamic therapy technique of janus nanoparticles on breast cancer. Artif Cells Nanomed Biotechnol 52(1):270–277. https://doi.org/10.1080/21691401.2024.2347369

Murali G, Kwon B, Kang H, Modigunta JKR, Park S, Lee S, Lee H, Park YH, Kim J, Park SY, Kim Y-J, In I (2022) Hematoporphyrin photosensitizer-linked carbon quantum dots for photodynamic therapy of cancer cells. ACS Appl Nano Mater 5(3):4376–4385. https://doi.org/10.1021/acsanm.2c00443

Nasseri B, Alizadeh E, Bani F, Davaran S, Akbarzadeh A, Rabiee N, Bahadori A, Ziaei M, Bagherzadeh M, Saeb MR, Mozafari M, Hamblin MR (2022) Nanomaterials for photothermal and photodynamic cancer therapy. Appl Phys Rev 9(1):011317. https://doi.org/10.1063/5.0047672

Niculescu A-G, Grumezescu AM (2021) Photodynamic therapy—an up-to-date review. Appl Sci 11(8):3626. https://doi.org/10.3390/app11083626

Pratavieira S, Requena MB, Stringasci MD, Ayala ETP, Bagnato VS (2022) The physics of light and sound in the fight against skin cancer. Braz J Phys 52(4):106. https://doi.org/10.1007/s13538-022-01121-8

Santhakumar H, Nair RV, Govindachar DM, Periyasamy G, Jayasree RS (2023) Engineering of tripeptide-stabilized gold nanoclusters with inherent photosensitizing property for bioimaging and photodynamic therapy. ACS Sustain Chem Eng 11(6):2102–2114. https://doi.org/10.1021/acssuschemeng.2c04857

Shangguan L, Chen Q, Shi B, Huang F (2017) Enhancing the solubility and bioactivity of anticancer drug tamoxifen by water-soluble pillar[6]arene-based host–guest complexation. Chem Commun 53(70):9749–9752. https://doi.org/10.1039/C7CC05305C

Sikder A, Chaudhuri A, Mondal S, Singh NDP (2021) Recent advances on stimuli-responsive combination therapy against multidrug-resistant bacteria and biofilm. ACS Appl Bio Mater 4(6): 4667–4683. https://doi.org/10.1021/acsabm.1c00150

Situ J, Yang Y, Zhang L, Yan H, Cheng Y (2024) Integration of O 2-economised tumour-targeted photosensitive magnetic nanomaterials in the diagnosis and therapy of gastric cancer. RSC Adv 14(14):9920–9932. https://doi.org/10.1039/D4RA00497C

Tabrizi L, McGarry R, Turzanska K, Varvarezos L, Fallon M, Brannigan R, Costello JT, Fitzgerald-Hughes D, Pryce MT (2024) Porphyrin-polymer as a photosensitizer prodrug for antimicrobial photodynamic therapy and biomolecule binding ability. Biomacromolecules 25(12): 7736–7749. https://doi.org/10.1021/acs.biomac.4c01011

Tamtaji M, Tyagi A, You CY, Galligan PR, Liu H, Liu Z, Karimi R, Cai Y, Roxas AP, Wong H, Luo Z (2021) Singlet oxygen photosensitization using graphene-based structures and immobilized dyes: a review. ACS Appl. Nano Mater 4(8):7563–7586. https://doi.org/10.1021/acsanm.1c01436

Teng K-X, Niu L-Y, Yang Q-Z (2022) A host–guest strategy for converting the photodynamic agents from a singlet oxygen generator to a superoxide radical generator. Chem Sci 13(20): 5951–5956. https://doi.org/10.1039/D2SC01469F

Tian M, Chen W, Wu Y, An J, Hong G, Chen M, Song F, Zheng W, Peng X (2022) Liposome-based nanoencapsulation of a mitochondria-stapling photosensitizer for efficient photodynamic therapy. ACS Appl Mater Interfaces 14(10):12050–12058. https://doi.org/10.1021/acsami.1c23156

Truong DH, Tran PTT, Tran TH (2024) Nanoparticles as carriers of photosensitizers to improve photodynamic therapy in cancer. Pharm Dev Technol 29(3):221–235. https://doi.org/10.1080/10837450.2024.2322570

Udrea AM, Smarandache A, Dinache A, Mares C, Nistorescu S, Avram S, Staicu A (2023) Photosensitizers-loaded nanocarriers for enhancement of photodynamic therapy in melanoma treatment. Pharmaceutics 15(8):2124. https://doi.org/10.3390/pharmaceutics15082124

Vafadar A, Tajbakhsh A, Hosseinpour-Soleimani F, Savardshtaki A, Hashempur MH (2024) Phytochemical-mediated efferocytosis and autophagy in inflammation control. Cell Death Discov 10(1):1–23. https://doi.org/10.1038/s41420-024-02254-2

Vinita NM, Devan U, Durgadevi S, Anitha S, Govarthanan M, Antony Joseph Velanganni A, Jeyakanthan J, Arul Prakash P, Mohamed Jaabir MS, Kumar P (2023) Impact of surface charge-tailored gold nanorods for selective targeting of mitochondria in breast cancer cells using photodynamic therapy. ACS Omega 8(37):33229–33241. https://doi.org/10.1021/acsomega.2c06731

Wu H, Du X, Xu J, Kong X, Li Y, Liu D, Yang X, Ye L, Ji J, Xi Y, Zhai G (2023) Multifunctional biomimetic nanoplatform based on photodynamic therapy and DNA repair intervention for the synergistic treatment of breast cancer. Acta Biomater 157:551–565. https://doi.org/10.1016/j.actbio.2022.12.010

Xiao Y-F, Chen W-C, Chen J-X, Lu G, Tian S, Cui X, Zhang Z, Chen H, Wan Y, Li S, Lee C-S (2022) Amplifying free radical generation of AIE photosensitizer with small singlet–triplet splitting for hypoxia-overcoming photodynamic therapy. ACS Appl Mater Interfaces 14(4): 5112–5121. https://doi.org/10.1021/acsami.1c23797

Xie J, Wang Y, Choi W, Jangili P, Ge Y, Xu Y, Kang J, Liu L, Zhang B, Xie Z, He J, Xie N, Nie G, Zhang H, Kim JS (2021) Overcoming barriers in photodynamic therapy harnessing nano-formulation strategies. Chem Soc Rev 50(16):9152–9201. https://doi.org/10.1039/D0CS01370F

Xu Y, Chen H, Xu S, Liu J, Chen Y, Gui L, Li H, Li R, Yuan Z, Li B (2022) β-Lactamase-responsive probe for efficient photodynamic therapy of drug-resistant bacterial infection. ACS Sens 7(5):1361–1371. https://doi.org/10.1021/acssensors.1c02485

Xu C, Yuan D, Cheng S, Lan X, Wang Q, Lin X, Liu W (2023) A preliminary study on the evaluation of photodynamic diagnosis for acne inflammatory skin lesions. J Biophotonics 16(12):e202300160. https://doi.org/10.1002/jbio.202300160

Yi X, Hu J-J, Dai J, Lou X, Zhao Z, Xia F, Tang BZ (2021) Self-guiding polymeric prodrug micelles with two aggregation-induced emission photosensitizers for enhanced chemo-photodynamic therapy. ACS Nano 15(2):3026–3037. https://doi.org/10.1021/acsnano.0c09407

Zhang Y, Yu W, Chen M, Zhang B, Zhang L, Li P (2023) The applications of nanozymes in cancer therapy: based on regulating pyroptosis, ferroptosis and autophagy of tumor cells. Nanoscale 15(29):12137–12156. https://doi.org/10.1039/D3NR01722B

Zhao W, Wang L, Zhang M, Liu Z, Wu C, Pan X, Huang Z, Lu C, Quan G (2024) Photodynamic therapy for cancer: mechanisms, photosensitizers, nanocarriers, and clinical studies. Med Comm 5(7):e603. https://doi.org/10.1002/mco2.603

Zheng X, Sun W, Ju M, Wu J, Huang H, Shen B (2022) A chemical biology toolbox to overcome the hypoxic tumor microenvironment for photodynamic therapy: a review. Biomater Sci 10(17): 4681–4693. https://doi.org/10.1039/D2BM00776B

Zhu TC, Finlay JC (2008) The role of photodynamic therapy (PDT) physics. Med Phys 35(7 Part 1): 3127–3136. https://doi.org/10.1118/1.2937440

Zhuang J, Wang B, Chen H, Zhang K, Li N, Zhao N, Tang BZ (2023) Efficient NIR-II type-I AIE photosensitizer for mitochondria-targeted photodynamic therapy through synergistic apoptosis–ferroptosis. ACS Nano 17(10):9110–9125. https://doi.org/10.1021/acsnano.2c12319

Progress in Magneto-Luminescent Nanomaterials for Cancer Theragnosis

9

Kanchan Chauhan, Pamela Vanessa Meixueiro Aguilar, Gustavo A. Hirata, and Prakhar Sengar

Abstract

Magnetic and luminescent properties within a single nanosystem provide a promising platform for cancer theragnostic, combining multiple functionalities in one. The magnetic component permits applications such as magnetic resonance imaging (MRI) and magnetic hyperthermia, improving diagnostic precision and offering a means for localized cancer treatment. The luminescent component aids optical imaging and phototherapy, offering supplementary diagnostic and therapeutic functionality. Together, these properties create a versatile tool that enhances diagnostic accuracy and sensitivity, potentially leading to earlier detection and more effective cancer treatment. Recently, magneto-luminescent nanosystems have revolutionized multimodal imaging and image-guided tumor therapy. This chapter elaborates on the latest advancements in magneto-luminescent nanomaterials, emphasizing their synthesis strategies and applications in cancer diagnostics and treatment. The discussion begins by introducing luminescent and magnetic nanomaterials and their properties used for biomedical applications. Then, examine the methods employed for designing magneto-luminescent nanomaterials and how the magnetic and optical properties interact and vary when coexisting in a single nanosystem. The chapter further explores their functional role in cancer diagnosis, highlighting their use as multimodal imaging probes and therapeutic agents for cancer. Furthermore, recent magneto-luminescent nanomaterials design advancements are discussed, along with the potential challenges associated with clinical translation, providing insights for future research.

K. Chauhan · P. V. M. Aguilar · G. A. Hirata · P. Sengar (✉)
Centro de Nanociencias y Nanotecnología, Universidad Nacional Autónoma de México, Tijuana Enscnada, Baja California, Mexico
e-mail: prakharsengar@ens.cnyn.unam.mx

R. Prasad, M. Mohan Yallapu (eds.), *Advancements in Cancer Theranostics*,
https://doi.org/10.1007/978-981-95-0938-6_9

Keywords

Magnetic nanoparticles · Luminescent nanoparticles · Cancer therapy and bioimaging · Targeted combination therapy · Theranostics

9.1 Introduction

Recently, the rapid advancement of nanofabrication has enabled the development of high-performing nanosystems. Formerly, developed single-component nanomaterials show limitations in functionality and performance, thus impeding their applicability. Additionally, single-component systems are often prone to aggregation, making them lose efficiency (Behrens and Appel 2016). Single-component nanomaterials have one property that is insufficient to meet multiple requirements simultaneously. Therefore, nanosystems that possess multiple properties receive much attention due to their ability to combine many functionalities through synergistic conjugation. Magneto-luminescent nanosystems are a class of multicomponent nanomaterials that have luminescence and magnetic properties at the nanoscale. These nanosystems can be fabricated by combining various magnetic and luminescent components via various synthesis approaches (Sengar et al. 2024; Zou et al. 2017; Li et al. 2020; Yang et al. 2019). These nanofabrication methods used for magneto-luminescent nanomaterials involve structural strategies such as core-shell, hybrid, vesicle structure, and heterodimer (Fig. 9.1). This combination of luminescent and magnetic properties enhances dispersion, sensitivity, and chemical stability.

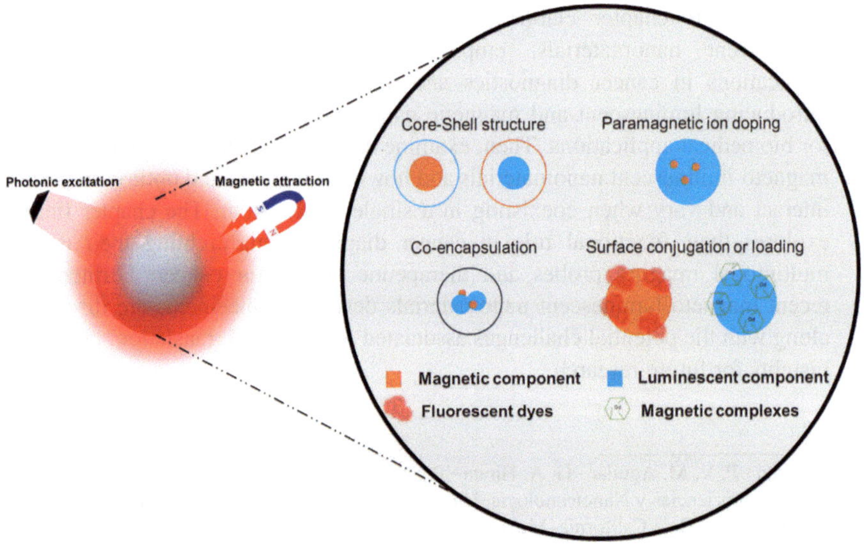

Fig. 9.1 Schematic representations of various nanofabrication strategies for magneto-luminescent nanomaterials

Now, in comparison to conventional disease diagnostic and therapeutic nanosystems, magneto-luminescent nanosystems demonstrate the following advantages:

1. They can be applied for multiple imaging simultaneously, thus overcoming the limitation of individual imaging modality and reaching better spatial resolution and accuracy (Liu et al. 2016; Wang et al. 2018a).
2. These nanosystems, besides passive targeting and tumor accumulation via enhanced permeation and retention effect (EPR), can be actively and magnetically guided for specific drug delivery (Liu et al. 2019).
3. High surface area due to the nanosize also allows the loading or functionalization of the disease-specific drug and their delivery via stimuli-responsive behavior using luminescence /magnetic heating stimuli (Yang et al. 2019; Zhang et al. 2016a).

Various magneto-luminescent nanomaterials have been developed and explored for their prospects in biomedical applications, for example, cell separation, disease diagnosis (e.g., cancer), biosensing, and fluorescent labeling. Many of these nanosystems have shown considerable potential, particularly for cancer theranostics, paving the way for clinical translation. However, some major challenges are still preventing the clinical translation of this new class of magneto-luminescent nanomaterials. One of the issues that can arise when combining magnetic and luminescent components, is the precise optimization of nanosystems properties to achieve the desired functionality at the nanoscale. Additionally, magnetic characteristics in a magneto-luminescent nanosystem will provide a natural tendency to accumulate, which could lead to clustering. This aggregation could block the capillaries during intravenous administration, posing a critical concern in clinical applications. Biocompatibility is another crucial factor for the clinical translation of these new magneto-luminescent nanosystems and must be addressed at multiple levels including cellular, tissue, hematologic, and immune compatibility. Therefore, there is a growing need to develop more versatile and stable magneto-luminescent nanosystems with high biocompatibility for application in clinical nanotheranostics.

This book chapter focuses on recent advances in magneto-luminescent nanomaterials synthesis procedures at the nanoscale level and their use in cancer diagnosis and treatment. We will begin by understanding the principles of luminescent and magnetic properties and how they influence biomedical applicability. Following that, we will explore the concept of magneto-luminescent nanomaterials, focusing on their unique properties, synthesis strategies, and recent advancements in their applications toward cancer theranostics. In conclusion, the current issues related to synthesis, stability, and properties of magneto-luminescent nanomaterials are examined. We will also examine the future directions, focusing on approaches to overcome these obstacles and improve the clinical capability and efficiency of magneto-luminescent nanosystems.

9.2 Luminescence Materials and Principle of Photoluminescence

Luminescence materials can convert energy to electromagnetic radiation, particularly photons, with wavelengths of ultraviolet (UV) to near infrared (NIR). The energy used to excite these luminescent materials can come from various sources, like high-energy electrons (cathodoluminescence), electric voltage (electroluminescence), mechanical energy (triboluminescence), chemical reaction (chemiluminescence), X-rays (radioluminescence), or low-energy photons in UV, visible, and NIR range (photoluminescence). The photoluminescence phenomenon occurs when a material absorbs photons causing its electrons to be energized to higher-energy levels. This excited state is unstable, and as the electrons return to their lower-energy states, they release the excess energy in the form of photons (Fig. 9.2) (Teets 2021). Some excitation energy is lost as heat through non-radiative processes; hence, the emitted photons usually have lower energy than the absorbed excitation photons. Photoluminescent materials are broadly categorized into fluorescent and phosphorescent. In fluorescent materials, the emission of photons stops at once after

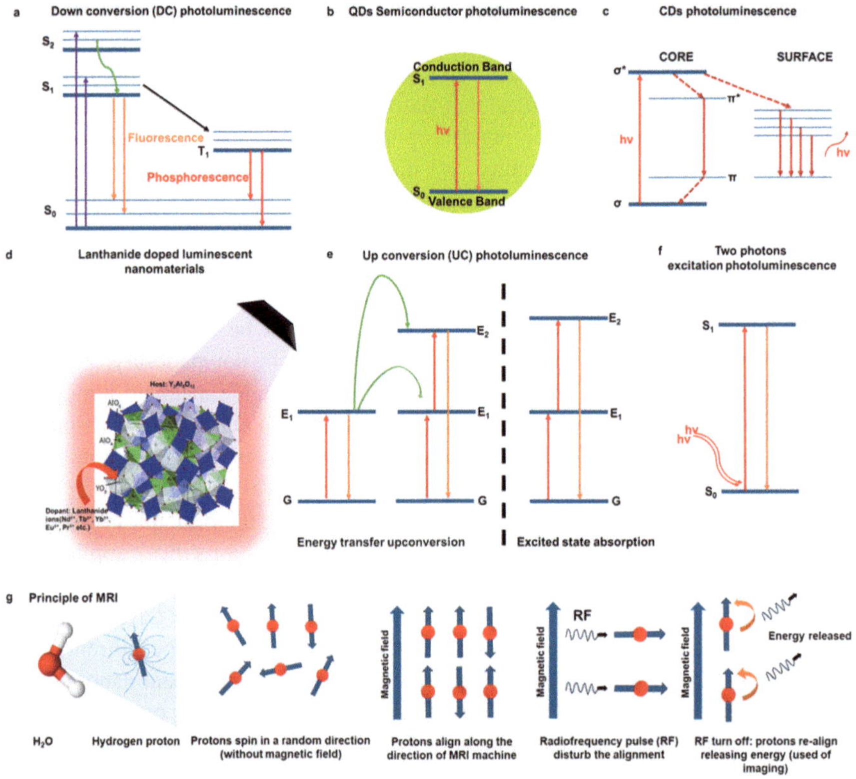

Fig. 9.2 Schematic illustration of various photoluminescent mechanisms and magnetic resonance imaging (MRI) principles in magneto-luminescent materials

excitation is removed (Fig. 9.2a). In contrast, phosphorescent/persistent luminescence materials can still emit photons for milliseconds to hours even after the excitation source is removed (Fig. 9.2a). Based on their composition, photoluminescent materials can be mainly divided into these types:

1. **Quantum dots (QDs)** are semiconductor nanoparticles, typically 2–10 nm in size, and display size-dependent luminescence behavior due to band gap transitions (Fig. 9.2b) (García de Arquer et al. 2021; Martynenko et al. 2017; Bányai and Koch 1993).
2. **Carbon dots (CDs)** are a newer class of carbon-based photoluminescent nanomaterials, usually smaller than 10 nm. Due to their carbon composition, they own intrinsic biocompatibility, making them desirable for bioimaging applications. The correct mechanism of CDs photoluminescence is still under investigation. However, current understanding suggests that photoluminescence is due to the combinatorial effect of surface state emission, defect-state emission, and π–π transitions of the carbon core (Fig. 9.2c) (Sengar et al. 2022; Goryacheva et al. 2017).
3. **Organic dyes** are also carbon-based small molecules that show photoluminescence due to a conjugated π-electron system. But they often lack photostability, which limits their utility (Wu et al. 2022).
4. **Lanthanide-doped phosphors** consist of an inorganic host matrix doped with lanthanide ions such as neodymium (Nd), terbium (Tb), ytterbium (Yb), or europium (Eu) (Fig. 9.2d) (Chávez-García et al. 2021, Sengar et al. 2019, 2020; Jain et al. 2018a). For example, in yttrium oxide doped with europium (Y_2O_3: Eu^{3+}), the host is Y_2O_3, and the activator is Eu^{3+}. By adjusting dopant ions, luminescent properties can be tuned. For instance, $Y_3Al_5O_{12}$ (YAG) doped with praseodymium (Pr^{3+}) emits UV photons, while the same host doped with neodymium (Nd^{3+}) emits NIR light upon NIR excitation (Sengar et al. 2016; Tejeda et al. 2018).

Because these materials can absorb and emit photons at different wavelengths, they are used in a wide range of applications, including bioimaging, sensors, lighting, and solar cells (Dubey et al. 2020; Huang et al. 2012; Khalid et al. 2023). Initially, photoluminescent materials working in the UV and visible regions were applied for bioimaging. However, UV and visible photons cannot travel deeply into human tissue, as these wavelengths are either absorbed or scattered by the tissue (Ash et al. 2017). Recently, the development of photoluminescent materials that can be excited and emit in far-red to NIR regions (650–1350 nm), often referred to as optical biological windows, has attracted significant attention. This spectral region allows photons to penetrate deeper into tissues, making these materials particularly valuable for in vivo bioimaging applications (Qu et al. 2024). Consequently, various NIR to NIR photoluminescent materials have been developed and implemented for such applications (Jackson et al. 2021). Photoluminescent nanomaterials can be synthesized by various synthesis routes depending on the type of the material, as summarized in Table 9.1.

Table 9.1 Methods involved in the synthesis of photoluminescent and magnetic nanoparticles

Type of nanomaterials	Synthesis routes	References
Photoluminescent component		
Lanthanide-doped nanomaterials	Sol–gel, combustion, hydro/solvo thermal, and coprecipitation	Jain et al. (2018a), Wen and Wang (2014)
Quantum dots (QDs)	Hydro/solvo thermal, colloidal synthesis, microemulsion, template-assisted, thermal decomposition, and electrochemical synthesis	Nagpal and Gusain (2022)
Carbon dots (CDs)	Hydro/solvothermal, microwave-assisted, electrochemical, and thermal decomposition	Sengar et al. (2022)
Organic florescent dyes	Self-assembly and chemical reactions	Fei and Gu (2009), Zhang et al. (2016b)
Magnetic component		
Magnetic nanoparticles	Coprecipitation, sol–gel, microemulsion, hydro/solvo thermal, microwave-assisted, and combustion	Ali et al. (2021)
Magnetic complexes	Direct chemical reaction of paramagnetic ions with ligands such as EDTA	Woodruff et al. (2013)

The use of luminescent nanomaterials for NIR → photonic conversion for bioimaging and therapy can be achieved through several excitation strategies. One common strategy is downconversion, where high-energy photons excite the nanomaterial, and it emits lower-energy photons (Fig. 9.2a). This photoluminescence process is energy-efficient and shows a high quantum yield. Luminescent materials doped with lanthanides such as neodymium (Nd^{3+}), erbium (Er^{3+}), ytterbium (Yb^{3+}), thulium (Tm^{3+}), and holmium (Ho^{3+}) provide appropriate transition levels for NIR → NIR downconversion (Sengar et al. 2023; Avram et al. 2018; Yang et al. 2021). Alternatively, NIR emission can also be accomplished through upconversion, where two low-energy NIR photons are absorbed to produce one higher-energy photon in the NIR range (Fig. 9.2e) (Jackson et al. 2021). However, single-doped upconversion luminescent materials typically demonstrate low quantum yield. To enhance the upconversion process, two lanthanide ions are used—one as a sensitizer and the other as an activator. For example, the widely studied upconversion systems β-$NaYF_4{}^{3+}$-Er^{3+} and β-$NaYF_4{}^{3+}$-Tm^{3+} use ytterbium ions (Yb^{3+}) as sensitizers, which absorb NIR photons (around 980 nm) and transfer energy to activator ions such as erbium (Er^{3+}), thulium (Tm^{3+}), or holmium (Ho^{3+}), resulting in strong NIR emission (energy transfer upconversion) (Fig. 9.2e). Various photoluminescent materials can convert visible photons to NIR photons via downconversion, limiting their use for deep-tissue bioimaging. To address this limitation about visible → NIR photoluminescent materials, new ultrafast femtosecond laser systems can generate two simultaneous NIR photons. Two NIR photons combined energy can excite the luminescent materials in similar ways as one visible photon, producing NIR emission (Fig. 9.2f) (Sengar et al. 2020; Pan et al. 2016; Jiang et al. 2024). These properties enhance their potential for in vivo bioimaging applications, making them strong candidates for developing magneto-luminescent nanosystems.

9.3 Magnetic Nanoparticles (MNPs)

Magnetic nanoparticles (MNPs) have demonstrated their applicability in many areas including energy storage, catalysis, magnetic inks/fluids, environmental, electronics, and biomedical. MNPs exhibiting paramagnetic and superparamagnetic behavior at room temperature are well-suited for magnetic resonance imaging (MRI) and therapy (Lee et al. 2014; Yang et al. 2022a; Blanco-Andujar et al. 2018). Paramagnetic and superparamagnetic nanoparticles both magnetize in the presence of an external magnetic field but lose their magnetization completely once the field is removed. Superparamagnetic nanoparticles exhibit stronger magnetization under the field due to their nanometric size (John 2021; Li et al. 2017). Additionally, the absence of magnetic domain formation in both paramagnetic and superparamagnetic nanomaterials prevents the retention of the residual magnetization when the external field is removed. This property is vital, as it prevents the aggregation of magnetic nanoparticles, making them suitable for biomedical applications. MNPs are primarily composed of pure metals such as cobalt (Co), iron (Fe), nickel (Ni), titanium (Ti), gadolinium (Gd), and manganese (Mn), as well as metal oxides like iron oxide (Fe_2O_3, Fe_3O_4), copper ferrite ($CuFe_2O_4$), and magnetic composites (Ali et al. 2021). Among these, iron oxide, along with complexes of Gd, are often used for biomedical applications due to their low toxicity (Li et al. 2016). MNPs can be synthesized via various methods such as template synthesis, coprecipitation, hydro/solvothermal, thermal decomposition, polyol process, and microemulsion as summarized in Table 9.1 (Li et al. 2016).

9.4 Application of MNPs for Magnetic Resonance Imaging (MRI)/Magnetic Hyperthermia

Magnetic resonance imaging (MRI) is a widely used noninvasive diagnostic modality in medicine due to its ability to generate high-resolution images of soft tissue (Grover et al. 2015). It is a highly effective technique for observing the central nervous system, heart, brain, and tumor. The MRI works by exposing the body to a powerful magnetic field, ranging from 0.2 to 7 Tesla. The basic principle of MRI is depicted in Fig. 9.2g; in the human body, the hydrogen atoms are mainly present in water molecules, acting as tiny magnets due to the magnetic moment of their proton. When exposed to a high magnetic field, the protons of hydrogen align their spin either parallel or antiparallel to the field direction. Thereafter, MRI equipment emits a radiofrequency pulse to disturb the magnetic alignment. After the pulse is turned off, the hydrogen nuclei return to their original state, releasing electromagnetic energy in the process. The MRI machine detects this released energy to create contrast images based on the difference in the water content between various tissues.

There are two main relaxation pathways that hydrogen nuclei follow when disturbed by radio waves. The first is T1 or longitudinal relaxation, which refers to the process of returning to the original aligned state. The second is T2 or transverse relaxation, which involves the hydrogen nuclei losing synchronicity with one another after the radiofrequency pulse stops. MRI images are generated based on T1/T2 contrast, allowing differentiation between tissues and tumor regions.

To enhance the quality of MRI scans, contrast agents are used to produce sharper, more detailed images. Gadolinium-based chelates, like Gd-DTPA, have been commonly employed as MRI contrast agents in clinical practice (Iyad et al. 2023). Moreover, superparamagnetic nanoparticles, mainly iron oxide nanoparticles (IONPs) with biocompatible encapsulations, have been clinically approved under commercial names as a contrast for MRI (Thakor et al. 2016). The process of generating contrast in MRI differs between various contrast agents. MNPs that influence the T1 and T2 relaxation processes are classified as T1 and T2 contrast agents. Paramagnetic contrast agents are usually employed for T1-weighted MRI (Lee et al. 2014), while superparamagnetic IONPs generate T2 contrast (Yang et al. 2022a). However, IONPs can also be tuned for T1 and T2 contrast abilities for MRI (Huang et al. 2014; Jung et al. 2014). Owing to their nanostructure, the surface of MNPs can be functionalized with therapeutic agents, allowing drug delivery to disease sites, which can also be directed via magnetic field influence. Additionally, superparamagnetic MNPs such as IONPs, due to their high magnetization, can function as nanoheaters when exposed to alternating magnetic fields. This heating effect, which can raise the temperature to over 42 °C, is sufficient to induce cytotoxicity in the cancer cells. This therapeutic modality is known as magnetic hyperthermia, and it is both noninvasive and specific to the disease region (Liu et al. 2020).

9.5 Magneto-Luminescent Nanomaterials

Magneto-luminescent nanomaterials are a hybrid of luminescent components (ln-doped nanoparticles, carbon dots, quantum dots, fluorescent dyes, etc.) and magnetic components (paramagnetic or superparamagnetic). As discussed previously, magnetic properties can be introduced by the incorporation of elements such as iron (Fe), gadolinium (Gd), nickel (Ni), manganese (Mn), and cobalt (Co) or their compounds like IONPs and metal-ferrites. Considering the advantages of NIR photons in deep bioimaging, NIR → NIR based luminescent components are preferred for magneto-luminescent nanohybrids. Due to the expansion of this area of research, many magneto-luminescent nanosystems have been formed and have shown bright prospects for cancer theragnosis.

9.6 IONP-Based Magneto-Luminescent Nanosystems

Iron oxide nanoparticles (IONPs) come in the category of ferromagnetic materials and, according to their chemical composition, can be differentiated into maghemite (γ-Fe_2O_3), magnetite (Fe_3O_4), and mixed ferrite (MFe_2O_4). Additionally, IONPs demonstrate remarkable biocompatibility and high magnetic relaxivity for MRI even in ultra-low field (Yang et al. 2022a; Jung et al. 2014; Yin et al. 2018). IONPs, due to their strong magnetization and size tunability, permit applicability in MRI, magnetic hyperthermia, and magnetically targeted drug delivery (Jiang et al. 2021). IONPs with size ~20 nm and smaller show superparamagnetic behavior at room temperature and high colloidal stability (Dadfar et al. 2020). Therefore, IONPs have been widely used as a prime magnetic component to design new nanosystems.

Various luminescent materials such as lanthanide-doped nanomaterials, carbon dots (CDs), semiconductor quantum dots (QDs), and fluorescent dyes have been conjugated with IONPs, which provides numerous advantages in cancer theranostics. The photoluminescence characteristics of QDs such as broad excitation wavelengths with a wide range of strong photonic emission and excellent photostability are suitable for optical bioimaging. QDs have a particle size of 2–10 nm, which is important to limit the size of the final magneto-luminescence system in the nanometric range. Combining QDs with IONPs can create a functional magneto-luminescent nanosystem that can be used in vivo and in vitro for optical imaging and therapy. Parks et al. synthesized CdSeTeS QDs encapsulated hollow Fe_2O_3 spheres (Ganganboina et al. 2020). The hollow magnetic sphere encapsulation over QDs was performed in three steps. QD encapsulation in the hollow magnetic sphere did not quench the fluorescence as charge and energy transfers were restricted owing to a sufficient gap between the Fe_2O_3 and QDs. This unique encapsulation strategy can be adapted with other luminescent nanomaterials to form quenching-free magneto-luminescent nanosystems. As most of the QDs are composed of toxic metals such as cadmium, lead, and selenium, their use in biomedical applications has been hindered. The biocompatible coating can somewhat solve the toxicity issue of QDs; for instance, Zhao et al. encapsulated SPIONs, QDs, and cilengitide within PEGylated liposomes to develop a theranostics magneto-luminescent nanosystem (Xu et al. 2018). The nanoplatform demonstrated magnetic targeting capability and precisely directed the surgical resection of glioma via MRI/FLI combined imaging techniques. In another report, SPIONs and NIR emitting PbS/CdS QDs were embedded in thiol-functionalized mesoporous silica with large pores (Yang et al. 2019).

Carbon dots (CDs) are another type of carbon-based luminescent nanomaterials that have been recently explored due to their intrinsic biocompatibility and ease of synthesis. Photoluminescence properties and size of CDs are almost identical to QDs, which makes them equally suitable for bioimaging and sensing. CDs also possess surface functional groups like carboxylic and amine which can be used for chemical conjugation. For example, graphene-based CDs were surface functionalized on the silica-coated hollow magnetic sphere, forming a multimodal theranostics nano platform HMNS/SiO_2/GQDs (Fig. 9.3a) (Wo et al. 2016). This

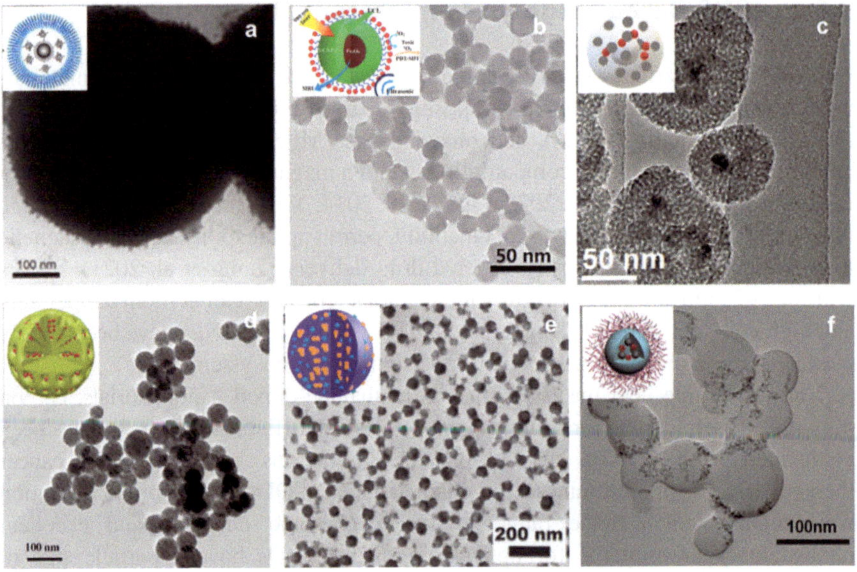

Fig. 9.3 Transmission electron microscopy (TEM) images of magneto-luminescent liposome coated HMNS/SiO_2/GQDs-DOX, Fe_3O_4@$NaGdF_4$:Yb:Er-HMME, HAp:Nd + SPIONs@$mSiO_2$, $mSiO_2$@$Gd_3Ga_5O_{12}$:Cr^{3+}, Nd^{3+} ($mSiO_2$@GGO), HSA-GGD-ICG, TB/SPIO@PS-PEG: (**a**) Reproduced with permission from reference Wo et al. (2016), Copyright 2016, Ivyspring International. (**b**) Reproduced with permission from reference Wang et al. (2019), Copyright 2019, Elsevier. (**c**) Reproduced with permission from reference Sengar et al. (2024) Copyright 2024, Elsevier. (**d**) Reproduced with permission from reference Shi et al. (2018), Copyright 2018, Elsevier. (**e**) Reproduced with permission from reference Wang et al. (2018a), Copyright 2018, Wiley-VCH GmbH. (**f**) Reproduced with permission from reference Meng et al. (2019), Copyright 2019, Royal Society of Chemistry]

nanoplatform was further loaded with doxorubicin (DOX) and coated with liposomes to provide therapeutic capabilities against cancer. CDs can be doped with metal ions during the hydrothermal process which alters the electronic, optical, and magnetic properties of the CDs (Li et al. 2022). Considering this attribute of CDs, Das and coworkers synthesized Fe-doped CDs through a hydrothermal approach, which presented high biocompatibility, fluorescence, and superparamagnetic properties (Das et al. 2019).

Another extensively researched category of photoluminescence nanomaterials is lanthanide-doped phosphors. This class of photoluminescence materials has very high photostability, emission tunability from visible to NIR photons, and long persistent luminescence capability. Majorly, lanthanide-doped luminescent nanomaterials with spectral conversion capability in the optical biological window (650–1350 nm) are more suitable for in vivo optical bioimaging. To attain NIR → NIR spectral conversion various lanthanide-doped nanophosphors have been studied, and various excitation strategies have been devised. Based on excitation strategies, lanthanides-doped nanophosphors can either be downconversion nanoparticles (DCNPs) or upconversion nanoparticles ($UCNP_S$). The UCNPs are composed of the host matrix with doped lanthanide ions such as Yb^{3+}, which mainly acts as a sensitizer, and Tm^{3+}, Er^{3+}, and Ho^{3+}, playing the role of emitter. The $UCNP_S$ follows an anti-Stokes process by absorbing two or more lower-energy photons and emitting one high-energy photon (Chen et al. 2014). The blend of UCNPs with IONPs is among the best-suited magneto-luminescent materials due to intense NIR upconversion and the exceptional photostability for in vivo bioimaging, surpassing clinical limitations. One such nanohybrid Fe_3O_4@$NaGdF_4$:Yb:Er-HMME was synthesized using a core-shell strategy where $NaGdF_4$:Yb:Er UCNPs were synthesized over the Fe_3O_4 nanoparticles by solvothermal process (Fig. 9.3b) (Wang et al. 2019). However, the loss in magnetization was observed after the surface coating with UCNPs. In other work, magneto-luminescent nanochains were formed by self-assembly of pheophorbide-A modified UCNPs $NaYF_4$: Yb, Ho, and IONPs (**Fu et al.** 2022). The nanosystem was able to produce high NIR emission and T2-MR image simultaneously. NIR → NIR emitting DCNPs are more efficient as excitation with one photon generates one photon, resulting in a high quantum yield, making them ideal for magnetic-luminescent nanomaterials. Our group recently reported a facile synthesis approach to co-encapsulated SPIONs with hydroxyapatite doped with neodymium within mesoporous silica (HAp: Nd + SPIONs@$mSiO_2$) (Fig. 9.3c) (Sengar et al. 2024). The synthesis involved mixing the two components during the formation of mesoporous silica, followed by hydrothermal treatment. This hybrid nanosystem exhibited NIR → NIR photoluminescent properties and retained the superparamagnetic nature, producing a magnetic heating effect.

Nevertheless, inorganic luminescent materials, by their size range, can have higher accumulation in the major organs, leading to toxicity in the long run. Like CDs, organic fluorescent dyes have structures with carbon-carbon double bonds or aromatic rings. Therefore, some organic dyes can emit in the NIR range with high quantum efficiency (Wu et al. 2022). The organic dye, based on its structure, can either be co-encapsulated or functionalized through chemical conjugation with the magnetic component, creating a magneto-luminescent nanohybrid with high clinical value in the field of diagnosis. Indocyanine green (ICG) fluorescent dye is a clinically approved NIR → NIR dye with photothermal and photodynamic cancer therapeutic capabilities (Wang et al. 2018b; Shirata et al. 2017). Also, with the NIR-activated photothermal effect, ICG can produce localized thermal expansion, producing photoacoustic waves, which can be detected to generate photoacoustic images. The unique properties make ICG an appropriate fluorescence probe that can be coupled with a magnetic component to form a magneto-luminescent nanohybrid with multimodal theranostics capabilities. One such nanohybrid was prepared using ICG-conjugated SPIONs nanoworm using EDC-NHS amide coupling (Yang et al. 2022b). Moreover, the simple but efficient magneto-luminescent nanohybrid was fabricated by co-encapsulation of ICG and magnetic contrast agent, glycyrrhetinic-acid-modified gadolinium(III)-1,4,7,10-tetraazacyclododecane-1,4,7,10-tetraacetate (GGD) in human serum albumin (HSA) (Fig. 9.3e) (Wang et al. 2018a). This nanohybrid, due to its clinically approved components, possesses a high clinical significance for cancer theranosis.

9.7 Other Magneto-Luminescent Nanosystems

Paramagnetic properties can also be introduced in the magneto-luminescent nanosystems by incorporating metallic ions such as gadolinium (Gd^{3+}), manganese (Mn^{2+}), cobalt (Co^{2+}), and nickel (Ni^{2+}) through doping or chemical conjugation. Among these metal ions, Gd^{3+} is the most comprehensively utilized to develop magneto-luminescent nanosystems. Gadolinium ions with seven unpaired electrons in the 4f orbital generate a significant magnetic moment to alter the magnetic relaxation of the water proton during MRI-producing contrast images (Iyad et al. 2023). Various luminescent nanomaterials such as CDs and lanthanide-doped nanomaterials have been combined with gadolinium to provide paramagnetic properties to the hybrid nanosystems. Magnetic garnet hosts like $Gd_3(Al, Ga)_5O_{12}$ doped with lanthanide ions are known to produce high luminescence output due to high thermal stability, low phonon energy, wavelength transparency, and facile energy transfer within the host (Sengar et al. 2016; Ueda and Tanabe 2019; Castaing et al. 2022). Moreover, due to the band gap tuning by altering the molar ratio of Ga and Al ions in the host, persistent luminescence can also be produced (Sengar et al. 2019, 2020; Ueda and Tanabe 2019; Castaing et al. 2022). For instance, our group synthesized magneto-luminescent $Gd_3(Al, Ga)_5O_{12}$: (Ce, Pr, Cr) nanoparticles by sol–gel and hydrothermal methods (Sengar et al. 2019, 2020). $Gd_3(Al, Ga)_5O_{12}$ host due to the presence of Gd ions in the host matrix demonstrated paramagnetic character. Moreover, $Gd_{2.99}Ce_{0.01}Al_{1.995}Cr_{0.005}Ga_3O_{12}$ nanoparticles presented NIR to NIR two-photon excitable persistent luminescence. However, garnet-based photoluminescent nanocrystals often lack regular morphology and poor dispersity, which hinders its ability to be used in bioimaging. Therefore, new synthesis approaches and polymer stabilization have been used to overcome these limitations. One such technique was used by Shi et al. to design a versatile magneto-luminescent nanosystem using mesoporous silica nanoparticles as a template (Fig. 9.3d) (Shi et al. 2018). For the synthesis, the ionic solution of the $Gd_3Ga_5O_{12}:Cr^{3+}$, Nd^{3+} precursors was embedded within the pores of mesoporous silica, followed by calcination, forming $Gd_3Ga_5O_{12}:Cr^{3+}$, Nd^{3+} nanocrystals.

Apart from the garnet host, other Gd-incorporated hosts like $NaGdF_4$ and Gd_2O_3 provide a high crystal field, which enhances the photoluminescence of the doped lanthanide ions and provides significant paramagnetic property for T1-weighted MRI (Zou et al. 2017; Lu et al. 2017; Jain et al. 2018b). Alternatingly, Gd ions can also be conjugated to other nanosystems via complexation, as shown by Carron et al., where Gd ions were complexed on the surface of UCNPs silica encapsulated $NaLnF_4:Yb^{3+}$, Tm^{3+} using diethylenetriamine pentaacetic acid (Carron et al. 2015). Furthermore, the Gd-doped CDs were synthesized utilizing a templated approach where luminescence and magnetic properties were size-dependent (Chen et al. 2016). The Gd@C-dots with 3 nm size showed the highest R1 relaxivity (10 mM^{-}

^{1} s^{-1}) with a fluorescence quantum efficiency of 30.2%. Manganese ion-based MRI contrast agents have recently gained importance for MRI owing to their low toxicity and cost-effectiveness (Pan et al. 2011). Some research groups successfully doped Mn ions in carbon dots via a simple hydro/solvothermal method to create biomaterial-based magneto-luminescent nanosystems (Stepanidenko et al. 2023; Ali et al. 2024).

9.8 Application of Magneto-Luminescent Nanosystems for Multimodal Imaging and Therapy of Cancer

Magneto-luminescent nanosystem primarily integrates two bioimaging modalities, magnetic resonance imaging (MRI)/fluorescence imaging (FLI), in one platform. MRI is capable of producing high-resolution three-dimensional images without tissue attenuation. On the contrary, FLI provides high sensitivity and the ability for real-time imaging, which are limited in MRI. This combinatorial imaging superiority has gained significant attention in designing magneto-luminescent nanosystems. Dual imaging can precisely localize cancerous tissue for image-guided therapeutics. Moreover, magneto-luminescent systems can be integrated with other therapeutic modalities such as drugs, and dyes capable of photon-induced hyperthermia. Additionally, the magneto-luminescent systems incorporating SPIONs as magnetic components can be heated via alternating magnetic field thus causing hyperthermia-mediated localized cell death. Multifunctional theragnostic capabilities of magneto-luminescent nanosystems led to the designing of various nanoconjugates demonstrating great potential in in vivo cancer imaging and therapy. Li et al. incorporated NIR emitting dye ICG, which is FDA-approved, and perfluorocarbons with ^{19}F, which is a stable isotope of fluorine capable of T2-weighted MR imaging in poly(lactic-co-glycolic acid) PLGA (Li et al. 2020). The versatile nanosystem showed time-dependent tumor accumulation along with background-free MR images. The presence of ICG enabled the localization of the magneto-luminescent nanosystem via NIR-FLI along with photoacoustic visualization. ICG also produced enough localized heat upon NIR excitation, leading to the successful treatment of tumor-bearing mice. In another biocompatible self-assembling magneto-luminescent nanohybrid, HSA-GGD-ICG was used to effectively image tumor sites via simultaneous MR and FL imaging (Fig. 9.4a, b) (Wang et al. 2018a). The nanohybrid accumulated at the tumor site after administration and was able to effectively treat it through ICG-mediated photothermal effect with a high thermal conversion of 85% (Fig. 9.4c, d). Instead of encapsulation of Gd-based MRI probes, Zao et al. constructed a core-shell nanocomposite consisting of NIR emitting persistent luminescent phosphor $ZnGa_2O_4$:Cr^{3+}, Sn^{4+} impregnated on mesoporous silica nanoparticles as a core (Zou et al. 2017). The resulting magneto-luminescent nanoplatform not only presents high NIR persistent luminescence in mice, but also exhibits high longitudinal relaxation due to the Gd_2O_3 shell. In another similar approach, multishell upconversion nanoparticle of $NaYF_4$ was developed. Each shell consisted of a different dopant combination Yb^{3+}, Er^{3+}, and Nd^{3+}, and the

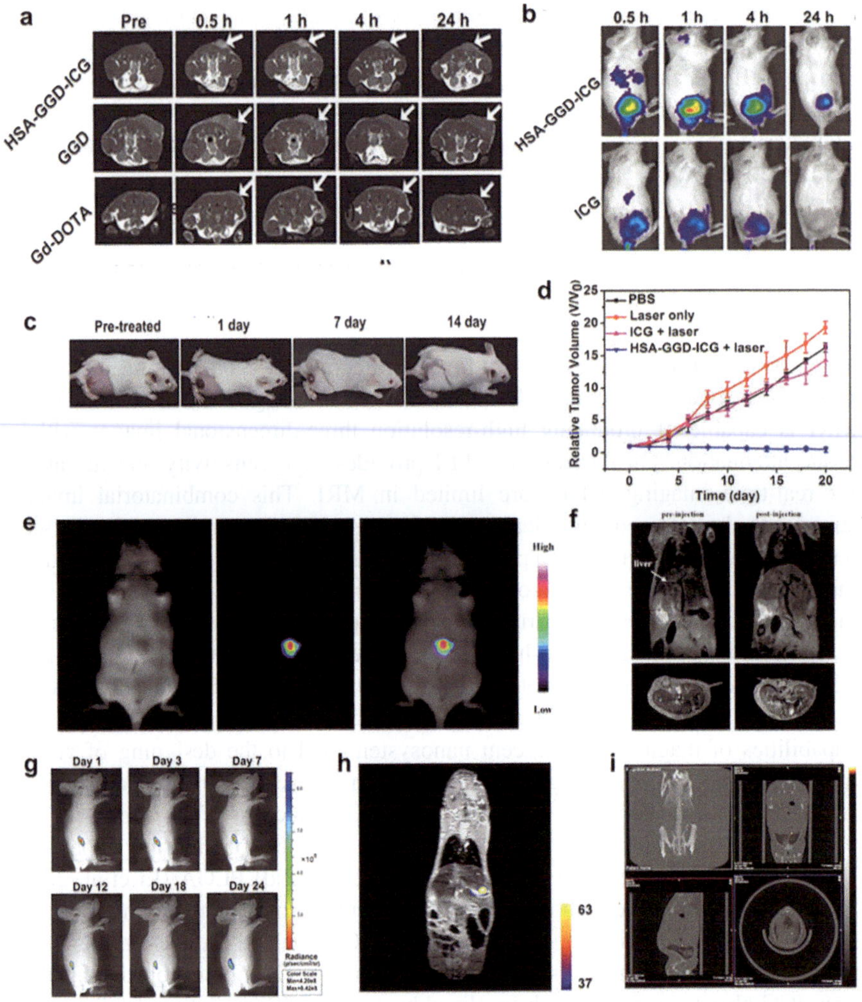

Fig. 9.4 In vivo theragnostic applications of magneto-luminescent nanomaterials: (**a**) In vivo MR imaging of H22 tumor-bearing mice taken at different time points after intravenous (i.v.) injection of HSA-GGD-ICG NPs, GGD, and Gd-DOTA at the same dose of 4.7 mg Gd per kg body weight (i.e., 1 g NPs, 32 mg GGD,20 mg ICG per kg). (**b**) In vivo fluorescence images of H22 tumor-bearing mice taken at different time points after intravenous (i.v.) injection of HSA-GGD-ICG NPs or ICG at the same dose of 0.2 mg ICG per kg body weight (i.e., 10 mg NPs and 0.32 mg GGD per kg). (**c**) Representative digital photographs of one mouse from HSA-GGD-ICG + laser group taken at various time points. (**d**) Corresponding tumor growth curves. (**e**) In vivo T_1-weighted MR imaging of mice before and after intravenous injection of $mSiO_2$@GGO NPs (top: coronal plane, bottom: transverse plane). (**f**) In vivo NIR-II luminescence imaging of a normal mouse after subcutaneous injection of $mSiO_2$@GGO NPs.g) In vivo fluorescence images of long-term cell tracing by TSP NPs ($n = 3$). (**g**) Representative in vivo fluorescence images of the mouse subcutaneously injected with 4104 of HuH-7 cells from day 0 to day 24. (**h**) Three-dimensional MPI-CT fusion images in vivo. (**i**) MPI-MRI fusion images in vivo. [**a–c** reproduced with permission from reference Wang et al. (2018a), copyright 2018, Wiley-VCH GmbH. **e**, **f** reproduced with permission from reference Shi et al. (2018), copyright 2018, Elsevier. **g–i** reproduced with permission from reference Meng et al. (2019), copyright 2019, Royal Society of Chemistry]

final shell was made of $NaGdF_4$ (Liu et al. 2016). It is worth noticing that even after several layers of coating, the authors were able to limit the size of the final nanocomposite to ~47 nm. By introducing Nd^{3+} in the luminescent system, the upconversion efficiency was enhanced ten-fold upon 800 nm excitation. In addition, the presence of Gd ion in the final shell enabled high R1 relaxivity for in vivo MR imaging. Garnet-based magneto-persistent luminescent materials with spherical morphology were synthesized via the ion impregnation method on mesoporous silica by Shi and Coworkers (2018). The synthesized nanocomposite presented in vivo persistent luminescence in NIR fluorescence emission upon excitation with UV and 808 nm photons, respectively (Fig. 9.4e). Also, nanocomposites showed high-contrast MR images with an R1 value of 10.70 mM^{-1} s^{-1} (Fig. 9.4f).

Incorporating Gd ions to design magneto-luminescent nanomaterials only provides the ability for MR imaging contrast capability. However, SPIONs present in magneto-luminescent nanomaterials not only act as MR contrast agents but can also provide cancer therapy through magnetic hyperthermia. Yang and his collaborator demonstrated how the incorporation of SPIONs and NIR emitting PbS/CdS quantum dots in large pores of mesoporous silica has high potential in cancer theragnostic (Yang et al. 2019). The nanoplatform was able to create T2 contrast MR images with an extremely high R2 value of 142 mM^{-1} s^{-1}. The large pores were further loaded with doxorubicin, a drug showing pH-responsive release. Additionally, the presence of SPIONs in the nanosystem generates a magneto-thermal effect, which speeds up the drug release. Polystyrene-polyethylene glycol self-assembly was used in another report to encapsulate NIR emitting dye TPABDFN along with SPIONs (Fig. 9.3f) (Meng et al. 2019). The magneto-luminescent nanoassembly exhibits high biocompatibility and stability, making it capable of long-term in vivo monitoring via FL imaging for 24 days (Fig. 9.4g). Magnetic particle imaging (MPI) is a new in vivo imaging modality that can locate and quantify SPIONs (Gleich and Weizenecker 2005). The nanoassembly was able to generate good contrast for MR imaging along with magnetic particle imaging (MPI) simultaneously (Fig. 9.4h). Also, it could be used for simultaneous computed tomography (CT) and MPI imaging (Fig. 9.4i). The research conducted in the development of several magneto-luminescent nanosystems employing various synthesis strategies has shown promising capabilities for multimodal cancer diagnosis and treatment (Table 9.2).

9.9 Conclusion and Future Perspective

Integrating magnetic and luminescence attributes in a single nanosystem provides applicability in various areas of physics, chemistry, and biomedicine. In the area of biomedicine, magneto-luminescent nanosystems have shown promising results related to the diagnosis and treatment of diseases such as cancer. Most of the research efforts have been made to design novel interaction strategies for efficient, biocompatible, and nanosized magneto-luminescent systems, subsequently validating the designed nanosystems for MRI and FLI dual imaging along with the

Table 9.2 Overview of magneto-luminescent nanosystems for multimodal imaging and cancer therapy applications

Magneto-luminescent nanosystem	Strategy	Luminescent Component	Excitation	Emission	Magnetic Component	Contrast Type	Relaxivity (r1/r2)	Imaging Type	Reference
Nd-doped $HAp/Fe_3O_4@mSiO_2$	Co-encapsulation of HAp and iron oxide in mesoporous silica	$Ca_{10}(PO_4)_6(OH)_2$: Nd	808 nm	880, 1064 nm	Fe_3O_4	MRI	–	MRI	Sengar et al. (2024)
AM-PP@ICGNPs	Encapsulated in PLGA, functionalized with A549 membrane	Indocyanine green (ICG)	745 nm	840 nm	Fe_3O_4	T1-T2 contrast	–	MR/PA/FL imaging	Li et al. (2020)
$mSiO_2@PbS/CdS-Fe_3O_4$ NPs	Loading on the large pore mesoporous silica	PbS/CdS (NIR)	806 nm	1200 nm	Fe_3O_4	T2 contrast	r2: 142 mM^{-1} s^{-1}	MRI	Yang et al. (2019)
HSA-GGD-ICG NPs	Co-encapsulation in human serum albumin	Indocyanine green (ICG)	808 nm	840 nm	Gd-based MRI contrast	T1 contrast	–	MRI/FLI	Wang et al. (2018a)
GAGG:Ce,Pr nanophosphor	Hydrothermal synthesis	$Gd_3Al_{5-x}Ga_xO_{12}:Ce^{3+}, Pr^{3+}$	455 nm	550 nm	Gd-doping	T1-T2 contrast	–	MRI	Sengar et al. (2019)
QD@MHS NPs	Hollow iron oxide shell with CdSeTeS QD core	CdSeTeS QDs		630 nm	Iron oxide shell	FLI & Electrochemical dual probe	–	FLI	Ganganboina et al. (2020)

QSC-lip	Cilengitide (CGT) encapsulated in liposomes	Carbon dots (CDs)	UV	Red	SPIONs	T1/T2 contrast	r1: 1.56 $mM^{-1}\ s^{-1}$, r2: 172.6 $mM^{-1}\ s^{-1}$	MRI	Xu et al. (2018)
FeCD	Microwave synthesis	Carbon dots (CDs)	485 nm	530 nm	SPIONs	T2 contrast	r2: 118.3 $mM^{-1}\ s^{-1}$	MRI	Das et al. (2019)
$mSiO_2$@GGO NPs	Template synthesis using mesoporous silica	$Gd_3Ga_5O_{12}:Cr^{3+}$, Nd^{3+}	254 nm	745 nm	Gd-doping	T1 contrast	r1: 10.70 $mM^{-1}\ s^{-1}$	MRI	Shi et al. (2018)
FA-conjugated $Gd_2O_3:Eu^{3+}$ NPs	APTMS and FA coating	$Gd_2O_3:Eu^{3+}$ NPs	254 nm	613 nm	Gd-doping	–	–	FLI	Jain et al. (2018b)
$NaGdF_4:Yb^{3+},Tm^{3+}$ NPs	Amide-linked Gd-chelates to $NaGdF_4:Yb^{3+}$, Tm^{3+}	$NaGdF_4:Yb^{3+}$, Tm^{3+}	980 nm	800 nm	Gd-doping	T1 contrast	r1: 25 $mM^{-1}\ s^{-1}$	MRI/optical imaging	Lu et al. (2017)
Gd@Cdots	Template synthesis using mesoporous silica	Gd-dots	350 nm	410 nm	Gd-doping	T1 contrast	r1: 10 $mM^{-1}\ s^{-1}$	MRI/ fluorescence microscopy	Chen et al. (2016)
Mn-doped CDs	Direct synthesis (hydrothermal)	Carbon dots	418 nm	560 nm	Mn-doping	T1 contrast	r1: 7.4 $mM^{-1}\ s^{-1}$	MRI	Stepanidenko et al. (2023)
TB/SPIO@PS-PEG (TSP NPs)	PS-PEG coating	Terbium (Tb)	480 nm	655 nm	SPIONs	T2 contrast	r2: 51.82 $mM^{-1}\ s^{-1}$	MRI/FMI/ MPI	Ali et al. (2024)

therapeutic functionality in in vivo models. The chapter covers the fundamentals of luminescence and magnetism, focusing on their biomedical applicability. Additionally, different strategies of magneto-luminescent nanomaterials design have been discussed extensively. Moreover, the chapter also includes various instances where the magneto-luminescent materials have demonstrated promising functionality in vivo related to the dual-modal imaging and therapy for cancer theranostics.

From the literature survey, it can be determined that mesoporous silica, biopolymers like (HSA, PLGA, etc.), and chemical functionalization of organic functional moieties are the most used strategies to create magneto-luminescent nanosystems. The use of magnetic and luminescent components that are already FDA-approved or under clinical trials for designing magneto-luminescent nanoplatforms is undoubtedly the most effective approach. This will allow faster clinical adaptation of the efficient magneto-luminescent nanosystems in the clinics. Many efficient nanosystems are only at the prototype stage, which will require further optimization for human applications and a better understanding of biodistribution/long-term biocompatibility. Additionally, most of the research lacks reproducibility in terms of material synthesis, which is very important for functional nanoproduct development. Furthermore, the use of complicated combination strategies may also lead to no reproducibility along with the high cost of industrial production. Another factor that is also crucial for the future development of magneto-luminescent materials is the dosage. For example, fluorescent materials being more sensitive are required in lower quantities than magnetic nanomaterials. Therefore, it is important to consider the dose ratio of each component for the optimal performance of the complete nanosystem. Future efforts should focus on establishing clinical systems that can evaluate the real-time efficacy of the magneto-luminescent nanosystems. The luminescent property intrinsic in magneto-luminescent nanoplatforms enables noninvasive visualization with high sensitivity, providing fast feedback, thus potentially influencing the real-time assessment of efficacy in clinics. In conclusion, magneto-luminescent nanoplatforms have demonstrated significant potential at the prototype stage for combined imaging and therapeutic applications in cancer. While challenges remain to be addressed, we believe that through collaborative efforts across various research fields, these nanoplatforms can achieve their full potential for clinical application.

References

Ali A, Shah T, Ullah R, Zhou P, Guo M, Ovais M, Tan Z, Rui YK (2021) Review on recent progress in magnetic nanoparticles: synthesis, characterization, and diverse applications. Front Chem 9: 629054. https://doi.org/10.3389/FCHEM.2021.629054/BIBTEX

Ali V, Kefayati H, Shafiee Ardestani M, Pourahmad A (2024) Synthesis and evaluation of new magneto-fluorescent carbon dot based on manganese citrate for MRI imaging. MAGMA 37: 139–148. https://doi.org/10.1007/s10334-023-01117-8

Ash C, Dubec M, Donne K, Bashford T (2017) Effect of wavelength and beam width on penetration in light-tissue interaction using computational methods. Lasers Med Sci 32:1909–1918. https://doi.org/10.1007/S10103-017-2317-4/FIGURES/8

Avram D, Tiseanu I, Vasile BS, Florea M, Tiseanu C (2018) Near infrared emission properties of Er doped cubic sesquioxides in the second/third biological windows. Sci Rep 8(1):1–12. https://doi.org/10.1038/s41598-018-36639-y

Bányai L, Koch SW (1993) Semiconductor quantum dots, vol 2. https://doi.org/10.1142/2019

Behrens S, Appel I (2016) Magnetic nanocomposites. Curr Opin Biotechnol 39:89–96. https://doi.org/10.1016/j.copbio.2016.02.005

Blanco-Andujar C, Teran FJ, Ortega D (2018) Current outlook and perspectives on nanoparticle-mediated magnetic hyperthermia. In: Iron oxide nanoparticles for biomedical applications, pp 197–245. https://doi.org/10.1016/B978-0-08-101925-2.00007-3

Carron S, Li QY, Vander Elst L, Muller RN, Parac-Vogt TN, Capobianco JA (2015) Assembly of near infra-red emitting upconverting nanoparticles and multiple Gd(iii)-chelates as a potential bimodal contrast agent for MRI and optical imaging. Dalton Trans 44:11331–11339. https://doi.org/10.1039/c5dt00919g

Castaing V, Lozano G, Míguez H (2022) Transparent phosphor thin films based on rare-earth-doped garnets: building blocks for versatile persistent luminescence materials. Adv Photonics Res 3. https://doi.org/10.1002/adpr.202100367

Chávez-García D, Sengar P, Juárez-Moreno K, Flores DLL, Calderón I, Barrera J, Hirata GAA (2021) Luminescence properties and cell uptake analysis of Y2O3:Eu, bi nanophosphors for bio-imaging applications. J Mater Res Technol 10:797–807. https://doi.org/10.1016/j.jmrt.2020.11.071

Chen G, Qiu H, Prasad PN, Chen X (2014) Upconversion nanoparticles: design, nanochemistry, and applications in theranostics. Chem Rev 114:5161–5214. https://doi.org/10.1021/CR400425H/ASSET/IMAGES/LARGE/CR-2013-00425H_0033.JPEG

Chen H, Wang GD, Sun X, Todd T, Zhang F, Xie J, Shen B (2016) Mesoporous silica as nanoreactors to prepare Gd-encapsulated carbon dots of controllable sizes and magnetic properties. Adv Funct Mater 26:3973–3982. https://doi.org/10.1002/adfm.201504177

Dadfar SM, Camozzi D, Darguzyte M, Roemhild K, Varvarà P, Metselaar J, Banala S, Straub M, Güvener N, Engelmann U, Slabu I, Buhl M, Van Leusen J, Kögerler P, Hermanns-Sachweh B, Schulz V, Kiessling F, Lammers T (2020) Size-isolation of superparamagnetic iron oxide nanoparticles improves MRI, MPI and hyperthermia performance. J Nanobiotechnol 18. https://doi.org/10.1186/s12951-020-0580-1

Das B, Girigoswami A, Dutta A, Pal P, Dutta J, Dadhich P, Srivas PK, Dhara S (2019) Carbon nanodots doped super-paramagnetic iron oxide nanoparticles for multimodal bioimaging and osteochondral tissue regeneration via external magnetic actuation. ACS Biomater Sci Eng 5: 3549–3560. https://doi.org/10.1021/acsbiomaterials.9b00571

Dubey V, Som S, Kumar V (2020) Luminescent materials in display and biomedical applications, luminescent materials in display and biomedical applications. CRC Press, Taylor & Francis Group, Boca Raton, FL. https://doi.org/10.1201/9780429025334/LUMINESCENT-MATERIALS-DISPLAY-BIOMEDICAL-APPLICATIONS-VIKAS-DUBEY-SUDIPTA-SOM-VIJAY-KUMAR

Fei X, Gu Y (2009) Progress in modifications and applications of fluorescent dye probe. Prog Nat Sci 19:1–7. https://doi.org/10.1016/J.PNSC.2008.06.004

Fu Q, Feng H, Liu L, Li Z, Li J, Hu J, Hu C, Yan X, Yang H, Song J (2022) Spatiotemporally controlled formation and rotation of magnetic nanochains in vivo for precise mechanotherapy of tumors. Angew Chem Int Ed 61. https://doi.org/10.1002/anie.202213319

Ganganboina AB, Chowdhury AD, Khoris IM, Doong RA, Li TC, Hara T, Abe F, Suzuki T, Park EY (2020) Hollow magnetic-fluorescent nanoparticles for dual-modality virus detection. Biosens Bioelectron 170:112680. https://doi.org/10.1016/J.BIOS.2020.112680

García de Arquer FP, Talapin DV, Klimov VI, Arakawa Y, Bayer M, Sargent EH (2021) Semiconductor quantum dots: technological progress and future challenges. Science 373. https://doi.org/10.1126/SCIENCE.AAZ8541/ASSET/C3AC7CAD-9110-4FD1-85BA-03C145F887FC/ASSETS/IMAGES/LARGE/SCIENCE.AAZ8541-F6.JPG

Gleich B, Weizenecker J (2005) Tomographic imaging using the nonlinear response of magnetic particles. Nature 435:1214–1217. https://doi.org/10.1038/nature03808

Goryacheva IY, Sapelkin AV, Sukhorukov GB (2017) Carbon nanodots: mechanisms of photoluminescence and principles of application. TrAC Trends Anal Chem 90:27–37. https://doi.org/10.1016/J.TRAC.2017.02.012

Grover VPB, Tognarelli JM, Crossey MME, Cox IJ, Taylor-Robinson SD, McPhail MJW (2015) Magnetic resonance imaging: principles and techniques: lessons for clinicians. J Clin Exp Hepatol 5:246–255. https://doi.org/10.1016/J.JCEH.2015.08.001

Huang X, Han S, Huang W, Liu X (2012) Enhancing solar cell efficiency: the search for luminescent materials as spectral converters. Chem Soc Rev 42:173–201. https://doi.org/10.1039/C2CS35288E

Huang G, Li H, Chen J, Zhao Z, Yang L, Chi X, Chen Z, Wang X, Gao J (2014) Tunable T1 and T2 contrast abilities of manganese-engineered iron oxide nanoparticles through size control. Nanoscale 6:10404–10412. https://doi.org/10.1039/C4NR02680B

Iyad N, Ahmad MS, Alkhatib SG, Hjouj M (2023) Gadolinium contrast agents-challenges and opportunities of a multidisciplinary approach: literature review. Eur J Radiol Open 11. https://doi.org/10.1016/j.ejro.2023.100503

Jackson CT, Jeong S, Dorlhiac GF, Landry MP (2021) Advances in engineering near-infrared luminescent materials. IScience 24. https://doi.org/10.1016/J.ISCI.2021.102156/ASSET/7FAB2C6C-1393-4049-9B95-6B1A61EBBAC3/MAIN.ASSETS/GR4_LRG.JPG

Jain A, Sengar P, Hirata GA (2018a) Rare-earth-doped Y 3 Al 5 O 12 (YAG) nanophosphors: synthesis, surface functionalization, and applications in thermoluminescence dosimetry and nanomedicine. J Phys D Appl Phys 51:303002. https://doi.org/10.1088/1361-6463/aaca49

Jain A, Fournier PGJ, Mendoza-Lavaniegos V, Sengar P, Guerra-Olvera FM, Iñiguez E, Kretzschmar TG, Hirata GA, Juárez P (2018b) Functionalized rare earth-doped nanoparticles for breast cancer nanodiagnostic using fluorescence and CT imaging. J Nanobiotechnol 16:26. https://doi.org/10.1186/s12951-018-0359-9

Jiang K, Zhang L, Bao G (2021) Magnetic iron oxide nanoparticles for biomedical applications. Curr Opin Biomed Eng 20. https://doi.org/10.1016/j.cobme.2021.100330

Jiang W, An W, Huang Z, Xu C, Shen Q, Pu C, Zhang S, Wu Q, Li L, Yu C (2024) A near-infrared fluorescent probe with two-photon excitation for in situ imaging of NQO1 in human colorectum cancer tissue. Talanta 274:126018. https://doi.org/10.1016/J.TALANTA.2024.126018

John SP (2021) Determination of ferrimagnetic and superparamagnetic components of magnetization and the effect of particle size on structural, magnetic and hyperfine properties of Mg0.5Zn0.5Fe2O4 nanoparticles. J Alloys Compd 869:159242. https://doi.org/10.1016/J.JALLCOM.2021.159242

Jung H, Park B, Lee C, Cho J, Suh J, Park JY, Kim YR, Kim J, Cho G, Cho HJ (2014) Dual MRI T1 and T2(*) contrast with size-controlled iron oxide nanoparticles. Nanomedicine 10:1679–1689. https://doi.org/10.1016/j.nano.2014.05.003

Khalid W, Zulfiqar A, Shehzad T, Haider SA, Abid Z, Altaf M, Ashraf RS, Shahid M (2023) Luminescent and photonic materials for sensing applications. Org Inorg Mater Based Sensors 3: 789–813. https://doi.org/10.1002/9783527834266.CH34

Lee SH, Kim BH, Bin Na H, Hyeon T (2014) Paramagnetic inorganic nanoparticles as T1 MRI contrast agents. Wiley Interdiscip Rev Nanomed Nanobiotechnol 6:196–209. https://doi.org/10.1002/wnan.1243

Li X, Wei J, Aifantis KE, Fan Y, Feng Q, Cui FZ, Watari F (2016) Current investigations into magnetic nanoparticles for biomedical applications. J Biomed Mater Res A 104:1285–1296. https://doi.org/10.1002/JBM.A.35654

Li Q, Kartikowati CW, Horie S, Ogi T, Iwaki T, Okuyama K (2017) Correlation between particle size/domain structure and magnetic properties of highly crystalline Fe3O4 nanoparticles. Sci Rep 7:1–7. https://doi.org/10.1038/s41598-017-09897-5

Li S, Jiang W, Yuan Y, Sui M, Yang Y, Huang L, Jiang L, Liu M, Chen S, Zhou X (2020) Delicately designed cancer cell membrane-camouflaged nanoparticles for targeted 19 F MR/PA/

FL imaging-guided photothermal therapy/. ACS Appl Mater Interfaces 12:57290–57301. https://doi.org/10.1021/acsami.0c13865

Li X, Fu Y, Zhao S, Xiao JF, Lan M, Wang B, Zhang K, Song X, Zeng L (2022) Metal ions-doped carbon dots: synthesis, properties, and applications. Chem Eng J 430:133101. https://doi.org/10.1016/J.CEJ.2021.133101

Liu Y, Kang N, Lv J, Zhou Z, Zhao Q, Ma L, Chen Z, Ren L, Nie L (2016) Deep photoacoustic/luminescence/magnetic resonance multimodal imaging in living subjects using high-efficiency upconversion nanocomposites. Adv Mater 28:6411–6419. https://doi.org/10.1002/adma.201506460

Liu JF, Jang B, Issadore D, Tsourkas A (2019) Use of magnetic fields and nanoparticles to trigger drug release and improve tumor targeting. Wiley Interdiscip Rev Nanomed Nanobiotechnol 11: e1571. https://doi.org/10.1002/WNAN.1571

Liu X, Zhang Y, Wang Y, Zhu W, Li G, Ma X, Zhang Y, Chen S, Tiwari S, Shi K, Zhang S, Fan HM, Zhao YX, Liang XJ (2020) Comprehensive understanding of magnetic hyperthermia for improving antitumor therapeutic efficacy. Theranostics 10:3793–3815. https://doi.org/10.7150/thno.40805

Lu Z, Deng R, Zhen M, Li X, Zou T, Zhou Y, Guan M, Zhang Y, Wang Y, Yu T, Shu C, Wang C (2017) Size-tunable NaGdF4 nanoparticles as: T 2 contrast agents for high-field magnetic resonance imaging. RSC Adv 7:43125–43131. https://doi.org/10.1039/c7ra08303c

Martynenko IV, Litvin AP, Purcell-Milton F, Baranov AV, Fedorov AV, Gun'Ko YK (2017) Application of semiconductor quantum dots in bioimaging and biosensing. J Mater Chem B 5: 6701–6727. https://doi.org/10.1039/C7TB01425B

Meng L, Ma X, Jiang S, Ji G, Han W, Xu B, Tian J, Tian W (2019) High-efficiency fluorescent and magnetic multimodal probe for long-term monitoring and deep penetration imaging of tumors. J Mater Chem B 7:5345–5351. https://doi.org/10.1039/C9TB00638A

Nagpal R, Gusain M (2022) Synthesis methods of quantum dots. In: Graphene, nanotubes and quantum dots-based nanotechnology: fundamentals and applications, pp 599–630. https://doi.org/10.1016/B978-0-323-85457-3.00006-2

Pan D, Schmieder AH, Wickline SA, Lanza GM (2011) Manganese-based MRI contrast agents: past, present, and future. Tetrahedron 67:8431–8444. https://doi.org/10.1016/j.tet.2011.07.076

Pan L, Sun S, Zhang L, Jiang K, Lin H (2016) Near-infrared emissive carbon dots for two-photon fluorescence bioimaging. Nanoscale 8:17350–17356. https://doi.org/10.1039/C6NR05878G

Qu J, Golovynska I, Liu J, Qu J, Golovynskyi S (2024) Optical transparency windows in near-infrared and short-wave infrared for the skin, skull, and brain: fluorescence bioimaging using PbS quantum dots. J Biophoton 17:e202400171. https://doi.org/10.1002/JBIO.202400171

Sengar P, Hirata GA, Farias MH, Castillón F (2016) Morphological optimization and (3-aminopropyl) trimethoxy silane surface modification of Y3Al5O12:Pr nanoscintillator for biomedical applications. Mater Res Bull 77:236–242. https://doi.org/10.1016/j.materresbull.2016.01.045

Sengar P, García-Tapia K, Can-Uc B, Juárez-Moreno K, Contreras-López OE, Hirata GA (2019) Simultaneous paramagnetic and persistence-luminescence in GAGG:Ce,Pr nanoparticles synthesized by sol-gel for biomedical applications. J Appl Phys 126:083107. https://doi.org/10.1063/1.5098788

Sengar P, Flores DL, Chauhan K, Can-Uc B, Juarez-Moreno K, Contreras OE, Digman MA, Hirata GA (2020) Visible/near infrared emitting garnet-based paramagnetic-persistence luminescent nanocrystals for two-photon bioimaging. Cryst Growth Des 20. https://doi.org/10.1021/acs.cgd.0c00577

Sengar P, Chauhan K, Hirata GA (2022) Progress on carbon dots and hydroxyapatite based biocompatible luminescent nanomaterials for cancer theranostics. Transl Oncol 24:101482. https://doi.org/10.1016/J.TRANON.2022.101482

Sengar P, Rojas-Baldivia K, Rodríguez-Hernández AG, Hirata GA (2023) Synthesis and optimization of Nd3+doped hydroxyapatite as NIR-to-NIR nanothermometers. Mater Today Commun 37:107117. https://doi.org/10.1016/J.MTCOMM.2023.107117

Sengar P, Chauhan K, De la Cruz W, Rodríguez-Hernández AG, Hirata GA (2024) Discovering the potential of biomaterial-based mesoporous and magneto-luminescent nanohybrid (Nd-doped hydroxyapatite/Fe_3O_4) for cancer theragnosis. Mater Today Commun 41:110299. https://doi.org/10.1016/j.mtcomm.2024.110299

Shi J, Sun X, Zheng S, Li J, Fu X, Zhang H (2018) A new near-infrared persistent luminescence nanoparticle as a multifunctional nanoplatform for multimodal imaging and cancer therapy. Biomaterials 152:15–23. https://doi.org/10.1016/j.biomaterials.2017.10.032

Shirata C, Kaneko J, Inagaki Y, Kokudo T, Sato M, Kiritani S, Akamatsu N, Arita J, Sakamoto Y, Hasegawa K, Kokudo N, Shirata C, Inagaki Y, Kiritani S, Kokudo T, Sato M, Hasegawa K, Kokudo N, Kaneko J, Akamatsu N (2017) Near-infrared photothermal/photodynamic therapy with indocyanine green induces apoptosis of hepatocellular carcinoma cells through oxidative stress. Sci Rep 7:13958. https://doi.org/10.1038/s41598-017-14401-0

Stepanidenko EA, Vedernikova AA, Badrieva ZF, Brui EA, Ondar SO, Miruschenko MD, Volina OV, Koroleva AV, Zhizhin EV, Ushakova EV (2023) Manganese-doped carbon dots as a promising nanoprobe for luminescent and magnetic resonance imaging. Photonics 10. https://doi.org/10.3390/photonics10070757

Teets TS (2021) Photoluminescence. https://doi.org/10.1021/ACSINFOCUS.7E5014

Tejeda EM, Arámburo M, Vargas E, Contreras OE, Hirata GA (2018) Novel bifunctional Nd:YAG/Fe3O4 nanocomposite as nanothermometer/nanoheater for potential biomedical applications. J Phys D Appl Phys 51:40LT01. https://doi.org/10.1088/1361-6463/AAD7E5

Thakor AS, Jokerst JV, Ghanouni P, Campbell JL, Mittra E, Gambhir SS (2016) Clinically approved nanoparticle imaging agents. J Nucl Med 57:1833–1837. https://doi.org/10.2967/JNUMED.116.181362

Ueda J, Tanabe S (2019) Review of luminescent properties of Ce3+-doped garnet phosphors: new insight into the effect of crystal and electronic structure. Opt Mater X(1). https://doi.org/10.1016/j.omx.2019.100018

Wang L, Lin H, Chi X, Sun C, Huang J, Tang X, Chen H, Luo X, Yin Z, Gao J (2018a) A self-assembled biocompatible nanoplatform for multimodal MR/fluorescence imaging assisted photothermal therapy and prognosis analysis. Small 14:1801612. https://doi.org/10.1002/smll.201801612

Wang H, Li X, Tse BWC, Yang H, Thorling CA, Liu Y, Touraud M, Chouane JB, Liu X, Roberts MS, Liang X (2018b) Indocyanine green-incorporating nanoparticles for cancer theranostics. Theranostics 8:1227–1242. https://doi.org/10.7150/thno.22872

Wang Z, Liu C, Zhao Y, Hu M, Ma D, Zhang P, Xue Y, Li X (2019) Photomagnetic nanoparticles in dual-modality imaging and photo-sonodynamic activity against bacteria. Chem Eng J 356:811–818. https://doi.org/10.1016/j.cej.2018.09.077

Wen H, Wang F (2014) Lanthanide-doped nanoparticles: synthesis, property, and application. In: Nanocrystalline materials: their synthesis-structure-property relationships and applications. Elsevier B.V, pp 121–160. https://doi.org/10.1016/B978-0-12-407796-6.00004-X

Wo F, Xu R, Shao Y, Zhang Z, Chu M, Shi D, Liu S (2016) A multimodal system with synergistic effects of magneto-mechanical, photothermal, photodynamic and chemo therapies of cancer in graphene-quantum dot-coated hollow magnetic nanospheres. Theranostics 6:485–500. https://doi.org/10.7150/thno.13411

Woodruff DN, Winpenny REP, Layfield RA (2013) Lanthanide single-molecule magnets. Chem Rev 113:5110–5148. https://doi.org/10.1021/CR400018Q/ASSET/CR400018Q.FP.PNG_V03

Wu J, Shi Z, Zhu L, Li J, Han X, Xu M, Hao S, Fan Y, Shao T, Bai H, Peng B, Hu W, Liu X, Yao C, Li L, Huang W (2022) The design and bioimaging applications of NIR fluorescent organic dyes with high brightness. Adv Opt Mater 10:2102514. https://doi.org/10.1002/ADOM.202102514

Xu HL, Yang JJ, ZhuGe DL, Lin MT, Zhu QY, Jin BH, Tong MQ, Shen BX, Xiao J, Zhao YZ (2018) Glioma-targeted delivery of a theranostic liposome integrated with quantum dots, superparamagnetic iron oxide, and Cilengitide for dual-imaging guiding cancer surgery. Adv Healthc Mater 7. https://doi.org/10.1002/adhm.201701130

Yang F, Skripka A, Tabatabaei MS, Hong SH, Ren F, Huang Y, Oh JK, Martel S, Liu X, Vetrone F, Ma D (2019) Magnetic photoluminescent nanoplatform built from large-pore mesoporous silica. Chem Mater 31:3201–3210. https://doi.org/10.1021/acs.chemmater.9b00028

Yang Y, Tu D, Zhang Y, Zhang P, Chen X (2021) Recent advances in design of lanthanide-containing NIR-II luminescent nanoprobes. iScience. https://doi.org/10.1016/j.isci

Yang H, Wang H, Wen C, Bai S, Wei P, Xu B, Xu Y, Liang C, Zhang Y, Zhang G, Wen H, Zhang L (2022a) Effects of iron oxide nanoparticles as T 2-MRI contrast agents on reproductive system in male mice. J Nanobiotechnol 20. https://doi.org/10.1186/s12951-022-01291-2

Yang CW, Liu K, Yao CY, Li B, Juhong A, Qiu Z, Huang X (2022b) Indocyanine green-conjugated superparamagnetic iron oxide nanoworm for multimodality breast cancer imaging. ACS Appl Nano Mater 5:18912–18920. https://doi.org/10.1021/acsanm.2c04687

Yin X, Russek SE, Zabow G, Sun F, Mohapatra J, Keenan KE, Boss MA, Zeng H, Liu JP, Viert A, Liou SH, Moreland J (2018) Large T1 contrast enhancement using superparamagnetic nanoparticles in ultra-low field MRI. Sci Rep 8(1):11863. https://doi.org/10.1038/s41598-018-30264-5

Zhang R, Cheng K, Antaris AL, Ma X, Yang M, Ramakrishnan S, Liu G, Lu A, Dai H, Tian M, Cheng Z (2016a) Hybrid anisotropic nanostructures for dual-modal cancer imaging and image-guided chemo-thermo therapies. Biomaterials 103:265–277. https://doi.org/10.1016/j.biomaterials.2016.06.063

Zhang M, Zheng ST, Liu XJ, Long Y, Yang BQ (2016b) A series of novel NIR fluorescent dyes: synthesis, theoretical calculations and fluorescence imaging applications in living cells. Dyes Pigments 125:220–228. https://doi.org/10.1016/J.DYEPIG.2015.10.028

Zou R, Gong S, Shi J, Jiao J, Wong K-L, Zhang H, Wang J, Su Q (2017) Magnetic-NIR persistent luminescent dual-modal ZGOCS@MSNs@Gd_2O_3 core–shell nanoprobes for in vivo imaging. Chem Mater 29:3938–3946. https://doi.org/10.1021/acs.chemmater.7b00087

Nanomaterial-Based Contrast and Therapeutic Agents

10

Sasvat Sayee Ram Ramesh, Paloma Patra, and Aravind Kumar Rengan

Abstract

Nanomaterials have emerged as powerful tools in biomedical imaging and therapy, offering unparalleled advantages as contrast and therapeutic agents. Their unique physicochemical properties, including high surface-to-volume ratios, tunable optical and magnetic characteristics, and functionalization capabilities, enable enhanced imaging precision across multiple modalities such as MRI, CT, ultrasound, and fluorescence imaging. Beyond their diagnostic applications, nanomaterials play a pivotal role in targeted drug delivery systems, improving therapeutic efficacy while minimizing systemic toxicity. This chapter explores their integration into advanced therapeutic strategies, such as photothermal and photodynamic therapies, where light-responsive and heat-generating nanoparticles selectively eradicate diseased tissues. Additionally, the concept of theranostics is examined, highlighting the potential of nanomaterials to function simultaneously as diagnostic and therapeutic agents, facilitating personalized and combination treatment approaches. Despite their promising applications, key challenges remain, including biocompatibility, toxicity, regulatory barriers, and scalability for clinical translation. Addressing these hurdles is crucial for the widespread adoption of nanomaterial-based medical technologies. This chapter provides a comprehensive review of the latest advancements, challenges, and future directions in the field, underscoring the transformative potential of nanomaterials in modern medicine.

S. S. R. Ramesh · P. Patra · A. K. Rengan (✉)
Department of Biomedical Engineering, Indian Institute of Technology, Hyderabad, India
e-mail: aravind@bme.iith.ac.in

R. Prasad, M. Mohan Yallapu (eds.), *Advancements in Cancer Theranostics*,
https://doi.org/10.1007/978-981-95-0938-6_10

Keywords

Nanomedicine · Photothermal therapy of cancer · Breast cancer · Theranostics · Nanoimaging · Clinical translational

10.1 Introduction

Nanomaterials are revolutionizing biomedicine by offering innovative solutions for disease diagnosis, treatment, and prevention. Their unique properties, including high surface area, tunable size, and functionalization capabilities, make them highly versatile for applications such as drug delivery, cancer immunotherapy, antiviral treatments, and tissue engineering. In cancer immunotherapy, nanomaterials serve as drug delivery platforms and immunomodulators, facilitating immunogenic cell death (ICD) and regulating immune responses to enhance anticancer effects (Li et al. 2022). In breast cancer, nanomedicine helps overcome immune escape mechanisms, improving the retention of immunostimulatory agents and modulating the tumor microenvironment for better tumor (Navarro-Ocón et al. 2022). Nanomaterials also demonstrate strong antiviral potential. $CuInP_2S_6$ (CIPS) nanosheets, for instance, bind to the spike protein of SARS-CoV-2, preventing viral entry and promoting elimination by macrophages (Zhang et al. 2022a). Virus-like particles (VLPs) are used in vaccine development, generating cross-reactive antibodies to combat infectious diseases, including COVID-19 (Hills and Howarth 2022).

In drug delivery, polymer-based nanomaterials improve the stability, bioavailability, and targeted transport of biomacromolecular therapeutics, such as proteins and nucleic acids (Machtakova et al. 2022). They are also used in antifungal drug delivery, enhancing drug loading efficiency and targeting precision, though clinical translation remains challenging (Nagasa and Belete 2022). Tissue engineering benefits from electroactive nanomaterials, which promote cell culture and regeneration, particularly in nerve and cardiac tissue applications (Xiong et al. 2022a). Carbon-based nanomaterials contribute to biofuel cells for biomedical devices by enhancing electron transfer and energy generation, which is essential for powering implantable medical technologies (ul Haque et al. 2023). Nanotechnology is also advancing diagnostics and therapeutics. In acute pancreatitis, nanoparticles enhance biomarker detection, imaging, and anti-inflammatory treatments (Jia et al. 2022). For respiratory diseases, nanomedicine plays a key role in drug delivery and vaccine development, particularly in mRNA vaccines and nanoparticle-based strategies to prevent conditions like COVID-19 (Yang et al. 2022). Through these diverse applications, nanomaterials continue to redefine biomedical research and clinical practice.

Nanomaterials exhibit distinct physical properties that make them exceptionally well-suited for both contrast enhancement and therapeutic applications in medical imaging and treatment. Their size, surface characteristics, and interactions with biological systems can be precisely engineered to optimize imaging contrast and facilitate the effective delivery of therapeutic agents. These properties play a crucial role in determining their performance in biomedical applications. One of the most significant attributes of nanoparticles is their high surface area-to-volume ratio, which enhances their reactivity and ability to interact with biological molecules. This characteristic is particularly valuable in drug delivery and imaging, as it allows for an increased loading capacity of therapeutic agents or contrast materials (Pavitra et al. 2022). Additionally, their nanoscale dimensions enable them to penetrate biological barriers, such as the blood-brain barrier, which is critical for targeting neurological conditions like gliomas. A notable example is Ti@FeAu core-shell nanoparticles, which have demonstrated improved permeability and enhanced contrast in glioma imaging (Thirumurugan et al. 2022).

Nanoparticles can also be engineered to exhibit specific optical and magnetic properties that are highly advantageous in biomedical imaging. For instance, up-conversion luminescence, a phenomenon observed in certain nanoparticles, makes them useful for optical thermometry and bioimaging. AgGd(MoO4)2:Er3+/Yb3+ nanoplates, for example, have been shown to possess enhanced luminescence and temperature-sensing capabilities, making them ideal for multimodal imaging (Pavitra et al. 2022). In the context of MRI, magnetic nanoparticles can be tailored to improve contrast by modifying their magnetic properties. Superparamagnetic nanoparticles, in particular, contribute to the enhancement of the negative contrast effect in MRI, thereby enabling more precise tumor localization (Thirumurugan et al. 2022).

Ensuring the biocompatibility of nanoparticles is essential for their safe application in medicine. Surface modifications, such as coating nanoparticles with biocompatible materials or functionalizing them with targeting ligands, can enhance their compatibility and reduce toxicity. For example, angiopep-2-decorated nanoparticles have demonstrated improved targeting and uptake by glioma cells, thereby enhancing their selective cytotoxicity (Thirumurugan et al. 2022). Furthermore, functionalization facilitates the integration of therapeutic agents, allowing nanoparticles to serve as both diagnostic and therapeutic tools, a concept known as theranostics. This dual functionality is exemplified by nanoparticles that not only deliver drugs but also provide imaging contrast, thereby enabling real-time monitoring of therapeutic effects (Zhang et al. 2022b). The interaction of nanoparticles with biological systems, such as pulmonary surfactants, significantly influences their distribution and therapeutic efficacy. A thorough understanding of these interactions is crucial for designing nanoparticles that minimize potential adverse effects while optimizing therapeutic outcomes (Liu et al. 2022a). Additionally, nanoparticles have the capacity to modulate immune responses, a property that is particularly relevant in immunotherapy. By tailoring their physicochemical characteristics, nanoparticles can be engineered to either activate or suppress the immune system, depending on the specific therapeutic requirement (Deng et al. 2022).

10.2 Importance of Contrast and Therapeutic Agents in Diagnostics and Treatment

The integration of contrast and therapeutic agents in medical diagnostics and treatment enhances imaging accuracy and therapeutic efficacy, playing a crucial role in personalized medicine. Contrast agents improve the visibility of internal structures in imaging techniques like MRI, PET, and ultrasound, while therapeutic agents facilitate targeted treatment. The development of advanced nanoscale and organic contrast agents has expanded imaging capabilities, leading to improved disease management and outcomes. NIR-II organic contrast agents offer deep tissue penetration and reduced light scattering, making them effective for optical imaging applications such as tumor and brain imaging (Jiang et al. 2022). In MRI, nanomaterial-based contrast agents, including metal oxide nanoparticles and quantum dots, enhance diagnostic accuracy and enable MRI-guided therapies like targeted drug delivery and hyperthermia (Wang et al. 2024a). Gadolinium-based agents also contribute to PET/MR imaging without significantly affecting PET quantification (Allen et al. 2022). Ultrasound contrast agents, particularly nanoscale formulations, facilitate molecular imaging and drug delivery, broadening their biomedical applications (Wang et al. 2024b). Theranostic approaches leverage iron oxide nanoparticles (IONPs) as both contrast agents and drug carriers, enabling real-time monitoring of treatment response and personalized therapies (Rahman 2023). MRI contrast agents further enhance radiotherapy precision by improving target delineation and treatment planning (Boldrini et al. 2023). Hybrid contrast agents, such as DOTA-gold nanoparticles, demonstrate both imaging and therapeutic potential, proving effective in cancer treatment under physiological conditions (Khan et al. 2022a).

Despite these advancements, challenges remain in optimizing nanosystems for imaging performance, as both large and small systems have distinct advantages (Zhou et al. 2022). Additionally, innovative MRI contrast agents, like doped CaF2 nanoparticles, are being explored for radiation dose measurement, potentially enhancing the safety and efficacy of radiation-based therapies (Frencken et al. 2024). However, issues such as toxicity, stability, and the need for standardized guidelines hinder clinical translation. Further research is essential to integrate these agents into personalized medicine strategies effectively. Nevertheless, ongoing developments in contrast and therapeutic agents hold great promise for advancing precision medicine and improving healthcare outcomes.

10.3 Why Nanomaterials Are Ideal Candidates for These Applications

Nanomaterials are increasingly recognized as ideal candidates for contrast and therapeutic agents due to their unique physicochemical properties, enabling enhanced imaging and targeted therapy. Their nanoscale size allows better penetration and interaction with biological systems, making them highly effective in precision medicine by improving diagnostic accuracy and therapeutic efficacy. These materials offer high sensitivity and specificity in diagnostic applications, with metal nanoparticles enhancing early disease detection due to their high surface area and optical properties (Jouyandeh et al. 2022). Magnetic nanoparticles (MNPs) provide superior contrast in MRI through their superparamagnetic properties, improving imaging resolution without ionizing radiation risks (Mohapatra et al. 2023). Fluorine-19 MRI benefits from nanocarriers encapsulating fluorinated compounds, increasing local contrast agent concentration, and enhancing diagnostic accuracy (Joseph et al. 2022). Beyond imaging, nanomaterials play a crucial role in targeted therapeutic delivery. They enable controlled drug release, minimizing systemic side effects and ensuring precise treatment, particularly in conditions like thrombosis (Zhao et al. 2012). Functionalization enhances biocompatibility and enables targeting of specific biological pathways, such as the spike protein of SARS-CoV-2, improving their potential against viral infections (Zhang et al. 2022a). Theranostic platforms integrate diagnostic and therapeutic functions within a single nanomaterial, facilitating simultaneous disease detection and treatment, ultimately optimizing patient outcomes (Zhang et al. 2022a; Zhao et al. 2012).

Nanomaterials also overcome biological barriers by offering novel administration routes that enhance bioavailability and therapeutic efficacy, ensuring site-specific targeting where needed (Zhang et al. 2022b). Surface modifications further improve interactions with biological systems, promoting better penetration and retention at target sites (Mohapatra et al. 2023). Despite their advantages, challenges remain in clinical implementation, including issues of bioavailability, immunotoxicity, and pharmacokinetics, along with regulatory hurdles and safety evaluations that limit widespread adoption (Zhang et al. 2022b; Souri et al. 2022). Nevertheless, ongoing research and development continue to drive advancements in nanomaterial-based diagnostics and therapeutics, promising safer and more effective medical solutions in the future. This chapter provides an in-depth exploration of nanomaterials as contrast agents, highlighting their unique properties that enhance imaging precision across various biomedical applications. It delves into their versatility in therapeutic settings, examining their role in targeted drug delivery systems that improve treatment efficacy while minimizing systemic side effects. Additionally, the review discusses the potential of nanomaterials in advanced therapeutic strategies, including photothermal therapy, where heat generated by nanoparticles selectively destroys diseased tissues, and photodynamic therapy, which utilizes light-activated nanomaterials to produce reactive oxygen species for targeted cell destruction. Furthermore, the concept of theranostics is explored, emphasizing how nanomaterials can simultaneously function as both diagnostic and therapeutic agents, enabling integrated approaches for more effective combination therapies.

Beyond their promising applications, this review also addresses the critical challenges and opportunities associated with nanomaterial-based medical technologies. Key concerns such as biocompatibility and potential safety risks, including toxicity and long-term biological effects, are thoroughly examined. Additionally, the review discusses regulatory and translational hurdles that must be overcome to facilitate the clinical adoption of nanomedicine, emphasizing the need for rigorous evaluation and standardization. Finally, it highlights the challenges related to scalability and large-scale manufacturing, underscoring the necessity of developing cost-effective and reproducible production methods to ensure the successful transition of nanomaterials from laboratory research to real-world medical applications.

10.4 Nanomaterial-Based Contrast Agents

10.4.1 Iron Oxide Nanoparticles for MRI

The magnetic characteristics, biocompatibility, and functionalization potential of iron oxide nanoparticles (IONPs) have made them attractive contrast agents for magnetic resonance imaging (MRI). By changing the relaxation periods of adjacent water protons, these nanoparticles improve MRI contrast and increase the precision of the diagnosis. Size, surface coating, and magnetic characteristics all affect how well they work; ultrasmall IONPs show improved relaxivities. Polyvinylpyrrolidone (PVP) coatings and other surface changes enhance stability and biocompatibility while also changing interactions with biological molecules, which impacts imaging and cellular uptake. Iron oxide nanoparticles (IONPs) were developed successfully in an experiment utilizing a pulsed laser ablation approach. The particles were produced in a solution of polyvinylpyrrolidone (PVP) and deionized water, and they came in a variety of sizes and dispersions. The IONPs were described as having a spherical form and exceptional dispersibility as FeO_3 and FeO_4. When evaluated as MRI contrast agents, these nanoparticles showed a notable increase in contrast, especially at higher dosages. Their promise for medical imaging was confirmed by their strong signal-to-noise ratio and 221.875% enhancement for T1-weighted images at 2.31 mM (Khumaeni et al. 2023a). Another investigation enhanced the stability and biocompatibility of polyvinylpyrrolidone (PVP)-coated superparamagnetic iron oxide nanoparticles (SPIONs) for MRI contrast applications by developing a one-pot production process. The nanoparticles demonstrated perfect magnetic characteristics, including a high magnetization of 48.4 emu/g and an optimum size of 11.61 nm, when PVP with a molecular weight of 40,000 g/mol was used. In comparison to uncoated SPIONs, the PVP40K-SPIONs also showed a greater relaxivity ratio (72.55), which increased their potency as T2-weighted MRI contrast agents. These findings demonstrate the potential of PVP-coated SPIONs for improvements in medical imaging (Alzoubi et al. 2023).

The possibility of dual-modality imaging, which combines MRI with PET or SPECT, improves their diagnostic potential even more. Radiolabeled iron oxide nanoparticles (RIONs) have the potential to be dual-modality contrast agents for SPECT/MRI and PET/MRI, combining excellent sensitivity and spatial resolution for early illness identification, according to recent research. Applications such as iodine-125 RIONs for monitoring stem cells in ischemic stroke, 99mTc-labeled RIONs for tumor targeting, and 68Ga-based RIONs for sentinel lymph node localization were investigated. Significant liver and spleen uptake were demonstrated in in vivo tests, indicating efficient biodistribution for targeted imaging. These findings were underscored the necessity for more studies to improve the targeting and therapeutic potential of RIONs (Karageorgou et al. 2023). IONPs are also being investigated for theranostic and medication delivery uses in personalized medicine. However, for clinical translation, issues including toxicity, long-term stability, and the complexity of nanoparticle production need to be resolved. Notwithstanding these challenges, IONP technology is being improved by continuing research, which increases its potential for use in medical diagnosis and therapy.

10.4.2 Advances in T1 and T2 Contrast Agents

Recent advancements in T1 and T2 nanocontrast agents have significantly improved the capabilities of magnetic resonance imaging (MRI), enhancing diagnostic precision and expanding therapeutic possibilities. Among these, metal oxide nanoparticle-based T1-T2 dual-mode MRI contrast agents stand out, as they effectively function in both T1 and T2 imaging modalities. These agents enhance MRI contrast and detail, surpassing traditional single-mode agents in providing a comprehensive view of disease states (Szpak et al. 2014). Paramagnetic silica-coated nanocrystals represent another breakthrough in this domain. These nanoparticles feature an inorganic core encased in a silica shell, covalently linked to chelated $Gd3+$ paramagnetic ions. This configuration boosts T1 and T2 relaxation times while improving biocompatibility and functionalization potential, making them ideal for targeted imaging and therapeutic applications (Frochot et al. 2015; Shirshahi and Soltani 2015; Gerion et al. 2006). Theranostic formulations, such as magneto-liposomes loaded with ω-3 polyunsaturated fatty acids, further illustrate the integration of therapeutic functions into imaging agents. These formulations not only act as contrast agents but also target inflammatory diseases, enabling simultaneous imaging and therapy. Similarly, stealth liposomes with pH-sensitive nanovalves release therapeutic cargo specifically in acidic tumor microenvironments, enhancing imaging and treatment outcomes (Calle et al. 2018).

The use of gadolinium compounds and manganese oxides as T1 contrast agents is gaining attention due to their paramagnetic properties, high relaxivity, and biocompatibility, making them promising alternatives in MRI applications (Haroyan and Ginat 2015). Overall, these advancements highlight a shift toward multifunctional nanocontrast agents that improve imaging quality while facilitating targeted therapies, thereby enhancing clinical outcomes. Continued research in this field is set to refine these technologies further, driving more effective diagnostic and therapeutic strategies in medicine. According to research, superparamagnetic iron oxide nanoparticles (SPIONs) have a considerably better longitudinal relaxivity than conventional 3 T agents, making them effective positive T1 contrast agents for low-field MRI at 64 mT. When it came to low field strengths, SPIONs performed better than gadolinium-based agents. The findings also show that the size of SPIONs affects their relaxivity, with larger particles having higher relaxivity. Additionally, T1-weighted imaging is favored by their balanced transverse to longitudinal relaxivity ratio. According to these results, SPIONs offer improved T1 contrast together with better safety and performance, making them attractive options for point-of-care low-field MRI systems (Oberdick et al. 2023). Another work additionally demonstrates how protein engineering may be used to create ultrasmall iron oxide nanoparticles (IONPs) with adjustable characteristics, building on earlier discoveries. The researchers were able to successfully alter the size and properties of the nanoparticles by creating proteins with different metal-coordination sites. Strong T1 and T2 relaxivity was demonstrated by the resultant protein-stabilized IONPs (Prot-IONPs), increasing their potential as MRI contrast agents. These modified Prot-IONPs exhibit potential for enhancing medical imaging methods, especially for in vivo MRI applications, due to their exceptional stability and biocompatibility (Aires et al. 2022). Adding to the earlier findings, another study developed pH-responsive extremely small iron oxide nanoparticles (ESIONPs) functionalized with polyethylene glycol (PEG) and a stimuli-responsive polymer (PSDM). These nanoparticles have switchable T1/T2 contrast capabilities, switching from T1 to T2 in response to the acidic environment of tumor tissues, hence increasing the T2 contrast signal at the tumor location. In vivo MRI tests demonstrated that these ESIONPs selectively target tumors, making them intriguing candidates for more accurate and precise tumor imaging and diagnostics (He et al. 2022). Another study reported nickel-doped iron oxide nanoparticles (NiIO NPs), which were developed as 7 nm monodispersed, biocompatible, and superparamagnetic particles. These nanoparticles efficiently converted standard T2 contrast agents into T1 agents, resulting in dramatically improved imaging clarity. NiIO NPs showed 2.4 times greater longitudinal relaxivity (9.1 mM^{-1} s^{-1}) compared to typical therapeutic T1 agents. Their dual-mode MRI capacity allows for higher-resolution T1 and T2 imaging, and their exceptional colloidal stability and biosafety make them potential candidates for clinical applications in deep tissue diagnostics (Lu et al. 2022).

10.5 Quantum Dots: Properties and Applications

Quantum dots (QDs) are nanoscale imaging agents with distinct optical features such as configurable emission wavelengths, high fluorescence intensity, and photobleaching resistance, making them ideal for biomedical applications. Their wide excitation and narrow emission spectra facilitate multiplexed imaging, while functionalized QDs improve precision by targeting specific cancer cells (Huang and Huang 2024; Khaledian et al. 2023; Guo et al. 2024). They are widely employed in cancer diagnostics and offer high-resolution imaging for early detection and therapy monitoring (Guo et al. 2024; Hamidu et al. 2023). Carbon-based QDs for integrated imaging and treatment are one example of an innovation that emphasizes their flexibility. Recent breakthroughs in fluorescence efficiency and safety continue to broaden their clinical and research applications (Jovanović et al. 2023). A paper reports the development of a novel bi-modal imaging contrast agent, I@MPA-Mn::ZnSe, which blends quantum dots (QDs) with the commercial contrast agent Iohexol to improve diagnostic imaging. The agent fluoresces at 585 nm, allowing for dual CT and fluorescence imaging capabilities. It also significantly improves CT contrast, with a four-fold increase in Hounsfield units (HU) compared to Iohexol, indicating its higher CT contrast efficiency (Alhasan et al. 2023). A novel T1 nanocontrast agent, Gd(DTPA)—GQDs, was developed by functionalizing graphene quantum dots with poly(ethylene glycol) bis(amine) and combining them with Gd-DTPA. The Gd(DTPA) – GQDs had a much greater longitudinal proton relaxivity (r1) of 10.90 mM^{-1} s^{-1}, exceeding commercial Gd-DTPA with an r1 of 4.18 mM^{-1} s^{-1}. These compounds were also very biocompatible, with no toxicity and good results in hemolysis and in vivo safety tests. MRI investigations indicated greater signal amplification in animal models, with obvious tumor visibility 2 h after injection. Gd(DTPA)—GQDs have great colloidal stability, making them suitable for advanced MRI applications (Li et al. 2023a).

Following previous progress, another work presents gadolinium-doped carbon quantum dots (Gd-doped CQDs) synthesized using a hydrothermal technique with starch as a precursor. The CQDs, functionalized with C-N and N-H bonds, had a size distribution ranging from 2 to 8 nm and exhibited significant photoluminescence, with a quantum yield of 13.2%. Notably, these CQDs exhibited stable fluorescence even under varied ionic conditions and temperatures up to 65 °C. The high longitudinal and transverse relaxivity values of 218.28 and 364.68 mM^{-1} s^{-1}, respectively, make them suitable MRI contrast agents. These findings indicate that Gd-doped CQDs might be useful multifunctional imaging probes, providing fluorescence and MRI capabilities for advanced biological applications (Molaei 2022). Despite the advancements in quantum dot (QD) technology, concerns about the potential toxicity of heavy metal-containing QDs, such as those with cadmium, remain. This has prompted ongoing research to ensure their safe application in medical imaging. While quantum dots present significant advantages as contrast agents, their use in clinical settings requires careful evaluation of their biocompatibility and potential toxicity (Li et al. 2023a; Naylor-Adamson et al. 2024). Efforts in the synthesis and functionalization of QDs continue to address these issues, which are vital for expanding their medical imaging applications. However, further research is essential to fully comprehend and mitigate any adverse effects associated with their use.

10.6 Gold and Silver Nanoparticles: Plasmonic Imaging

Gold and silver nanoparticles have emerged as effective plasmonic imaging agents owing to their unique optical characteristics, which improve imaging contrast and allow for accurate visualization of biological structures. These nanoparticles have been utilized in a range of imaging modalities, including optical, photoacoustic, and computed tomography (CT) imaging, resulting in substantial advances in diagnosis and therapies. Gold nanoparticles (AuNPs), with their significant light scattering and absorption, improve optical imaging techniques such as optical coherence tomography (OCT) and photoacoustic imaging (PAI), which transform absorbed light into sound waves to produce high-resolution pictures. Gold superclusters (AuSC) are minuscule gold nanoparticle clusters that have been optimized for intravascular optical coherence tomography (IV-OCT). These superclusters, coated with polymer for stability and functionality, target P-selectin, an early sign of vascular inflammation. In a rat model, developed AuSC improved IV-OCT molecular imaging by linking signal strength with inflammation severity and moving IV-OCT from anatomical imaging to molecular characterization (Calvert et al. 2023). Another study combines polarization-sensitive and spectral contrast optical coherence tomography (PS-OCT and SC-OCT) to identify gold nanobipyramids (GNBPs) from surrounding tissues in vivo animal models. By identifying GNBPs' distinctive fingerprints as exogenous contrast agents, this approach improves molecular imaging capabilities, visualizes cellular processes, and shows promise for larger applications in advanced OCT diagnostics (Keahey et al. 2021).

In CT imaging, their high atomic number promotes X-ray attenuation, making them effective contrast agents for tumor visualization. O. Ibrahim et al. developed spherical thiol-capped gold nanoparticles (AuNPs) averaging 20 nm in size, which produced strong contrast in spectral photon-counting CT (SPCCT) imaging proportionate to their concentration. The minimum detectable concentrations for simulated lesions as tiny as 1 mm were identified, and a unique phantom was developed to optimize high-Z nanoparticle contrast in X-ray imaging, highlighting AuNPs' potential to improve cancer lesion visibility (Ibrahim et al. 2024). Similarly, silver nanoparticles (AgNPs) are used in optoacoustic imaging to improve tumor detection by utilizing their plasmon resonance. In order to provide reliable photoacoustic imaging, research highlights silver nanoaggregates (AgNAs) with improved near-infrared optical responses, blackbody-like absorption, and good photothermal stability. AgNAs showed targeted in vivo cancer imaging after being functionalized with tumor-targeting peptides, indicating their promise as long-lasting and effective contrast agents for cutting-edge biomedical applications (Kim et al. 2024). Hybrid nanoparticles, such as Ag@Au complexes, combine the characteristics of gold and silver to improve CT contrast and tumor imaging. Synthesized by pulsed laser ablation in an iodine medium, gold, silver, and mixed gold-silver (Au-Ag) nanoparticles demonstrated superior CT contrast properties. In vitro and in vivo, Au-Ag nanoparticles significantly increased their Hounsfield Unit (HU) values and enhanced CT contrast by 196% compared to standard iodine agents. These results indicate Au-Ag combinations as viable substitutes for cutting-edge medical imaging (Khumaeni et al. 2023b).

10.7 Silica Nanoparticles

Silica nanoparticles (SiNPs) are versatile contrast agents in medical imaging, offering high surface area, tunable pore size, and biocompatibility for efficient delivery of contrast agents. In CT, barium-doped and iodinated mesoporous SiNPs demonstrate superior X-ray attenuation and stability. A research reports the effective fabrication of barium-doped mesoporous silica nanoparticles (SiO2-Ba NPs) with diameters less than 50 nm, marking their first use as CT contrast agents. These nanoparticles have a

4.5-fold greater X-ray attenuation efficacy than standard agents such as iopromide, as well as outstanding structural stability with little barium ion release. These findings highlight SiO2-Ba NPs as attractive options for safer, more effective CT imaging (Mazhit et al. 2023). In MRI, silica-based manganese and gadolinium-loaded nanoparticles enhance relaxivity for T1 and T2 imaging. A study developed an innovative T1-weighted MRI contrast agent utilizing porous silica nanoparticles loaded with cationic Gd(III) complexes, resulting in excellent relaxivity increases. The relaxivity rose from 9.05 mM^{-1} s^{-1} to 118.32 mM^{-1} s^{-1} due to electrostatic confinement and interaction with serum albumin. This durable, biocompatible nanomaterial has promise for diagnostic applications in T1-weighted MRI (Mallik et al. 2022). Their multifunctional capabilities make SiNPs valuable for advanced diagnostic applications. A research investigation demonstrates the effective design of mesoporous silica nanorods with tailored lengths and aspect ratios, which include magnetic iron oxide nanoparticles into their structure. These nanorods displayed excellent performance as T2-weighted MRI contrast agents, as well as multimodal imaging capabilities via surface functionalization with fluorophores. Their biocompatibility and multifunctional design make them intriguing candidates for enhanced medical imaging and therapeutic applications (Grzelak et al. 2022).

10.8 Ultrasound Contrast Agents

Ultrasound nanodroplets are being used as improved contrast agents for ultrasound imaging and targeted drug administration. These drugs increase the echogenicity of tissues, which improves the contrast resolution of ultrasound images. Nanodroplets have various benefits over typical microbubbles, including smaller size, longer circulation durations, and the ability to undergo phase changes, which improves imaging and therapeutic applications. Nanodroplets typically have a perfluorocarbon core, such as perfluorohexane or perfluoropentane, surrounded by a stabilizing shell (Luke 2024; Van et al. 2021). When exposed to ultrasound, the core goes through a liquid-to-gas phase change, creating microbubbles that increase echogenicity. Shell materials can vary, with lipid- or polymer-based shells being utilized to improve biocompatibility and minimize absorption by the reticulo-endothelial system (RES) (Li et al. 2023b). These nanodroplets are used in contrast-enhanced ultrasound imaging to reveal features such as tumor vasculature and other tissues. They may be designed to target specific biological markers, hence increasing imaging specificity (Li et al. 2023b; Liang et al. 2024). Furthermore, nanodroplets show potential for drug delivery since they allow therapeutic chemicals to be encapsulated and released at the target spot when stimulated by ultrasound. Their dual-mode capabilities have also been investigated, with nanodroplets serving as contrast agents for both ultrasound and magnetic resonance imaging (MRI), offering full diagnostic data (Li et al. 2023b; Riaz et al. 2022).

10.9 Hybrid Nanomaterials for Improved Contrast Agents

Hybrid nanomaterials have emerged as innovative platforms for improving contrast agents in medical imaging, particularly in ultrasound (US) and magnetic resonance imaging (MRI). These materials exploit the unique properties of nanoparticles to enhance imaging performance and therapeutic potential. A prominent example is poly(vinyl alcohol)-shelled microbubbles, which combine the advantages of polymeric shells with imaging functionalities. Incorporating superparamagnetic iron oxide nanoparticles (SPIONs) into these microbubbles, either through physical or chemical methods, significantly boosts their imaging performance, with the structural design of these hybrid probes playing a critical role in their effectiveness across imaging modalities (Kothapalli 2013). The versatility of hybrid nanoparticles extends beyond microbubbles, encompassing nanoscale metal-organic frameworks, lipid-based nanoparticles, dendrimers, and micelles. These nanoparticles are increasingly explored for cancer imaging and therapy, as they can encapsulate imaging agents and therapeutic drugs, enabling simultaneous diagnosis and treatment—an approach known as theranostics (Xue and Liang 2014). Their nanoscale size, along with the ability to modify their surface, facilitates the development of multifunctional systems that improve the specificity and sensitivity of imaging techniques (Xue and Liang 2014; He and Lin 2015).

A study presents a novel methodology for engineering protein-stabilized iron oxide nanoparticles (Prot-IONPs) as targeted MRI contrast agents, leveraging the engineerability of proteins to enhance nanoparticle stability and magnetic properties essential for effective imaging. By functionalizing Prot-IONPs with alendronate (Ald), the nanoparticles were designed to specifically target atherosclerotic plaques through the chemical diversity of proteins, enabling disease-specific biomarker binding. In a proof-of-concept application, Prot-IONPs-Ald provided positive contrast imaging of atherosclerotic plaques in vivo, marking a significant improvement over traditional negative contrast agents. Tested in an atherosclerotic mouse model (ApoE−/− mice on a high-fat diet), the nanoparticles demonstrated successful targeting, stability, and enhanced imaging capabilities. This innovative approach highlights the potential of protein–nanoparticle hybrids as versatile platforms for developing targeted MRI contrast agents, offering improved diagnostic accuracy and sensitivity for detecting and monitoring various medical conditions (Guedes et al. 2024).

One study by Zhigao Yi et al. demonstrated the development of hybrid lanthanide nanoparticles as binary contrast agents for T1/T2 dual-weighted MRI, enabling comprehensive imaging for enhanced tumor diagnosis. These nanoparticles were synthesized and tested for dual functionality, improving both T1 and T2 contrast and providing more detailed imaging data, a significant leap for accurate tumor characterization (Yi et al. 2023). Similarly, the synthesis of hybrid gold nanoparticles combined with Dotarem and stabilized by lactose-modified chitosan (Chitlac) introduced a unique platform for MRI imaging and photothermal therapy (PTT). These nanoparticles exhibited high stability, non-toxicity, and remarkable diagnostic and therapeutic potential across multiple cancer cell lines, indicating their utility as dual-function agents in nanomedicine (Khan et al. 2022b).

In another study, an enzyme-responsive hybrid nanoplatform composed of manganese oxide and hyaluronic acid-conjugated mesoporous silica nanoparticles (HA-MnO@MSN) was developed for dual-targeted MRI and photoacoustic imaging. By catalyzing the conversion of hydrogen peroxide into oxygen, the nanoparticles improved tumor oxygenation, enhancing imaging contrast. In vivo tests demonstrated effective targeting of CD44-expressing cancer cells and prolonged imaging performance, making them a promising tool for tumor diagnosis and therapy (Sivasubramanian et al. 2022). Adding to this, a novel hybrid material combining carbon quantum dots (CQDs) and rhodium nanoparticles (Rh NPs) was synthesized using a microwave-assisted hydrothermal method. This hybrid system demonstrated exceptional biocompatibility and multimodal imaging capabilities, integrating optical fluorescence and X-ray fluorescence properties for enhanced bioimaging applications (Saladino et al. 2021).

Lastly, another approach combined manganese oxide (MnO) with lipid liquid crystalline nanoparticles (LLCNPs) to create hybrid MRI contrast agents. This integration leveraged the magnetic properties of MnO and the biocompatibility of lipid-based systems, resulting in safe and efficient imaging agents. The study emphasized their potential to overcome the limitations of traditional contrast agents, offering improved imaging quality and safety for clinical applications (Flak et al. 2023). Collectively, these methodologies underscore the transformative potential of hybrid nanomaterials in medical imaging, advancing the field through multifunctionality, enhanced targeting, and improved biocompatibility.

10.10 Nanomaterial-Based Therapeutic Agents

Nanomaterial-based therapeutic agents signify a ground-breaking advancement in modern medicine, particularly within drug delivery systems. These agents, encompassing biodegradable nanoparticles, liposomes, dendrimers, and various nanomaterials, are designed to improve the efficacy and specificity of therapeutic interventions. Biodegradable nanoparticles stand out for enabling controlled drug delivery while reducing toxicity, as their ability to degrade within the body supports a more sustainable therapeutic strategy (Kurul et al. 2025). Similarly, liposomes act as nanocarriers that encapsulate drugs, ensuring targeted delivery to specific tissues, which enhances therapeutic outcomes and minimizes side effects (Kurul et al. 2025; Gilmore et al. 2008; Slavcev et al. 2011). Dendrimers, highly branched macromolecules, offer notable benefits in drug delivery and gene therapy, as their structural versatility allows the incorporation of multiple functional groups for tailored therapeutic applications (Ruggiero et al. 2010). Metal nanoparticles and carbon nanomaterials are also increasingly recognized for their capacity to be engineered for specific conditions, enhancing their role as therapeutic agents. Magnetic nanomaterials bring additional sophistication, serving dual functions in targeted drug delivery and imaging, enabling precise delivery and real-time monitoring of therapeutic efficacy (Haleem et al. 2023). As research progresses, the transformative potential of nanomaterials in addressing complex diseases like cancer and neurodegenerative disorders becomes increasingly evident. Thus, nanomaterial-based therapeutic agents are pioneering medical innovation, offering robust solutions for targeted drug delivery and heightened therapeutic efficacy, shaping the future of medicine (Haleem et al. 2023).

10.10.1 Drug Delivery Systems

Nanomaterial-based drug delivery agents represent a revolutionary approach in medicine, utilizing the distinctive properties of nanoparticles to enhance therapeutic efficacy while reducing side effects. These systems, including liposomes, biodegradable nanoparticles, and dendrimers, are designed to deliver drugs directly to targeted tissues, improving bioavailability and minimizing toxicity (Kurul et al. 2025; Ruggiero et al. 2010; Azad and Rojanasakul 2006; Kim et al. 2014). A critical feature of these systems is the integration of targeting and therapeutic components within the nanoparticles. The targeting component ensures specific binding to diseased cells, enabling receptor-mediated uptake, while the therapeutic component is released directly within the target cell, ensuring precise drug action (Kurul et al. 2025).

This dual functionality is essential for achieving targeted drug delivery, which minimizes side effects and optimizes treatment outcomes (Kim et al. 2014; Sci et al. 2011; Ponnappan and Chugh 2015). Nanocarriers, such as polymeric nanoparticles and liposomes, further enhance therapeutic efficacy by enabling controlled drug release over time, increasing bioavailability, and reducing toxicity risks associated with conventional methods (Betancourt et al. 2009). Additionally, the Enhanced Permeability and Retention (EPR) effect facilitates nanoparticle accumulation in tumor tissues, leveraging the leaky vasculature of tumors to improve drug delivery to cancer sites (Fang et al. 2011).

Beyond oncology, these nanotechnology-based systems have diverse applications, including the delivery of macromolecular agents and vaccines. Encapsulation within nanostructured materials stabilizes therapeutic agents and enables site-specific accumulation, making them particularly advantageous for treating complex diseases (Azad and Rojanasakul 2006; Sci et al. 2011). In conclusion, nanomaterial-based drug delivery systems are at the forefront of modern medicine, offering innovative solutions to longstanding challenges in drug delivery. By combining advanced materials with engineering principles, these agents significantly enhance treatment efficacy while reducing adverse effects, marking a major advancement in therapeutics (Thassu et al. 2007).

10.10.2 Lipid-Based Nanoparticles: Liposomes and SLNs

Lipid-based nanoparticles (LNPs) have emerged as a versatile and effective platform for drug delivery, addressing challenges associated with traditional methods. These nanoparticles enhance solubility, bioavailability, and controlled drug release, making them suitable for diverse therapeutic applications. Their adaptability allows for the encapsulation of both hydrophilic and lipophilic drugs, improving absorption across biological barriers, including the gut and ocular tissues (Rehman et al. 2024; Baig et al. 2024). LNPs are particularly beneficial for Biopharmaceutics Classification System (BCS) class II and IV drugs, which have low solubility and permeability (Gangavarapu et al. 2024). Encapsulation in lipid matrices further enhances gastrointestinal absorption and solubilization through lymphatic pathways (Mehrdadi 2024). Variants such as solid lipid nanoparticles (SLNs) and nanostructured lipid carriers (NLCs) enable controlled drug release, maintaining therapeutic levels over extended periods (Munir et al. 2024). Functionalization with targeting moieties, like peptides, improves specificity and reduces off-target effects, enhancing therapeutic outcomes (Aschmann et al. 2024). In ocular drug delivery, cationic nanostructured lipid carriers (CNLCs) exhibit superior bio-adhesion to ocular tissues, enabling more effective drug delivery (Baig et al. 2024). LNPs have shown significant promise across various therapeutic areas. They enhance the stability and bioavailability of bioactive compounds and facilitate controlled release (Patel et al. 2024). In cancer therapy, their ability to penetrate deep skin layers and mimic lipoprotein structures aids in the targeted delivery of drugs, including siRNA, through receptor-mediated transport (Golestani 2024; Yaghmur et al. 2023). The advancement of lipid nanoparticle (LNP) technologies has driven significant progress in drug delivery systems through innovative methodologies and novel findings across diverse studies. One study introduced 30-n-LNPs, a formulation enriched with 30% distearoylphosphatidylcholine (DSPC), deviating from traditional compositions. These LNPs demonstrated targeted biodistribution to inflamed colon tissues in a dextran sodium sulfate (DSS)-induced colitis mouse model, bypassing clearing organs such as the liver and spleen. When loaded with interleukin-10-encoding mRNA, the 30-n-LNPs effectively alleviated colitis, underscoring the potential for finely tuned LNP compositions in disease-specific therapies beyond inflammatory bowel diseases (Rampado et al. 2025). Complementary to this, a study utilizing multinuclear solid-state MAS NMR spectroscopy, enhanced by dynamic nuclear polarization (DNP), provided unprecedented insights into the spatial organization of siRNA-loaded LNPs. This method revealed a core-shell structure with siRNA at the core, targeting agents near the surface, and sugars on the outermost layer, enabling optimized designs for targeted siRNA delivery. The novel structural understanding was paired with significant improvements in functional outcomes, potentially broadening applications for multi-component biological formulations (Porat-Dahlerbruch et al. 2025). In a reformulation effort, researchers explored LNP designs omitting cholesterol, previously considered essential, and optimized degradable-core ionizable cationic lipids. This resulted in organ-specific mRNA distribution, with targeted accumulation and expression in the lungs and liver. The findings challenge prevailing paradigms and demonstrate that streamlined LNP designs can enhance delivery efficacy while reducing unwanted accumulation in off-target organs, presenting opportunities for diverse mRNA therapeutics (Su et al. 2024).

Another study focused on the functionalization of LNPs with fusogenic coiled-coil peptides to improve cytosolic delivery. These peptides bypass the endolysosomal pathway, facilitating efficient payload delivery while reducing side effects. Surface modifications with targeting moieties further enhanced cell-specific interactions, demonstrating the potential for precision drug delivery with reduced off-target effects (Aschmann et al. 2024). In the realm of respiratory disease treatment, an innovative approach employed palmitoylethanolamide (PEA) to develop inhalable LNPs (PEA-LNPs). These nanoparticles demonstrated effective aerosol performance, improved solubility of PEA, and rapid macrophage internalization while exhibiting no cytotoxicity. This novel formulation offers promising therapeutic potential by maximizing drug concentration at the target site while minimizing systemic exposure (Maretti et al. 2024). Finally, a study mimicked Comirnaty drug-free LNPs using a simple preparation method, enabling structural comparison through SAXS/PCS and additional spectroscopy techniques. Contrary to previous classifications as solid lipid nanoparticles, these LNPs were found to exhibit a liquid-state structure with a narrow size distribution, offering new perspectives on their drug delivery mechanisms and associated medical benefits (Unruh et al. 2024). Collectively, these studies highlight the transformative potential of LNP technologies, showcasing advancements in targeted delivery, structural optimization, and functional performance. These findings pave the way for more effective, tailored therapies across a wide spectrum of diseases.

10.10.3 Polymer-Based Nanoparticles: Dendrimers and Micelles

Polymer-based nanoparticles represent a versatile and promising technology in drug delivery, offering enhanced stability, bioavailability, and targeted delivery of therapeutic agents across various biomedical applications, including cancer therapy, tissue engineering, and disease treatment. Their biocompatibility and biodegradability reduce toxicity and immune responses, enabling their use in non-invasive delivery routes such as nasal, oral, and ocular applications. These nanoparticles can be engineered for controlled drug release and targeted delivery, with mucoadhesive polymers like chitosan derivatives tailored for stomach targeting and polyethylene glycol functionalization for small intestine targeting. The ability to design nanoparticles with specific sizes, shapes, and surface characteristics allows them to overcome biological barriers, such as the blood-brain barrier, through surfactant or apolipoprotein coatings (Feczkó 2021).

The synthesis of polymer-based nanoparticles has been advanced through microfluidic technologies, which provide precise control over their properties and enable consistent quality for testing in biomimetic models, including organ-on-a-chip systems (Fabozzi et al. 2021). Innovations in polymer science have also led to functionalized nanoparticles and polymer-hybrid systems that combine polymers with inorganic or organic materials, resulting in improved targeting, stability, and drug release kinetics (Ferreira Soares et al. 2020; Sur et al. 2019). Applications of these nanoparticles are particularly prominent in cancer therapy, where they deliver chemotherapeutic agents directly to tumor sites, minimizing systemic toxicity and improving therapeutic efficacy (Sur et al. 2019). In tissue engineering, polymeric nanoparticles integrated into fibers have demonstrated potential for wound dressings and regenerative medicine by providing controlled drug release and enhancing bioactivity (Domingues et al. 2023). Additionally, protein-based nanoparticles, such as those derived from silk and collagen, offer biodegradability and bioavailability, further broadening their suitability for medical applications (De Frates et al. 2018). This combination of adaptability, functionalization, and targeted efficiency solidifies polymer-based nanoparticles as a cornerstone of emerging nanomedicine technologies.

Multiple studies collectively advance the field of drug delivery systems through innovative methodologies and compelling results, each introducing distinct contributions. One study presented a novel ultrasonication approach for synthesizing methylene polyethylene glycol-polystyrene (mPEG-PS) diblock copolymer-based solid lipid nanoparticles (SLNs), specifically targeting HIV-1. Using response surface methodology, the SLNs were optimized to achieve a particle size of 198 nm and an entrapment efficiency of 67.42%. Extensive analytical evaluations, including quantitative PCR and caspase-3 assays, demonstrated significant cellular responses, underscoring the nanoparticles' therapeutic potential in HIV treatment (Sreeharsha et al. 2024). Similarly, another study successfully synthesized lipid-polymer hybrid nanoparticles (LPHNPs) using lipid monolayer shells and PLGA cores for the targeted delivery of the anticancer drug etoposide. Optimization through Response Surface Methodology (RSM) revealed improved encapsulation efficiency and sustained drug release of 14% over 80 h. Notably, these LPHNPs exhibited enhanced antitumor activity against lung cancer cell lines, demonstrating their versatility for hydrophobic drug delivery in cancer therapy (Zare Kazemabadi et al. 2021). Combining lipid-polymer systems with pH-responsiveness, another study developed amphiphilic polymer prodrugs that leverage acidic tumor environments to trigger precise drug release. This targeted delivery minimizes damage to healthy cells and integrates real-time monitoring of drug release via imaging. The prodrugs' selective, responsive nature highlights their promise for improved cancer treatment (Kayal et al. 2024). Complementing these findings, the ROS-responsive iRGD-BDOX@CPNs system enhanced precision cancer therapy through dual chemo–/photodynamic approaches. Its active targeting via integrin $\alpha v\beta 3$ receptors and fluorescence-guided real-time monitoring further improved therapeutic outcomes. Efficient ROS-triggered drug release from this system underscores its clinical potential for advanced tumor treatment (Zhang et al. 2021). Another notable study focused on interpenetrating polymer network (IPN) nanoparticles synthesized from carboxymethyl guar gum and chitosan for controlled metformin hydrochloride delivery. These nanoparticles exhibited an optimized particle size of 440.6 nm and an entrapment efficiency of up to 88.23%. In diabetic rat models, IPN formulations showed sustained hypoglycemic effects, demonstrating their efficacy in prolonged drug delivery (Roy et al. 2024).

Lastly, the synthesis of lipid-polymer core-shell hybrid nanoparticles using PLGA and lecithin showcased high encapsulation efficiency (92%) and colloidal stability. Their controlled drug release over 16 days and cytocompatibility emphasized the platform's utility, particularly in neurosurgery, for sustained therapeutic delivery of vasodilators like nicardipine hydrochloride (Soomherun et al. 2024). Together, these studies highlight ground-breaking methodologies and underscore the transformative potential of nanoparticle-based drug delivery in treating diverse medical conditions.

10.10.4 Inorganic Carriers: Mesoporous Silica and Gold

Inorganic carriers have emerged as a promising strategy to enhance the efficacy and specificity of therapeutic agents, particularly in cancer treatment. These carriers provide significant advantages, including high drug loading capacity, stability, and biocompatibility, making them ideal for developing smart drug delivery systems. Among the prominent inorganic materials studied are mesoporous silica nanoparticles (MSNs), gold nanoparticles, metal-organic frameworks (MOFs), calcium carbonate particles, and emerging materials like graphene oxide and black phosphorus. MSNs stand out for their large surface area and tunable pore size, enabling high drug loading and controlled release. Their excellent biocompatibility and stability allow for the delivery of various therapeutic agents, and functionalization with targeting ligands enhances their selectivity for tumor cells, minimizing side effects on healthy tissues. Similarly, gold nanoparticles offer unique optical properties and ease of surface modification, enabling targeted drug delivery. These carriers can respond to external stimuli such as light or heat, facilitating controlled drug release at the target site, and their ability to penetrate biological barriers adds to their effectiveness in delivering chemotherapeutics (Shi et al. 2020). MOFs, crystalline materials composed of metal ions coordinated to organic ligands, provide a highly porous structure suitable for drug adsorption. Their surface can be tailored for specific applications, and they have been shown to improve the cytotoxicity, stability, and release profile of anticancer drugs (Tarasi et al. 2022). Calcium carbonate particles also offer pH-sensitive drug release, high surface area, and a porous structure, making them particularly effective in protecting drugs from degradation in the gastrointestinal tract. These particles can be hybridized with organic materials like alginate to enhance biocompatibility and drug release properties (Huang et al. 2022; Zelenikhin et al. 2023).

Emerging materials such as graphene oxide and black phosphorus exhibit high surface areas and the ability to load a variety of drugs. Their engineered properties improve drug stability and facilitate targeted delivery, further establishing their potential in cancer therapy (Shi et al. 2020). While inorganic carriers offer numerous benefits, challenges such as toxicity, long-term stability, and scalability remain critical to address. Integrating inorganic carriers with organic materials to form hybrid systems may offer a balanced solution, combining the strengths of both for more effective drug delivery systems.

Metal-organic frameworks (MOFs) feature prominently, demonstrating their versatility and effectiveness as drug carriers. One study utilized MIL-100(Fe) and MIL-53(Fe) MOFs to encapsulate poorly soluble drugs, hydrochlorothiazide and dapsone. The high porosity of these MOFs enabled remarkable drug loading capacities of 23.1% and 27.6%, respectively, with sustained release over 72 h under physiological conditions, confirming their potential for prolonged and controlled drug delivery (Yadav et al. 2023). Similarly, another study introduced TMU-6 and TMU-21 MOFs, tailored with amine pillars to manipulate hydrophobicity and optimize the adsorption of the anticancer drug Nimesulide. Hirshfeld surface analysis revealed that structural modifications, such as replacing benzene with naphthalene cores, significantly enhanced drug interaction dynamics, resulting in higher cytotoxicity against cancer cell lines compared to free drugs (Tarasi et al. 2022). Together, these findings underscore the adaptability of MOFs for precise drug delivery applications.

Further innovation was presented in an FeMn-based MOF synthesized from $Fe_2Mn(\mu_3\text{-}O)$ clusters, which combined dual-stimuli responsiveness (pH and hydrogen sulfide) with ferromagnetic properties. Encapsulating 5-fluorouracil, this system achieved a high drug loading capacity of 43.8 wt% and demonstrated rapid release in tumor-mimicking environments. Mathematical modeling confirmed controlled release kinetics, while cytotoxicity assessments highlighted the MOF's biocompatibility, positioning it as a strong candidate for cancer therapy (Akbar et al. 2022). Complementing MOF-based strategies, another study developed mucin-based nanoparticles capable of co-delivering hydrophobic and hydrophilic drugs. The system maintained drug stability across varying pH levels and achieved controlled, sustained release, highlighting its potential for complex therapeutic regimens. This co-delivery approach could streamline treatments requiring multiple drugs with differing solubility profiles, significantly improving patient outcomes (Lutz et al. 2022).

Expanding on material innovation, genetically engineered carriers and carbon allotropes showcased exciting potential. Engineered platforms like virus-like nanoparticles enabled gene-drug delivery to tumor tissues, enhancing therapeutic efficacy and minimizing off-target effects. Magnetic-responsive systems, combining elastin-like proteins and superparamagnetic iron oxide nanoparticles, facilitated precise drug release and monitoring. Additionally, carbon allotropes such as graphene oxide and fullerenes were explored for their high surface area and unique properties, laying the groundwork for future breakthroughs in nanocarrier technology (Hajebi et al. 2022).

These studies collectively emphasize the transformative potential of novel materials and methodologies in drug delivery systems. By addressing key challenges such as targeted delivery, dual drug compatibility, and tumor-specific release, these innovations pave the way for more effective and patient-centered therapeutic strategies.

10.10.5 Photothermal Therapy (PTT)

Photothermal therapy (PTT) is an innovative, non-invasive cancer treatment that utilizes laser-based techniques to selectively destroy tumor cells by converting light energy into heat through tumor-specific photo-absorbers (Ou et al. 2016; Shah et al. 2008). The method's efficacy is significantly enhanced by photothermal agents (PTAs), substances that absorb light and induce localized tumor cell death by generating heat (Navarro-Ocón et al. 2022). Nanomaterials play a critical role in improving photothermal conversion efficiency and enabling enhanced accumulation of therapeutic agents within tumors, a phenomenon attributed to the enhanced permeability and retention (EPR) effect caused by the leaky vasculature of tumor tissues (Fang et al. 2011; Maeda et al. 2000; Wang 2015). Among these, gold nanoparticles (AuNPs) are particularly notable due to their unique optical properties that allow effective absorption of near-infrared (NIR) light, converting it into heat to target cancer cells (Zhang et al. 2014; Radwan and Azzazy 2009; Duncan et al. 2010). Their tunable surface properties further enhance targeting and stability, making them ideal for PTT applications. The PTT process typically uses a continuous wave laser to deliver light energy to tumors, enabling efficient heat conversion. To optimize treatment, identifying the presence and spatial distribution of photo-absorbers within the tumor prior to therapy is crucial, which is effectively achieved through photoacoustic imaging. This technique detects nanoparticles and guides the PTT process while ensuring precision. Additionally, ultrasound imaging complements photoacoustic imaging by computing thermal maps and monitoring temperature changes during treatment, ensuring effective therapy delivery (Shah et al. 2008). The following table summarizes key details from recent studies exploring innovative materials and methodologies in cancer therapy. The studies primarily focus on various forms of photothermal therapy (PTT), photodynamic therapy (PDT), and combinations of these with additional therapeutic strategies like immunotherapy, ferroptosis, and chemotherapy. Notable advancements include the development of novel nanomaterials, dual-action therapies, and unique methodologies for enhanced tumor targeting, photothermal efficiency, and therapeutic outcomes. Table 10.1 highlights key materials, therapy types, and the main findings from each study, offering valuable insights into the ongoing evolution of cancer treatment strategies.

PTT represents a promising cancer treatment approach, leveraging the unique properties of nanomaterials and advanced imaging techniques to enhance therapeutic efficacy. The integration of photoacoustic and ultrasound imaging facilitates precise tumor targeting and real-time treatment monitoring. With ongoing advancements in nanotechnology and imaging modalities, PTT holds significant potential to improve patient outcomes in cancer therapy (Ou et al. 2016; Shah et al. 2008).

Table 10.1 Summary of recent advancements in nanomaterial-based cancer therapies: photothermal, photodynamic, and combination approaches

Key Material/Agent	Therapy Type(s)	Photothermal Efficiency	Tumor Targeting	Key Innovation	Notable Outcomes	Reference
Indian ink (tumor simulant)	Photothermal Therapy	Focuses on converting light into heat to target tumors	Tumor simulant for experimental research	Circular array-based photoacoustic photothermal system for real-time, precise temperature monitoring	Improved treatment outcomes with real-time temperature imaging and 2D temperature mapping, significant for treatment planning	Qin et al. (2024)
Pt/Au nanotubes (encapsulated with docetaxel and phase change materials)	Photothermal + Chemotherapy	NIR-induced phase change releases chemotherapy	Triple-negative breast cancer	Integration of photothermal therapy with chemotherapy via controlled release triggered by NIR	75.4% tumor inhibition rate, improved survival, and effective combination of therapies	Yang et al. (2024)
Gold and silver noble-metal PTAs (nanoparticles)	Photothermal Therapy	Effective light-to-heat conversion	Squamous cell carcinoma	Numerical analysis to optimize photothermal therapy parameters	Optimization of volume fraction, laser intensity, and nanoparticle radius for better PTT efficacy	Kim and Kim (2024)
Fucoidan-based CuS nanoparticles, hyaluronic acid microneedles, doxorubicin	Photothermal + Chemotherapy + Chemodynamic Therapy	Enhanced by photothermal-induced ROS generation	Melanoma	Microneedle system with CO2 bubble generation for enhanced drug delivery	Significant melanoma growth inhibition in vivo, reduced side effects due to combined therapies	Tao et al. (2023)
Biodegradable metal-gallate nanoparticles (MnGA)	Mild-Temperature Photothermal Therapy	Uses NIR to achieve mild temperatures (43 °C)	Osteosarcoma	Pyroptosis induction via nanoparticles to enhance therapy	Enhanced therapeutic efficacy and reduced side effects due to combination with chemotherapy	Liu et al. (2023)
Thermoresponsive hydrogel	Photothermal Therapy	NIR-triggered phase transition	Bacterial skin infections	Thermoresponsive hydrogel for controlled heat regulation and safety	Accelerated wound healing, effective bacterial eradication, and safety in photothermal treatment	Fu et al. (2023)

(continued)

Table 10.1 (continued)

Key Material/Agent	Therapy Type(s)	Photothermal Efficiency	Tumor Targeting	Key Innovation	Notable Outcomes	Reference
Programmable nanophotosensitizer	Photothermal Therapy	Optimized for NIR absorption and heat generation	Cancer (general)	Multistage delivery system and imaging-guided therapy	Enhanced efficacy, better tumor reduction, and fewer side effects through precise targeting	Ma et al. (2023)
Glucose Oxidase (GOX), Indocyanine Green (IR820), α-Cyano-4-hydroxycinnamic Acid (CHC)	Photothermal-Starvation Therapy	Dual action with photothermal heating and metabolic starvation	Colorectal cancer	Coassembly of materials for dual therapy strategy	Significant suppression of CRC recurrence and metastasis, promising new therapeutic avenue	Duan et al. (2023)
FRET Nanobullet (HMME + PANI-ES)	Photodynamic Therapy (PDT), Photothermal Therapy (PTT)	High	Targeted using PEGA-pVEC peptide	Synergistic PDT and PTT, FRET-based amplification of imaging and therapy	80% tumor volume reduction, enhanced therapeutic efficacy, and imaging	Tan et al. (2023)
MnBV (Biliverdin Manganese Complex)	Photothermal Therapy (PTT), Immunotherapy	High	Dual-targeting in tumors	Combination of photothermal and immunotherapy with biomolecular assembly	Inhibition of TNBC metastasis and recurrence, promising dual-target therapy for aggressive cancers	Zhao et al. (2024a)
Functionalized Injection Needle	Electrothermal Therapy (ETT)	N/A	Circulating Tumor Cells (CTCs)	Electrothermal method for CTC capture and elimination	More effective than traditional methods, improved cancer treatment targeting metastasis	Zhang et al. (2024a)
Pt-1 (Pt(IV) Prodrug + BODIPY)	Photodynamic Therapy (PDT), Photothermal Therapy (PTT), Photoactivated Chemotherapy (PACT)	42.10%	Tumor accumulation in vivo	Combined PDT, PTT, and PACT with photoinduced cisplatin release	First Pt(IV) prodrug active in both PTT and PDT, improved tumor treatment and imaging	Spector et al. (2024)

PFeM (Iron + Mangiferin Complex)	Photothermal Therapy (PTT), Ferroptosis	N/A	Tumor cells	Self-healing nanotherapeutics combining PTT and ferroptosis	Enhanced tumor therapy, reduced inflammation, and heat tolerance	Zheng et al. (2024)
NDS-BPN NP (Naphthalene Diamine-Based Nano-Agent)	Photothermal Therapy (PTT)	55.05%	Tumor targeting	AIE-based nanoparticle synthesis, NIR emission for deeper tissue penetration	Successful photothermal ablation, promising imaging-guided tumor therapy	Zhang et al. (2024b)
LP-CuS Nanoparticles (LP-CuS NPs)	Photothermal Therapy (PTT), Chemotherapy	High	Tumor targeting	LP-functionalized CuS NPs for improved biocompatibility and drug release	Synergistic chemo-photothermal therapy, enhanced DOX cytotoxicity, and cellular uptake	Gupta et al. (2024)
Ag2S/g-C3N4 Nanoparticles (NPs)	Photothermal Therapy (PTT), Photodynamic Therapy (PDT)	31.28%	Tumor targeting	Hydrothermal synthesis for tailored nanoparticle properties	Long-term tumor control with combined PTT and PDT, no tumor regrowth for 30 days	Wang et al. (2024c)

10.10.6 Photodynamic Therapy (PDT)

Photodynamic Therapy (PDT) is a non-invasive treatment that utilizes light-sensitive compounds called photosensitizers, which produce reactive oxygen species (ROS) upon activation by specific light wavelengths to destroy targeted cells. It has shown potential in treating conditions like acne, cancer, and other diseases involving abnormal cell growth, with its precision in targeting diseased tissues while minimizing damage to healthy tissues. The integration of advanced technologies, such as nanotechnology and X-ray-induced PDT, has further enhanced its therapeutic potential. PDT operates by administering a photosensitizer that accumulates in target cells, generating ROS like singlet oxygen, superoxide anion radicals, and hydroxyl radicals upon light activation, leading to cell death through apoptosis or necrosis. Its efficacy depends on light dosimetry, photosensitizer concentration, and tissue oxygenation, with pleural PDT studies demonstrating its effectiveness (Sun et al. 2024). Nanoparticles have revolutionized PDT for cancer treatment by improving drug delivery, target specificity, and reducing side effects. They enhance photosensitizer solubility, stability, and targeted delivery, optimizing therapeutic outcomes. Porphyrin-based nanoparticles improve the delivery and activation of photosensitizers, reducing side effects (Glowacka-Sobotta et al. 2024). Metal-based nanoparticles, functionalized with folic acid, target cancer cells with high precision by binding to folate receptors, boosting the effectiveness of PDT and photothermal therapy (PTT) (Chandra et al. 2024; Mariano et al. 2024). Hybrid nanoparticles, such as those combining up-conversion nanoparticles and chlorin e6, show superior tumor accumulation and reduced side effects compared to traditional photosensitizers (Kawasaki et al. 2024).

10.10.7 Photosensitizers and Their Delivery via Nanocarriers

Photosensitizers (PSs) are essential for photodynamic therapy (PDT), a treatment that uses light-activated compounds to generate reactive oxygen species (ROS) for selectively destroying cancer cells. However, their clinical utility is often hindered by issues such as poor solubility, stability, and off-target effects. Nanocarriers have emerged as a promising solution, enhancing the delivery and efficacy of PSs in PDT. Traditional nanocarriers like liposomes and hydrogels offer high encapsulation efficiency, sustained release, biocompatibility, and localized drug depots, improving PS stability and solubility (Romero et al. 2024). Extracellular vesicles (EVs), naturally occurring nanoparticles, provide advantages such as prolonged circulation, immune evasion, and tumor homing, effectively delivering PSs while minimizing off-target effects (Lv et al. 2024a). Inorganic and organic nanoparticles, including metal-organic and covalent organic frameworks, offer structural stability, high drug loading, and stimuli-responsive release of PSs in the tumor microenvironment (Romero et al. 2024; Sasaki et al. 2024). Innovative strategies are also advancing PDT outcomes. Active targeting and surface modification of nanocarriers with ligands enable precise delivery to intracellular organelles like mitochondria or lysosomes, addressing tumor-specific challenges such as hypoxia and vascular disruption (Modi et al. 2025). Carrier-free approaches, where PSs self-assemble with other agents like siRNA, create hybrid systems with high loading ratios and enhanced therapeutic effects (Li et al. 2024a). Peptide-based nanocarriers facilitate the co-delivery of PSs and antisense oligonucleotides, inducing synergistic cytotoxicity through targeted cellular pathways, with nuclear localization signals further enhancing delivery efficiency (Li et al. 2024a). Despite these advancements, challenges remain, including the complexity of synthesizing multifunctional nanocarriers, ensuring biocompatibility and safety, and addressing PDT limitations such as light penetration into deep tissues and oxygen dependency (Zhao et al. 2024b; Truong et al. 2024).

10.10.8 Applications in Cancer Therapy

Recent advancements in photodynamic therapy (PDT) have introduced a variety of novel materials designed to enhance therapeutic efficacy, improve targeting, and overcome critical limitations such as hypoxia, oxidative resistance, and nonspecific phototoxicity. These studies collectively highlight innovative strategies for optimizing PDT through material engineering and precise modulation of singlet oxygen ($^1O_2$) generation, reactive oxygen species (ROS) production, and immune response induction. One of the most significant breakthroughs is the introduction of the photo-triggered porphyrin polyelectrolyte nano-assembling (PPN) strategy. This system leverages charge interactions and metal-ligand coordination to create a unique microenvironment that enables controlled $^1O_2$ generation in three distinct stages—pre-PDT quenching, during-PDT activation, and post-PDT silencing. This approach achieved a remarkable 123-fold enhancement in photodynamic efficiency, offering unprecedented remote control over $^1O_2$ production to mitigate nonspecific phototoxicity and improve therapeutic precision (Yao et al. 2025). Complementing this approach, a novel conjugate system integrating nitric oxide (NO) donors with silicon phthalocyanine (SiPc-NO@RGD NPs) demonstrated (Fig. 10.1a) exceptional PDT efficiency in breast cancer models by generating high levels of ROS and enabling precise tumor targeting (Aikelamu et al. 2024).

Building upon targeting strategies, quantum dot-functionalized nanoparticles (QDs-RGD) exhibited dual functionalities by not only enhancing PDT efficiency in pancreatic cancer but also inducing a potent antitumor immune response through immunogenic cell death. This immune activation component suggests a promising avenue for long-term tumor suppression beyond direct cytotoxic effects (Li et al. 2024b). Similarly, lanthanide nanoparticle-based platforms (LCR nanoplatforms) showcased a dual-channel excitation mechanism in the NIR-II region, enabling deep tissue penetration and real-time fluorescence imaging, while achieving precise PDT activation via energy transfer mechanisms (Lv et al. 2024b).

Addressing tumor microenvironmental challenges, carrier-free hybrid nanoparticles were developed to combat hypoxia and oxidative resistance in oral carcinoma (Fig. 10.1b). These self-assembled nanoparticles effectively reversed the

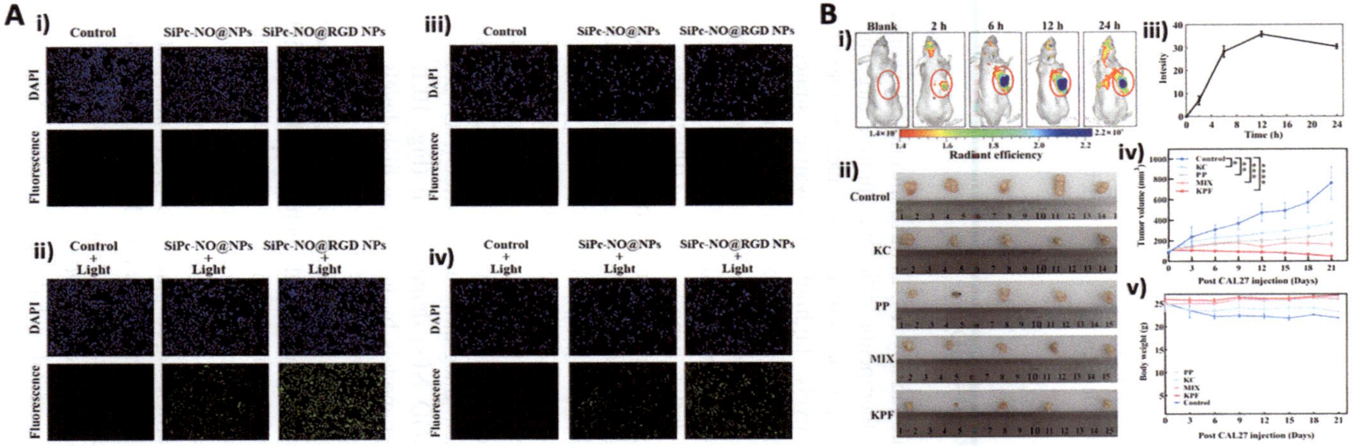

Fig. 10.1 In figure (**a**) (i) and (iii), no green fluorescence was detected in either the control or experimental groups of 4 T1 and MCF-7 cells in the absence of laser irradiation, indicating minimal intrinsic fluorescence. However, in figure (**a**) (ii) and (iv), cells treated with both DCFH-DA and SiPc–NO self-assembled nanoparticles exhibited distinct green fluorescence upon laser irradiation, confirming a light-triggered response. The fluorescence intensity correlated with cellular uptake results, suggesting that nanoparticle internalization influenced the observed signal. Notably, SiPc–NO@RGD nanoparticles generated significantly higher levels of reactive oxygen species (ROS) compared to SiPc–NO@NPs, likely due to enhanced cellular targeting and uptake, underscoring their potential as efficient phototherapeutic agents (Aikelamu et al. 2024). KPF nanoparticles (KPF NPs) accumulate at tumors via the enhanced permeability and retention (EPR) effect, enabling effective in vivo therapy. After injection into CAL-27 tumor-bearing mice with PP (5 mg/kg, 200 µL), fluorescence appeared within 2 h (**b** i), peaked at 12 h (Fig. 10.1b ii), and lasted 24 h. In post-photodynamic therapy (PDT), saline-treated tumors grew significantly (100 mm^3 to 760 mm^3, (**b**) iii, iv), while PP, KC, and their combination reduced growth to 400 mm^3, 250 mm^3, and 150 mm^3, respectively. KPF NPs showed the highest efficacy with 95% tumor inhibition. Without laser irradiation, tumors continued growing, confirming suppression mainly from singlet oxygen ($^\wedge 1O^2$) generation. These results align with in vitro findings. Additionally, while saline- and KC-treated mice lost weight, KPF NP treatment led to weight gain, indicating improved health (**b** v) (Li et al. 2024c). (*This figure is reproduced from an open-access article distributed under the terms of the Creative Commons Attribution 4.0 (CC BY-4.0) license*)

limiting conditions within tumors, enhancing the bioavailability of protoporphyrin IX and improving PDT outcomes (Li et al. 2024c). A complementary approach using a hyaluronan-coated metal-organic framework (ZMCH) nanoplatform successfully integrated proteolysis targeting chimeras (PROTACs) with PDT, resulting in significant tumor suppression in 4 T1 tumor-bearing mice and enhanced antitumor T-cell immunity (Zhu et al. 2024). This highlights the potential of combining PDT with targeted protein degradation to reshape the immunosuppressive tumor microenvironment.

Further optimization of PDT efficiency was demonstrated through hybrid nanoparticles incorporating up-conversion nanoparticles and chlorin e6-bearing pullulan. This combination yielded an astounding 400-fold increase in PDT activity compared to conventional photosensitizers, largely due to improved energy transfer and enhanced tumor accumulation through the enhanced permeability and retention (EPR) effect (Kawasaki et al. 2024). Similarly, a mitochondria-targeted smart nanosystem using Au25 gold clusters, catalase, and GSH-responsive mesoporous silica nanoparticles addressed tumor hypoxia while maximizing ROS production through NIR laser irradiation at 808 nm, resulting in superior in vivo tumor suppression (Gao et al. 2025).

Advancements in photosensitizer formulations have also contributed to PDT improvements. Purpurin-18-N-propylimide methyl ester (P18 N PI ME) solid lipid nanoparticles exhibited superior PDT effects compared to free photosensitizers, demonstrating controlled release and selective phototoxicity in cervical and lung cancer models (Yeo et al. 2024). Finally, water-soluble fluorescent organic nanoparticles functionalized with purpurin-18 and chlorin p6 (FONPs[Cp6]) facilitated enhanced intracellular ROS generation and mitochondrial apoptosis activation in colorectal cancer cells, underscoring the importance of subcellular targeting mechanisms for PDT efficacy (Chkair et al. 2024).

Together, these studies underscore the rapid evolution of PDT through the development of multifunctional nanoplatforms, advanced targeting strategies, and synergistic approaches that integrate PDT with immune modulation and tumor microenvironment adaptation. The potential for clinical translation of these novel materials is promising, offering improved specificity, reduced side effects, and enhanced therapeutic outcomes for various cancer types.

10.10.9 Theranostic Agents and Combination Therapies

Nanotheranostic agents have emerged as a transformative approach in medical research, particularly for addressing complex diseases such as cancer. These agents leverage nanoparticles as carriers for both therapeutic drugs and imaging agents, enabling concurrent diagnosis and treatment within a single platform. Their multifunctionality significantly enhances therapeutic precision, as they can be engineered for targeted delivery while facilitating real-time imaging, thereby optimizing treatment efficacy (Xue and Liang 2014). In oncology, combination therapies are increasingly regarded as crucial for improving therapeutic outcomes, as they integrate multiple treatment strategies to counteract drug resistance and enhance overall efficacy. By incorporating nanotheranostic agents alongside conventional anticancer drugs, such as Survivin antisense oligonucleotides, researchers aim to refine treatment regimens and overcome key limitations in cancer therapy. The success of these approaches hinges on the strategic design of delivery systems and the timing of administration to maximize their synergistic effects (Komarova and Wodarz 2014; Oudega et al. 2012). Various nanoparticle-based platforms, including dendrimers, liposomes, magnetic nanoparticles, and micelles, have been extensively explored for their potential in combination therapies. Each of these systems offers distinct advantages—dendrimers, for instance, allow for extensive surface modifications to enhance targeting specificity, while liposomes and lipid-based nanoparticles exhibit high biocompatibility and the capacity to encapsulate both hydrophilic and hydrophobic therapeutic agents, expanding their clinical applicability. The integration of nanotheranostic platforms into combination therapy not only improves drug pharmacokinetics but also enhances the overall therapeutic index by minimizing adverse effects and increasing patient compliance (Xue and Liang 2014). As advancements continue in nanotechnology and personalized medicine, the development of more sophisticated nanostructured systems will further refine treatment strategies, paving the way for more effective and individualized cancer therapies. Ultimately, the synergistic application of nanotheranostic agents in combination therapy represents a promising frontier in precision medicine, offering innovative solutions to the challenges of disease management (Komarova and Wodarz 2014; Oudega et al. 2012).

The development of innovative nanomaterials for theranostic applications has gained momentum, with various studies exploring novel approaches to cancer imaging and treatment. Lentinan-coated selenium nanoparticles (LNT-SeNPs) emerged as a promising theranostic agent, integrating selective anticancer properties with MRI detection via label-free chemical exchange saturation transfer (CEST) imaging. These nanoparticles demonstrated (Fig. 10.2a) enhanced biocompatibility and stability, facilitating efficient drug delivery when combined with TNF-α, as evidenced by increased MRI signals and improved therapeutic outcomes in murine models (Liu et al. 2022b). Similarly, a nanotherapeutic strategy leveraging engineered nanoparticles capable of dual cytotoxic and immunosuppressive activities was developed for post-transplantation cancer treatment. This approach achieved a balance between preventing organ rejection and maintaining sufficient immune activity to suppress cancer recurrence, significantly enhancing treatment outcomes (Yang et al. 2023).

Further advancements in theranostics were realized through the integration of inorganic photothermal agents (PTAs) in the near-infrared-II (NIR-II) bio-window, optimizing deep tissue penetration for superior photothermal therapy (PTT). These PTAs exhibited multifunctionality, enabling photoacoustic imaging and synergistic treatment with photodynamic therapy (PDT), chemodynamic therapy (CDT), and immunotherapy, marking a substantial improvement in multimodal cancer therapy (Zhang et al. 2022c). Another significant development involved iron-based micelles encapsulating indocyanine green (ICG) and cyclic tripeptide cRGD, forming

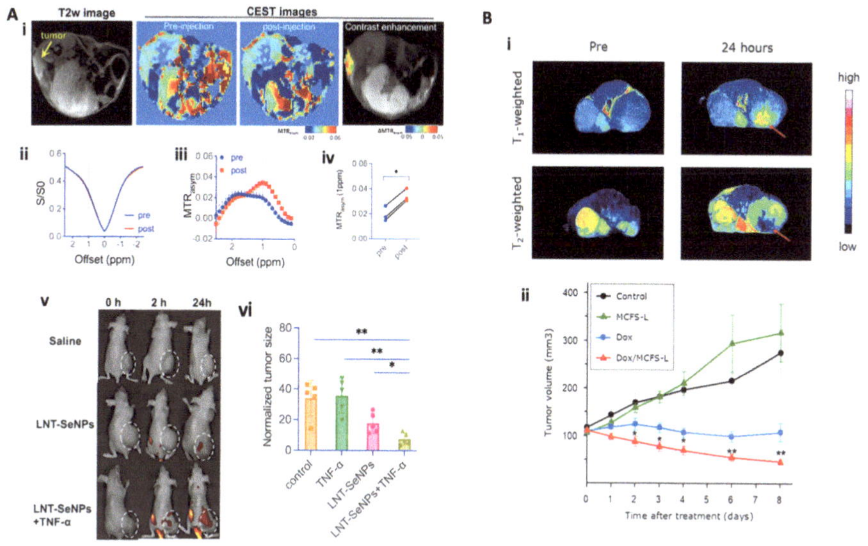

Fig. 10.2 In figure (**a**) (i–iv) When LNT-SeNPs and TNF-α were co-injected into CT26 tumor-bearing mice, CEST MRI showed a 75% increase in tumor contrast (0.020 ± 0.016 to 0.035 ± 0.005 at 1 h post-injection). As shown in Fig. A v, fluorescence appeared in the combination group at 2 h but not in the LNT-SeNPs-only group, becoming stronger in both by 24 h, though more intense with co-administration. Figure (**a**) (vi) shows that after 2 weeks, LNT-SeNPs + TNF-α induced the strongest tumor growth inhibition (normalized size increase: 7.6 vs. 33.9 in controls, $p = 0.0079$), while LNT-SeNPs alone had no significant effect (17.8, $p = 0.5560$), and TNF-α alone had no impact (35.6, $p = 0.6905$) (Liu et al. 2022b). (**b**) Magnetic targeting and therapeutic evaluation of MCFS-L. (i) T2- and T1-weighted MRI scans showing successful tumor targeting of MCFS-L in an MMTV-PyMT tumor-bearing mouse under a magnetic field, with MCFS nanoparticles appearing as a dark region on T2- and an enhanced signal on T1-weighted images 24 h post-injection. (ii) Antitumor efficacy of doxorubicin-loaded MCFS-L in an orthotopic mammary tumor model, demonstrating a 41% tumor volume reduction 8 days post-treatment, whereas systemically administered doxorubicin only achieved tumor stasis (Mikhaylov et al. 2022). (*This figure is reproduced from an open-access article distributed under the terms of the Creative Commons Attribution 4.0 (CC BY-4.0) license*)

ICG@SANPs-cRGD nanoparticles. These were utilized in a combined PTT/PDT strategy alongside Anti-PD-L1 immunotherapy, effectively inhibiting both primary tumor growth and metastasis while facilitating dual-mode fluorescence and MRI imaging (Kong et al. 2022).

Magnetic cobalt ferrite spinel (MCFS) nanoparticles demonstrated dual-contrast MRI capabilities as both T1- and T2-weighted contrast agents. These nanoparticles, encapsulated in liposomes for targeted doxorubicin delivery, exhibited superior relaxivity coefficients, improved imaging sensitivity, and enhanced drug delivery efficacy, making them a viable candidate for integrated imaging and treatment (Fig. 10.2b) (Mikhaylov et al. 2022). In parallel, nanocomposites incorporating samarium oxide, gadolinium oxide, and iron oxide onto carbon nanohorn (CNH) carriers functionalized with polyethylene glycol (PEG) showcased a high drug loading capacity. These nanocomposites enabled controlled drug release under acidic tumor conditions, facilitating combinatorial chemotherapy and radiotherapy while enhancing imaging capabilities (Rahmania et al. 2023).

A novel acetazolamide derivative, AZM-SH, was incorporated into a magneto-fluorescent nanosystem (NANO3) for colorectal cancer treatment. Functionalized with Cetuximab (CTX) and loaded with doxorubicin (DOX) and ofloxacin (OFLO), these nanoparticles exhibited increased cellular uptake, targeted drug delivery, and an enteric-coated oral delivery system ensuring precise colonic release (Marcelo et al. 2023). Another targeted approach involved trastuzumab-conjugated nanoparticles for HER2+ breast cancer treatment, co-delivering docetaxel and siHER2. These nanoparticles demonstrated superior efficacy in drug-resistant orthotopic HER2+ tumors and successfully crossed the blood-brain barrier to target trastuzumab-resistant brain metastases (Ngamcherdtrakul et al. 2022).

Mesoporous silica nanoparticles (MSNs) modified with curcumin and a cisplatin prodrug provided a synergistic approach to overcoming multidrug resistance in cancer therapy. The embedded curcumin acted as a photosensitizer, generating reactive oxygen species (ROS) under light irradiation, while the cisplatin prodrug ensured sustained chemotherapy, significantly enhancing therapeutic outcomes in resistant cancer cells (Busa et al. 2022). Polysialic acid (PSA) nanoparticles conjugated with the iRGD peptide and photosensitizer chlorin e6 (Ce6) facilitated tumor-targeted drug delivery. This nanoplatform effectively enhanced penetration and retention within tumor tissues, improving the efficacy of PDT and chemotherapy using doxorubicin (Kim et al. 2023).

Finally, a novel metal-organic framework (MOF)-based nanoplatform incorporating manganese-doped nanoparticles was developed for hypoxia-responsive imaging and therapy. These nanoparticles functioned as efficient radiosensitizers and oxygen generators, improving radiotherapy outcomes while serving as MRI contrast agents. Their ability to modulate the tumor microenvironment contributed to enhanced treatment efficacy with reduced systemic toxicity (Guo et al. 2022). Collectively, these advancements underscore the potential of multifunctional nanomaterials in theranostic applications, paving the way for more precise and effective cancer treatments.

10.11 Challenges and Opportunities

10.11.1 Biocompatibility and Safety Concerns

Nanotechnology has revolutionized the field of medicine, offering innovative solutions for disease diagnosis, drug delivery, and therapy. Nanomaterials-based contrast agents and therapeutic systems have demonstrated remarkable potential in imaging modalities like MRI, CT, and ultrasound, as well as in therapeutic techniques such as photodynamic therapy (PDT), sonodynamic therapy (SDT), and targeted drug delivery. However, the integration of these nanomaterials into clinical practice necessitates a comprehensive understanding of their biocompatibility and safety. This section explores the biocompatibility challenges and safety concerns associated with nanomaterials-based contrast and therapeutic agents, emphasizing the importance of optimizing these materials for clinical application (Nel et al. 2009).

10.11.1.1 Cytotoxicity

Nanomaterials can induce cytotoxic effects through mechanisms such as reactive oxygen species (ROS) generation, mitochondrial dysfunction, and apoptosis. For instance, quantum dots, composed of heavy metals like cadmium, may release toxic ions, affecting cellular health. Similarly, silver nanoparticles (AgNPs) are known to disrupt cell membranes and cause oxidative stress (Fadeel and Garcia-Bennett 2010).

10.11.1.2 Immunogenicity

Nanomaterials may provoke immune responses, leading to inflammation or hypersensitivity. It has been shown that liposomal nanoparticles, if improperly formulated, can activate the complement system. As also, SPIONs can also be recognized as foreign by macrophages, triggering an immune reaction (Mahmoudi et al. 2012).

10.11.1.3 Hemocompatibility

Nanomaterials used in intravenous applications must not interfere with blood components. However, carbon-based nanomaterials may aggregate, causing thrombosis or hemolysis. Positively charged nanoparticles tend to disrupt red blood cell membranes (Domingues et al. 2022).

10.11.1.4 Biodistribution and Accumulation

The small size and surface properties of nanomaterials allow them to penetrate tissues effectively, but this also results in potential unintended accumulation. For instance, gold nanoparticles tend to accumulate in the liver and spleen. Persistent nanomaterials, such as carbon nanotubes, can accumulate in organs, leading to chronic toxicity (Domingues et al. 2022).

10.11.1.5 Genotoxicity

Certain nanomaterials can interact with DNA, causing mutations or damage. Studies have shown that metal oxide nanoparticles (e.g., zinc oxide) can induce DNA strand breaks through ROS generation. Prolonged exposure to nanomaterials like titanium dioxide (TiO_2) may cause chromosomal aberrations (Oberdörster et al. 2005).

10.11.2 Safety Concerns in Therapeutic Applications

10.11.2.1 Dose-Dependent Toxicity

Nanomaterials often exhibit a narrow therapeutic window, where slight changes in dosage can lead to toxicity. For example, excessive doses of SPIONs can overwhelm iron metabolism pathways. High concentrations of photothermal agents like gold nanorods can cause localized tissue damage (Mahmoudi et al. 2012).

10.11.2.2 Degradation Products

Decomposition of nanomaterials within the body can release toxic substances. Biodegradable polymeric nanoparticles may produce acidic degradation byproducts. Metal-based nanoparticles, such as QDs, release ions that may disrupt cellular function (Yang and Ma 2010; Derfus et al. 2004).

10.11.2.3 Long-Term Effects

Chronic exposure to nanomaterials raises concerns about their potential carcinogenicity and endocrine-disrupting effects. For instance, prolonged use of silica nanoparticles has been linked to pulmonary fibrosis. Persistent nanostructures, like carbon nanotubes, may act as asbestos-like fibers in the lungs (Nel et al. 2006; Murugadoss et al. 2017).

10.11.2.4 Interaction with Biological Barriers

Nanomaterials must cross biological barriers to reach their target sites. However, their interaction with these barriers can lead to unintended effects. One such effect is the disruption of the blood-brain barrier by nanoparticles which can expose the brain to harmful substances. Nanoparticles with poor renal clearance may accumulate in the kidneys, causing nephrotoxicity (Yang and Ma 2010).

10.11.2.5 Environmental Impact

The disposal and environmental persistence of nanomaterials also pose risks. These risks are primarily accumulation in ecosystems which may affect aquatic life. Also, nanomaterials in wastewater can re-enter the food chain, leading to human exposure (Nel et al. 2006).

10.11.3 Strategies to Mitigate Safety Concerns

10.11.3.1 Surface Functionalization

Modifying the surface of nanoparticles with biocompatible polymers or ligands can enhance stability, reduce toxicity, and improve targeting. Examples include polyethylene glycol (PEG) coating to reduce immunogenicity, functionalizing SPIONs with dextran to improve biocompatibility, and coating nanomaterials with biocompatible polymers like polyethylene glycol (PEG) to enhance clearance via renal pathways (Gupta and Gupta 2005).

10.11.3.2 Biodegradable Nanomaterials

Using materials that degrade into non-toxic byproducts, such as polylactic acid (PLA) and poly (lactic-co-glycolic acid) (PLGA) for drug delivery. Calcium phosphate nanoparticles for bone regeneration are also one of the biodegradable materials that are widely used. Another strategy is to design smaller nanoparticles—

nanoparticles smaller than 5 nm are more efficiently excreted via renal pathways (Aillon et al. 2009; Buchman et al. 2019).

10.11.3.3 Dose Optimization and Periodic Monitoring

Using the minimum effective dose reduces the risk of long-term buildup. Also, long-term imaging studies can track nanomaterial biodistribution and accumulation in vivo (Cohen et al. 2015).

10.11.3.4 Controlled Release Mechanisms

Engineering nanocarriers to release therapeutic agents in response to specific stimuli, such as pH or temperature, minimizes off-target effects (Blum et al. 2015).

10.11.3.5 Thorough Preclinical Testing

Comprehensive in vitro and in vivo testing, including cytotoxicity assays using diverse cell lines and long-term animal studies to evaluate chronic toxicity and carcinogenicity (Khlebtsov and Dykman 2011).

10.11.3.6 Regulatory Oversight

Ensuring that nanomaterials undergo stringent evaluation by regulatory agencies like the FDA and EMA to confirm their safety and efficacy (Foulkes et al. 2020).

10.11.4 Short-Term Toxicity

Nanomaterials interact with biological systems in complex ways, often leading to acute toxic effects. The primary mechanisms include:

10.11.4.1 Oxidative Stress and ROS Generation

Many nanomaterials, particularly metal-based ones like zinc oxide (ZnO) and silver nanoparticles (AgNPs), induce oxidative stress by generating reactive oxygen species (ROS). Elevated ROS levels can damage cellular components, including lipids, proteins, and DNA, resulting in cytotoxicity. For instance, quantum dots (QDs) produce ROS under UV or visible light exposure, causing oxidative damage to nearby cells (Derfus et al. 2004; Sruthi et al. 2018).

10.11.4.2 Inflammation and Immune Activation

Nanoparticles can activate immune cells, leading to the release of inflammatory cytokines and acute inflammation. Lipid-based nanoparticles used in drug delivery may trigger complement activation-related pseudo-allergy (CARPA), resulting in systemic inflammation.

10.11.4.3 Membrane Disruption

The high surface energy of nanoparticles often results in their interaction with cell membranes, causing structural disruptions. Positively charged gold nanoparticles interact with negatively charged cell membranes, leading to cell lysis (Khlebtsov and Dykman 2011).

10.11.4.4 Protein Corona Formation

Upon entering the biological milieu, nanoparticles interact with serum proteins, forming a "protein corona" that can alter their behavior and cause acute toxic effects, such as unintentional cellular uptake or immune activation (Buchman et al. 2019).

10.11.4.5 Metal Ion Release

Metallic nanoparticles, including silver, cadmium, and zinc oxide, can release toxic ions, leading to short-term toxicity. For instance, the release of cadmium ions from quantum dots is known to disrupt cellular homeostasis and trigger cell death (Derfus et al. 2004; Sruthi et al. 2018).

10.11.5 Types of Short-Term Toxic Effects

10.11.5.1 Cytotoxicity

Nanomaterials can induce acute cytotoxic effects, leading to cell death through necrosis or apoptosis. This is particularly concerning for therapeutic agents administered in high doses. Superparamagnetic iron oxide nanoparticles (SPIONs), used in MRI, were shown to cause dose-dependent cytotoxicity in hepatocytes during in vitro studies.

10.11.5.2 Hematological Toxicity

Intravenously administered nanomaterials may interact with blood components, leading to hematological issues such as hemolysis, coagulation disturbances, or platelet aggregation. Carbon nanotubes have been reported to induce platelet activation, increasing the risk of thrombosis in animal models.

10.11.5.3 Pulmonary Toxicity

Inhaled nanomaterials, such as silica or titanium dioxide nanoparticles, can cause acute respiratory issues, including inflammation, oxidative damage, and reduced lung function. For instance, short-term exposure to titanium dioxide nanoparticles in animal studies resulted in pulmonary inflammation and fibrosis.

10.11.5.4 Hepatotoxicity

The liver is a primary organ for nanoparticle metabolism and clearance. Many nanoparticles accumulate in the liver, leading to acute hepatotoxicity characterized by inflammation, oxidative stress, and impaired liver function. Gold nanoparticles administered to mice showed significant liver enzyme elevation and histopathological changes within 24 h.

10.11.5.5 Renal Toxicity

The kidneys play a critical role in nanoparticle excretion. Nanoparticles that are not adequately cleared can accumulate, leading to nephrotoxicity. Zinc oxide nanoparticles have been shown to cause acute renal toxicity in rodents by inducing oxidative damage to kidney cells.

10.11.5.6 Neurotoxicity

Nanoparticles that cross the blood-brain barrier can cause short-term neurotoxic effects, including neuronal inflammation and oxidative stress. Silver nanoparticles were found to induce acute neuroinflammation and neuronal cell death in animal models.

10.11.5.7 Gastrointestinal Toxicity

Orally administered nanoparticles, such as those used for drug delivery, may irritate the gastrointestinal tract, causing inflammation and epithelial damage. Zinc oxide nanoparticles caused acute damage to the intestinal lining in animal studies (Nel et al. 2006, 2009; Mahmoudi et al. 2012; Oberdörster et al. 2005; Derfus et al. 2004; Murugadoss et al. 2017; Aillon et al. 2009; Buchman et al. 2019; Khlebtsov and Dykman 2011; Sruthi et al. 2018; Mahmoudi et al. 2011; Poland et al. 2008; Xia et al. 2008).

10.11.6 Factors Influencing Short-Term Toxic Effects

Several factors determine the severity and nature of short-term toxic effects caused by nanomaterials.

10.11.6.1 Physicochemical Properties

Size: Smaller nanoparticles penetrate tissues more easily, potentially causing more damage.

Shape: Rod-shaped nanoparticles tend to induce more toxicity than spherical ones due to their higher aspect ratio.

Surface Charge: Positively charged nanoparticles are more likely to interact with negatively charged cell membranes, increasing toxicity.

10.11.6.2 Dose and Concentration

Higher doses of nanomaterials are associated with more pronounced toxic effects, as demonstrated in various in vitro and in vivo studies.

10.11.6.3 Route of Administration

The mode of administration significantly affects toxicity. Intravenous routes may lead to systemic effects, while inhalation predominantly affects the lungs.

10.11.6.4 Agglomeration

Nanoparticles tend to aggregate in biological environments, which can affect their biodistribution and toxicity.

10.11.6.5 Surface Modifications

Functionalized nanoparticles, such as PEGylated liposomes, exhibit reduced toxicity compared to their unmodified counterparts. QDs used for fluorescence imaging release cadmium ions, which induce oxidative stress and cytotoxicity within hours of exposure. Coating QDs with biocompatible polymers has been shown to reduce these toxic effects. AgNPs, widely used for their antimicrobial properties, cause acute oxidative stress and inflammation in skin cells. Recent studies suggest that capping agents like citrate can mitigate this toxicity. SPIONs are effective MRI contrast agents but can cause oxidative stress in liver cells shortly after administration. Functionalization with dextran or silica has been demonstrated to reduce this effect (Sukhanova et al. 2018; Abbasi et al. 2023).

10.12 Long-Term Accumulation Effects

10.12.1 Mechanisms of Long-Term Accumulation

10.12.1.1 Reduced Biodegradability

Many nanomaterials, particularly those based on metals (e.g., gold, silver, and titanium dioxide), exhibit low or negligible biodegradability, leading to persistent retention in tissues. It has been shown that gold nanoparticles used in imaging have been detected in the liver and spleen months after administration without significant excretion. Gold nanoparticles resist enzymatic degradation and are primarily sequestered in the liver and spleen due to their interaction with the mononuclear phagocyte system (MPS). Chronic accumulation has been shown to induce hepatic inflammation, oxidative stress, and altered metabolic function.

10.12.1.2 Inefficient Clearance Pathways

Nanomaterials often evade natural clearance mechanisms like renal excretion or macrophage-mediated phagocytosis due to their size, shape, or surface properties. Carbon nanotubes (CNTs), due to their high aspect ratio, resist macrophage engulfment and accumulate in the lungs.

10.12.1.3 Preferential Tissue Uptake

Specific tissues, such as the liver, spleen, and lymph nodes, tend to accumulate nanomaterials due to their role in filtering blood and lymphatic fluid. For instance, quantum dots injected intravenously were predominantly sequestered in the liver and spleen in rodent studies.

10.12.1.4 Bio-corona Formation

The formation of protein or lipid coronas around nanoparticles, known as opsonization, alters their biokinetics, leading to prolonged retention in certain tissues.

10.12.1.5 Repeated or Chronic Exposure

In industrial or environmental settings, repeated exposure to nanoparticles like silica or titanium dioxide can result in a cumulative buildup in organs (Bourquin et al. 2018; Gaillet and Rouanet 2015; Gunduz et al. 2017; Xiong et al. 2022b).

10.12.2 Biological Consequences of Long-Term Accumulation

10.12.2.1 Chronic Inflammation

Persistent accumulation can stimulate the immune system, leading to low-grade chronic inflammation. This is mediated by the continuous activation of macrophages and the release of pro-inflammatory cytokines. Titanium dioxide nanoparticles cause chronic pulmonary inflammation when accumulated in the lungs after repeated inhalation.

10.12.2.2 Fibrosis and Tissue Remodeling

Long-term nanomaterial accumulation in tissues like the lungs, liver, or kidneys can trigger fibrosis due to excessive deposition of extracellular matrix proteins. Carbon nanotubes have been shown to induce lung fibrosis in animal models after chronic exposure.

10.12.2.3 Organ Dysfunction

Accumulated nanoparticles can disrupt cellular and organ function by inducing oxidative stress, mitochondrial damage, and lipid peroxidation. Accumulation of silver nanoparticles in the liver has been linked to impaired hepatic enzyme activity and liver dysfunction in animal studies.

10.12.2.4 Neurotoxicity

Some nanoparticles can cross the blood-brain barrier (BBB) and accumulate in the brain, potentially leading to neurodegenerative diseases. Repeated exposure to manganese oxide nanoparticles in animal models resulted in neuronal damage and symptoms resembling Parkinson's disease.

10.12.2.5 Carcinogenic Potential

Chronic inflammation and oxidative stress induced by accumulated nanoparticles can promote DNA damage and cellular transformation, increasing cancer risk. Prolonged exposure to multi-walled carbon nanotubes in rodents has been associated with mesothelioma-like conditions (Bourquin et al. 2018; Gaillet and Rouanet 2015; Gunduz et al. 2017; Xiong et al. 2022b).

10.13 Future Directions

10.13.1 Machine Learning for Toxicity Prediction

Advances in computational biology and machine learning are enabling predictive models to assess nanomaterial toxicity, reducing reliance on animal studies.

10.13.2 Dynamic Toxicity Screening

High-throughput screening systems that mimic dynamic biological environments can provide more accurate assessments of short-term toxic effects.

10.13.3 Regulatory Oversight

Stringent regulatory frameworks are essential to ensure that nanomaterials undergo rigorous safety testing before clinical approval.

The long-term accumulation of nanomaterials poses significant risks to human health, including chronic inflammation, fibrosis, neurotoxicity, and potential carcinogenic effects. These risks are influenced by the physicochemical properties of nanomaterials, their biodegradability, and clearance efficiency. By understanding these mechanisms and implementing strategies like biodegradable designs and surface functionalization, the safety of nanomaterials in clinical and industrial applications can be improved. Continued research, coupled with stringent regulatory oversight, is essential to balance the benefits of nanotechnology with its potential long-term risks.

Emerging approaches, such as machine learning and artificial intelligence, are being employed to predict the biocompatibility and toxicity of nanomaterials. Advanced computational models can simulate nanoparticle interactions with biological systems, reducing the reliance on animal testing. Additionally, interdisciplinary collaborations among material scientists, biologists, and toxicologists are essential to developing safer nanomaterials.

Nanomaterials-based contrast and therapeutic agents hold immense promise for revolutionizing modern medicine. However, their clinical success hinges on addressing biocompatibility and safety challenges. Through surface modifications, biodegradable designs, controlled release systems, and rigorous testing protocols, researchers can mitigate the associated risks. The development of safe and effective nanomaterials will not only enhance therapeutic outcomes but also pave the way for their widespread adoption in clinical practice (Kumar et al. 2024; Sajid et al. 2015).

10.14 Conclusion

Nanomaterials have emerged as powerful tools in biomedicine, offering significant advancements in both diagnostic imaging and therapeutic applications. Their unique physicochemical properties, including tunable size, high surface-to-volume ratio, and functionalization capabilities, enable enhanced contrast in imaging techniques such as MRI, CT, ultrasound, and optical imaging. Additionally, nanomaterial-based therapeutic platforms, including drug delivery systems, photothermal and photodynamic therapies, gene delivery, and theranostic agents, have revolutionized precision medicine by improving treatment efficacy and minimizing systemic toxicity.

Despite these remarkable advancements, several challenges remain. Issues related to biocompatibility, long-term toxicity, regulatory approval, and large-scale manufacturing hinder the widespread clinical adoption of nanomaterials. The need for standardized safety evaluations and rigorous regulatory frameworks is critical to ensuring their safe and effective translation from laboratory research to real-world medical applications. Future research should focus on developing next-generation multifunctional nanomaterials that integrate diagnostic and therapeutic capabilities for personalized medicine. Innovations in stimuli-responsive nanomaterials, targeted delivery mechanisms, and improved biocompatibility will further enhance their clinical potential. Additionally, efforts to improve cost-effectiveness and scalability will be essential for broader industrial adoption.

In conclusion, nanomaterials hold immense promise for revolutionizing modern medicine by enabling more precise diagnostics and effective treatments. While challenges persist, continuous advancements in nanotechnology, material design, and translational research will drive the integration of nanomaterials into mainstream clinical applications, ultimately improving patient outcomes and shaping the future of healthcare.

Author Contributions
Sasvat Sayee Ram Ramesh—conceptualization, investigation, data curation, formal analysis, original draft writing (Sect. 1, 2, 3, 5); Paloma Patra—investigation, writing the draft (Sect. 4), review and editing; Aravind Kumar Rengan—conceptualization, review and editing of the draft, valuable suggestion, and approval of the final manuscript. All the authors have read and approved the final manuscript.

Conflict of Interest Statement The authors declare no conflicts of interest.

Data Availability Statement Not applicable.

Ethics Statement Not applicable.

References

Abbasi R, Shineh G, Mobaraki M, Doughty S, Tayebi L (2023) Structural parameters of nanoparticles affecting their toxicity for bisomedical applications: a review. J Nanoparticle Res 25(3):1–35

Aikelamu K, Bai J, Zhang Q, Huang J, Wang M, Zhong C (2024) Self-assembled nanoparticles of silicon (IV)–NO Donor phthalocyanine conjugate for tumor photodynamic therapy in red light. Pharmaceutics 16:1166

Aillon KL, Xie Y, El-Gendy N, Berkland CJ, Forrest ML (2009) Effects of nanomaterial physicochemical properties on in vivo toxicity. Adv Drug Deliv Rev 61:457–466

Aires A, Fernández-Afonso Y, Guedes G, Guisasola E, Gutiérrez L, Cortajarena AL (2022) Engineered protein-driven synthesis of tunable iron oxide nanoparticles as T1 and T2 magnetic resonance imaging contrast agents. Chem Mater 34:10832–10841

Akbar MU, Badar M, Zaheer M (2022) Programmable drug release from a dual-stimuli responsive magnetic metal-organic framework. ACS Omega 7:32588–32598

Alhasan A, Sani SA, Tajuddin HA, Ali TH, Hisham S, Ung NM, Azhar NA, Said NA, Abd Jamil AH, Bradley DA (2023) Synthesis of I@MPA-Mn:ZnSe as an efficient contrast agent for CT/fluorescence bi-modal imaging application. Radiat Phys Chemi 209:110947

Allen TJ, Henze Bancroft LC, Kumar M, Bradshaw TJ, Strigel RM, McMillan AB, Fowler AM (2022) Gadolinium-based contrast agent attenuation does not impact PET quantification in simultaneous dynamic contrast enhanced breast PET/MR. Med Phys 49:5206–5215

Alzoubi FY, Abu Noqta O, Al Zoubi T, Al-Khateeb HM, Alqadi MK, Abuelsamen A, Makhadmeh GN (2023) A novel one-pot synthesis of PVP-coated iron oxide nanoparticles as biocompatible contrast agents for enhanced T2-weighted MRI. J Comp Sci 7:131

Aschmann D, Knol RA, Kros A (2024) Lipid-based nanoparticle functionalization with coiled-coil peptides for in vitro and in vivo drug delivery. Acc Chem Res 57:1098–1110

Azad N, Rojanasakul Y (2006) Nanobiotechnology in drug delivery. Am J Drug Deliv 4(2):79–88

Baig MS, Karade SK, Ahmad A, Khan MA, Haque A, Webster TJ, Faiyazuddin M, Al-Qahtani NH (2024) Lipid-based nanoparticles: innovations in ocular drug delivery. Front Mol Biosci 11: 1421959

Betancourt T, Doiron A, Homan KA, Brannon-Peppas L (2009) Controlled release and nanotechnology. Nanotechnol Drug Deliv, pp 283–312

Blum AP, Kammeyer JK, Rush AM, Callmann CE, Hahn ME, Gianneschi NC (2015) Stimuli-responsive nanomaterials for biomedical applications. J Am Chem Soc 137:2140–2154

Boldrini L, Alongi F, Romano A et al (2023) Current practices and perspectives on the integration of contrast agents in MRI-guided radiation therapy clinical practice: A worldwide survey. Clin Transl Radiat Oncol 40:100615

Bourquin J, Milosevic A, Hauser D, Lehner R, Blank F, Petri-Fink A, Rothen-Rutishauser B (2018) Biodistribution, clearance, and long-term fate of clinically relevant nanomaterials. Adv Mater. https://doi.org/10.1002/ADMA.201704307

Buchman JT, Hudson-Smith NV, Landy KM, Haynes CL (2019) Understanding nanoparticle toxicity mechanisms to inform redesign strategies to reduce environmental impact. Acc Chem Res 52:1632–1642

Busa P, Kankala RK, Deng JP, Liu CL, Lee CH (2022) Conquering cancer multi-drug resistance using curcumin and cisplatin prodrug-encapsulated mesoporous silica nanoparticles for synergistic chemo- and photodynamic therapies. Nano 3693(12):3693

Calle D, Ballesteros P, Cerdán S (2018) Advanced contrast agents for multimodal biomedical imaging based on nanotechnology. Methods Mol Biol 1718:441–457

Calvert ND, Baxter J, Torrens AA et al (2023) Highly NIR-II scattering gold superclusters for intravascular optical coherence tomography molecular imaging. Nat Nanotechnol 20(2): 276–285. https://doi.org/10.26434/CHEMRXIV-2023-HNR3G-V2

Chandra J, Nasir N, Wahab S, Sahebkar A, Kesharwani P (2024) Harnessing the power of targeted metal nanocarriers mediated photodynamic and photothermal therapy. Nanomedicine. https://doi.org/10.1080/17435889.2024.2419820

Chkair R, Couvez J, Brégier F, Diab-Assaf M, Sol V, Blanchard-Desce M, Liagre B, Chemin G (2024) Activity of hydrophilic, biocompatible, fluorescent, organic nanoparticles functionalized with purpurin-18 in photodynamic therapy for colorectal cancer. Nano 14:1557

Cohen JM, DeLoid GM, Demokritou P (2015) A critical review of *in vitro* dosimetry for engineered nanomaterials. Nanomedicine 10:3015–3032

De Frates K, Markiewicz T, Gallo P, Rack A, Weyhmiller A, Jarmusik B, Hu X (2018) Protein polymer-based nanoparticles: fabrication and medical applications. Int J Mol Sci 19:1717

Deng J, Wang J, Shi J, Li H, Lu M, Fan Z, Gu Z, Cheng H (2022) Tailoring the physicochemical properties of nanomaterials for immunomodulation. Adv Drug Deliv Rev 180:114039

Derfus AM, Chan WCW, Bhatia SN (2004) Probing the cytotoxicity of semiconductor quantum dots. Nano Lett 4:11–18

Domingues C, Santos A, Alvarez-Lorenzo C, Concheiro A, Jarak I, Veiga F, Barbosa I, Dourado M, Figueiras A (2022) where is nano today and where is it Headed? A review of nanomedicine and the dilemma of nanotoxicology. ACS Nano 16:9994–10041

Domingues JM, Miranda CS, Homem NC, Felgueiras HP, Antunes JC (2023) Nanoparticle synthesis and their integration into polymer-based fibers for biomedical applications. Biomedicine 11:1862

Duan X, Tian H, Zheng S et al (2023) Photothermal-starvation therapy nanomodulator capable of inhibiting colorectal cancer recurrence and metastasis by energy metabolism reduction. Adv Healthc Mater 12:2300968

Duncan B, Kim C, Rotello VM (2010) Gold nanoparticle platforms as drug and biomacromolecule delivery systems. J Controlled Rel 148:122–127

Fabozzi A, Della Sala F, Di Gennaro M, Solimando N, Pagliuca M, Borzacchiello A (2021) Polymer based nanoparticles for biomedical applications by microfluidic techniques: from design to biological evaluation. Polym Chem 12:6667–6687

Fadeel B, Garcia-Bennett AE (2010) Better safe than sorry: Understanding the toxicological properties of inorganic nanoparticles manufactured for biomedical applications. Adv Drug Deliv Rev 62:362–374

Fang J, Nakamura H, Maeda H (2011) The EPR effect: unique features of tumor blood vessels for drug delivery, factors involved, and limitations and augmentation of the effect. Adv Drug Deliv Rev 63:136–151

Feczkó T (2021) Polymeric nanotherapeutics acting at special regions of body. J Drug Deliv Sci Technol 64:102597

Ferreira Soares DC, Domingues SC, Viana DB, Tebaldi ML (2020) Polymer-hybrid nanoparticles: current advances in biomedical applications. Biomed Pharmacother 131:110695

Flak D, Zalewski T, Fiedorowicz K et al (2023) Hybrids of manganese oxide and lipid liquid crystalline nanoparticles (LLCNPs@MnO) as potential magnetic resonance imaging (MRI) contrast agents. J Mater Chem B 11:8732–8753

Foulkes R, Man E, Thind J, Yeung S, Joy A, Hoskins C (2020) The regulation of nanomaterials and nanomedicines for clinical application: current and future perspectives. Biomater Sci 8:4653–4664

Frencken AL, Richtsmeier D, Leonard RL et al (2024) X-ray-sensitive doped CaF2-based MRI contrast agents for local radiation dose measurement. ACS Appl Mater Interfaces 16:13453–13465

Frochot C, Sève A, Vanderesse R, Toussaint M, Barberi-Heyob M (2015) Silica-based nanoparticles for photodynamic therapy. Appl Nanosci Photomed, pp 449–464

Fu H, Xue K, Zhang Y, Xiao M, Wu K, Shi L, Zhu C (2023) Thermoresponsive hydrogel-enabled thermostatic photothermal therapy for enhanced healing of bacteria-infected wounds. Adv Sci 10:2206865

Gaillet S, Rouanet JM (2015) Silver nanoparticles: their potential toxic effects after oral exposure and underlying mechanisms—a review. Food Chem Toxicol 77:58–63

Gangavarapu A, Tapia-Lopez LV, Sarkar B, Pena-Zacarias J, Zayed A, Badruddoza M, Nurunnabi M (2024) Lipid nanoparticles for enhancing oral bioavailability. Nanoscale 16:18319–18338

Gao F, Liu Y, Zhu L, Zhang J, Chang Y, Gao W, Ma G, Ma X, Guo Y (2025) An intelligent triple assisted gold cluster-based nanosystem for enhanced tumor photodynamic therapy. Chin J Chem 43:175–183

Gerion D, Herberg J, Gjersing E, Ramon E, Maxwell R, Gray JW, Budinger TF, Chen FF (2006) Enhancement of T1 and T2 relaxation by paramagnetic silica-coated nanocrystals. J Phys Chem C 111:540–551

Gilmore JL, Yi X, Quan L, Kabanov AV (2008) Novel nanomaterials for clinical neuroscience. J Neuroimmune Pharmacol 3:83–94

Glowacka-Sobotta A, Czarczynska-Goslinska B, Ziental D, Wysocki M, Michalak M, Güzel E, Sobotta L (2024) Versatile porphyrin arrangements for photodynamic therapy—a review. Nano 14:1879

Golestani P (2024) Lipid-based nanoparticles as a promising treatment for the skin cancer. Heliyon 10:e29898

Grzelak J, Gázquez J, Grayston A, Teles M, Herranz F, Roher N, Rosell A, Roig A, Gich M (2022) Magnetic mesoporous silica nanorods loaded with ceria and functionalized with fluorophores for multimodal imaging. ACS Appl Nano Mater 5:2113–2125

Guedes G, Uribe KB, Martínez-Parra L, Aires A, Beraza M, Ruiz-Cabello J, Cortajarena AL (2024) Engineering protein-nanoparticle hybrids as targeted contrast agents. ACS Appl Mater Interfaces 16:59861

Gunduz N, Ceylan H, Guler MO, Tekinay AB (2017) Intracellular accumulation of gold nanoparticles leads to inhibition of macropinocytosis to reduce the endoplasmic reticulum stress. Sci Rep 7(1):40493

Guo W, Gao H, Li H, Ge S, Zhang F, Wang L, Shi H, Han A (2022) Self-assembly of a multifunction DNA tetrahedron for effective delivery of aptamer PL1 and Pcsk9 siRNA potentiate immune checkpoint therapy for colorectal cancer. ACS Appl Mater Interfaces 14: 31634–31644

Guo W, Song X, Liu J, Liu W, Chu X, Lei Z (2024) Quantum dots as a potential multifunctional material for the enhancement of clinical diagnosis strategies and cancer treatments. Nano 14: 1088

Gupta AK, Gupta M (2005) Cytotoxicity suppression and cellular uptake enhancement of surface modified magnetic nanoparticles. Biomaterials 26:1565–1573

Gupta S, Dutta B, Shelar SB, Gangwar A, Bhattacharyya K, Bairwa KK, Hassan PA, Barick KC (2024) Polyphosphate-mediated crystallographic and colloidal stabilization of CuS nanoparticles: enhanced NIR-responsive chemo-photothermal efficacy. ACS Appl Bio Mater 7:6641–6655

Hajebi S, Yousefiasl S, Rahimmanesh I et al (2022) Genetically engineered viral vectors and organic-based non-viral nanocarriers for drug delivery applications. Adv Healthc Mater 11: 2201583

Haleem A, Javaid M, Singh RP, Rab S, Suman R (2023) Applications of nanotechnology in medical field: a brief review. Glob Health J 7:70–77

Hamidu A, Pitt WG, Husseini GA (2023) Recent breakthroughs in using quantum dots for cancer imaging and drug delivery purposes. Nano 13:2566

Haroyan H, Ginat DT (2015) Gadolinium-based contrast agents. Neuroimaging Pharma:105–110

He C, Lin W (2015) Hybrid nanoparticles for cancer imaging and therapy. Cancer Treat Res 166: 173–192

He Y, Mao Z, Lu Z, Yan J, Zhang Y, Bianco A, Cao Y, Pei R (2022) Extremely small iron oxide nanoparticles with pH-dependent solubility transition as T1/T2 switchable contrast agents for MRI. ACS Appl Nano Mater 5:15826–15836

Hills RA, Howarth M (2022) Virus-like particles against infectious disease and cancer: guidance for the nano-architect. Curr Opin Biotechnol 73:346–354
Huang S, Huang G (2024) The utilization of quantum dot labeling as a burgeoning technique in the field of biological imaging. RSC Adv 14:20884–20897
Huang Y, Cao L, Parakhonskiy BV, Skirtach AG (2022) Hard, soft, and hard-and-soft drug delivery carriers based on CaCO3 and alginate biomaterials: synthesis, properties, pharmaceutical applications. Pharmaceutics 14:909
Ibrahim YO, Maalej N, Masood Pirzada B, Younis Raja A, Anjum DH, Jan N, Behouch A, Ul Haq Qurashi A (2024) Gold nanoparticles spectral CT imaging and limit of detectability in a new materials contrast-detail phantom. Phys Med 120:103326
Jia WL, Xu LF, Xu WJ, Yang M, Zhang YW (2022) Application of nanotechnology in the diagnosis and treatment of acute pancreatitis. Nanoscale Adv 4:1949–1961
Jiang Z, Ding Y, Lovell JF, Zhang Y (2022) Design and application of organic contrast agents for molecular imaging in the second near infrared (NIR-II) window. Photoacoustics 28:100426
Joseph JM, Gigliobianco MR, Firouzabadi BM, Censi R, Di Martino P (2022) Nanotechnology as a versatile tool for 19F-MRI agent's formulation: a glimpse into the use of perfluorinated and fluorinated compounds in nanoparticles. Pharmaceutics 14:382
Jouyandeh M, Sajadi SM, Seidi F et al (2022) Metal nanoparticles-assisted early diagnosis of diseases. OpenNano 8:100104
Jovanović S, Marković Z, Budimir M, Prekodravac J, Zmejkoski D, Kepić D, Bonasera A, Marković BT (2023) Lights and dots toward therapy—carbon-based quantum dots as new agents for photodynamic therapy. Pharmaceutics 15:1170
Karageorgou MA, Bouziotis P, Stiliaris E, Stamopoulos D (2023) Radiolabeled iron oxide nanoparticles as dual modality contrast agents in SPECT/MRI and PET/MRI. Nano 13:503
Kawasaki R, Eto T, Kono N, Ohdake R, Yamana K, Hirano H, Kawamura S, Tarutani N, Katagiri K, Ikeda A (2024) Photodynamic therapy using hybrid nanoparticles comprising of upconversion nanoparticles and chlorin e6-bearing pullulan. Biomater Sci 12:5766–5774
Kayal S, Kola P, Pal J, Mandal M, Dhara D (2024) Self-indicating polymer prodrug nanoparticles for pH-responsive drug delivery in cancer cells and real-time monitoring of drug release. ACS Appl Bio Mater 7:5810–5822
Keahey P, Si P, Razavi M, Yu S, Lippok N, Villiger M, Padera TP, De La Zerda A, Bouma B (2021) Spectral- and polarization-dependent scattering of gold nanobipyramids for exogenous contrast in optical coherence tomography. Nano Lett 21:8595–8601
Khaledian S, Abdoli M, Fatahian R, Zahabi SS, Khaledian S, Abdoli M, Fatahian R, Zahabi SS (2023) Quantum dots in cancer cell imaging. In: Quantum dots—recent advances, new perspectives and contemporary applications. https://doi.org/10.5772/INTECHOPEN.107671
Khan M, Liu H, Sacco P, Marsich E, Li X, Djaker N, Spadavecchia J (2022a) DOTAREM (DOTA)–gold-nanoparticles: design, spectroscopic evaluation to build hybrid contrast agents to applications in nanomedicine. Int J Nanomed 17:4105–4118
Khan M, Liu H, Sacco P, Marsich E, Li X, Djaker N, Spadavecchia J (2022b) DOTAREM (DOTA)–gold-nanoparticles: design, spectroscopic evaluation to build hybrid contrast agents to applications in nanomedecine. Int J Nanomedicine 17:4105–4118
Khlebtsov N, Dykman L (2011) Biodistribution and toxicity of engineered gold nanoparticles: a review of in vitro and in vivo studies. Chem Soc Rev 40:1647–1671
Khumaeni A, Kurniawan G, Hidayanto E (2023a) Iron oxide nanoparticles produced by a low-energy Nd: YAG Laser ablation technique and their application as contrast agent for magnetic resonance imaging. J Nano Res 80:65–80
Khumaeni A, Budi WS, Avicenna S, Muniroh M, Kusumaningrum N, Damayanti O, Fitria S (2023b) Gold and silver nanoparticles as computed tomography (CT) contrast agents produced by a pulsed laser ablation technique: study in-vitro and in-vivo. Rasayan J Chem 16:502–508
Kim D, Kim H (2024) Effectiveness of photothermal therapy using various noble-metal photothermal agents. Int J Thermal Sci 200:108998

Kim J, Piao Y, Hyeon T, Malley W, Kubeil M (2014) Nanomaterials for drugs delivery. Soft Nanosci Lett 4:51–52

Kim GL, Park B, Jang EH, Gu J, Seo SR, Cheung H, Lee HJ, Lee S, Kim JH (2023) Dual effect of chemo-PDT with tumor targeting nanoparticles containing iRGD peptide. Pharmaceutics 15: 614

Kim M, Kubelick KP, Yu AM, VanderLaan D, Jhunjhunwala A, Nikolai RJ, Cadena M, Kim J, Emelianov SY (2024) Regulating interparticle proximity in plasmonic nanosphere aggregates to enhance photoacoustic response and photothermal stability. Adv Funct Mater 34:2313963

Komarova NL, Wodarz D (2014) Combination therapies: short-term versus long-term strategies. In: Targeted cancer treatment in silico: small molecule inhibitors and oncolytic viruses. New York, NY, Springer New York, pp 89–106

Kong C, Xu B, Qiu G, Wei M, Zhang M, Tang J, Li L, Liu J, Bao S (2022) Multifunctional nanoparticles-mediated PTT/PDT synergistic immune activation and antitumor activity combined with Anti-PD-L1 immunotherapy for breast cancer treatment. Int J Nanomedicine 17: 5391–5411

Kothapalli SVVN (2013) Ultrasound contrast agents loaded with magnetic nanoparticles: acoustic and mechanical characterization. KTH Royal Institute of Technology

Kumar R, Dutt S, Tripathi AD, Singh AK, Chaturvedi VK, Singh SK (2024) Navigating safety and toxicity challenges in nanomedicine: strategies, assessment, and mitigation. Nanotechnol Life Sci Part F 3808:15–37

Kurul F, Turkmen H, Cetin AE, Topkaya SN (2025) Nanomedicine: How nanomaterials are transforming drug delivery, bio-imaging, and diagnosis. Next Nanotechnol 7:100129

Li Q, Shi Z, Zhang F, Zeng W, Zhu D, Mei L (2022) Symphony of nanomaterials and immunotherapy based on the cancer–immunity cycle. Acta Pharm Sin B 12:107–134

Li Z, Qi G, Shi G, Zhang M, Hu H, Hao L (2023a) Engineered graphene quantum dots as a magnetic resonance signal amplifier for biomedical imaging. Molecules 28:2363

Li Y, Chen Y, Liu R, Duan S, Chen L, Sun J, Zhang L (2023b) Study on contrast-enhanced ultrasound imaging and anti-tumor effects of drug-loaded nanodroplets with tumor targeting and ultrasound sensitivity. Front Biosci Landmark 28:115

Li N, Dong F, Sun L, Qian Y, Zhang L, Wang G, Yuan L, Liu H, Jiang Y, Wang Y (2024a) Carrier-free delivery of nucleic acid and photosensitizer nanoparticles for enhanced photodynamic and gene antitumour therapy. Fundam Res. https://doi.org/10.1016/J.FMRE.2024.03.014

Li MM, Zhang Y, Sun F, Huai MX, Zhang FY, Pan JX, Qu CY, Shen F, Li ZH, Xu LM (2024b) Photodynamic therapy using RGD-functionalized quantum dots elicit a potent immune response in a syngeneic mouse model of pancreatic. Cancer Int J Nanomedicine 19:9487–9502

Li X, Li Z, Su Y, Zhou J, Li Y, Zhao Q, Yang X, Shi L, Shen L (2024c) Carrier-free hybrid nanoparticles for enhanced photodynamic therapy in oral carcinoma via reversal of hypoxia and oxidative resistance. Pharmaceutics 16:1130

Liang Y, Zhang S, Wang D et al (2024) Dual-functional nanodroplet for tumor vasculature ultrasound imaging and tumor immunosuppressive microenvironment remodeling. Adv Healthc Mater 13:2401274

Liu Q, Guan J, Song R, Zhang X, Mao S (2022a) Physicochemical properties of nanoparticles affecting their fate and the physiological function of pulmonary surfactants. Acta Biomater 140: 76–87

Liu G, Ling J, He L, Xu Y, Chen T, Shi C, Luo L (2022b) Theranostic cancer treatment using lentinan-coated selenium nanoparticles and label-free CEST MRI. Pharmaceutics 15:120

Liu K, Zhang L, Lu H, Wen Y, Bi B, Wang G, Jiang Y, Zeng L, Zhao J (2023) Enhanced mild-temperature photothermal therapy by pyroptosis-boosted ATP deprivation with biodegradable nanoformulation. J Nanobiotechnol 21:1–17

Lu C, Xu X, Zhang T, Wang Z, Chai Y (2022) Facile synthesis of superparamagnetic nickel-doped iron oxide nanoparticles as high-performance T1 contrast agents for magnetic resonance imaging. J Mater Chem B 10:1623–1633

Luke GP (2024) Perfluorocarbon nanodroplets for ultrasound imaging and drug delivery. J Acoust Soc Am 155:A286
Lutz TM, Kimna C, Lieleg O (2022) A pH-stable, mucin based nanoparticle system for the co-delivery of hydrophobic and hydrophilic drugs. Int J Biol Macromol 215:102–112
Lv B, Zhao Y, Liang Y, Cao J (2024a) Extracellular vesicles: an advanced delivery platform for photosensitizers in tumor photodynamic therapy. Chem Eng J 498:155438
Lv K, Wang H, Fu X, Chen S, Zhang R, Zhou Y, Feng J, Zhang H (2024b) An integrated nanoplatform via dual channel excitation for both precise fluorescence imaging and photodynamic therapy of orthotopic breast tumor in NIR-II region. Small 20:2404007
Ma H, Mu X, Tang Y, Li C, Wang Y, Lu Y, Zhou X, Li Z (2023) Programmable multistage small-molecule nano-photosensitizer for multimodal imaging-guided photothermal therapy. Acta Biomater 157:408–416
Machtakova M, Thérien-Aubin H, Landfester K (2022) Polymer nano-systems for the encapsulation and delivery of active biomacromolecular therapeutic agents. Chem Soc Rev 51(1): 128–152
Maeda H, Wu J, Sawa T, Matsumura Y, Hori K (2000) Tumor vascular permeability and the EPR effect in macromolecular therapeutics: a review. J Controlled Rel 65:271–284
Mahmoudi M, Sant S, Wang B, Laurent S, Sen T (2011) Superparamagnetic iron oxide nanoparticles (SPIONs): Development, surface modification and applications in chemotherapy. Adv Drug Deliv Rev 63:24–46
Mahmoudi M, Hofmann H, Rothen-Rutishauser B, Petri-Fink A (2012) Assessing the in vitro and in vivo toxicity of superparamagnetic iron oxide nanoparticles. Chem Rev 112:2323–2338
Mallik R, Khannam M, Saha M, Marandi S, Kumar S, Mukherjee C (2022) The electrostatic confinement of aquated monocationic Gd(III) complex-molecules within the inner core of porous silica nanoparticles creates a highly efficient T1 contrast agent for magnetic resonance imaging. Dalton Trans 51:14138–14149
Marcelo GA, Montpeyó D, Galhano J, Martínez-Máñez R, Capelo-Martínez JL, Lorenzo J, Lodeiro C, Oliveira E (2023) Development of new targeted nanotherapy combined with magneto-fluorescent nanoparticles against colorectal cancer. Int J Mol Sci 24:6612
Maretti E, Gioia F, Rustichelli C, Molinari S, Leo E (2024) Inflammatory-targeted lipid carrier as a new nanomaterial to formulate an inhaled drug delivery system. Molecules 29:1616
Mariano S, Carata E, Calcagnile L, Panzarini E (2024) Recent advances in photodynamic therapy: metal-based nanoparticles as tools to improve cancer therapy. Pharmaceutics 16:932
Mazhit A, Zhumanova K, Duisenbayeva B, Atabaev TS (2023) Highly efficient CT contrast agent based on barium-doped mesoporous silica nanoparticles (SiO2-Ba NPs). Mater Lett 353:135318
Mehrdadi S (2024) Lipid-based nanoparticles as oral drug delivery systems: overcoming poor gastrointestinal absorption and enhancing bioavailability of peptide and protein therapeutics. Adv Pharm Bull 14:48–66
Mikhaylov G, Mikac U, Butinar M, Turk V, Turk B, Psakhie S, Vasiljeva O (2022) Theranostic applications of an ultra-sensitive T1 and T2 magnetic resonance contrast agent based on cobalt ferrite spinel nanoparticles. Cancers (Basel) 14:4026
Modi SK, Mohapatra P, Bhatt P, Singh A, Parmar AS, Roy A, Joshi V, Singh MS (2025) Targeting tumor microenvironment with photodynamic nanomedicine. Med Res Rev 45:66–96
Mohapatra J, Nigam S, George J, Arellano AC, Wang P, Liu JP (2023) Principles and applications of magnetic nanomaterials in magnetically guided bioimaging. Mater Today Phys 32:101003
Molaei MJ (2022) Gadolinium-doped fluorescent carbon quantum dots as MRI contrast agents and fluorescent probes. Sci Rep 12:1–11
Munir M, Zaman M, Waqar MA, Khan MA, Alvi MN (2024) Solid lipid nanoparticles: a versatile approach for controlled release and targeted drug delivery. J Liposome Res 34:335–348
Murugadoss S, Lison D, Godderis L, Van Den BS, Mast J, Brassinne F, Sebaihi N, Hoet PH (2017) Toxicology of silica nanoparticles: an update. Arch Toxicol 91:2967–3010
Nagasa GD, Belete A (2022) Review on nanomaterials and nano-scaled systems for topical and systemic delivery of antifungal drugs. J Multidiscip Healthc 15:1819–1840

Navarro-Ocón A, Blaya-Cánovas JL, López-Tejada A et al (2022) Nanomedicine as a promising tool to overcome immune escape in breast cancer. Pharmaceutics 14(3):505

Naylor-Adamson L, Price TW, Booth Z, Stasiuk GJ, Calaminus SDJ (2024) Quantum dot imaging agents: haematopoietic cell interactions and biocompatibility. Cells 13:354

Nel A, Xia T, Mädler L, Li N (2006) Toxic potential of materials at the o level. Science (1979) 311: 622–627

Nel AE, Mädler L, Velegol D, Xia T, Hoek EMV, Somasundaran P, Klaessig F, Castranova V, Thompson M (2009) Understanding biophysicochemical interactions at the nano–bio interface. Nat Mater 8:543–557

Ngamcherdtrakul W, Bejan DS, Cruz-Muñoz W et al (2022) Targeted nanoparticle for co-delivery of HER2 siRNA and a taxane to mirror the standard treatment of HER2+ breast cancer: efficacy in breast tumor and brain metastasis. Small 18:2107550

Oberdick S, Jordanova K, Lundstrom J, Parigi G, Poorman M, Zabow HG, Keenan K (2023) Iron oxide nanoparticles as positive T1 contrast agents for low-field magnetic resonance imaging at 64 mT. Sci Rep 13(1):11520. https://doi.org/10.21203/RS.3.RS-2485292/V1

Oberdörster G, Oberdörster E, Oberdörster J (2005) Nanotoxicology: an emerging discipline evolving from studies of ultrafine particles. Environ Health Perspect 113:823–839

Ou G, Li Z, Li D, Cheng L, Liu Z, Wu H (2016) Photothermal therapy by using titanium oxide nanoparticles. Nano Res 9:1236–1243

Oudega M, Bradbury EJ, Ramer MS (2012) Combination therapies. Handb Clin Neurol 109:617–636

Patel P, Garala K, Singh S, Prajapati BG, Chittasupho C (2024) Lipid-based nanoparticles in delivering bioactive compounds for improving therapeutic efficacy. Pharmaceuticals 17:329

Pavitra E, Lee H, Hwang SK, Park JY, Varaprasad GL, Rao MVB, Han YK, Raju GSR, Huh YS (2022) Cooperative ligand fields enriched luminescence of AgGd(MoO4)2:Er3+/Yb3+@mSi core–shell upconversion nanoplates for optical thermometry and biomedical applications. Appl Surf Sci 579:152166

Poland CA, Duffin R, Kinloch I, Maynard A, Wallace WAH, Seaton A, Stone V, Brown S, MacNee W, Donaldson K (2008) Carbon nanotubes introduced into the abdominal cavity of mice show asbestos-like pathogenicity in a pilot study. Nat Nanotechnol 3:423–428

Ponnappan N, Chugh A (2015) Nanoparticle-mediated delivery of therapeutic drugs. Pharmaceut Med 29:155–167

Porat-Dahlerbruch G, Sergeyev IV, Quinn CM et al (2025) Spatial organization of lipid nanoparticle siRNA delivery systems revealed by an integrated magnetic resonance approach. Small Methods 9:2400622

Qin Z, Lai P, Sun M (2024) Photoacoustic thermal-strain measurement towards noninvasive and accurate temperature mapping in photothermal therapy. Photoacoustics 40:100651

Radwan SH, Azzazy HME (2009) Gold nanoparticles for molecular diagnostics. Expert Rev Mol Diagn 9:511–524

Rahman M (2023) Magnetic resonance imaging and iron-oxide nanoparticles in the era of personalized medicine. Nano 7:424–449

Rahmania FJ, Huang YS, Workie YA et al (2023) Preparation of functional nanoparticles-loaded magnetic carbon nanohorn nanocomposites towards composite treatment. Nano 13:839

Rampado R, Somu Naidu G, Karpov O et al (2025) Lipid nanoparticles with fine-tuned composition show enhanced colon targeting as a platform for mRNA therapeutics. Adv Sci 12:2408744

Rehman M, Tahir N, Sohail MF, Qadri MU, Duarte SOD, Brandão P, Esteves T, Javed I, Fonte P (2024) Lipid-based nanoformulations for drug delivery: an ongoing perspective. Pharmaceutics 16:1376

Riaz R, Waqar H, Ahmad NM, Abbas SR (2022) Novel magnetic elastic phase-change nanodroplets as dual mode contrast agent for ultrasound and magnetic resonance imaging. Polymers (Basel) 14:2915

Romero MP, González MA, Romero MP, González MA (2024) Drug-delivery based on encapsulation for photodynamic therapy and photothermal therapy. Biomater Microencapsul. https://doi.org/10.5772/INTECHOPEN.1005277

Roy C, Gandhi A, Manna S, Jana S (2024) Fabrication and evaluation of carboxymethyl guar gum-chitosan interpenetrating polymer network (IPN) nanoparticles for controlled drug delivery. Med Nov Technol Devices 22:100300

Ruggiero C, Pastorino L, Herrera OL (2010) Nanotechnology based targeted drug delivery. In: 2010 Annual International Conference of the IEEE Engineering in Medicine and Biology Society, EMBC'10, pp 3731–3732

Sajid M, Ilyas M, Basheer C, Tariq M, Daud M, Baig N, Shehzad F (2015) Impact of nanoparticles on human and environment: review of toxicity factors, exposures, control strategies, and future prospects. Environ Sci Pollut Res 22:4122–4143

Saladino GM, Kilic NI, Brodin B, Hamawandi B, Yazgan I, Hertz HM, Toprak MS (2021) Carbon quantum dots conjugated rhodium nanoparticles as hybrid multimodal contrast agents. Nano 11: 2165

Sasaki I, Brégier F, Chemin G, Daniel J, Couvez J, Chkair R, Vaultier M, Sol V, Blanchard-Desce M (2024) Hydrophilic biocompatible fluorescent organic nanoparticles as nanocarriers for biosourced photosensitizers for photodynamic therapy. Nano 14:216

Sci IJT, Singh S, Pandey VK, Prakash Tewari R, Agarwal V (2011) Nanoparticle based drug delivery system: advantages and applications. Indian J Sci Technol 4:177–184

Shah J, Park S, Aglyamov SR, Larson T, Ma L, Sokolov KV, Johnston KP, Milner TE, Emelianov SY (2008) Photoacoustic imaging and temperature measurement for photothermal cancer therapy. J Biomed Optics 13(3):034024. https://doi.org/10.1117/12940362

Shi Z, Zhou Y, Fan T, Lin Y, Zhang H, Mei L (2020) Inorganic nano-carriers based smart drug delivery systems for tumor therapy. Smart Mater Med 1:32–47

Shirshahi V, Soltani M (2015) Solid silica nanoparticles: applications in molecular imaging. Contrast Media Mol Imaging 10:1–17

Sivasubramanian M, Chu CH, Cheng SH, Chen NT, Chen CT, Chuang YC, Yu H, Chen YL, De LL, Lo LW (2022) Multimodal magnetic resonance and photoacoustic imaging of tumor-specific enzyme-responsive hybrid nanoparticles for oxygen modulation. Front Bioeng Biotechnol 10:910902

Slavcev RA, Wettig S, Kaur T, Slavcev RA, Wettig S, Kaur T (2011) Nanomedicine based approaches to cancer diagnosis and therapy. Non-Viral Gene Ther. https://doi.org/10.5772/18304

Soomherun N, Kreua-ongarjnukool N, Niyomthai ST, Chumnanvej S (2024) Lipid-polymer hybrid nanoparticles synthesized via lipid-based surface engineering for a robust drug delivery platform. Colloids Surf B Biointerfaces 237:113858

Souri M, Chiani M, Farhangi A, Mehrabi MR, Nourouzian D, Raahemifar K, Soltani M (2022) Anti-COVID-19 nanomaterials: directions to improve prevention, diagnosis, and treatment. Nano 12:783

Spector DV, Bykusov V, Zharova A et al (2024) Nanoformulation of the photoactive cisplatin prodrug for combined photothermal therapy and bioimaging. ACS Appl Nano Mater. https://doi.org/10.1021/ACSANM.4C04680/ASSET/IMAGES/LARGE/AN4C04680_0007.JPEG

Sreeharsha N, Kunigal Sridhar S, Bhuvanahalli Rangappa A, Goudanavar P, Karadigere Nagaraju P, Raghavendra Naveen N, Narayanappa Shiroorkar P, Haq Asif A, Meravanige G, Swaroop Duddi Sreehari K (2024) Ultrasonication-mediated synthesis of diblock polymer-based nanoparticles for advanced drug delivery systems: insights and optimization. Ultrason Sonochem 111:107137

Sruthi S, Ashtami J, Mohanan PV (2018) Biomedical application and hidden toxicity of Zinc oxide nanoparticles. Mater Today Chem 10:175–186

Su K, Shi L, Sheng T et al (2024) Reformulating lipid nanoparticles for organ-targeted mRNA accumulation and translation. Nat Commun 15:1–12

Sukhanova A, Bozrova S, Sokolov P, Berestovoy M, Karaulov A, Nabiev I (2018) Dependence of nanoparticle toxicity on their physical and chemical properties. Nanoscale Res Lett 13:1–21
Sun H, Ong Y, Kim MM, Dimofte A, Singhal S, Cengel KA, Yodh AG, Zhu TC (2024) A comprehensive study of reactive oxygen species explicit dosimetry for pleural photodynamic therapy. Antioxidants 13:1436
Sur S, Rathore A, Dave V, Reddy KR, Chouhan RS, Sadhu V (2019) Recent developments in functionalized polymer nanoparticles for efficient drug delivery system. Nano-Struct Nano-Objects 20:100397
Szpak A, Fiejdasz S, Prendota W, Strączek T, Kapusta C, Szmyd J, Nowakowska M, Zapotoczny S (2014) T1–T2 Dual-modal MRI contrast agents based on superparamagnetic iron oxide nanoparticles with surface attached gadolinium complexes. J Nanopart Res 16:1–11
Tan M, Li X, Zhang H, Zheng M, Xiong J, Cao Y, Cao G, Wang Z, Ran H (2023) Förster resonance energy transfer nanobullet for photoacoustic imaging and amplified photothermal-photodynamic therapy of cancer. Adv Healthc Mater 12:2202943
Tao J, Wang B, Dong Y, Chen XY, Li S, Jiang T, Zhao X (2023) Photothermal and acid-responsive fucoidan CuS bubble pump microneedles for combined CDT/PTT/CT treatment of melanoma. ACS Appl Mater Interfaces 15:40267–40279
Tarasi S, Ramazani A, Morsali A, Hu ML, Ghafghazi S, Tarasi R, Ahmadi Y (2022) Drug delivery using hydrophilic metal-organic frameworks (MOFs): effect of structure properties of MOFs on biological behavior of carriers. Inorg Chem 61:13125–13132
Thassu D, Pathak Y, Deleers M (2007) Nanoparticulate drug-delivery systems: an overview. Nanoparticulate Drug Deliv Syst:1–31
Thirumurugan S, Dash P, Liu X et al (2022) Angiopep-2-decorated titanium–alloy core–shell magnetic nanoparticles for nanotheranostics and medical imaging. Nanoscale 14:14789–14800
Truong DH, Tran PTT, Tran TH (2024) Nanoparticles as carriers of photosensitizers to improve photodynamic therapy in cancer. Pharm Dev Technol 29:221–235
ul Haque S, Nasar A, Duteanu N, Inamuddin (2023) Carbon based-nanomaterials used in biofuel cells—a review. Fuel 331:125634
Unruh T, Götz K, Vogel C, Fröhlich E, Scheurer A, Porcar L, Steiniger F (2024) Mesoscopic structure of lipid nanoparticle formulations for mRNA drug delivery: comirnaty and drug-free dispersions. ACS Nano 18:9746–9764
Van NA, Jandhyala S, Jordan T, Luke GP (2021) Repeated acoustic vaporization of perfluorohexane nanodroplets for contrast-enhanced ultrasound imaging. IEEE Trans Ultrason Ferroelectr Freq Control 68:3497–3506
Wang AZ (2015) EPR or no EPR? Sci Transl Med, The billion-dollar question. https://doi.org/10.1126/SCITRANSLMED.AAC8108
Wang W, Shang S, Wang Y, Xu B (2024a) Utilization of nanomaterials in MRI contrast agents and their role in therapy guided by imaging. Front Bioeng Biotechnol 12:1484577
Wang J, Wang Y, Zhong L, Yan F, Zheng H (2024b) Nanoscale contrast agents: a promising tool for ultrasound imaging and therapy. Adv Drug Deliv Rev 207:115200
Wang H, You J, Shi D, Cai L, Shi H, Shi H (2024c) Multifunctional Ag2S/g-C3N4 Nanoparticles for photodynamic and photothermal cancer therapy under near-infrared excitation. ACS Appl Nano Mater 7:19904–19914
Xia T, Kovochich M, Liong M, Zink JI, Nel AE (2008) Cationic polystyrene nanosphere toxicity depends on cell-specific endocytic and mitochondrial injury pathways. ACS Nano 2:85–96
Xiong W, Wang S, Wei Z, Cai Y, Li B, Lin F, Xia D (2022a) Knowledge domain and hotspots predict concerning electroactive biomaterials applied in tissue engineering: a bibliometric and visualized analysis from 2011 to 2021. Front Bioeng Biotechnol 10:904629
Xiong P, Huang X, Ye N, Lu Q, Zhang G, Peng S, Wang H, Liu Y (2022b) Cytotoxicity of metal-based nanoparticles: from mechanisms and methods of evaluation to pathological manifestations. Adv Sci 9:2106049
Xue X, Liang X-J (2014) Multifunctional nanoparticles for theranostics and imaging. In: Nanomedicine: principles and perspectives. New York, NY, Springer New York, pp 101–115

Yadav P, Bhardwaj P, Maruthi M, Chakraborty A, Kanoo P (2023) Metal–organic framework based drug delivery systems as smart carriers for release of poorly soluble drugs hydrochlorothiazide and dapsone. Dalton Trans 52:11725–11734

Yaghmur A, Østergaard J, Mu H (2023) Lipid nanoparticles for targeted delivery of anticancer therapeutics: recent advances in development of siRNA and lipoprotein-mimicking nanocarriers. Adv Drug Deliv Rev 203:115136

Yang K, Ma Y-Q (2010) Computer simulation of the translocation of nanoparticles with different shapes across a lipid bilayer. Nat Nanotechnol 5:579–583

Yang Y, Luo D, Inam M, Hu J, Zhou Y, Xu C, Chen W (2022) A scientometrics study of the nanomedicines assisted in respiratory diseases. Front Bioeng Biotechnol 10:1053653

Yang Z, Xie H, Wan J et al (2023) A nanotherapeutic strategy that engages cytotoxic and immunosuppressive activities for the treatment of cancer recurrence following organ transplantation. EBioMedicine 92:104594

Yang H, Zhang Q, Dai L, Wang Y, Zheng G, Zhang X, Zheng D, Ji X, Sang Y, Nie Z (2024) Docetaxel-encapsulated catalytic Pt/Au nanotubes for synergistic chemo-photothermal therapy of triple-negative breast cancer. Adv Healthc Mater 13:2400662

Yao Y, Chen S, Yan C, Wang J, Liu J, Zhu WH, Fan C, Guo Z (2025) Photo-triggered fluorescence polyelectrolyte nanoassemblies: manipulate and boost singlet oxygen in photodynamic therapy. Angew Chem Int Ed 64:e202416963

Yeo S, Wu H, Yoon I et al (2024) Enhanced photodynamic therapy efficacy through solid lipid nanoparticle of purpurin-18-N-propylimide methyl ester for cancer treatment. Int J Mol Sci 25: 10382

Yi Z, Li X, Lu W, Liu H, Zeng S, Hao J (2023) Correction: hybrid lanthanide nanoparticles as a new class of binary contrast agents for in vivo T1/T2 dual-weighted MRI and synergistic tumor diagnosis. J Mater Chem B 11:9765–9765

Zare Kazemabadi F, Heydarinasab A, Akbarzadehkhiyavi A, Ardjmand M (2021) Development, optimization and in vitro evaluation of etoposide loaded lipid polymer hybrid nanoparticles for controlled drug delivery on lung cancer. Chem Methodol 5:135–152

Zelenikhin PV, Galeeva AG, Islamova RR, Lopatin ON, Yarullin RS, Ilinskaya ON (2023) Hybrid organomineral carriers for therapeutic proteins. Russ J Bioorg Chem 49:298–305

Zhang J, Chen G, Guay D, Chaker M, Ma D (2014) Highly active PtAu alloy nanoparticle catalysts for the reduction of 4-nitrophenol. Nanoscale 6:2125–2130

Zhang Z, Lu Z, Yuan Q, Zhang C, Tang Y (2021) ROS-Responsive and active targeted drug delivery based on conjugated polymer nanoparticles for synergistic chemo–/photodynamic therapy. J Mater Chem B 9:2240–2248

Zhang G, Cong Y, Liu FL et al (2022a) A nanomaterial targeting the spike protein captures SARS-CoV-2 variants and promotes viral elimination. Nat Nanotechnol 17(9):993–1003

Zhang P, Li Y, Tang W, Zhao J, Jing L, McHugh KJ (2022b) Theranostic nanoparticles with disease-specific administration strategies. Nano Today 42:101335

Zhang L, Forgham H, Huang X, Shen A, Davis TP, Qiao R, Guo B (2022c) All-in-one inorganic nanoagents for near-infrared-II photothermal-based cancer theranostics. Mater Today Adv 14: 100226

Zhang B, Wang F, Kong X, Zhang X, Chen T, Li X (2024a) Electrothermal therapy for the in vivo elimination of circulating tumor cells by using a functionalized injection needle. Appl Phys Lett 125:203701

Zhang Q, Li E, Zhang Y, Chen Y, Wang D, Wang S (2024b) Aggregation-induced emission-active organic nanoagent with high photothermal conversion efficiency for near-infrared imaging-guided tumor photothermal therapy. ACS Biomater Sci Eng. https://doi.org/10.1021/ACSBIOMATERIALS.4C00959/ASSET/IMAGES/LARGE/AB4C00959_0005.JPEG

Zhao Q, Wang L, Cheng R, Mao L, Arnold RD, Howerth EW, Chen ZG, Platt S (2012) Magnetic nanoparticle-based hyperthermia for head & neck cancer in mouse models. Theranostics 2:113–121

Zhao K, Shen G, Liu Y et al (2024a) Dual-targeted assembled nanodrugs for near-infrared photothermal immunotherapy of triple-negative breast cancer. ACS Appl Mater Interfaces 16: 67532–67544

Zhao W, Wang L, Zhang M, Liu Z, Wu C, Pan X, Huang Z, Lu C, Quan G (2024b) Photodynamic therapy for cancer: mechanisms, photosensitizers, nanocarriers, and clinical studies. MedComm (Beijing) 5:e603

Zheng Y, Zheng F, Xu R, Sun X, Yu J, Chen H, Gao Y (2024) Self-healing photothermal nanotherapeutics for enhanced tumor therapy through triple ferroptosis amplification and cascade inflammation inhibition. ACS Appl Mater Interfaces 16:51994–52007

Zhou Y, Liu R, Shevtsov M, Gao H (2022) When imaging meets size-transformable nanosystems. Adv Drug Deliv Rev 183:114176

Zhu H, Gao F, Li Y et al (2024) MOF-based nanoparticles for tumor-targeted protein degradation and photodynamic therapy induce enhanced anti-tumor immunity. Nano Today 56:102308

Cancer Imaging and Scope of Early-Stage Diagnosis: Molecular Imaging Modalities

11

Aseem Setia, Komal Rani, Nandini Vinodrao Randhave, Nidhi Verma, Ankit Kumar Malik, Vikas Kumar, Vaishali, and Madaswamy S. Muthu

Abstract

Cancer has been an enormous issue for a long time. The ability to precisely localize cancer at an early stage, crucial for effective treatment, has been made possible by recent developments in molecular imaging, which have radically altered cancer research. Molecular imaging non-invasiveness, speed, safety, and user-friendliness make it a priceless asset in the fight against cancer in its early stages. Imaging technologies including positron emission tomography (PET), magnetic resonance imaging (MRI), and computed tomography (CT) enable the early diagnosis of tiny or hidden malignancies by offering high-resolution, real-time molecular visualization of tumors. Clinicians can detect tumors before they become noticeable or cause symptoms, which allows for better treatment results and early intervention. When it comes to cancer detection and therapy, molecular imaging is crucial. Modern molecular imaging techniques allow for a more exact visualization of malignant cells and tissues; they include fluorescence imaging, MRI, PET, and CT. The advancements in cancer diagnosis, treatment response monitoring in real-time, and our overall knowledge of cancer biology have opened up exciting new possibilities for targeted treatment and early intervention. Artificial intelligence (AI) and machine learning (ML) are transforming molecular imaging by making data interpretation, pattern recognition, and picture analysis much better. These advancements in technology pave the way for more precise anomaly identification, better tumor segmentation, and individualized treatment strategies. Imaging techniques may be optimized by AI-driven algorithms, which speed up cancer diagnostics while decreasing the likelihood of a human mistake.

A. Setia · K. Rani · N. V. Randhave · N. Verma · A. K. Malik · V. Kumar · Vaishali
M. S. Muthu (✉)
Department of Pharmaceutical Engineering and Technology, Indian Institute of Technology (IIT) BHU, Varanasi, Uttar Pradesh, India
e-mail: msmuthu.phe@itbhu.ac.in

R. Prasad, M. Mohan Yallapu (eds.), *Advancements in Cancer Theranostics*,
https://doi.org/10.1007/978-981-95-0938-6_11

This chapter provides a comprehensive discussion of various molecular imaging modalities for early-stage cancer detection. In addition, we incorporated our discussion with AI and machine learning concepts for imaging modalities.

Keywords

Cancer diagnosis · Molecular imaging and therapy · Theranostics · Nanomedicine · Translational · Imaging

11.1 Introduction

Cancer remains a foremost global health concern, ranking as the second leading cause of worldwide morbidity and mortality. According to cancer statistics, the American Cancer Society estimates that in 2025, there will be 2.04 million new cancer cases and 618,120 cancer deaths projected to occur in the United States (Siegel et al. 2025). It is a multiplex disease attributed to uncontrolled cell growth and abnormal proliferation, often resulting in widespread metastasis. Diagnosing cancer at its early stages is particularly challenging due to its multifactorial nature, which makes timely detection and effective treatment a critical concern for researchers and clinicians. Hence, treatment options for advanced-stage cancers often prove to be less effective (Chacko and Ankri 2024). Traditional diagnostic approaches such as biopsy and histopathological analysis are often invasive and may not detect cancer in its early stage. Recent advancements in medical imaging technologies including nanotheranostics imaging, significantly enhanced the ability to diagnose, treat, and monitor cancer (Randhave et al. 2024; Vikas et al. 2023). It provides essential information on tumor location, size, and metastasis, allowing for better diagnosis and treatment strategies. Traditional anatomical imaging techniques such as ultrasound, CT, and MRI are commonly used, but particularly molecular imaging modalities are gaining prominence for their ability to target specific molecular markers associated with cancer (Prusty et al. 2023). The traditional anatomical imaging has significant limitations, such as inadequate tissue resolution, exposure to radiation, high costs, and discomfort for patients (Fass 2008). Utilizing molecularly focused agents offers a more precise and sensitive approach to detecting cancer, particularly at early stages when tumors are still microscopic and not visible through conventional imaging methods. Molecular imaging techniques, including PET, SPECT, MRI, and CT, can track the movement of contrast agents within the body, revealing cellular and molecular changes and offering a non-invasive and highly sensitive means of identifying early-stage cancer, evaluating treatment responses, and guiding targeted therapies (Bai et al. 2023; Corradini et al. 2019). Optical molecular imaging is an advanced technique that effectively combines optical detection with molecular imaging to specifically target biomarkers found in tumor cells and their microenvironment. This powerful method decisively distinguishes tumor tissues from surrounding normal tissues, delivering high-resolution, non-invasive, and safe monitoring. With its ability to detect early-stage cancer, optical molecular imaging is essential for facilitating continuous monitoring during and after surgical procedures (Li et al. 2024). Additionally, hybrid imaging modalities combine anatomical imaging (like CT or MRI) with functional imaging (such as SPECT or PET), allowing for a comprehensive view of both the structure and function of tissues. These systems enable the visualization of anatomical details alongside physiological processes in a single image, which improves diagnostic accuracy (Corradini et al. 2019). For example, combining CT with PET (PET/CT) provides both high-resolution anatomical images and functional data about metabolic activity, making it invaluable for oncology (Patton et al. 2009).

Nanotheranostics imaging, which combines therapeutic and diagnostic capabilities at the nanoscale, further revolutionizes cancer care by enabling targeted treatment delivery while simultaneously monitoring therapeutic efficacy (Setia et al. 2024a). This integrated approach holds great promise for the early diagnosis, personalized treatment, and effective management of cancer (Mura and Couvreur 2012). Recent advancements in technology have increased the need for more accurate cancer detection and analysis methods. However, the complex and heterogeneous nature of cancer data makes clinical interpretation challenging. The integration of artificial intelligence (AI) and machine learning (ML) into cancer imaging is further enhancing the scope and accuracy of early-stage diagnosis (Hunter et al. 2022). AI and ML algorithms can process vast amounts of imaging data, identify subtle patterns, and predict outcomes with a level of precision that surpasses human capability. In response, machine learning techniques, especially deep learning (DL), are gaining prominence in clinical oncology (Khandakar et al. 2024). DL, an advanced subset of ML, is particularly well-suited for image processing tasks, including medical image analysis, due to its ability to automatically learn and identify patterns in large, complex datasets (Asif et al. 2024). By automating image interpretation, AI not only accelerates the diagnostic process but also minimizes human error, ensuring earlier and more accurate detection of malignancies. Furthermore, AI-driven predictive models are being developed to personalize treatment plans, offering tailored approaches for patients based on their unique molecular imaging profiles (Huang et al. 2020; Chen et al. 2021).

11.2 Nanotheranostics Imaging Modalities

11.2.1 Positron Emission Tomography (PET) for Breast Cancer

Positron Emission Tomography, or PET, is a well-known and common way of screening the whole body in medical diagnostics. Imaging using contrast agents comprising PET radionuclide producing pairs of γ-rays is necessary with this imaging approach. (Shokeen et al. 2008). The most widely used radioactive contrast agents include ^{13}N, ^{18}F, ^{64}Cu, ^{74}As, ^{62}Cu, ^{76}Br, ^{82}Rb, ^{15}O, ^{86}Y, ^{11}C, ^{89}Zr, ^{68}Ga, ^{44}Sc, ^{72}As, and ^{124}I (Bar-Shalom et al. 2000; Adam and Wilbur 2005). The fast clearance efficiency and strong ionizing radiation exposure of patients are limitations of the tiny radionuclides. Nanotheranostics imaging has replaced these radio atoms as the gold standard in PET scans due to its superior biodistribution and ability to accumulate imaging contrast agents at specific target sites (Gupta et al. 2024).

For instance, Chen et al. developed a renally-clearable, 89Zr-labeled anti-HER2-targeted ultrasmall silica nanoparticle immunoconjugate with unique biological characteristics for the detection and therapy of breast tumor. A versatile optical materials platform, fluorescent core-shell silica nanoparticles which is known as Cornell dots or C dots with a size less than 10 nm can contain near-infrared (NIR) dyes inside, as well as a number of targeting ligands with contrast-producing radiolabels on the outside. This design has been successfully applied in clinical settings. The goal of this research was to develop water-synthesized, next-generation ultrasmall prime C dots (C′ dots) with enhanced size control, increased particle brightness, and the ability to modularly and orthogonally functionalize the PEG shell. Following surface amination, small C′ dots containing Cy5 dye and are PEGylated (PEG-Cy5-C′ dots) modified by radiolabels chelators, and anti-HER2 scFv fragments. An efficient click chemistry bioconjugation technique is achieved through area-specific engineering of anti-HER2 scFv fragments. The ultrasmall particle-based design tackles present issues in nanoparticles and immune targeting of HER2+ breast cancer (Fig. 11.1) (Chen et al. 2018).

11.2.2 Computed Tomography (CT) and Dual-Modality Imaging for Lung Cancer

Computed tomography (CT) imaging is essential for the precise diagnosis, localization, and staging of lung tumors, as it provides high-resolution anatomical details (Verschakelen et al. 2002). It facilitates the precise assessment of disease severity and tumor size, as well as the planning of treatment. Nevertheless, dual-modality imaging systems, such as PET-CT, combine the advantageous characteristics of PET with the comprehensive anatomy information of CT. This combination enables physicians to obtain both structural and functional information in a single imaging session, thereby enhancing the precision of diagnosis, the staging of diseases, and the planning of interventions. It is crucial to have a precise understanding of the tumor location and its progression in lung cancer (Lardinois et al. 2003). Dual-modality imaging facilitates the visualization of the tumor's size, activity, and progression, providing a comprehensive understanding of the condition. PET scans are ideal for detecting tumors and other anomalous tissue growth because they offer information on metabolic activity. However, it lacks the anatomical context necessary for accurate localization. In contrast, CT provides precise images of structures, but it does not provide information regarding their functionality. The combination of these methodologies provides a more comprehensive understanding of the disease, thereby enabling physicians to make more informed decisions regarding treatment and staging (Greenwood et al. 2024). The utilization of PET and CT is further enhanced by the advancement of nanotheranostics, which integrates nanotechnology with diagnostic and treatment methodologies. Nanotheranostics can be customized to target specific cancer indications, thereby

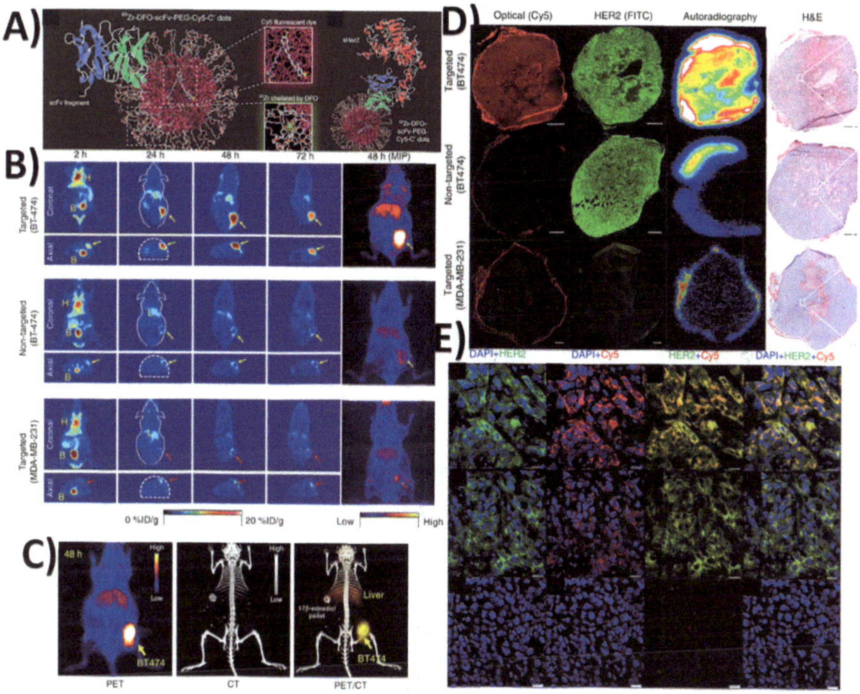

Fig. 11.1 (**a**) The synthesis of 89Zr-DFO-scFv-PEG-Cy5-C′ dot bonded to sHER2 and site-specific engineering of anti-HER scFv fragments. (**b**) In vivo PET imaging of HER2-targeted breast cancer in xenograft models. (**c**) PET, CT, AND PET/CT imaging in tumor mice. (**d**) Ex vivo tumor histopathology that is correlative and (**e**) confocal microscopy of random areas from corresponding tumor tissue specimens. Reproduced with permission from reference (Chen et al. 2018) Figs. 1, 3 and 5 (Springer Nature©2018)

increasing the sensitivity and precision of the dual imaging system (Armstrong et al. 2024). A recent preclinical investigation by Badea et al. revealed the application of a nanomedicine-based liposome loaded with iodinated CT contrast agent for distinguishing tumors depending on their rate of growth. This was accomplished by the utilization of 2D clinical mammography of mammary cancer in a rat model. The study, aimed to assess the efficacy of the liposomes loaded with iodinated CT contrast agents in the characterization of pulmonary nodules that are either slowly growing or fastly growing in genetically engineered mouse models with primary non-small cell lung cancer. The research illustrated that tumors that were growing swiftly exhibited a higher vascular permeability to nanomedicine contrast agents than those that were growing slowly (Fig. 11.2) (Badea et al. 2012).

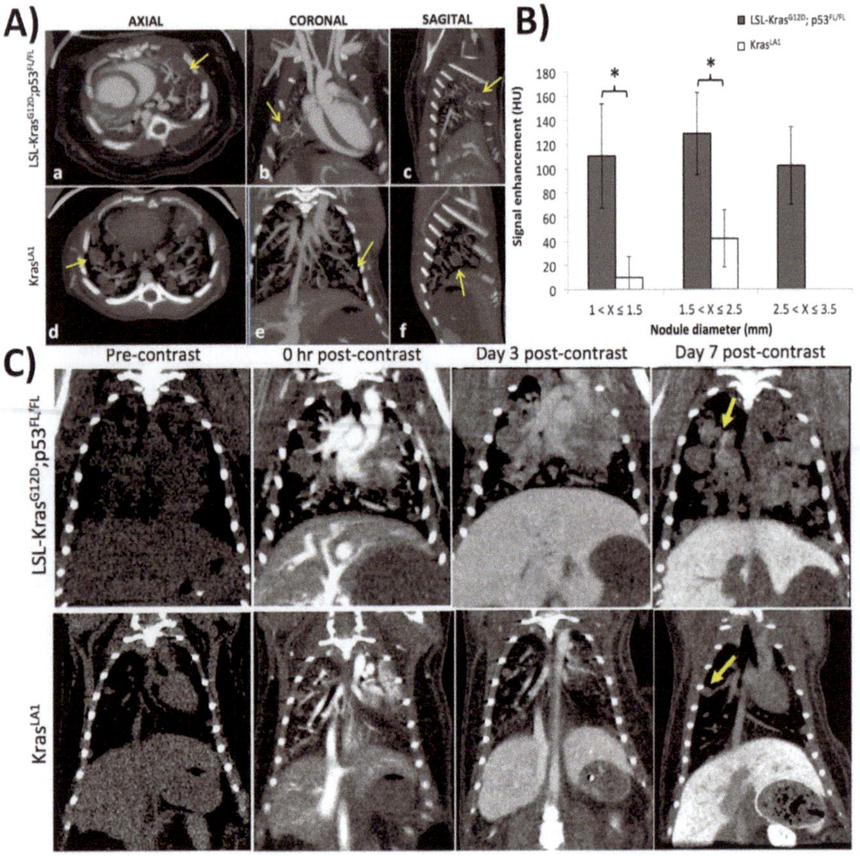

Fig. 11.2 (**a**) Maximum intensity projection images of the orthogonal thick slab showing the blood flow to nodules in lung cancer models. (**b**) On the seventh day, the two main nodules of lung cancer showed differential signal amplification. (**c**) Photographs of micro-CT images of mice taken previously and at 0 h, 3 days, and 7 days after injection of liposomal contrast agent. Reproduced with permission from reference (Badea et al. 2012) Figs. 4, 5, and 6 (PLOS One©2012)

11.2.3 Optical Imaging for Brain Cancer

Optical imaging, which provides high-quality image, real-time data without the need of radiation or external agents, is a potential technique for assessing brain tumors (Zhang et al. 2022; Muthu et al. 2011). These highly detailed techniques circumvent the limitations of PET (Czerner et al. 2023), CT (Ohlerth and Scharf 2007), and magnetic resonance imaging (MRI) (Ohlerth and Scharf 2007) by offering detailed resolution and more accessible imaging options. The main optical imaging techniques are fluorescence imaging, optical coherence tomography (OCT), bioluminescence imaging, photoacoustic imaging, and Raman imaging. Optical imaging in nanotheranostics refers to the use of optical imaging techniques in combination with nanomaterials for therapeutic and diagnostic purposes at the nanoscale (Murar et al. 2022; Setia et al. 2024b).

For specific orthotopic glioma theranostics, Lv et al. have developed a core-shell yttrium vanadate nanoparticle having YVO_4:Nd^{3+} particles as the core and MnO_2 nanosheets as the shell. Furthermore, lactoferrin (LF) was used to functionalize the nanoparticles, and hematoporphyrinmonomethyl ether (HMME), a sonosensitizer, was incorporated into them. This allowed them to perform bimodality of magnetic resonance imaging (MRI) and near-infrared fluorescence imaging (NIR-II), traverse the blood-brain barrier, and offer extremely effective sonodynamic treatment (SDT) for orthotopic gliomas (Fig. 11.3I). To find out if the nanoparticles may be employed as a bimodal contrast agent for MRI and NIR-II fluorescence imaging, more investigation was conducted. The nanoparticles demonstrated NIR-II emission from the vessels for 5 s after injection, suggesting that they functioned effectively in the NIR-II range of luminescence (Fig. 11.3III (a)). After blood flow, it was easier to see the network of vessels, and Fig. 11.3III (b) shows the exceptional spatial resolution for the intensity profile with a width of 0.59 mm. The nanoparticles capacity to identify malignancies was then assessed in vivo. When NIR-II emission was used to illuminate the tumors in both groups 6 h after the nanoparticles were injected, the group treated with lactoferrin-modified nanoparticles showed a noticeably stronger signal, suggesting that the addition of lactoferrin molecules enhances the nanoparticles' capacity to target gliomas in Fig. 11.3I (c). Additionally, the produced nanoparticles were evaluated for their ability to respond to the tumor microenvironment-weighted MRI contrast agent. The signals in the simulated TME increased as the manganese content increased, in contrast to normal conditions (Fig. 11.3II (d)). Furthermore, as displayed in Fig. 11.3II (e), the longitudinal relativity was five times greater than it would be in normal conditions. Additionally, four groups of rats were randomly selected: the control group, the US group, the YHM group, and the YHM plus US group. Through treatment, the tumors in the control, US, and YHM groups rapidly grew in size. Tumor size was assessed using T2-weighted MRI. Nonetheless, the YHM + US group was successful in restricting the growth of the tumor, which further illustrated the beneficial effects of SDT as a treatment (Fig. 11.3III) (Lv et al. 2022).

11.2.4 Hybrid Imaging Systems for Gastric Cancer

The purpose of hybrid imaging systems for gastric cancer, a rapidly developing field of diagnostic and therapeutic approach, is to combine many imaging modalities to provide a more accurate, detailed, and complete understanding of cancers of the digestive tract (Bockisch et al. 2009). These technologies allow the integration of cellular, structural, and functional data in one examination by leveraging the benefits of many imaging modalities. The hybrid imaging provides both structural and cellular insights, it is highly beneficial for the diagnosis, staging, and therapy response monitoring of cancer of the gut (Le Fur et al. 2020). They are especially

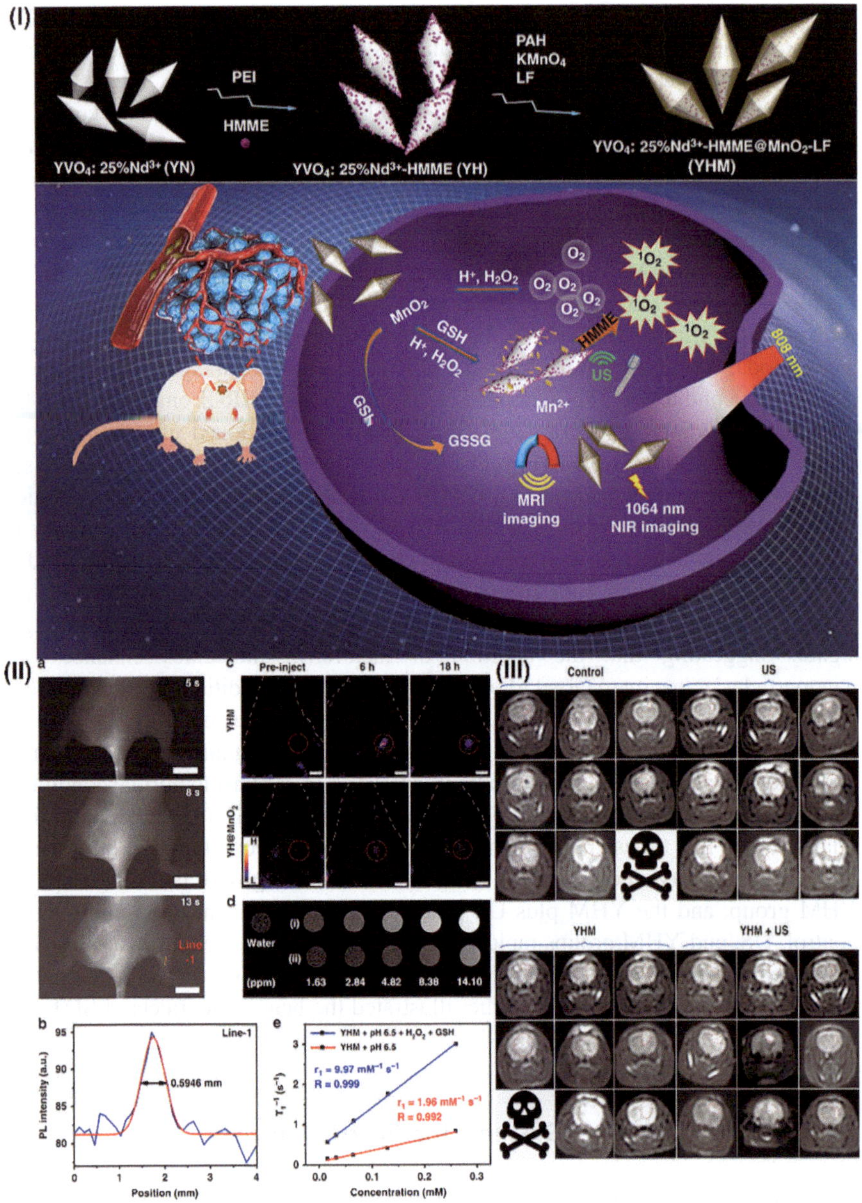

Fig. 11.3 (I) Diagrammatic representation of the preparation, multimodal imaging, and sonodynamic therapy of the yttrium vanadate nanoparticles. (II) NIR-II imaging and MRI of the yttrium vanadate nanoparticles. (III) T2-weighted MRI images of the tumors at 0, 5, and 8 days. Reproduced with permission from reference (Lv et al. 2022). Scheme 1, Figs. 5 and 6 (Nature©2022)

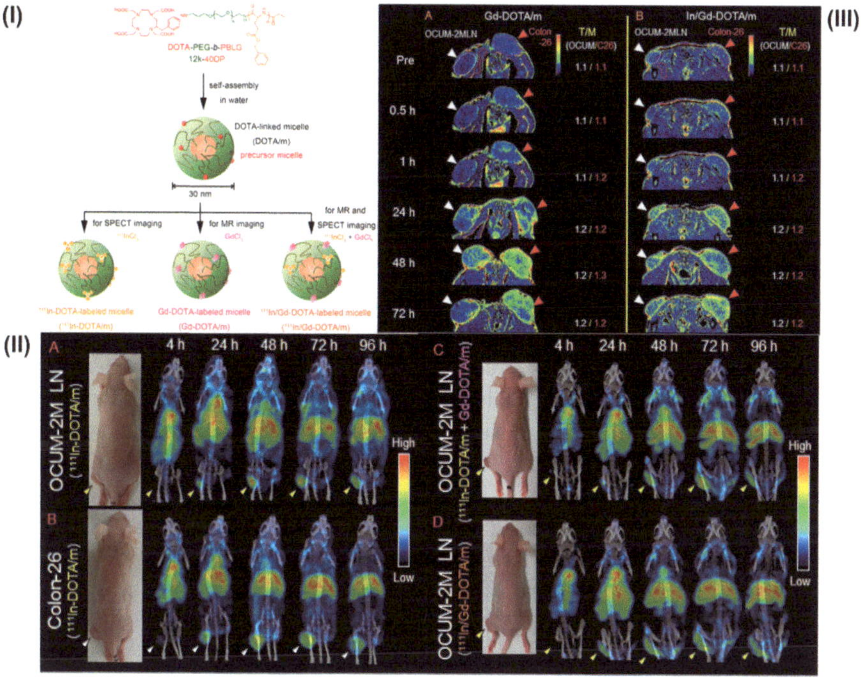

Fig. 11.4 (I) Schematic representation of the preparation of DOTA-micelles. (II) SPECT/CT imaging of the micelles in the mice bearing OCUM-2 M LN and Colon 26 tumors. (III) MRI of the micelles in the mice bearing OCUM-2 M LN and Colon 26 tumors. Reproduced with permission from reference (Miura et al. 2015). Figures 1, 8, and 9 (ACS Publication©2015)

useful in nanotheranostics because they combine therapeutic and diagnostic modalities, allowing for real-time examination of therapy response. Examples of hybrid imaging systems include PET/MRI, SPECT/CT, PA/MRI, optical imaging/CT, and optical imaging/MRI (Li et al. 2016).

For multimodal imaging of scirrhous gastric cancer, Miura et al. designed a polymeric micelles functionalized with DOTA-PEG-b-PBLG. As shown in Fig. 11.4I, the micelles were labeled with ^{111}In for SPECT and Gd for MRI imaging, respectively. The biodistribution of the micelles was evaluated using SPECT/CT imaging in the mice with OCUM-2 M LN and Colon-26 tumors. After 4 h of injection of ^{111}In-DOTA/m, a high level of radioactivity was detected in the blood, but neither tumor displayed a discernible increase in signal. After 48 h, the buildup in both tumors gradually increased before decreasing. After a day, the tumors were finally visible on SPECT/CT imaging (Fig. 11.4II (a, b)). Similar to SPECT/CT imaging with 111In-DOTA/m, OCUM-2 M LN tumors could be seen 24–96 h after injection using a combination of 111In-DOTA/m and Gd-DOTA/m and the 111In/Gd-DOTA/m (Fig. 11.4II (c, d)). To confirm that the generated micelles could be used for multimodal imaging, Gd-DOTA/m and In/Gd-DOTA/m were further evaluated for MRI imaging in mice with tumors. Both tumors showed a progressive micelle buildup and an increase in the MRI signal over time. The tumor's signal strength dramatically increased after 24–72 h of injection (Fig. 11.4III (a, b)) (Miura et al. 2015).

11.3 Artificial Intelligence in Molecular Imaging

Molecular imaging confers a non-invasive methodology for the investigation and quantitation of the molecules intricate in biological processes, providing comprehensive perceptions that correspond to conventional structural imaging approaches such as CT and MRI (Yin et al. 2024). Structural imaging is an imperative approach for the recognition of abnormal structures, while molecular imaging focuses on the molecular conformation of tissues that directs the types of molecules present, with a strong emphasis on precision (Cassidy and Radda 2005). Artificial intelligence (AI) encompasses the mimicking of complex cognitive functions like reasoning and perception. However, machine learning (ML) enables algorithms to learn from data, forming models that validate the patterns attributed to specific tasks (Herskovits 2021). AI/ML modeling is an invaluable endowment to molecular imaging, particularly through the use of deep learning and generative adversarial networks (GANs) (Medhi et al. 2024). This advanced technology is enhancing the anatomization of medical images, offering opportunities for improved accuracy and efficiency in diagnostics (Litjens et al. 2017). The journey of AI/ML began in the middle of the 20th century, with key tactics like backpropagation evolving in the 1960s. The idea of convolutional neural networks (CNNs) was also established at that moment. Since the 1980s, Japan has been a leader in adopting AI/ML to robotics, with a substantial sweep in occurrences around 2010 due to enhanced computational power (Hind et al. 2024). However, obligations like the ambiguity of AI decision-making and model validity remain, affecting trust in clinical settings (Minoshima and Cross 2022).

11.3.1 Role of AI in Early-Stage Cancer Diagnosis

Cancer, with its diverse forms, poses a significant and life-threatening challenge to global health (Chauhan et al. 2024). This condition stems from a complex interplay of genetic and epigenetic mutations and often presents diagnostic hurdles, especially at the early stages when symptoms may not be readily apparent or when cancer cells are still in the process of mutation. In oncology, early detection is crucial, as early-stage conditions are often curable. By capitalizing on this opportunity, patient outcomes may improve, unnecessary treatments can be eliminated, and survival rates may increase (Sufyan et al. 2023). AI plays a pivotal role in the early-stage diagnosis of cancer, offering significant promise in overcoming the crucial challenges in oncology with its advanced computational and analytical power by enhancing diagnostic accuracy, improving the identification of biomarkers, and providing critical insights into treatment planning (Foser et al. 2024). Furthermore, it aids in distinguishing between benign and malignant lesions, determining tumor margins during surgery, and monitoring the tumor's biological progression over time. It can rapidly process high-resolution imaging data, such as CT scans, MRIs, and X-rays, far more efficiently than human clinicians alone (Alshuhri et al. 2024). In combination with ML algorithms, AI can detect subtle changes in tissue and cell structure that may indicate the presence of cancer, even before these changes are noticeable to the human eye. This enables earlier intervention and a more accurate understanding of the location, size, and characteristics of tumors. Moreover, through advanced pattern recognition techniques, it can analyze complex data from Next-Generation Sequencing (NGS) and identify mutations or genetic markers that are relevant for cancer diagnosis. This capability not only aids in detecting cancer earlier but also helps in personalizing treatment plans based on an individual's unique genetic makeup. Personalized medicine, powered by AI, can optimize drug delivery, improve patient outcomes, and minimize adverse side effects by tailoring therapies to the specific biological characteristics of the cancer (Iqbal et al. 2021).

For instance, Muhammad et al. developed an artificial neural network (ANN) model to detect pancreatic cancer at its early stage. The ANN model was trained using health data from 800,114 participants, which comprises 898 pancreatic cancer cases. The model incorporated 18 features to assess cancer risk for individuals. With sensitivity and specificity rates exceeding 80%, and the area covered by the receiver operating characteristic curve measuring 0.86 for the training group and 0.85 for the testing group, the constructed model demonstrated excellent performance. The subsequent findings demonstrated that the ANN can accurately forecast the likelihood of pancreatic cancer, providing a possible means of identifying those who could potentially benefit from early detection and treatment (Muhammad et al. 2019).

11.3.2 Machine Learning for Molecular Imaging Modalities

ML, particularly deep learning (DL), has become a transformative force in molecular imaging, a critical medical imaging technique that provides high-resolution images of molecular and cellular processes within the body. The rapid advancements in imaging technologies and the exponential growth of medical data have highlighted the need for more accurate, automated, and quantitative approaches to process and analyze this vast amount of information. In this context, combining ML algorithms with molecular imaging has become a viable way to improve therapeutic and diagnostic approaches (Jimenez-Mesa et al. 2023). For example, Yi et al. investigated whether ML-based radiomics models from ^{68}Ga-PSMA PET/CT images could predict intraprostatic lesions not visible on PET scans in primary prostate cancer patients. They conducted a retrospective analysis using data from two institutions, with 64 patients in the training set and 36 in the test set. Three random forest models were created using features from standard and delayed PET images and compared their performance to PSA density (PSAD). The results showed that these radiomics models outperformed PSAD in predicting hidden intraprostatic lesions, indicating a promising non-invasive approach for identifying such lesions in these patients (Yi et al. 2022).

ML and DL algorithms offer several benefits for molecular imaging, including improving the precision and efficiency of data analysis. To help with prompt and precise diagnosis, this expertise is essential for finding disease-specific biomarkers and creating predictive models for early detection and prognosis. Furthermore, sophisticated insights into disease mechanisms and treatment outcomes are made possible by the ability of ML and DL models to identify intricate patterns and relationships in the data. The DL approach excels in feature extraction, which enables more precise and informative image interpretation. Furthermore, these algorithms can enhance image reconstruction by reducing noise and mitigating artifacts, leading to better image quality and more reliable diagnostic results (Yin et al. 2021). For instance, Zhang et al. introduced a GAN-based post processing technique that employs total gradient loss to improve the quality and resolution of fluorescence images. Ultimately, this approach led to the sharper images with clearer boundaries by aligning the gradient of the generated image with that of the original. The results highlighted it outperforms SRGAN by reducing unrealistic textures, though false features remain a challenge due to network hallucinations (Zhang et al. 2019).

The support vector machine (SVM) stands out as one of the most widely utilized ML algorithms in molecular imaging due to its exceptional capability in managing nonlinear relationships and processing large feature spaces effectively. It has shown significant promise in medical imaging tasks, delivering robust and accurate outcomes. These advantages make SVM an excellent tool for analyzing complex medical images, such as those used in molecular imaging, and for supporting clinical decision-making (Gao et al. 2015).

11.4 Case Study and Clinical Findings

Malignant lesion detection, illness progression assessment, and treatment response evaluation are all aided by advanced imaging techniques. These methods are crucial for locating distant metastases and directing treatment choices (Pulumati et al. 2023). Several nanomedicines that have received FDA/EMA approval have shown better pharmacokinetics and toxicity profiles than their traditional chemotherapeutic counterparts. Hence, there is an urgent need for a clinical tool for patient stratification.

In a study, Do et al., compared the ability of AI-CAD, an AI-based computer-aided diagnosis system, to identify problematic microcalcification on mammography with that of trained breast radiologists. From June 2003 to November 2019, 420 patients who underwent biopsy for suspected microcalcification had 435 unilateral mammographies reviewed retrospectively. Malignancy scores were computed by using commercial AI-CAD for the mammography pictures. Under the receiving operator characteristics curve (AUC), radiologists and AI-CAD were evaluated for their diagnostic performance. All three of the adjusted AUCs for AI-CAD were considerably higher than those for radiologists alone. There was no need to conduct superfluous biopsies as none of the 27 instances that were reduced to category 3 with a 2% threshold were found to be malignant on pathological investigation. According to the results, AI-CAD performed as well as radiologists when it came to identifying suspected microcalcification on mammography, suggesting it might help with clinical decision-making for treating breast microcalcification (Fig. 11.5) (Do et al. 2021).

Moreover, Voortman et al. reported a case study of a lung cancer patient. The case report of a lung cancer patient with no prior cardiac history before receiving an experimental combination of chemotherapy and bortezomib, who developed severe but reversible congestive heart failure. It was suggested that pre-existing (subclinical) left ventricular dysfunction was present by the retrospective measurement of elevated levels of N-terminal pro-B-type natriuretic peptide (NT-proBNP) in archival serum samples. As a potential risk factor for cardiac complications caused by systemic proteasome suppression, they propose that the patient may have had subclinical cardiomyopathy at baseline, which is defined by dysregulation of the ubiquitin-proteasome system. Protease inhibition medication, including bortezomib, requires thorough monitoring of patients with cardiac disease or risk factors for it. Patients with lung cancer should exercise caution since they frequently have cardiac comorbidities (Voortman and Giaccone 2006).

11.5 Scope of Early-Stage Cancer Detection Using Molecular Imaging

The rapidly emerging areas in the preclinical drug development process for cancer treatment are non-invasive imaging and molecular imaging (O'Farrell et al. 2013). These imaging techniques include optical techniques like fluorescence and bioluminescence imaging, as well as SPECT, MRI, PET, and ultrasound. The capacity to objectively and non-invasively track pharmacodynamic alterations at the molecular and cell level in living animals is one of the many benefits that these molecular imaging technologies offer over conventional screening techniques (Vermeulen et al. 2022). Molecular imaging is required to enhance early cancer detection and attain better results. The use of molecular imaging in the diagnosis of early-stage cancer has grown in popularity as a result of developments in imaging methods, tumor biology, and technology (Frangioni 2008; Bonlawar et al. 2024). Tumor biomarkers can be found in vivo by molecular imaging technologies before a tumor can be physically recognized with conventional techniques like CT or MRI (Smith et al. 2003). Even when the tumor is extremely small to be seen with other imaging techniques, PET combined with radiolabelled tracers, such as fluorodeoxyglucose (FDG), can assist in identifying early metabolic alterations in

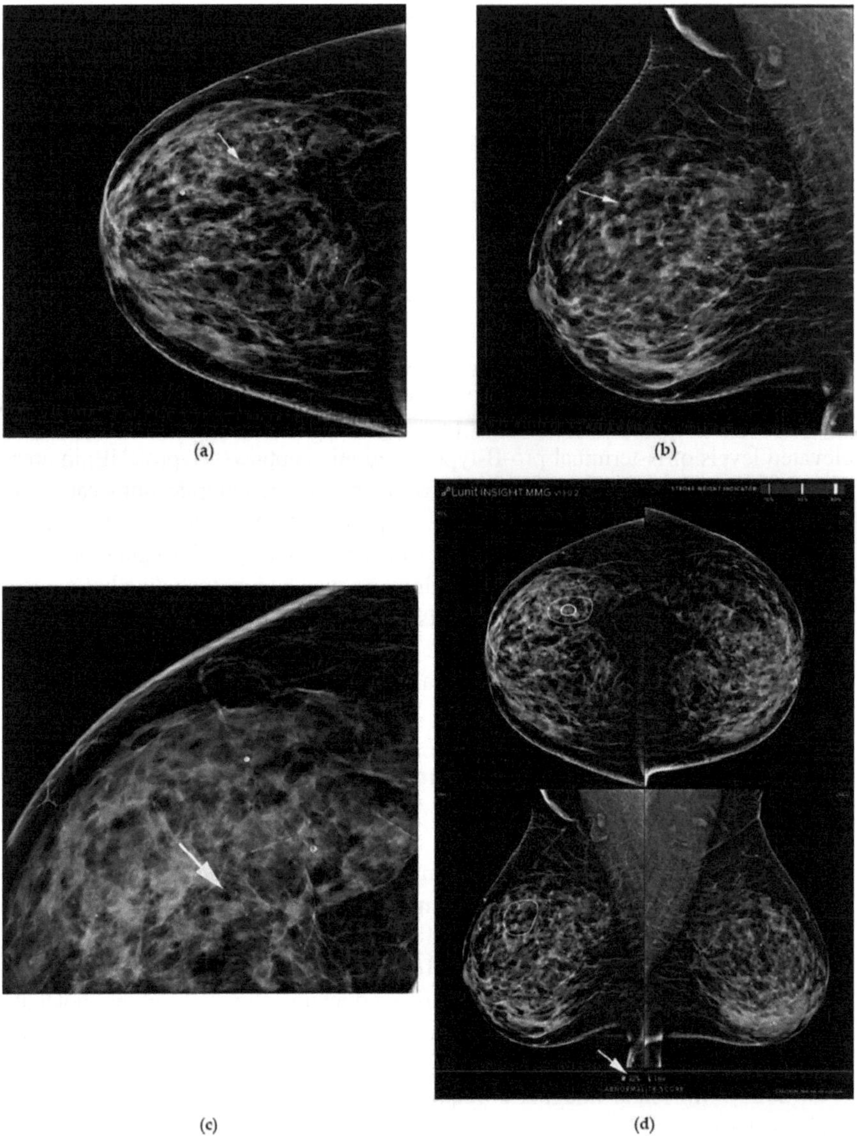

Fig. 11.5 A 64-year-old lady had screening mammography, where AI-CAD found microcalcifications. (**a**, **b**) Clusters of microcalcifications were revealed by routine mammography, which included the right craniocaudal and mediolateral oblique views. (**c**) In the compression view and at a craniocaudal magnification of the right side, clusters of amorphous microcalcifications were visible. (**d**) There was a 92% malignancy rate as determined by AI-CAD. Reproduced with permission from reference (Do et al. 2021). Figure 2 (MDPI©2020)

cancer cells (Schwenck et al. 2023). Because tailored contrast agents directly bind to cancer biomarkers, they can improve MRI and enable more accurate detection of tiny lesions. Additionally, SPECT detects tissue changes using tracers that release gamma rays, which can be crucial for early cancer detection (Keidar et al. 2003). The tumor oxygen levels, acidic environment, circulation, and expression of genes can all be better understood with the use of molecular imaging (Liu et al. 2017). Furthermore, frequently before more conventional techniques do, molecular imaging often gives doctors insight into the course of cancers and the efficacy of therapies like immunotherapy, chemotherapy, and targeted drugs (Rowe and Pomper 2022). Additionally, in vivo molecular imaging has significant promise for influencing medicine through the use of targeted therapy and the provision of personalized medicine. Ultrasound imaging with molecularly-targeted microbubbles and optical imaging with fluorescent probes are both appealing techniques due to their ability to provide images with spatial accuracy, real-time imaging, affordability, and radiation protection (Bortot et al. 2023). A single scan can offer both anatomical and functional imaging by combining multiple molecular imaging modalities (such as PET/MRI and PET/CT) (Padmanabhan et al. 2017). However, molecular imaging has a developing and promising role in early cancer diagnosis. It's anticipated that developments in technology, imaging techniques (Dhamija et al. 2024), and molecular biology will transform personalized cancer diagnostics by enabling earlier, more precise, and less invasive cancer detection and treatment (Rai 2024).

11.6 Conclusion

Imaging for cancer is essential for early identification and diagnosis, which in turn improves patient outcomes through the ability to respond promptly. Researchers are turning to molecular imaging modalities like SPECT, PET, and MRI to see and track cellular and molecular biological processes. These methods enable the detection of curable malignancies in their earliest stages by producing high-resolution pictures of tumors in real-time. A landmark development in cancer treatment is the application of AI and ML to cancer imaging. These innovations improve the accuracy of diagnoses, which in turn allows for earlier tumor diagnosis and the development of more tailored treatment plans. There is a new hope for cancer patients throughout the world owing to the enormous potential of AI and ML to transform cancer care through better outcomes and faster clinical decision making.

Author Contribution A.S., K.R., and N.V.R. have prepared the first outline and initial draft. N.V. R., N.V., and A.K.M. have done the literature. V.K. and V have prepared an overview of AI-ML and incorporated it into the chapter. M.S.M. prepared initial draft, has done the revision and final editing of the chapter.

References

Adam MJ, Wilbur DS (2005) Radiohalogens for imaging and therapy. Chem Soc Rev 34(2): 153–163

Alshuhri MS, Al-Musawi SG, Al-Alwany AA, Uinarni H, Rasulova I, Rodrigues P, Alkhafaji AT, Alshanberi AM, Alawadi AH, Abbas AH (2024) Artificial intelligence in cancer diagnosis: opportunities and challenges. Pathol Res Pract 253:154996

Armstrong MF, Burkett BJ, O'Byrne TJ, Gottlich HC, Yin LX, Tasche KK, Price DL, Moore EJ, Routman DM, Gamez M, Lester SC (2024) Cancer in patients with incidental asymmetric oropharynx positron emission tomography uptake. JAMA Otolaryngol Head Neck Surg 150(12):1078–1086

Asif S, Wenhui Y, ur-Rehman S, ul-ain Q, Amjad K, Yueyang Y, Jinhai S, Awais M (2024) Advancements and prospects of machine learning in medical diagnostics: unveiling the future of diagnostic precision. Arch Comput Methods Eng:1–31

Badea CT, Athreya KK, Espinosa G, Clark D, Ghafoori AP, Li Y, Kirsch DG, Johnson GA, Annapragada A, Ghaghada KB (2012) Computed tomography imaging of primary lung cancer in mice using a liposomal-iodinated contrast agent. PLoS One 7(4):e34496

Bai JW, Qiu SQ, Zhang GJ (2023) Molecular and functional imaging in cancer-targeted therapy: current applications and future directions. Signal Transduct Target Ther 8(1):89

Bar-Shalom R, Valdivia AY, Blaufox MD (2000) PET imaging in oncology. Semin Nucl Med 30(3):150–185

Bockisch A, Freudenberg LS, Schmidt D, Kuwert T (2009) Hybrid imaging by SPECT/CT and PET/CT: proven outcomes in cancer imaging. Semin Nucl Med 39(4):276–289

Bonlawar J, Setia A, Challa RR, Vallamkonda B, Mehata AK, Viswanadh MK, Muthu MS (2024) Targeted nanotheransotics: integration of preclinical MRI and CT in the molecular imaging and therapy of advanced diseases. Nano 8(3):401–426

Bortot B, Mangogna A, Di Lorenzo G, Stabile G, Ricci G, Biffi S (2023) Image-guided cancer surgery: a narrative review on imaging modalities and emerging nanotechnology strategies. J Nanobiotechnol 21(1):155

Cassidy PJ, Radda GK (2005) Molecular imaging perspectives. J R Soc Interface 2(3):133–144

Chacko N, Ankri R (2024) Non-invasive early-stage cancer detection: current methods and future perspectives. Clin Exp Med 25(1):17

Chauhan M, Shekhar S, Yadav B, Garg V, Dutt R, Mehata AK, Goswami P, Koch B, Muthu MS, Singh RP (2024) AS1411 aptamer/RGD dual functionalized theranostic chitosan-PLGA nanoparticles for brain cancer treatment and imaging. Biomater Adv 160:213833

Chen F, Ma K, Madajewski B, Zhuang L, Zhang L, Rickert K, Marelli M, Yoo B, Turker MZ, Overholtzer M, Quinn TP, Gonen M, Zanzonico P, Tuesca A, Bowen MA, Norton L, Subramony JA, Wiesner U, Bradbury MS (2018) Ultrasmall targeted nanoparticles with engineered antibody fragments for imaging detection of HER2-overexpressing breast cancer. Nat Commun 9(1):4141

Chen ZH, Lin L, Wu CF, Li CF, Xu RH, Sun Y (2021) Artificial intelligence for assisting cancer diagnosis and treatment in the era of precision medicine. Cancer Commun 41(11):1100–1115

Corradini S, Alongi F, Andratschke N, Belka C, Boldrini L, Cellini F, Debus J, Guckenberger M, Hörner-Rieber J, Lagerwaard FJ, Mazzola R, Palacios MA, Philippens MEP, Raaijmakers CPJ, Terhaard CHJ, Valentini V, Niyazi M (2019) MR-guidance in clinical reality: current treatment challenges and future perspectives. Radiat Oncol 14(1):92

Czerner CP, Derlin T, Bengel FM, Weiberg D (2023) Positron emission tomography in oncology. Radiologie (Heidelb) 63(7):549–560

Dhamija P, Mehata AK, Tamang R, Bonlawar J, Vaishali, Malik AK, Setia A, Kumar S, Challa RR, Koch B, Muthu MS (2024) Redox-sensitive poly(lactic-co-glycolic acid) nanoparticles of palbociclib: development, ultrasound/photoacoustic imaging, and smart breast cancer therapy. Mol Pharm 21(6):2713–2726

Do YA, Jang M, Yun B, Shin SU, Kim B, Kim SM (2021) Diagnostic performance of artificial intelligence-based computer-aided diagnosis for breast microcalcification on mammography. Diagnostics (Basel, Switzerland) 11(8)

Fass L (2008) Imaging and cancer: a review. Mol Oncol 2(2):115–152

Foser S, Maiese K, Digumarthy SR, Puig-Butille JA, Rebhan C (2024) Looking to the future of early detection in cancer: liquid biopsies, imaging, and artificial intelligence. Clin Chem 70(1): 27–32

Frangioni JV (2008) New technologies for human cancer imaging. J Clin Oncol 26(24):4012–4021

Gao X, Chu C, Li Y, Lu P, Wang W, Liu W, Yu L (2015) The method and efficacy of support vector machine classifiers based on texture features and multi-resolution histogram from (18)F-FDG PET-CT images for the evaluation of mediastinal lymph nodes in patients with lung cancer. Eur J Radiol 84(2):312–317

Greenwood HE, Barber AR, Edwards RS, Tyrrell WE, George ME, Dos Santos SN, Baark F, Tanc M, Khalil E, Falzone A, Ward NP, DeBlasi JM, Torrente L, Soni PN, Pearce DR, Firth G, Smith LM, Vilhelmsson Timmermand O, Huebner A, Swanton C, Hynds RE, DeNicola GM, Witney TH (2024) Imaging NRF2 activation in non-small cell lung cancer with positron emission tomography. Nat Commun 15(1):10484

Gupta D, Roy P, Sharma R, Kasana R, Rathore P, Gupta TK (2024) Recent nanotheranostic approaches in cancer research. Clin Exp Med 24(1):8

Herskovits EH (2021) Artificial intelligence in molecular imaging. Ann Transl Med 9(9):824

Hind S, van der Vlist FN, Kanderske M (2024) Challenges as catalysts: how Waymo's open dataset challenges shape AI development. AI Soc:1–17

Huang S, Yang J, Fong S, Zhao Q (2020) Artificial intelligence in cancer diagnosis and prognosis: opportunities and challenges. Cancer Lett 471:61–71

Hunter B, Hindocha S, Lee RW (2022) The role of artificial intelligence in early cancer diagnosis. Cancer 14(6)

Iqbal MJ, Javed Z, Sadia H, Qureshi IA, Irshad A, Ahmed R, Malik K, Raza S, Abbas A, Pezzani R, Sharifi-Rad J (2021) Clinical applications of artificial intelligence and machine learning in cancer diagnosis: looking into the future. Cancer Cell Int 21(1):270

Jimenez-Mesa C, Arco JE, Martinez-Murcia FJ, Suckling J, Ramirez J, Gorriz JM (2023) Applications of machine learning and deep learning in SPECT and PET imaging: general overview, challenges and future prospects. Pharmacol Res 197:106984

Keidar Z, Israel O, Krausz Y (2003) SPECT/CT in tumor imaging: technical aspects and clinical applications. Semin Nucl Med 33(3):205–218

Khandakar S, Al Mamun M, Islam M, Hossain K, Melon M, Javed M (2024) Unveiling early detection and prevention of cancer: machine learning and deep learning approaches. Educ Adm Theory Pract 30(5):14614–14628

Lardinois D, Weder W, Hany TF, Kamel EM, Korom S, Seifert B, von Schulthess GK, Steinert HC (2003) Staging of non-small-cell lung cancer with integrated positron-emission tomography and computed tomography. N Engl J Med 348(25):2500–2507

Le Fur M, Zhou IY, Catalano O, Caravan P (2020) Toward molecular imaging of intestinal pathology. Inflamm Bowel Dis 26(10):1470–1484

Li X, Zhang XN, Li XD, Chang J (2016) Multimodality imaging in nanomedicine and nanotheranostics. Cancer Biol Med 13(3):339–348

Li Y, Li Z, Li Y, Gao X, Wang T, Ma X, Wu M (2024) Optical molecular imaging in cancer research: current impact and future prospect. Oncol Trans Med 10(5):212–222

Litjens G, Kooi T, Bejnordi BE, Setio AAA, Ciompi F, Ghafoorian M, van der Laak J, van Ginneken B, Sánchez CI (2017) A survey on deep learning in medical image analysis. Med Image Anal 42:60–88

Liu JN, Bu W, Shi J (2017) Chemical design and synthesis of functionalized probes for imaging and treating tumor hypoxia. Chem Rev 117(9):6160–6224

Lv Z, Jin L, Cao Y, Zhang H, Xue D, Yin N, Zhang T, Wang Y, Liu J, Liu X, Zhang H (2022) A nanotheranostic agent based on Nd(3+)-doped YVO(4) with blood-brain-barrier permeability for NIR-II fluorescence imaging/magnetic resonance imaging and boosted sonodynamic therapy of orthotopic glioma. Light Sci Appl 11(1):116

Medhi D, Kamidi SR, Mamatha Sree KP, Shaikh S, Rasheed S, Thengu Murichathil AH, Nazir Z (2024) Artificial intelligence and its role in diagnosing heart failure: a narrative review. Cureus 16(5):e59661

Minoshima S, Cross D (2022) Application of artificial intelligence in brain molecular imaging. Ann Nucl Med 36(2):103–110

Miura Y, Tsuji AB, Sugyo A, Sudo H, Aoki I, Inubushi M, Yashiro M, Hirakawa K, Cabral H, Nishiyama N, Saga T, Kataoka K (2015) Polymeric Micelle platform for multimodal tomographic imaging to detect scirrhous gastric cancer. ACS Biomater Sci Eng 1(11):1067–1076

Muhammad W, Hart GR, Nartowt B, Farrell JJ, Johung K, Liang Y, Deng J (2019) Pancreatic cancer prediction through an artificial neural network. Front Artif Intell 2:2

Mura S, Couvreur P (2012) Nanotheranostics for personalized medicine. Adv Drug Deliv Rev 64(13):1394–1416

Murar M, Albertazzi L, Pujals S (2022) Advanced optical imaging-guided nanotheranostics towards personalized cancer drug delivery. Nano 12(3)

Muthu MS, Kulkarni SA, Xiong J, Feng SS (2011) Vitamin E TPGS coated liposomes enhanced cellular uptake and cytotoxicity of docetaxel in brain cancer cells. Int J Pharm 421(2):332–340

O'Farrell AC, Shnyder SD, Marston G, Coletta PL, Gill JH (2013) Non-invasive molecular imaging for preclinical cancer therapeutic development. Br J Pharmacol 169(4):719–735

Ohlerth S, Scharf G (2007) Computed tomography in small animals—basic principles and state of the art applications. Vet J 173(2):254–271

Padmanabhan P, Nedumaran AM, Mishra S, Pandarinathan G, Archunan G, Gulyás B (2017) The advents of hybrid imaging modalities: a new era in neuroimaging applications. Adv Biosyst 1(8):e1700019

Patton JA, Townsend DW, Hutton BF (2009) Hybrid imaging technology: from dreams and vision to clinical devices. Nucl Med Semin 39(4):247–263

Prusty S, Dora L, Panda R, Agrawal S, Abraham A (2023) Molecular imaging analysis in cancer using deep learning: a review. Biomed Eng Res 39(4):1015–1032

Pulumati A, Pulumati A, Dwarakanath BS, Verma A, Papineni RVL (2023) Technological advancements in cancer diagnostics: Improvements and limitations. Cancer Rep (Hoboken) 6(2):e1764

Rai HM (2024) Cancer detection and segmentation using machine learning and deep learning techniques: a review. Multimed Tools Appl 83(9):27001–27035

Randhave N, Setia A, Challa RR, Vallamkonda B, Badgujar P, Verma N, Malik AK, Mehata AK, Muthu MS (2024) Autophagy targeted nanomedicines and nanotheranostics in cancer imaging and therapy. J Drug Deliv Technol 99:105945

Rowe SP, Pomper MG (2022) Molecular imaging in oncology: current impact and future directions. CA Cancer J Clin 72(4):333–352

Schwenck J, Sonanini D, Cotton JM, Rammensee HG, la Fougère C, Zender L, Pichler BJ (2023) Advances in PET imaging of cancer. Nat Rev Cancer 23(7):474–490

Setia A, Vallamkonda B, Challa RR, Mehata AK, Badgujar P, Muthu MS (2024a) Herbal theranostics: controlled, targeted delivery and imaging of herbal molecules. Nano 8(3):344–379

Setia A, Kumari P, Mehata AK, Malik AK, Mahto SK, Muthu MS (2024b) Cetuximab decorated redox sensitive D-alpha-tocopheryl-polyethyleneglycol-1000-succinate based nanoparticles for cabazitaxel delivery: formulation, lung targeting and enhanced anti-cancer effects. Int J Pharm 653:123881

Shokeen M, Fettig NM, Rossin R (2008) Synthesis, in vitro and in vivo evaluation of radiolabeled nanoparticles. Quart J Nucl Med Mol Imaging 52(3):267

Siegel RL, Kratzer TB, Giaquinto AN, Sung H, Jemal A (2025) CA Cancer J Clin 75(1):10–45

Smith JJ, Sorensen AG, Thrall JH (2003) Biomarkers in imaging: realizing radiology's future. Radiology 227(3):633–638

Sufyan M, Shokat Z, Ashfaq UA (2023) Artificial intelligence in cancer diagnosis and therapy: current status and future perspective. Comput Biol Med 165:107356

Vermeulen I, Isin EM, Barton P, Cillero-Pastor B, Heeren RMA (2022) Multimodal molecular imaging in drug discovery and development. Drug Discov Today 27(8):2086–2099

Verschakelen JA, Bogaert J, De Wever W (2002) Computed tomography in staging for lung cancer. Eur Respir J Suppl 35:40s–48s

Vikas AK, Mehata MK, Viswanadh AK, Malik A, Setia P, Kumari SK, Mahto MS (2023) Muthu, EGFR targeted redox sensitive chitosan nanoparticles of cabazitaxel: dual-targeted cancer therapy, lung distribution, and targeting studies by photoacoustic and optical imaging. Biomacromolecules 24(11):4989–5003

Voortman J, Giaccone G (2006) Severe reversible cardiac failure after bortezomib treatment combined with chemotherapy in a non-small cell lung cancer patient: a case report. BMC Cancer 6:129

Yi Z, Hu S, Lin X, Zou Q, Zou M, Zhang Z, Xu L, Jiang N, Zhang Y (2022) Machine learning-based prediction of invisible intraprostatic prostate cancer lesions on (68) Ga-PSMA-11 PET/CT in patients with primary prostate cancer. Eur J Nucl Med Mol Imaging 49(5): 1523–1534

Yin L, Cao Z, Wang K, Tian J, Yang X, Zhang J (2021) A review of the application of machine learning in molecular imaging. Ann Transl Med 9(9):825

Yin C, Hu P, Qin L, Wang Z, Zhao H (2024) The current status and future directions on nanoparticles for tumor molecular imaging. Int J Nanomed 19:9549–9574

Zhang C, Wang K, An Y, He K, Tong T, Tian J (2019) Improved generative adversarial networks using the total gradient loss for the resolution enhancement of fluorescence images. Biomed Opt Express 10(9):4742–4756

Zhang L, Liu Y, Huang H, Xie H, Zhang B, Xia W, Guo B (2022) Multifunctional nanotheranostics for near infrared optical imaging-guided treatment of brain tumors. Adv Drug Deliv Rev 190: 114536

Functional Biomedicine for Image-Guided Therapeutic of Solid Tumors 12

Deepak Kulkarni, Prabhakar Panzade, Santosh Shelke, Gayatri More, Kashish Mulani, and Prabhanjan Giram

Abstract

The development of functional biomedicine for image-guided therapy has revolutionized the treatment of solid tumors, offering a precise, noninvasive approach to both diagnosis and therapy.

Innovations in the design and development of biomedicines for image-guided surgery (IGS) have transformed the treatment of solid tumors. These image-guided surgeries feature precision and enhanced margins of safety for delicate cancerous tissues. This book chapter explores functional nanomaterials combined with imaging agents such as IRDye, lanthanides, actinides, and other metals. These functional biomedicines integrate with sophisticated imaging techniques, including computed tomography (CT), positron emission tomography (PET), cystoscopy, and magnetic resonance imaging (MRI), to improve patient treatment outcomes.

The functional nanomedicines discussed include inorganic metallic and silica nanomaterials, carbon-based nanotubes, and fullerenes, alongside widely studied organic nanocarriers such as liposomes and polymeric carriers. These materials aim to enhance pharmacokinetics and pharmacodynamics (PK/PD), ultimately ensuring safety and efficacy. This class of nanomaterials offers unique benefits for image-guided surgery, with some still in the early stages of preclinical and clinical studies.

D. Kulkarni · P. Panzade · S. Shelke · G. More · K. Mulani
Department of Pharmaceutics, Srinath College of Pharmacy, Bajajnagar, Chhatrapati Sambhajinagar, India

P. Giram (✉)
Department of Pharmaceutical Sciences, University at Buffalo, The State University of New York, Buffalo, NY, USA

R. Prasad, M. Mohan Yallapu (eds.), *Advancements in Cancer Theranostics*,
https://doi.org/10.1007/978-981-95-0938-6_12

Organic and polymer chemists have made significant advancements in nanomaterial design, biocompatibility, and safety. However, clinical translation and regulatory approval have faced considerable challenges, showing limited success. This chapter also discusses the opportunities and challenges for future research and development in functional biomedicine for image-guided surgery in solid tumors.

Keywords

Imaging · Solid tumor · Cancer · Therapeutic

12.1 Introduction

Latest developments in nanotechnology have revolutionized the field of medicine offering newer avenues for the diagnosis and therapy of cancer. One of such novel strategy is image-guided therapy in which imaging techniques are used to improve the precision and efficacy of therapeutic outcome especially in the cancer treatment (Dennahy et al. 2022). Image-guided therapy integrates imaging technique with therapeutic intervention permitting real-time visualization of tumors and adjacent tissues which aids in the treatment. Besides, therapeutic outcome improved significantly owing to reduced impact on healthy tissues and precise images of tumor location, size, and shape provided by image-guided therapy (Jiang et al. 2023). Advanced imaging techniques provide real-time images important for precision therapy amid various treatment options like surgery, radiotherapy, and chemotherapy. Therefore, image-guided therapy offers several advantages, including precise treatment, minimal side effects with improved patient safety, leading to enhanced therapeutic outcome in cancer therapy (Zeng et al. 2022).

Nanomaterials like nanoparticles, nanorods, nanospheres, nanostars, and nanoshells are chemical substances having nanoscale dimensions, unique physicochemical properties, and a small size relative to cancer cells (100–1000 times smaller). These are suitable for multimodal imaging and targeted delivery because of their high surface area, adjustable optical and magnetic properties, and functionalization potential. Due to advancements in nanotechnology, various nanomaterials have been combined with image-guided therapy for cancer treatment, attributed to their inherent imaging potential and theranostic applications. These nanomedicines can augment therapeutic efficiency by accumulating in tumor tissues to give targeted drug delivery. Moreover, nanomaterials can be tailored to include contrast agents, which provide imaging potential due to their massive surface area (Huang et al. 2012; Siddique and Chow 2020).

Nanomaterials are utilized in image-guided therapy predominantly due to their capability to act as contrast agents in several imaging modalities, including magnetic resonance imaging (MRI), positron emission tomography (PET) and photoacoustic imaging, and optical imaging. These materials enhance tissue permeation and thereby provide high-resolution images which aids in defining tumor boundaries that will assist in surgical resection. Additionally, nanoparticles can deliver therapeutic agents directly to tumor sites while concurrently permitting real-time monitoring of treatment response via imaging techniques (Xu et al. 2024; Liu et al. 2020). Nanomaterials display a pivotal role in medical imaging, such as magnetic resonance imaging (MRI), positron emission tomography (PET), and photoacoustic imaging. These materials work by amplifying signals in various imaging approaches such as magnetic resonance imaging (MRI), computed tomography (CT), and fluorescence imaging, leading to improved visualization of tumors, (Fernandez-Fernandez et al. 2011) for example, enhanced MRI contrast by iron oxide nanoparticles and real-time tracing of the distribution of nanoparticles in the body (Hu et al. 2018). Nanomaterials widely employed in cancer image-guided therapy include gold nanoparticles, quantum dots, liposomes, iron oxide nanoparticles, and silica-based platforms. Gold nanoparticles (AuNPs) are used in image-guided therapy of cancer due to their unique features such as excellent optical properties, biocompatibility, and ease of surface modification. These act as contrast agents in CT through increased X-ray absorption, aid plasmon resonance for photothermal imaging, and facilitate fluorescence imaging (Sargazi et al. 2022). Quantum dots, which display excellent photostability and modifiable fluorescence, offer high-resolution fluorescence imaging of tumors and help in multimodal imaging. Likewise, liposomes exhibit potential to encapsulate some imaging agents like fluorescent dyes to boost tumor visualization for imaging modalities like MRI, CT, PET, or fluorescence imaging while ensuring targeted delivery and minimizing systemic toxicity (Prasad et al. 2020). Palladium-based nanomaterials also showed promising theranostic applications in cancer imaging as contrast agents and therapy owing to their favorable properties, that is, intense absorption in the near-infrared region, superior photothermal conversion efficiency, exceptional photothermal stability along with biocompatibility. Besides, these can be surface functionalized with ligands, therapeutic agents, or imaging agents, leading to enhanced selectivity (Liu et al. 2020).

The various image-guided techniques and their applications in solid tumor diagnosis and therapeutics are given in Table 12.1.

12.1.1 Metallic Nanoparticles

Metallic nanoparticles play an important role as contrast agents in image-guided therapy for solid tumors. The role of metallic nanoparticles is important in the diagnosis of cancerous tumors and various pathological conditions. Several metallic nanoparticles have been used in previous studies for image-guided therapy of tumors.

12.1.1.1 Silver Nanoparticles

Silver nanoparticles represent one of the explored categories of metallic nanoparticles for new and improved biomedicine. It is a commercially available and clinically safe AgNP-based product, and sustainable studies have been

Table 12.1 Applications of image-guided techniques in tumor diagnosis and therapy

Imaging technology	Applications
Fluorescence imaging	• Provide real-time, high-resolution visualization of tumor tissue, enabling early detection, precise localization, and effective monitoring of therapeutic response • Monitor the delivery of drugs specifically to tumor tissues to ensure targeted delivery • Help surgeons in intraoperative imaging to visualize tumor tissue in real-time and guide the removal of tumors without affecting healthy tissue • Used to visualize angiogenesis in tumors, thus providing insight into the progression of cancer
Photoacoustic imaging	• Can detect cancer in the early stage and monitor tumor angiogenesis • Aids in biopsies and surgeries owing to its ability to provide real-time imaging
Magnetic resonance imaging (MRI)	• Potential to differentiate between benign and malignant tumors, considering tissue composition and vascularity • Helps in the detection of tumors in the brain, liver, breast, etc., by giving thorough anatomical images
Computed tomography (CT)	• Tumor detection and diagnosis by providing cross-sectional images • Precise cancer staging by identifying metastasis in lymph nodes and organs
Positron emission tomography (PET)	• Ability to detect increased glucose metabolism in cancer cells • Cancers not detected by CT or MRI can be identified by PET
Ultrasound imaging	• Superficial tumors detected by this imaging modality, like breast, thyroid • Detect abnormalities in different tissues and organs

conducted to assess the applicability of nanosilver for therapeutic activity and alternative strategies for cancer therapy.

Lee and Colleagues (2020) developed a composite of porous silicon microparticles (PSiMPs) and silver nanoparticles for cancer immunotherapy. These PSiMPs and silver nanoparticles were subjected to cancer therapy via CT as an image-guided technique by scanning the lungs as a localized agent. The grafting of silver nanoparticles onto porous silicon microparticles is performed via several techniques, such as scanning electron microscopy and X-ray diffraction. The composite has a high CT signal intensity of approximately 1000 HU. To produce and view images of the lungs, a strong CT signal is produced at the site of injection for up to several hours without spreading. It shows no toxicity or immune response, indicating its potential for treating lung cancer. Silver nanoparticles and PSiMPs are safe for mice and do not affect their immune response. Figure 12.1 illustrates the image-guided localization agent for computed tomography.

Nair and coreserchers (Joseph et al. 2017) explored biogenic nanochemobiotics fabricated from silver nanoparticles and galactoxyloglucan with effective biodistribution in solid tumors via SERS fingerprinting (surface-enhanced Raman scattering platform). Silver nanoparticles (SNPs) and naturally occurring

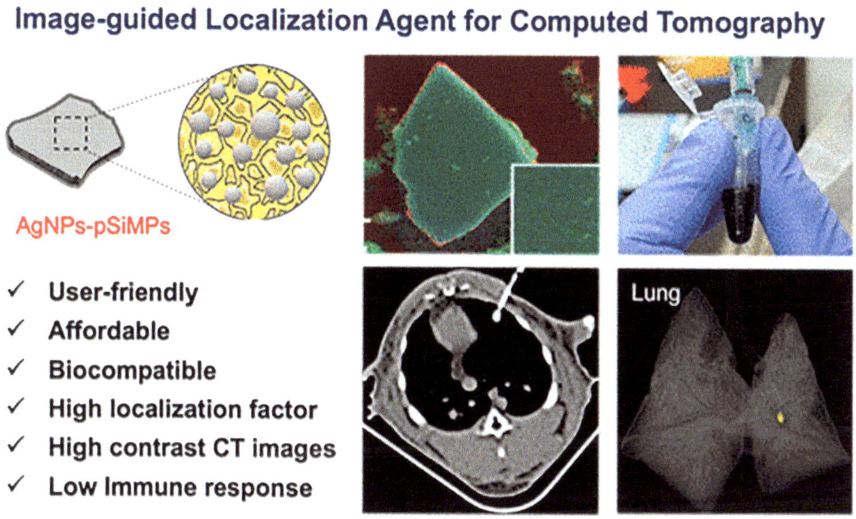

Fig. 12.1 Image-guided localization agent for computed tomography. Adapted with permission from reference Lee et al. (2020)

galactoxyglucan (PST001), a composite denoted as SNP@PST, have good biocompatibility. These nanocomposites are eco-friendly and biocompatible. SNP@PST plays a role in cancer therapy by minimizing selective cytotoxicity toward tumors while minimizing toxicity to normal cells, which are mainly used to treat cancer. A novel nanocomposite used for cancer treatment that combines both chemical and antibacterial effects that kill cancer cells while minimizing harm to normal cells. The nontoxic nature of SNP@PST was demonstrated in mice. Nonocomposites subjected to cancer therapy via SERS can be used to monitor the distribution of tumors and thereby improve therapeutic effectiveness.

With recent advancements in nanotechnology, several types of metal nanomaterials have been developed to increase the anticancer efficacy of radiotherapy. Among these nanoparticles, silver nanoparticles are highly applicable for therapeutic, cancer, and diagnostic conditions. These silver nanoparticles are known for their excellent radiosensitizing properties.

Lee and colleagues (2020) developed AgNPs functionalized with polyethylene glycol (PEG) and As1411 (AsNPs), which were assessed for their tumor-targeting properties for cancer treatment and radiosensitizing effects both in vitro and in vivo. These findings provide an important basis for the potential use of AsNPs as effective nanoradiosensitizers to target tumors and are applicable for the treatment of glioma. Malignant gliomas are cerebral tumors that can be treated with modified forms of AgNPs to target glioma via systemic administration.

12.1.1.2 Gold-Based NPs

Gold NPs are medically safe, have a surface that can be easily modified, and efficiently create Auger electrons when irradiated. Compared with other nanoparticles and their derivatives, gold NPs have longer blood circulation times, reduce nephrotoxicity, and have a greater X-ray absorption coefficient. Gold NPs can be easily fabricated via a wide range of solution-based synthetic approaches. Gold nanoparticles are toxic in nature, and their toxicity depends on size; these effects were confirmed via in vivo and in vitro experiments.

NPs are functionalized with different components, polymers, and drugs. This functionalization, in addition to enhancing biocompatibility and bioavailability, also promotes target selectivity and specificity and improves therapeutic efficiency.

Liu et al. (2013) developed acetylated folic acid (FA)-modified amine-terminated generation 5 poly(amidoamine) dendrimers ($G5{\cdot}NH_2$–FA) as stabilizers to prepare modified forms of AuNPs that are stabilized and demonstrate favorable cytocompatibility. These nanoparticles are resistant at pH values ranging from 6 to 8 under various aqueous conditions and are suitable at sufficient temperatures of up to 50 °C. The FA-modified form of nanoparticles was subjected to the diagnosis of tumor cells via computed tomography (CT) imaging for the treatment of cancer cells in vivo as well as xenograft tumor models in vivo. FA-mediated cellular uptake targets direct binding pathways and results in increased CT values in KB cells. These FA-modified AuNPs can be used for CT imaging of cancer cells in vivo and are used for the diagnosis and treatment of cancerous tumors. In addition, it is used for various biomedical applications in sensing, diagnosis, and therapeutics.

Yan Dou (2017) developed multifunctional nanotheranostic PB@AU core-state nanoparticles (CSNPs) based on Prussian blue nanoparticles (PNPs) and gold nanoparticles (AuNPs). PB@AUNP core saturation components were produced by combining both PBNPs and AUNPs, two different types of intrinsic theranostic NPs. These materials have good colloidal stability. The imaging techniques used for tumor detection are magnetic resonance (MR) imaging and computed tomography (CT) imaging. Both techniques are utilized to localize and detect cancer cells effectively. The use of multiple modified CSNPs enhances the imaging capabilities and improves the success rate of tumor detection. AUNPs offer CT enhancement capability and radiosensitization. By both MR and CT imaging, CBNPs achieve effective tumor localization through passive targeting, which allows them to accumulate at the tumor site after IV injection. Imaging techniques such as MR and CT have shown the efficient localization of CSNPs in tumors. These factors increase the success rate of cancer tumor detection.

12.1.1.3 Copper-Based NPs

Zhang and Colleagues (2016) discovered PEGylated Cu2xse NPs for image-guided cancer therapy (Fig. 12.2). The normal human needs copper for energy production, healthy blood vessels, and the well-being of the cardiovascular system. Both selenium and copper have excellent biodegradable properties. To improve their water solubility and colloidal properties, ultrasmall NPs are coated with HS-PEG-SH. This

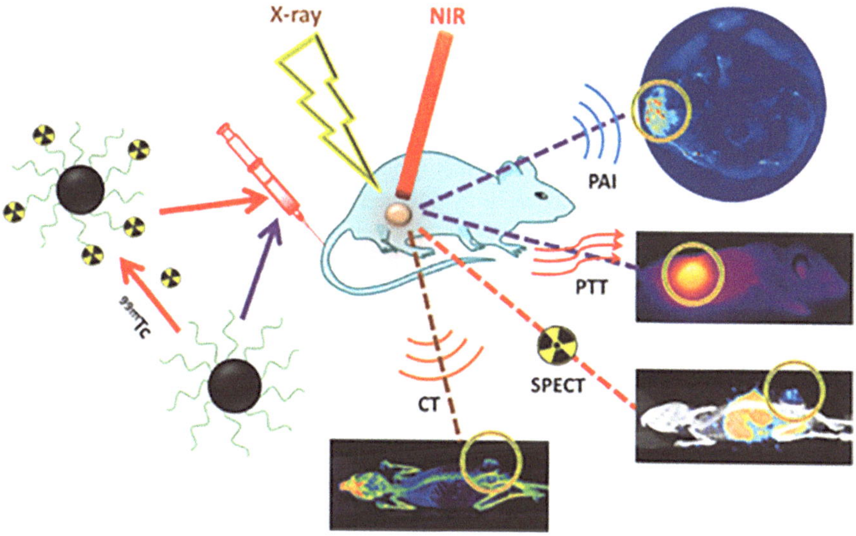

Fig. 12.2 PEGylated $Cu_{2-x}Se$ nanoparticles for image-guided cancer therapy. Adapted with permission from reference Zhang et al. (2016)

modified form of NP demonstrated low cytotoxicity toward 4 T1 mammary carcinoma cells at concentrations of up to 100 g/L. Copper nanoparticles were subjected to cancer immunotherapy via CT as an image-guided technique. For in vitro studies, cancer-bearing BALB/c mice were intravenously injected with a solution of modified NPs. After injection, CT images clearly revealed enhanced permeability and retention effects. At 10 h postinjection, the HU value was approximately 50. The purpose of this approach was to improve the success rate of cancer detection in the presence of Cu and Se.

12.1.1.4 Iron-Based Nanomaterials

Lu et al. (2018) used a dendrimer with small iron oxide NPs to stabilize gold nanoflowers. AuNPs are used to achieve a photoacoustic response, whereas FeO_3 results in improved CT imaging. FeO_3 also has strong colloidal stability, and its cellular absorption is also effective. The image-guided technique used to detect tumors is computed tomography (CT) to detect cancer cells easily and for the diagnosis of cancer patients. The modified NP studies were carried out with 4 T1 mammary carcinoma cells to examine their cytocompatibility. After continuous laser exposure, the cell viability was approximately 34.7%, which demonstrated the effectiveness of the Fe_3O_4/Au DSNPs as a PT ablation material. After injection, the CT visibility produced for the tumor increased by 1.7 times in 60 min. Additionally, better CT signal intensity was produced, indicating superior localized imaging capability.

12.1.1.5 Platinum-Based Nanomaterials

Wang et al. (2019) developed functionalized platinum nanoparticles for cancer therapy. Platinum is functionalized with a light-activatable dual prodrug polymer to produce dual prodrug polymer nanoparticles (DPP NPs). The platinum-based NPs were subjected to cancer therapy via drug-mediated computed tomography (DMCT) imaging. This technique facilitates real-time monitoring of the distribution and metabolism of activable dual prodrug polymer nanoparticles in vivo, facilitating the optimization of light irradiation parameters during cancer therapy. Dimethyl cantharidin has anticancer activity. The combination of DMC with platinum results in a drug-based chemotherapy effect. Different CT images were observed in mice for up to many hours compared with those of mice treated with cisplatin. Tumor growth is prevented in male KM mice harboring U14 xenograft tumors when they are subjected to the dual drug polymer (DPP) system under laser irradiation. Several mice treated with the dual prodrug polymer via laser irradiation guided by Pt DMCT were cured. The body weights and physiological parameters of the mice treated with DPP NPs remained constant. Compared with the other groups, the DPP NP group demonstrated increased anticancer efficacy and decreased toxicity.

12.1.1.6 Tungsten-Based Nanomaterials

Jakhmola et al. (2014) developed modified tungsten oxide (WO_3) NPs coated with polycaprolactone (PCL). To improve stability, they were combined with biodegradable polycaprolactone (PCL) and a PEGylated surfactant. The imaging technique used was CT imaging. The purpose of using tungsten oxide NPs is to enhance CT images while ensuring low toxicity. The 100-nm diameter NPs with a 0.1 m concentration displayed good contrast in the CT images. In in vivo experiments, it did not have any toxic effects and had a high CT signal, and was quickly released from the body. PCL-WO3 NPs prolong the blood circulation period to a minimum of hours; however, they are easily eliminated. They do not accumulate in the body, making them suitable candidates for CT angiography.

12.1.1.7 Silica Nanoparticles

In recent developments, silica-based nanoparticles have contributed to their action by researching tumors and acting as drug delivery vectors. Silica assists with image-guided techniques to detect the tumor at a particular site and is used for cancer diagnosis. SiNP-mediated drugs can be combined with many chemotherapeutic agents, photosensitizers, SiRNAs, and gene therapeutic agents and have one or more functional agents to improve the success rate of cancer detection.

Feng Chen (2013) and coresearchers developed functionalized mesoporous silica ($mSiO_2$) nanoparticles, which are biocompatible and applicable for many diagnostic activities and are used as drug delivery vehicles to target tumors. The functionalized $mSiO_2$ nanoparticles were subjected to positron emission tomography (PET) imaging to target tumors and were used for drug delivery in 4 T1 murine breast tumor-bearing mice. This mesoporous silica ($mSiO_2$) was functionalized with a thiol group, and PEG was added with the TRC105 antibody, which is specific for CD105/endoglin. The TRC105 antibody binds to CD105/endoglin and contributes to its function by targeting tumors, which are expressed on the tumor vasculature. This

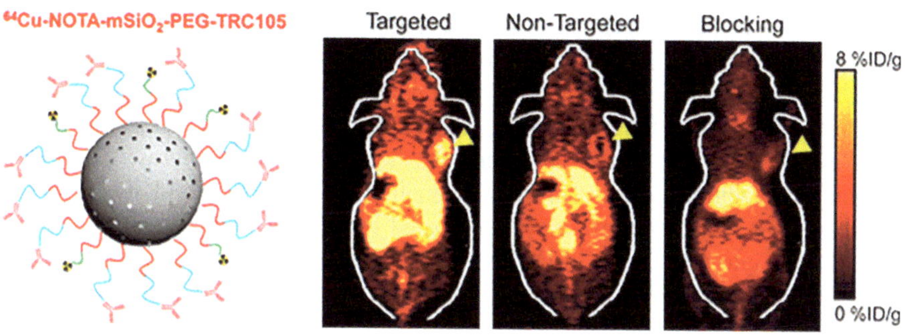

Fig. 12.3 In vivo tumor targeting with image-guided drug delivery. Adapted with permission from reference Chen et al. (2013)

targeted binding enhances the accumulation of these functionalized nanoparticles at the tumor site. In previous studies, doxorubicin (DOX) was used to deliver drugs to 4 T1 tumors through intravenous injection of DOX-loaded NOTA-mSiO2-PEG-TRC105, which enhanced the potential of image-guided drug delivery and targeted tumors at specific sites. Figure 12.3 illustrates the in vivo tumor targeting with image-guided drug delivery.

12.1.2 Carbon-Based Functional Biomolecules

Tumors exhibit uncontrolled growth, and their diagnosis and treatment are very critical and challenging. Hence, various new and unique techniques and biomedicines have been developed for diagnosis and therapy. One of the best techniques is the use of carbon-based functional biomedicines. They are widely used in the diagnosis of solid tumors. It possesses certain properties, such as high strength and excellent surface area. Various functionalizations can be performed to potentiate their application in cancer therapy.

12.1.2.1 Graphene Oxide

Graphene oxide nanomaterials play a vital role in image-guided techniques for the diagnosis and treatment of solid tumors. It is the most promising nanomaterial among the carbon family, as it has emerged as a polynomial functional tool. The use of graphene oxide has increased rapidly because of its excellent properties, such as high surface area and biocompatibility. It is used in the diagnosis and targeting of solid tumors via image-guided techniques. In many studies, graphene oxide nanomaterials or nanosheets are used for image-guided therapy of solid tumors.

Guo and coresearchers reported the development of a multifunctionalized graphene oxide (GO) nanocomposite for enhancing targeted imaging and therapy for prostate cancer. This method is based on the functionalization of graphene oxide with dendrimers containing amino groups, which are then used to attach gadolinium diethylene triamine pentaacetate (Gd-DTPA) and monoclonal antibodies that target prostate stem cell antigen (PSCA). The GO-DEN(Gd-DTPA)-mAb is constructed to target the Gd-DTPA-mAb that binds to the overexpressed PSCA on cancer cells and therefore increases the selectivity of the imaging and therapeutic agents. During magnetic resonance imaging (MRI), GO-DEN(Gd-DTPA)-mAb was used as a contrast agent, which markedly improved the T1-weighted contrast of prostate tumors and facilitated better in vivo evaluation of the cancer and its metastatic processes. The nanocomposites were prepared by loading doxorubicin hydrochloride (DOX) into GO-DEN(Gd-DTPA)-mAb, which enhanced targeted delivery to PC-3 cells. Tumor growth inhibition in the mouse models was appreciable during the in vivo studies. These findings demonstrated that this multifunctionalized graphene oxide nanocomposite has potential for simultaneous imaging and therapeutic application against prostate cancer (Guo et al. 2016).

The synthesis of GO nanoparticles with the intent to use them in breast cancer therapy targets the characteristic attributes of GO, such as its broad surface area, biocompatibility, and range of functionalization with several targeting agents, including but not limited to FSHR monoclonal figures, antibodies, and nucleic acids. For example, monoclonal antibodies that target specific receptors, such as follicle-stimulating hormone receptors, can be attached to enhance specificity and cellular uptake by tumor cells. These antibodies can also be used to fuse chemotherapeutic agents, such as siRNAs, to GO nanoparticles through static attachment or covalent interfacing, ensuring targeting without deleterious off-target effects. In other words, nanocarriers can be loaded with drugs or fusion molecules without perturbing each other through static or strong bonding forces. Additionally, the incorporation of chemotherapeutic agents, such as doxorubicin, facilitates controlled release depending on the stimuli that have been exerted on the nanoparticles, thus augmenting therapeutic influences on targeted cancer cells while limiting healthy tissues from sustaining damage. Image-aiding procedures are also typed into this category, with the aim of targeting the localization within the tumor via real-time understanding of the movement of the particles, thus improving the dispersal and effectiveness of the drug. This capability optimizes treatment protocols and improves the manner in which patients respond to treatment (Grilli et al. 2022).

Graphene oxide has gained significant attention in biomedicine because of its unique properties. Graphene oxide has superior optical absorption properties in the near-infrared region. Hence, graphene oxide is widely used in imaging techniques for targeting solid tumors, where it can absorb light and destroy cancer cells. Its ability to carry drugs and target specific cells enhances its therapeutic activity.

Jun and coresearchers synthesized folic acid–chitosan–graphene oxide conjugate (FA-CS-GO) as a multifunctional nanomaterial that allows the combination of several components, increasing the efficiency of cancer treatment (Fig. 12.4). Folic acid works as a targeting ligand that binds selectively to overexpressed folate receptors on the surfaces of cancer cells, thus enhancing the distribution of the nanomaterial to tumors. Chitosan, which is biocompatible and biodegradable, also contributes by providing some structural integrity and stability while enhancing cell uptake, whereas graphene oxide performs as a support or carrier of images and sunny therapeutic agents because of its diverse features, such as its large surface area and

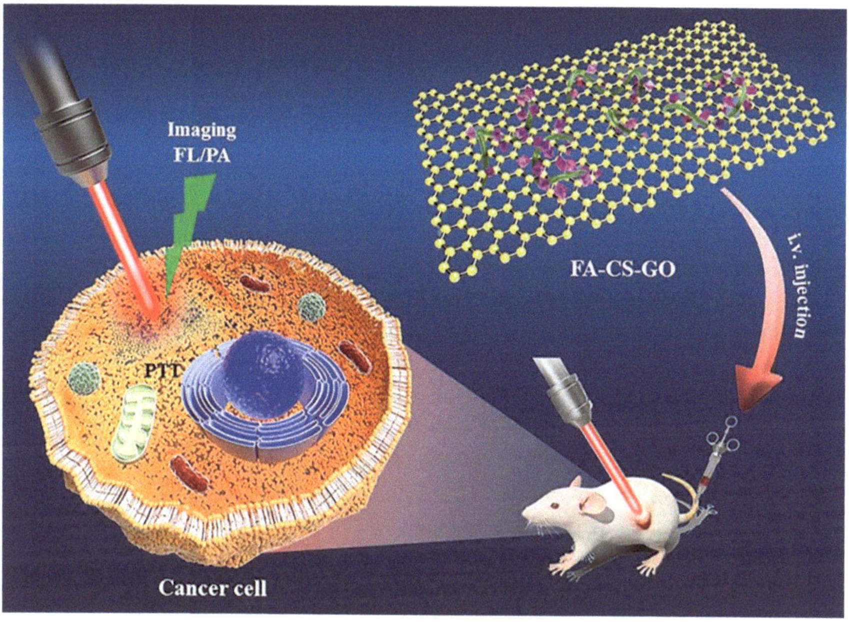

Fig. 12.4 Folic acid-conjugated chitosan-functionalized graphene oxide for image-guided tumor therapy. Adapted with permission from reference Jun et al. (2020)

wide range of optical characteristics, including high near-infrared (NIR) light absorption, which is useful in photothermal therapy. In this context, the use of image-guided techniques, particularly PAI and FL imaging, increases targeting efficacy. PAI works by using the NIR absorption of FA-CS-GO to emit ultrasound signals, which are used to form images. These images are high in resolution and are used to dynamically track the FA-CS-GO distribution in tumor tissues, whereas FL imaging captures the signals given off by the excited FA-CS-GO and covers the dynamic aspect of the molecule's localization in tissues and tumors. Receptor-mediated endocytosis is the major targeting mechanism of FA-CS-GO, where folic acid binds to folate receptors present on cancer cells and facilitates their cellular uptake, along with the enhanced permeability and retention (EPR) effect, which results in the accumulation of nanoparticles in tumor tissues (Jun et al. 2020). Additionally, the formulation is designed to prolong circulation time in the bloodstream, ensuring extended exposure to the tumor site before clearance. Collectively, these features contribute to the effective targeting of tumors, significantly enhancing the overall efficacy of cancer therapies.

12.1.2.2 Fullerene

Fullerene is a carbon allotrope that has been widely used in cancer diagnosis. Various functionalized fullerenes were prepared for targeting the tumor site via an image-guided technique. The specific physicochemical properties of fullerenes promoted the development of tumor theranostics because they have wide applications in imaging techniques. Fullerene also possesses antioxidant properties.

Chen et al. fabricated the use of functionalized fullerene in image-guided techniques. Fullerene is functionalized with polyhydroxy fullerene (PHF; C60 (OH) O Na) or a carboxy group (CF; C60(COOH)2)3) to enhance the effect of fullerene. The imaging technique used was photoacoustic imaging (PAI). Experiments were carried out on tumor-bearing nude mice, yielding promising results for tumor diagnosis via PAI after the functionalization of fullerene. This functionalization was carried out to potentiate the effect and facilitate easy diagnosis of the tumor (Chen et al. 2012). Figure 12.5 illustrates the treatment of a mouse tumor with polyhydroxy fullerenes.

Fullerenes represent a promising area of research in nanotechnology and medicine with potential applications in cancer therapy and imaging. Fullerene has mostly been applied in biological fields. The arrangement of carbon atoms endows fullerenes with unique physical and chemical properties, such as high stability, electrical conductivity, and ability to absorb light.

Di Giosia and coresearchers functionalized FWS with biomolecules such as proteins, peptides, and antibodies. This functionalization enhances the specificity of fullerene for image therapy for solid tumors. It also enhances the solubility and potentiates the photoactivity of fullerene. The imaging techniques used were optoacoustic imaging and third-harmonic generation (THG). This technique has wide applications in cancer treatment and diagnosis (Giosia et al. 2021).

Fullerene has the ability to directly target specific cells. Fullerene absorbs light and hence is widely used in imaging techniques, showing promising potential for the treatment and diagnosis of cancer.

Mauro and colleagues reported that a biocompatible coating on fullerene enhances the image-guided technique of fullerene for solid tumors. Fullerene is functionalized or coated with polymers such as PEG to potentiate its effects. The technique used was magnetic resonance imaging (MRI). This technique is used for brain and central nervous system imaging (Mauro et al. 2021).

12.1.2.3 Carbon Nanotubes

Carbon nanotubes are cylindrical nanostructures composed of carbon atoms arranged in a hexagonal lattice. It has good mechanical, electrical, and thermal properties. Owing to these excellent properties, it is widely used in research and biomedicine. A graphene sheet is rolled on it; hence, it also exhibits metallic and semiconductor behavior. It also possesses good strength, flexibility, and thermal conductivity properties and is therefore widely used in image-guided techniques.

Rubio et al. reported that functionalized CNTs are widely used in image-guided techniques. They functionalized carbon nanotubes with 1,3-dipolar cycloadditions and amidation reactions. They carried out experiments on mice that were implanted with CT-26 colon cancer target tumors. These functionalized carbon nanotubes were used for targeting colon cancer or colon tumors (Fig. 12.6). The functionalization enhances the specificity to the target site. The imaging technique used to target colon cancer is multiphoton luminescence imaging (Rubio et al. 2015).

Carbon nanotubes, especially single-wall-coated nanotubes, have various unique properties and are widely used in image-guided therapy. It is widely used in the

A

B

C

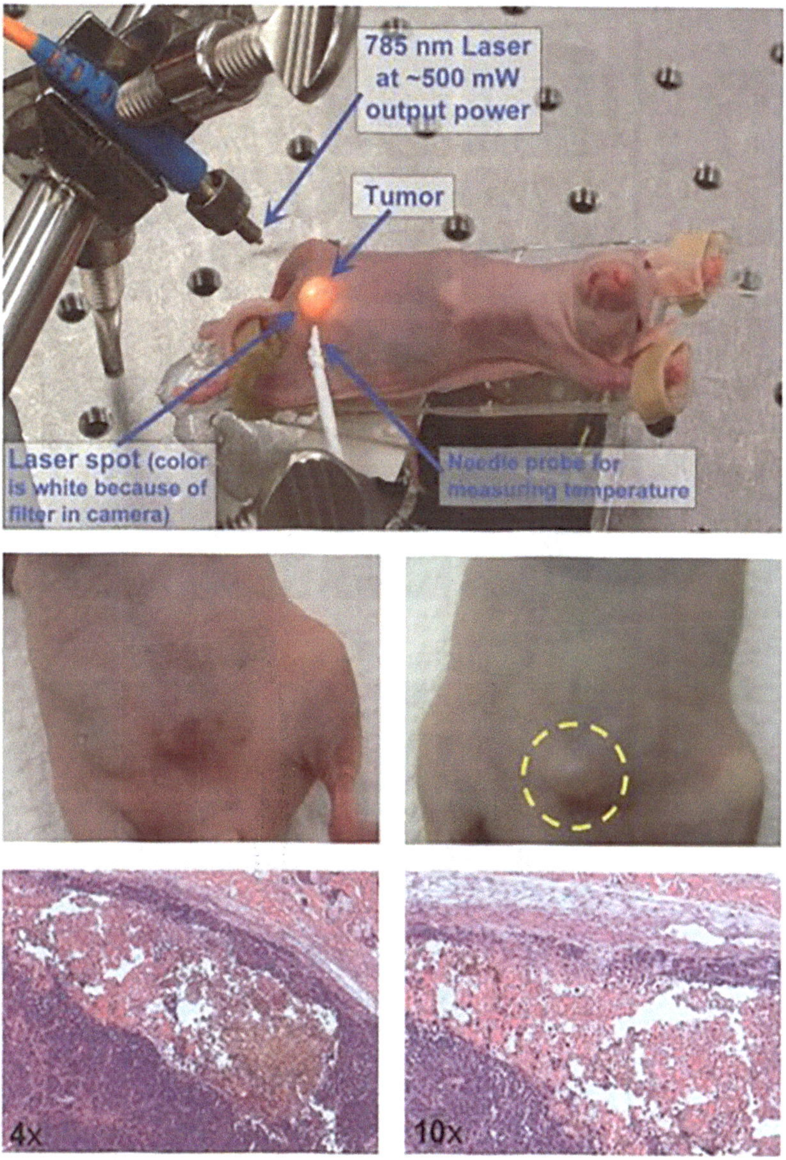

Fig. 12.5 Treatment of a mouse tumor with polyhydroxy fullerenes. Adapted with permission from reference Krishna et al. (2010)

diagnosis of tumors. The results show the application of these materials in heat management and thermal interface materials. It has wide applications in sensors, transistors, and conductive materials, and certain modifications have been made to single-walled carbon nanotubes to increase their effectiveness and enhance their use.

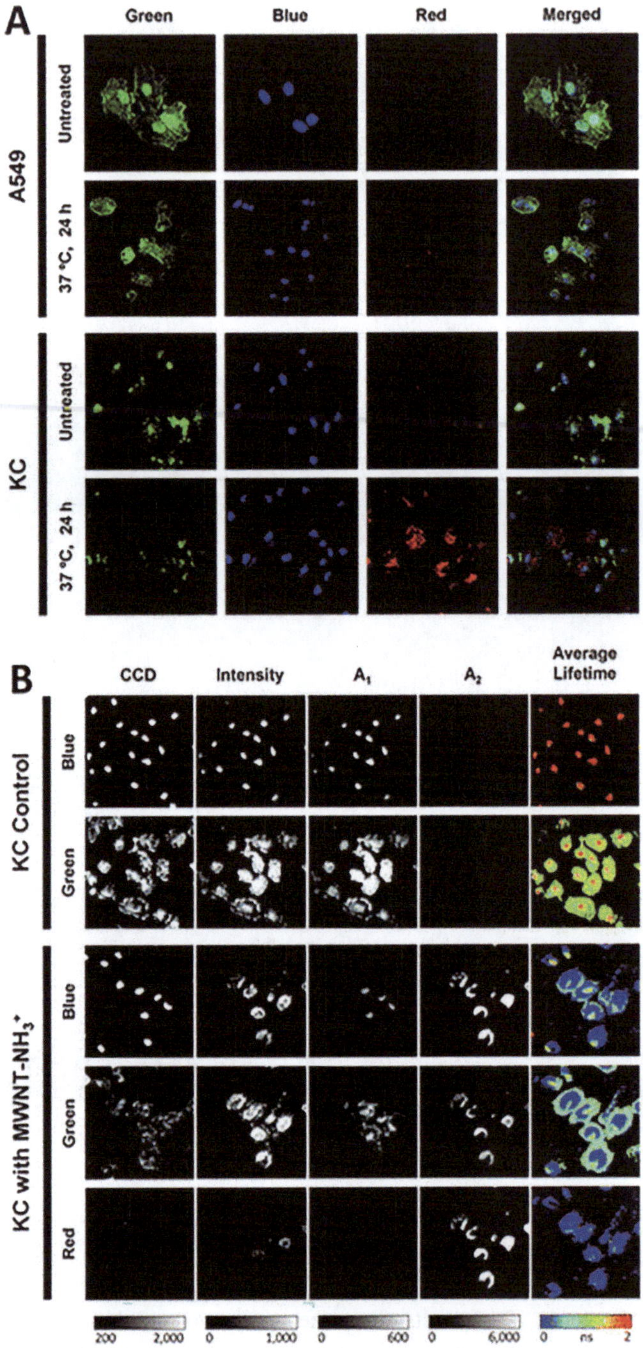

Fig. 12.6 Multiphoton luminescence imaging of chemically functionalized multiwalled carbon nanotubes in solid tumors. Adapted with permission from reference Rubio et al. (2015)

Zhao and coresearchers fabricated carbon nanotubes coated with polydopamine (PDA), which was further modified with polyethylene glycol (PEG) to increase the efficacy of SWCN. The imaging technique that was used was magnetic resonance imaging (MRI). This functionalization is a robust photothermal agent for photothermal therapy. This technique is used for breast tumors or breast cell lines. The experiment was carried out on the mice, and the tumors were diagnosed via magnetic resonance imaging (Zhao et al. 2016).

Sitharaman and his coresearchers investigated that single-wall-coated nanoparticles are functionalized with gadolinium (Gd-SWCNTs). This functionalization enhances the efficacy of carbon nanotubes. The experiment was carried out on the mouse legs, and the tumor was identified. The image-guided technique that was used for these functionalized nanotubes was magnetic resonance imaging (MRI). MRI is the most effective technique that was used for the diagnosis of tumors (Sitharaman et al. 2013).

12.2 Opportunities and Challenges

The use of functional biomedicines for image-guided therapy in solid tumors has made significant advancements in terms of improving precision and accuracy. However, challenges remain that need to be addressed to meet the unmet needs of patients and the community. As science progresses in image-guided surgery to fill existing gaps, researchers and the scientific community must deepen their understanding of tumor pathophysiology, the tumor microenvironment, metastasis, and the physicochemical properties of functional biomedicines. This includes the characteristics of the functional material itself, as well as the loaded cargo, such as contrast or imaging agents.

12.2.1 Opportunities

The molecular specificity of functional biomedicines can be enhanced through targeted therapy, which is crucial for improving tumor targeting while minimizing distribution to normal healthy cells. Understanding the pharmacokinetics (PK) and pharmacodynamics (PD) of both the parent imaging agent and the functionalized biomedicine is a key step in assessing the absorption, distribution, metabolism, and excretion (ADME) of these nanomedicines. Achieving optimal targeting and delivery strategies for functional nanomedicines in image-guided surgery requires collaborative efforts from material designers, formulation scientists, and biologists (Ferrari 2005; Allen and Cullis 2013; Petros and DeSimone 2010).

Another important aspect is safety, particularly biocompatibility and biodegradability, which significantly influence the safety profile of functional nanomedicines for image-guided applications. Polymer and material chemists strive to design nontoxic biomaterials and rigorously assess degradation and toxicity in preclinical animal models using short, intermediate, and long-term toxicity tests according to

OECD guidelines. Innovations in polymer design, using techniques like ATRP (Atom Transfer Radical Polymerization), RAFT (Reversible Addition-Fragmentation chain Transfer), and ROP (Ring-Opening Polymerization), have led to the development of diverse polymeric architectures. These architectures can respond to stimuli such as pH, temperature, redox conditions, enzymes, and hydrogen peroxide, thereby enhancing therapeutic efficacy and safety profiles (Ferrari 2005; Kamaly et al. 2016; Giram and Garnaik 2021; Giram et al. 2021, 2024a, b; Patil et al. 2023; Pardeshi et al. 2023).

Multimodal imaging has gained popularity in the field of image-guided surgery, as it integrates MRI, CT, and PET to visualize tumor signatures, providing clinically significant insights for grading and staging. This approach offers a pathway toward personalized and patient-specific treatments. Despite promising preclinical and clinical advances, the efficacy and safety of new functional biomedicines remain limited. Utilizing PK/PD models, physiologically based pharmacokinetic (PBPK) models, and AI can provide deeper insights, potentially enhancing their clinical applications (Kircher et al. 2012; Derendorf and Meibohm 1999; Jones and Rowland-Yeo 2013; Esteva et al. 2019).

The final hurdle involves navigating regulatory guidelines, particularly the stringent requirements imposed by bodies like the US FDA, the European Medicines Agency, and other international regulators. There is a need for a unified international framework and a comprehensive database for approval and amendments, which would simplify the process globally. National and international collaborations among academia, industry, and regulatory bodies are essential to streamline the development and clinical use of functional biomedicines (FDA and EMA) (Zerhouni and Hamburg 2016; Guidance for Industry n.d.).

12.2.2 Challenges

Functional biomedicines for image-guided surgery have demonstrated efficacy and safety in preclinical studies. However, global challenges remain in improving precision and targeting for deep tumor tissues and overcoming anatomical barriers. The design of functional biomedicines, from lab-scale to pilot-scale and large-scale industrial production, requires sophisticated techniques and analytical tools. Achieving the target product profile with the intended quality is feasible through the design of experiments for process optimization and production, following ICH Q8 guidelines.

One major challenge in effective image-guided surgery is tumor heterogeneity, which complicates the surgeon's ability to define a clear safety margin. Variability among patient populations, due to differences in genetics, immunological status, and tumor burden, also impacts biodistribution and patient outcomes. These issues can be mitigated through the design of personalized, functionalized biomedicines.

Regulatory hurdles can be addressed by creating harmonized, unified, clear, concise, and direct guidelines that are easy for researchers, the scientific community, and industrial innovators to follow. The workflow of image-guided surgery presents additional challenges, including patient compliance and discomfort, which need to be carefully managed.

12.3 Conclusion and Future Prospects

Functional biomedicine for image-guided surgery has achieved limited success due to several hurdles, including complex pharmacokinetics, short half-life, a challenging tumor microenvironment, and tumor heterogeneity. To forecast a brighter future for image-guided surgery, collaboration among material scientists, formulation experts, and oncologists is essential to develop a comprehensive strategy.

A significant issue in translating preclinical results to clinical settings is the poor correlation between in vitro and in vivo data. Often, results from preclinical animal models are not reproducible in clinical trials because these models fail to capture the complexities of human patients. Developing and utilizing primary human tumor xenografts and patient-derived xenograft models is crucial to streamline initial pharmacokinetic, safety, toxicity, and efficacy studies.

There is a real need for image-guided surgery, particularly for photodynamic therapy, radiation therapy, and photothermal therapy. Addressing this need requires a deeper understanding of the barriers to capturing accurate data. These barriers include not only the therapy itself but also the intrinsic properties of functional biomedicines. Current research and development should focus on using appropriate animal models and accurately extrapolating pharmacokinetic and pharmacodynamic data to clinical settings. Additionally, assessing the degradation and toxicity of functional biomedicines is vital for their successful application in image-guided surgery. Enhancing image processing and minimizing artifacts could also be achieved through the integration of AI for image analysis (Pagar et al. 2022).

In the future, it is expected that numerous functional biomedicines will be developed to address current challenges. Ideal materials should minimize off-target effects on normal tissues while maintaining high specificity for tumors. Targeting strategies using antibodies represent a promising area for achieving precise targeting of anatomical and pathological sites. These advancements could offer significant benefits for both patients and surgeons, enhancing precision and accuracy in surgical procedures (Gadade et al. 2023).

Acknowledgements Authors are thankful to affiliated institutes for providing facilities for the literature survey.

Conflict of Interest Authors have no conflict of interest.

References

Allen TM, Cullis PR (2013) Liposomal drug delivery systems: from concept to clinical applications. Adv Drug Deliv Rev 65(1):36–48. https://doi.org/10.1016/j.addr.2012.09.037

Chen Z, Ma L, Liu Y, Chen C (2012) Applications of functionalized fullerenes in tumor theranostics. Theranostics 2(3):238–250. https://doi.org/10.7150/thno.3509

Chen F, Hong H, Zhang Y, Valdovinos HF, Shi S, Kwon GS, Theuer CP, Barnhart TE, Cai W (2013) In vivo tumor targeting and image-guided drug delivery with antibody-conjugated, radiolabeled mesoporous silica nanoparticles. ACS Nano 7(10):9027–9039. https://doi.org/10.1021/nn403617j

Dennahy IS, Han Z, MacCuaig WM, Chalfant HM, Condacse A, Hagood JM, Claros-Sorto JC, Razaq W, Holter-Chakrabarty J, Squires R, Edil BH, Jain A, McNally LR (2022) Nanotheranostics for image-guided cancer treatment. Pharmaceutics 14(5):917. https://doi.org/10.3390/pharmaceutics14050917

Derendorf H, Meibohm B (1999) Pharm Res 16(2):176–185. https://doi.org/10.1023/A:1011907920641

Dou Y, Li X, Yang W, Guo Y, Wu M, Liu Y, Li X, Zhang X, Chang J (2017) PB@Au Core–satellite multifunctional nanotheranostics for magnetic resonance and computed tomography imaging in vivo and synergetic photothermal and radiosensitive therapy. ACS Appl Mater Interfaces 9(2):1263–1272. https://doi.org/10.1021/acsami.6b13493

Esteva A, Robicquet A, Ramsundar B, Kuleshov V, DePristo M, Chou K, Cui C, Corrado G, Thrun S, Dean J (2019) A guide to deep learning in healthcare. Nat Med 25(1):24–29. https://doi.org/10.1038/s41591-018-0316-z

Fernandez-Fernandez A, Manchanda R, McGoron AJ (2011) Theranostic applications of nanomaterials in cancer: drug delivery, image-guided therapy, and multifunctional platforms. Appl Biochem Biotechnol 165(7–8):1628–1651. https://doi.org/10.1007/s12010-011-9383-z

Ferrari M (2005) Cancer nanotechnology: opportunities and challenges. Nat Rev Cancer 5(3):161–171. https://doi.org/10.1038/nrc1566

Gadade DD, Jain N, Sareen R, Giram PS, Modi A (2023) Strategies for cancer targeting: novel drug delivery systems opportunities and future challenges. In: Malviya R, Sundram S (eds) Targeted cancer therapy in biomedical engineering. Biological and medical physics, biomedical engineering. Springer Nature Singapore, Singapore, pp 1–42. https://doi.org/10.1007/978-981-19-9786-0_1

Giosia MD, Alice S, Markus S, Andrea C, Fabio A, Nardella MI, Vincenzo M, Francesco V, Marco M, Francesco Z, Stefania R, Matteo C, Vasilis N (2021) A bio-conjugated fullerene as a subcellular-targeted and multifaceted phototheranostic agent. Adv Funct Mater 31(20):2101527. https://doi.org/10.1002/adfm.202101527

Giram PS, Garnaik B (2021) Evaluation of biocompatibility of synthesized low molecular weight PLGA copolymers using zinc L-proline through green route for biomedical application. Polym Adv Technol 32(11):4502–4515. https://doi.org/10.1002/pat.5452

Giram PS, Wang JT-W, Walters AA, Rade PP, Akhtar M, Han S, Faruqu FN, Abdel-Bar HM, Garnaik B, Al-Jamal KT (2021) Green synthesis of methoxy-poly(ethylene glycol)-*Block*-poly (L-lactide-*Co*-glycolide) copolymer using zinc proline as a biocompatible initiator for irinotecan delivery to colon cancer in vivo. Biomater Sci 9(3):795–806. https://doi.org/10.1039/D0BM01421D

Giram PS, Nimma R, Bulbule A, Yadav AS, Gorain M, Venkata Radharani NN, Kundu GC, Garnaik B (2024a) Poly(D,L-lactide-co-glycolide) surface-anchored biotin-loaded irinotecan nanoparticles for active targeting of colon cancer. ACS Omega:acsomega.3c07833. https://doi.org/10.1021/acsomega.3c07833

Giram PS, Nimma R, Bulbule A, Yadav AS, Gorain M, Venkata Radharani NN, Kundu GC, Garnaik B (2024b) Engineered PLGA core–lipid shell hybrid nanocarriers improve the efficacy and safety of irinotecan to combat colon cancer. ACS Biomater Sci Eng 10(10):6661–6676. https://doi.org/10.1021/acsbiomaterials.4c01260

Grilli F, Gohari PH, Zou S (2022) Characteristics of graphene oxide for gene transfection and controlled release in breast cancer cells. Int J Mol Sci 23(12):6802. https://doi.org/10.3390/ijms23126802

Guidance for Industry (n.d.). Considerations for the design of early-phase clinical trials of cellular and gene therapy products

Guo L, Shi H, Wu H, Zhang Y, Wang X, Wu D, An L, Yang S (2016) Prostate cancer targeted multifunctionalized graphene oxide for magnetic resonance imaging and drug delivery. Carbon 107:87–99. https://doi.org/10.1016/j.carbon.2016.05.054

Hu Y, Mignani S, Majoral J-P, Shen M, Shi X (2018) Construction of iron oxide nanoparticle-based hybrid platforms for tumor imaging and therapy. Chem Soc Rev 47(5):1874–1900. https://doi.org/10.1039/C7CS00657H

Huang Y, He S, Cao W, Cai K, Liang X-J (2012) Biomedical nanomaterials for imaging-guided cancer therapy. Nanoscale 4(20):6135. https://doi.org/10.1039/c2nr31715j

Jakhmola A, Anton N, Anton H, Messaddeq N, Hallouard F, Klymchenko A, Mely Y, Vandamme TF (2014) Poly-ε-caprolactone tungsten oxide nanoparticles as a contrast agent for X-ray computed tomography. Biomaterials 35(9):2981–2986. https://doi.org/10.1016/j.biomaterials.2013.12.032

Jiang Z, Zhang M, Li P, Wang Y, Fu Q (2023) Nanomaterial-based CT contrast agents and their applications in image-guided therapy. Theranostics 13(2):483–509. https://doi.org/10.7150/thno.79625

Jones H, Rowland-Yeo K (2013) Basic concepts in physiologically based pharmacokinetic modeling in drug discovery and development. CPT Pharmacomet Syst Pharmacol 2(8):1–12. https://doi.org/10.1038/psp.2013.41

Joseph MM, Nair JB, Adukkadan RN, Hari N, Pillai RK, Nair AJ, Maiti KK, Therakathinal T, S. (2017) Exploration of biogenic nano-chemobiotics fabricated by silver nanoparticle and galactoxyloglucan with an efficient biodistribution in solid tumor investigated by SERS fingerprinting. ACS Appl Mater Interfaces 9(23):19578–19590. https://doi.org/10.1021/acsami.7b03191

Jun SW, Manivasagan P, Kwon J, Nguyen VT, Mondal S, Ly CD, Lee J, Kang Y-H, Kim C-S, Oh J (2020) Folic acid-conjugated chitosan-functionalized graphene oxide for highly efficient photoacoustic imaging-guided tumor-targeted photothermal therapy. Int J Biol Macromol 155:961–971. https://doi.org/10.1016/j.ijbiomac.2019.11.055

Kamaly N, Yameen B, Wu J, Farokhzad OC (2016) Degradable controlled-release polymers and polymeric nanoparticles: mechanisms of controlling drug release. Chem Rev 116(4):2602–2663. https://doi.org/10.1021/acs.chemrev.5b00346

Kircher MF, Hricak H, Larson SM (2012) Molecular imaging for personalized cancer care. Mol Oncol 6(2):182–195. https://doi.org/10.1016/j.molonc.2012.02.005

Krishna V, Singh A, Sharma P, Iwakuma N, Wang Q, Zhang Q, Knapik J, Jiang H, Grobmyer SR, Koopman B, Moudgil B (2010) Polyhydroxy fullerenes for non-invasive cancer imaging and therapy. Small 6(20):2236–2241. https://doi.org/10.1002/smll.201000847

Lee EM, Lee J, Kim Y, Yi KS, Cho J, Kim J, An JM, Lee D, Kim SJ, An E, Hong YJ, Jo H, Lee SH, Jung Y, Choi C-H, Kang JS, Hur J, Kim D (2020) Hybrid composite of silver nanoparticle–porous silicon microparticles as an image-guided localization agent for computed tomography scan of the lungs. ACS Biomater Sci Eng 6(8):4390–4396. https://doi.org/10.1021/acsbiomaterials.0c00611

Liu H, Shen M, Zhao J, Zhu J, Xiao T, Cao X, Zhang G, Shi X (2013) Facile Formation of folic acid-modified dendrimer-stabilized gold–silver alloy nanoparticles for potential cellular computed tomography imaging applications. Analyst 138(7):1979–1987. https://doi.org/10.1039/C3AN36649A

Liu Y, Li J, Chen M, Chen X, Zheng N (2020) Palladium-based nanomaterials for cancer imaging and therapy. Theranostics 10(22):10057–10074. https://doi.org/10.7150/thno.45990

Lu S, Li X, Zhang J, Peng C, Shen M, Shi X (2018) Dendrimer-stabilized gold nanoflowers embedded with ultrasmall iron oxide nanoparticles for multimode imaging–guided combination therapy of tumors. Adv Sci 5(12):1801612. https://doi.org/10.1002/advs.201801612

Mauro N, Utzeri MA, Varvarà P, Cavallaro G (2021) Functionalization of metal and carbon nanoparticles with potential in cancer theranostics. Molecules 26(11):3085. https://doi.org/10.3390/molecules26113085

Pagar RR, Musale SR, Pawar G, Kulkarni D, Giram PS (2022) Comprehensive review on the degradation chemistry and toxicity studies of functional materials. ACS Biomater Sci Eng 8(6): 2161–2195. https://doi.org/10.1021/acsbiomaterials.1c01304

Pardeshi SR, Giram PS, More MP, Patil NB, Patil PB, Pardeshi CV, Deshmukh PK, Naik JB (2023) Gold nanoparticles: from synthesis to theranostic applications and clinical scenario. In: Pardeshi CV (ed) Nanomaterial-based drug delivery systems. Springer International Publishing, Cham, pp 269–294. https://doi.org/10.1007/978-3-031-30529-0_9

Patil RM, Shete PB, Giram PS, Somvanshi SB, Thorat ND (2023) In vivo mechanistic study of superparamagnetic materials. In: Thorat N, Sahu NK (eds) Superparamagnetic materials for cancer medicine. Nanomedicine and nanotoxicology. Springer International Publishing, Cham, pp 219–241. https://doi.org/10.1007/978-3-031-37287-2_11

Petros RA, DeSimone JM (2010) Strategies in the design of nanoparticles for therapeutic applications. Nat Rev Drug Discov 9(8):615–627. https://doi.org/10.1038/nrd2591

Prasad R, Jain NK, Yadav AS, Chauhan DS, Devrukhkar J, Kumawat MK, Shinde S, Gorain M, Thakor AS, Kundu GC, Conde J, Srivastava R (2020) Liposomal nanotheranostics for multimode targeted in vivo bioimaging and near-infrared light mediated cancer therapy. Commun Biol 3(1):1–14. https://doi.org/10.1038/s42003-020-1016-z

Rubio N, Hirvonen LM, Chong EZ, Wang JTW, Bourgognon M, Kafa H, Hassan HA, Al-Jamal WT, McCarthy D, Hogstrand C, Festy F, Al-Jamal KT (2015) Multiphoton luminescence imaging of chemically functionalized multi-walled carbon nanotubes in cells and solid tumors. Chem Commun 51(45):9366–9369. https://doi.org/10.1039/C5CC02675J

Sargazi S, Laraib U, Er S, Rahdar A, Hassanisaadi M, Zafar MN, Díez-Pascual AM, Bilal M (2022) Application of green gold nanoparticles in cancer therapy and diagnosis. Nano 12(7):1102. https://doi.org/10.3390/nano12071102

Siddique S, Chow JCL (2020) Application of nanomaterials in biomedical imaging and cancer therapy. Nano 10(9):1700. https://doi.org/10.3390/nano10091700

Sitharaman B, Jacobson BD, Wadghiri YZ, Bryant H, Frank J (2013) The magnetic, relaxometric, and optical properties of gadolinium-catalyzed single walled carbon nanotubes. J Appl Phys 113(13):134308. https://doi.org/10.1063/1.4796183

Wang Z, Kuang G, Yu Z, Li A, Zhou D, Huang Y (2019) Light-activatable dual prodrug polymer nanoparticle for precise synergistic chemotherapy guided by drug-mediated computed tomography imaging. Acta Biomater 94:459–468. https://doi.org/10.1016/j.actbio.2019.05.047

Xu H, Guo Z, Li M, Chaves HV, Pinto VDPT, Filho GC, Du M, Bezerra MM (2024) Copper-based nanomaterials for image-guided cancer therapy. BIO Integr 5(1). https://doi.org/10.15212/bioi-2024-0013

Zeng Z, Gao H, Chen C, Xiao L, Zhang K (2022) Bioresponsive nanomaterials: recent advances in cancer multimodal imaging and imaging-guided therapy. Front Chem 10:881812. https://doi.org/10.3389/fchem.2022.881812

Zerhouni E, Hamburg M (2016) The need for global regulatory harmonization: a public health imperative. Sci Transl Med 8(338):338ed6. https://doi.org/10.1126/scitranslmed.aaf1396

Zhang S, Sun C, Zeng J, Sun Q, Wang G, Wang Y, Wu Y, Dou S, Gao M, Li Z (2016) Ambient aqueous synthesis of ultrasmall PEGylated Cu2−Se nanoparticles as a multifunctional theranostic agent for multimodal imaging guided photothermal therapy of cancer. Adv Mater 28(40):8927–8936. https://doi.org/10.1002/adma.201602193

Zhao H, Chao Y, Liu J, Huang J, Pan J, Guo W, Wu J, Sheng M, Yang K, Wang J, Liu Z (2016) Polydopamine coated single-walled carbon nanotubes as a versatile platform with radionuclide labeling for multimodal tumor imaging and therapy. Theranostics 6(11):1833–1843. https://doi.org/10.7150/thno.16047

Multifacated Nanoimaging Agents and Preclinical/Clinical Examinations for Localized Theranostics

13

Navneet Kaur, Shweta Meena, and Ankush

Abstract

Nanoimaging agents have emerged as powerful tools for localized theranostics, including diagnostic imaging and targeted therapy. These nanoimaging agents are designed to deliver therapeutic payloads to targeted disease locations and improve contrast in imaging modalities such as MRI, PET, and CT. Hence, we can simultaneously diagnose, treat, and monitor disease progression. Recent preclinical studies have demonstrated the efficacy of multifunctional nanoparticles in applications such as cancer imaging and therapy. Several organic nanoparticles (NPs), inorganic NPs, and carbon-based NPs have been explored as localized cancer theranostics. They showed promising results with improved selectivity and sensitivity with fewer adverse effects. While many promising nanoimaging agents remain in preclinical development, several formulations have progressed to clinical trials or received regulatory approval. Clinical studies have highlighted both the potential benefits and challenges of translating nanoparticle-based theranostics into human use. Key areas of ongoing research include improving tissue specificity, enhancing detection sensitivity, and optimizing the safety profiles of nanoimaging agents. In this chapter, we will

N. Kaur (✉)
Department of Chemical and Biomedical Engineering, FAMU-FSU College of Engineering, Tallahassee, FL, USA

National High Magnetic Field Laboratory, FSU, Tallahassee, FL, USA
e-mail: Nk22u@fsu.edu

S. Meena
Department of Electronics and Communication Engineering, National Institute of Technology Kurukshetra, Kurukshetra, Haryana, India

Ankush
UNF Sustainable Solutions Laboratory, University of North Florida, Jacksonville, FL, USA

R. Prasad, M. Mohan Yallapu (eds.), *Advancements in Cancer Theranostics*,
https://doi.org/10.1007/978-981-95-0938-6_13

explore the development and application of multifaceted nanoimaging agents for both diagnostic imaging and targeted therapy. We will highlight the novel nanoparticle designs and surface modifications to extend circulation times, improve targeting, and enable multimodal imaging capabilities. Moreover, we will also outline the challenges and future directions for the clinical translation of multifaceted nanoimaging agents.

Keywords

Nanoimaging and therapy · Clinical · Nanotheranostics · Solid tumor · Nanoparticles and medicines · Targeted cancer therapy · Light-mediated therapy

13.1 Introduction

Multifaceted nanoimaging agents have emerged as a promising tool in the field of theranostics, offering a powerful combination of diagnostic and therapeutic capabilities within a single nanoplatform. These advanced nanoparticles are designed to improve personalized medicine by allowing accurate disease detection, targeted drug delivery, and real-time monitoring of treatment (Lee et al. 2012). These agents are engineered with a range of functionalities, including multimodal imaging capabilities that enable in-depth disease characterization across different spatial dimensions and resolution levels. For instance, a nanoparticle could provide contrast for magnetic resonance imaging (MRI), positron emission tomography (PET), and near-infrared fluorescence imaging (NIRFI), offering multiple views on disease progression and treatment. They can deliver therapeutic payloads specifically to affected areas, increasing treatment efficacy while reducing off-target side effects, especially beneficial in the field of cancer, where traditional chemotherapies often result in serious systemic toxicity (Ryu et al. 2012). Certain nanoparticles are designed to respond to external stimuli, such as light, to produce localized therapeutic effects. This includes photothermal therapy (PTT), where light exposure causes the nanoparticles to emit heat to destroy cancer cells, and photodynamic therapy (PDT), in which they release reactive oxygen species to initiate cell death. Several types of nanoimaging agents, such as magnetic nanoparticles, quantum dots, gold nanoparticles, and carbon-based nanostructures, have been explored for cancer treatment and have shown promising results in both preclinical and clinical studies. These agents enhance imaging capabilities, allowing for improved tumor detection, localization, and characterization. Their unique properties enable targeted drug delivery and real-time monitoring of therapeutic responses, significantly enhancing treatment efficacy (Ahmed et al. 2012). Also, studies have shown that these nanoimaging agents can minimize side effects by reducing the dosage of conventional therapies, ultimately contributing to more personalized and effective cancer treatment strategies. In this chapter, we will explore the synthesis and functionalization strategies of multifaceted nanoimaging agents, discussing their applications and efficacy across preclinical and clinical trials. It will also discuss the current challenges and future of localized theranostic applications, which have the potential to improve early diagnosis, treatment, and personalized care for patients.

13.2 Types of Nanoimaging Agents for Theranostics

There are the main types of nanoimaging agents for theranostics, each with unique properties that enable both imaging and targeted therapy, for more precise and effective treatment approaches (Lim et al. 2015) (Fig. 13.1).

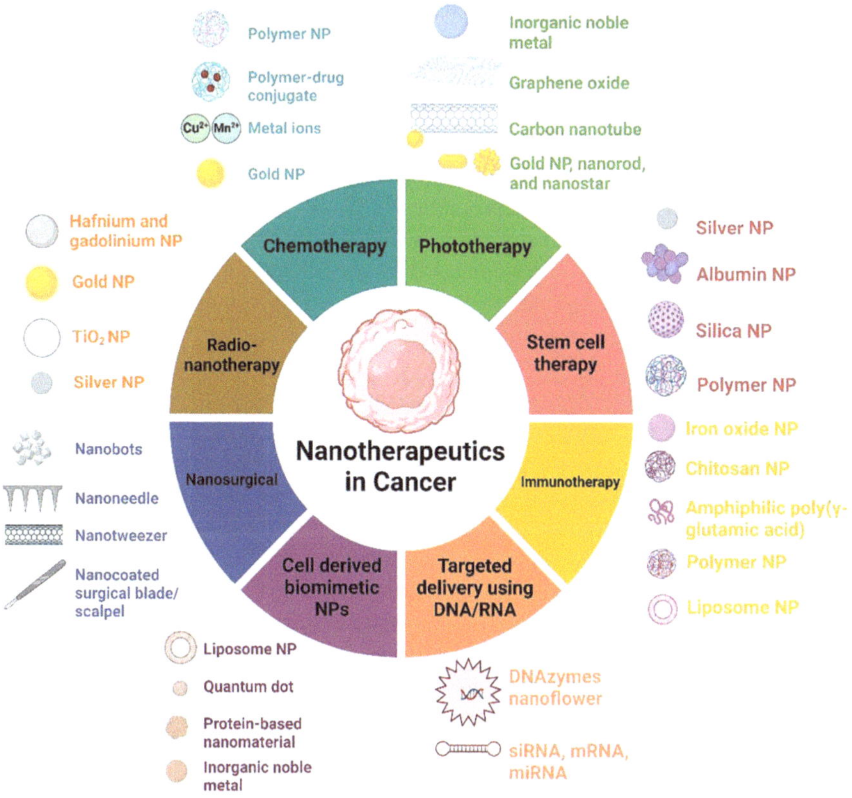

Fig. 13.1 Different types of nanoparticles and therapies for imaging and treatment of cancer. Created with biorender.com

13.2.1 Magnetic Nanoparticles (MNPs)

Magnetic nanoparticles (MNPs), including iron oxide nanoparticles like magnetite (Fe_3O_4) and maghemite (γ-Fe_2O_3), have emerged as tools in theranostics due to their unique magnetic properties and biocompatibility. These nanoparticles are particularly effective as contrast agents for Magnetic Resonance Imaging (MRI), where Superparamagnetic Iron Oxide Nanoparticles (SPIONs) are utilized to enhance imaging quality. They primarily influence T2 and T2* relaxation times, leading to negative contrast that aids in the visualization of tissues and organs, making them invaluable for detecting tumors and other abnormalities. Beyond imaging, MNPs can also serve as carriers for therapeutic agents. When an external magnetic field is applied, these nanoparticles can be directed to specific target sites within the body, facilitating localized drug delivery. This targeted approach is especially promising in cancer treatment, where drug-loaded MNPs can concentrate at tumor sites, enhancing therapeutic efficacy while minimizing systemic side effects. This adaptability, coupled with their responsiveness to magnetic fields, allows for targeted delivery and minimally invasive diagnostic and therapeutic applications, offering a promising alternative to conventional treatments. MNPs have been synthesized with different methods, including co-precipitation, thermal decomposition, and hydrothermal techniques. The size and surface properties of MNPs are crucial, as they directly influence cellular uptake, biodistribution, and clearance from the body. To enhance stability and functionalization potential, MNPs are often coated with biocompatible materials such as polyethylene glycol (PEG), dextran, or silica, which reduce aggregation and provide targeting sites for ligands, drugs, or imaging agents. A biocompatible copolymer, poly (oligo(ethylene glycol) methacrylate-co-methacrylic acid) (P(OEGMA-co-MAA)), was synthesized to coat superparamagnetic iron oxide nanoparticles. The nanoparticles, tunable in size (10–25 nm), demonstrated long-term colloidal stability and prolonged circulation time in vivo, enhancing magnetic resonance imaging applications (Lutz et al. 2006). Another study (Wang et al. 2006) developed a one-pot template-free method for synthesizing amine-functionalized magnetite nanoparticles and hollow nanospheres using $FeCl_3{\cdot}6H_2O$ as the sole iron source. These magnetic nanomaterials were characterized by various techniques and demonstrated potential applications in immunoassays and magnetic resonance imaging in live mice. Moreover, liposomes have been explored with magnetic particles as a theranostics agent. Maghemite (γ-Fe_2O_3) nanocrystals encapsulated within large unilamellar liposomes composed of egg phosphatidylcholine and DSPE-PEG (2000) were synthesized. These magnetic-fluid-loaded liposomes (MFLs) exhibited excellent physical stability, low toxicity, and superior T2 contrast agent properties in magnetic resonance imaging, demonstrating effective mobility under a magnetic field gradient. In vivo assessments confirmed their stealthy behavior and efficacy as contrast agents in mouse MR angiography (Martina et al. 2005). Targeting agents have been attached to these MNPs and were used for imaging and treatment. Human transferrin was covalently coupled to ultrasmall superparamagnetic iron oxide (USPIO) particles and evaluated in SMT/2A tumor-bearing rats. This formulation demonstrated a 40% reduction in tumor signal intensity 150 min postinjection, highlighting its potential for enhanced tumor detection via MR imaging compared to control USPIO formulations (Kresse et al. 1998). A nanoconjugate was developed as a targeted MRI contrast agent for detecting cancer cells that overexpress the folate receptor. Synthesized by coating superparamagnetic iron oxide nanoparticles with bifunctional PEG and conjugating with folic acid, it demonstrated a 12-fold higher uptake in HeLa adenocarcinoma cells compared to PEG- or dextran-coated nanoparticles. This specificity was confirmed through MR phantom imaging and a competitive inhibition assay, highlighting its potential for precise cancer detection (Sun et al. 2006). Furthermore, MNPs were also studied for clinical studies. In a pilot study with 15 patients, superparamagnetic iron oxide (ferrite) particles were administered intravenously at doses of 10–50 μmol/kg as a contrast agent for magnetic resonance (MR) imaging. Ferrite-enhanced images significantly improved hepatic lesion detection and reduced the threshold size to 3 mm ($P < 0.01$). These findings suggest ferrite could enhance abdominal MR imaging performance without chronic toxicity in humans, consistent with preclinical data (Stark et al. 1988). Another study investigated the targeting capabilities of multifunctional nanoparticles for cancer treatment, comparing passive targeting through the EPR effect, active targeting of $\alpha v\beta 3$ integrin via RGD grafting, magnetic targeting, and their combination. Using PLGA-based nanoparticles loaded with paclitaxel and superparamagnetic iron oxides, results showed that combined active and magnetic targeting significantly enhanced nanoparticle accumulation in tumors (eightfold increase), improved MRI contrast, and increased median survival from 13 to 22 days, highlighting the potential of dual-targeting strategies for personalized cancer therapy (Schleich et al. 2014).

13.2.2 Gold Nanoparticles (AuNPs)

Gold nanoparticles are highly versatile theranostic agents known for their unique optical properties, including surface plasmon resonance, where AuNPs absorb light and convert it into heat (Khoury and Vo-Dinh 2008). This capability allows them to accumulate in tumors and generate localized heat upon exposure to near-infrared light, effectively destroying cancer cells (Vigderman and Zubarev 2013). Also, AuNPs serve as excellent contrast agents for various imaging modalities, including computed tomography (CT) and photoacoustic imaging, due to their strong light scattering properties (Hauck and Chan 2007). Their biocompatibility makes them suitable for use in biological systems, and they can be easily functionalized with targeting ligands, drugs, or imaging agents. This versatility enables the design of multifunctional theranostic platforms that integrate both diagnostic and therapeutic functions, enhancing the effectiveness of cancer treatments. Various types of AuNPs have been used for theranostics, including spherical nanoparticles, gold nanorods, nanoshells, nanostars, and nanocages (Hauck and Chan 2007; Shanavas et al. 2018, 2019; Kaur et al. 2023; Liu et al. 2015; Coughlin et al. 2013; Choi et al. 2012; Tong et al. 2009). Cetuximab-conjugated polyethylene glycol (PEG)ylated gold nanorods (CET-PGNRs) were developed to target epithelial cancer cells using near-infrared light. Biocompatibility and specificity of nanoprobes were improved by replacing the cetyltrimethylammonium bromide bilayer on GNRs with heterobifunctional PEG. Functionalized with cetuximab through EDC-NHS chemistry, CET-PGNRs demonstrated excellent tumor targeting, optical properties, and strong potential for simultaneous absorption imaging and photothermal therapy of cancer cells (Choi et al. 2012). Another study presents a modification of GNPs with the transactivator of transcription (TAT) peptide (TAT-Au NP) that effectively crosses the BBB to deliver the anticancer drug doxorubicin (Dox) and Gd3+ contrast agents to brain tumors. Mice treated with pH-sensitive Dox-conjugated TAT-Au NPs showed significantly improved survival compared to free Dox, while TAT-Au NPs also enhanced brain tumor imaging with prolonged retention of Gd3+ chelates, highlighting their potential for improved therapy and noninvasive imaging (Cheng et al. 2014). Further, the combination of CT and photoacoustic (PA) imaging offered high resolution and 3D visualization for precise tissue localization, enhancing photothermal therapy (PTT) guidance. The study introduced gold nanoparticles (AuNPs) coated with Prussian blue (PB), which improved photostability and made them effective photoabsorbing agents for both PTT and PA imaging. The core/shell Au@PB nanoparticles provided significant CT contrast and enabled complete ablation of 100 mm^3 tumors in mice after NIR laser treatment, showcasing their potential for integrated cancer diagnosis and noninvasive therapy (Jing et al. 2014). Another work developed gadolinium-conjugated gold nanoshells that improved contrast in various imaging modalities and effectively converted near-infrared light to heat for cancer cell ablation, demonstrating their potential for image-guided photothermal therapy (Coughlin et al. 2013). Moreover, gold nanostars (GNS) have promising results for in vivo and in vitro studies for imaging and treatment. Gold nanostar was developed for multimodality theranostics, including SERS detection, CT imaging, two-photon luminescence (TPL) imaging, and photothermal therapy (PTT). The 30-nm GNS demonstrated superior tumor uptake and deeper interstitial penetration compared to larger particles, with effective photothermal ablation of tumors in vivo using a near-infrared laser, highlighting its promise for biosensing and image-guided therapy (Liu et al. 2015). Another study demonstrated the gold nanostars-based nanocomposites (MGSNs) through organosilica coating and Gd chelate conjugation. The MGSNs exhibited strong NIR absorbance, enhanced T1-MR imaging capability, and excellent stability, enabling effective in vivo MR and SERS imaging. With high photothermal ablation efficiency against tumor cells upon NIR irradiation, MGSNs demonstrated significant potential as a bimodal imaging-guided platform for focused photothermal tumor therapy (Gao et al. 2015).

13.2.3 Quantum Dots (QDs)

Quantum dots are semiconductor nanocrystals that act as fluorescent imaging agents due to their size-dependent optical properties. These properties allow QDs to exhibit tunable emission wavelengths based on their size, enabling the creation of a diverse array of colors from a single excitation source (Ji et al. 2014). This characteristic facilitates multiplexed imaging, which helps to visualize multiple targets simultaneously within biological systems. Furthermore, QDs can be functionalized with specific targeting ligands to enhance their specificity for tissues or cell types, improving both imaging contrast and therapeutic delivery (Karakoti et al. 2015). In addition to their imaging capabilities, quantum dots act as photosensitizers for photodynamic therapy, thereby providing a dual function that combines diagnosis and treatment in a single nanoplatform (Volkov 2015). Gadolinium-doped carbon dots (Gd-doped CDs) were developed to enhance MRI-guided radiotherapy. Gd-doped CDs exhibited high water dispersibility, stable photoluminescence, and excellent biocompatibility, with a relaxivity rate of 6.45 mM^{-1} s^{-1}. They showed prolonged circulation, efficient tumor targeting, and rapid renal clearance, making them promising T1 contrast agents and radiosensitizers for precise MRI-guided cancer therapy (Du et al. 2017). Besides radiotherapy, they have been used for photothermal and photodynamic therapy. Another study developed poly(acrylic acid)-coated $Cu_2(OH)PO_4$ quantum dots ($Cu_2(OH)PO_4$@PAA QDs) with strong near-infrared absorption, enabling effective photothermal and photodynamic therapies upon NIR irradiation. Tested in vitro on cervical cancer cells and in vivo on mice, $Cu_2(OH)PO_4$@PAA QDs showed significant tumor ablation and served as a photoacoustic imaging contrast, making them a promising multifunctional theranostic platform (Guo et al. 2017). Another study explored co-loaded quantum dots (CdTe/CdS/ZnS) and paclitaxel in nanostructured lipid carriers (NLC) for multifunctional cancer therapy. The co-loaded NLCs, with high encapsulation efficacy (80.7%), demonstrated effective tumor targeting and inhibition, along with imaging abilities for H22 tumors in vivo and ex vivo (Olerile et al. 2017). Magnetofluorescent carbon quantum dots (GdNS@CQDs) synthesized via a one-pot microwave method and conjugated with folic acid (FA-GdNS@CQDs) showed high stability, strong fluorescence, and magnetic properties, making them effective for dual fluorescence/MR imaging. Also, FA-GdNS@CQDs loaded with doxorubicin (FA-GdNS@CQDs-DOX) demonstrated targeted, pH-sensitive drug release, with high uptake and localization in cancer cells (Chiu et al. 2016).

13.2.4 Lipid-Based Nanoparticles

Lipid-based nanoparticles (LNPs) have emerged as a promising platform for theranostic applications, particularly in cancer treatment and imaging. LNPs are biocompatible, biodegradable, and can target cancer cells. They can incorporate contrast agents for MRI, PET, or optical imaging, while simultaneously delivering drugs or gene therapies to target sites. This dual functionality allows for image-guided drug delivery, stimuli-responsive treatment, and noninvasive monitoring of therapeutic efficacy. Various types of lipid-based nanoparticles, including liposomes, solid lipid nanoparticles, and nanostructured lipid carriers, have been explored as cancer theranostics. Tc-99 m radiolabeled, folate-targeted liposomes encapsulating paclitaxel and vinorelbine were developed, demonstrating promising efficacy through high cellular uptake, enhanced cytotoxicity, and effective tumor inhibition in vivo, with reduced toxicity and prevention of lung metastasis in NSCLC-bearing mice (Karpuz et al. 2021). Another study developed cationic liposomes encapsulated with imatinib and used as theranostic agents for breast cancer. The neutral liposomes demonstrated high cellular binding, controlled drug release, and over 90% radiolabeling efficiency. While cationic liposomes exhibited high nonspecific cytotoxicity, the neutral, 68Ga-radiolabeled, imatinib-loaded formulation showed dose-dependent cytotoxicity, served as a promising candidate for targeted breast cancer therapy (Karpuz et al. 2024). Doxorubicin-loaded LTSLs combined with MR-HIFU significantly increased tumor drug concentration and distribution in rabbit Vx2 tumors compared to free doxorubicin. This image-guided, noninvasive method demonstrated promise for precise drug delivery in solid tumors (Ranjan et al. 2011).To enhance tumor-targeted accumulation of near-infrared (NIR) theranostic agents for effective phototherapy, IR-780 iodide-loaded solid lipid nanoparticles (SLNs) conjugated with c(RGDyK) were developed. These cRGD-IR-780 SLNs demonstrated improved stability, monodispersity, and selective targeting of αvβ3 integrin-overexpressing cells. In vitro and in vivo studies showed effective tumor eradication in U87MG models under laser irradiation, highlighting cRGD-IR-780 SLNs as a promising imaging-guided, targeted PTT agent for cancer therapy (Kuang et al. 2017a). Further, the study evaluated the release of doxorubicin and [Gd (HPDO3A)(H_2O)] from low temperature-sensitive liposomes (LTSL) and traditional temperature-sensitive liposomes (TTSL). The LTSL system demonstrated greater doxorubicin leakage at 37 °C and a faster release at 42 °C, allowing for MRI imaging of the drug release near the melting phase transition temperature (de Smet et al. 2010). Moreover, solid lipid nanoparticles have been explored as theranostics. LDL-mimetic solid lipid nanoparticles (SLNs) incorporating quantum dots were developed for traceable anticancer theranostics, using a paclitaxel-siRNA combination for synergistic effects. These SLN/siRNA complexes showed efficient delivery and anticancer activity in lung carcinoma cells by inducing apoptosis, while quantum dots enabled visualization of intracellular SLN translocation (Olerile 2020). To enhance tumor accumulation of near-infrared (NIR) theranostic agents for effective phototherapy, cRGD-conjugated solid lipid nanoparticles (SLNs) were developed to deliver IR-780 iodide, an imaging-guided PTT agent with superior stability over traditional dyes. These cRGD-IR-780 SLNs displayed high tumor-targeting efficiency, especially in cells with αvβ3 integrin, and successfully eradicated U87MG tumors with laser irradiation, showing promise as a targeted NIR-guided phototherapy agent for cancer (Kuang et al. 2017b).

13.2.5 Polymeric Nanoparticles

Polymeric nanoparticles have emerged as promising theranostic platforms, typically consisting of a polymer core or shell that encapsulates therapeutic agents, along with imaging agents for diagnostics and targeting ligands for improved tumor accumulation. They offer several advantages, including biocompatibility, high drug loading capacity, controlled release, and multimodal imaging capabilities. Polymeric nanoparticles can deliver various therapeutics, including chemotherapy drugs and nucleic acids, while allowing real-time monitoring of their distribution and efficacy through imaging techniques like MRI or fluorescence imaging. Multifunctional gold-coated PLGA nanoshells (AuPLGA-NS) were synthesized using a two-step method to serve both therapeutic and imaging functions. These biodegradable nanoparticles showed 80% cancer cell death through photothermal therapy and provided X-ray/CT contrast comparable to iodine, demonstrating their potential as a simple, effective nanosystem for cancer theranostics (Shanavas et al. 2018). DOX–magnetic PLGA nanoparticles conjugated with antibodies were developed for breast cancer detection and therapy, showing strong cancer cell affinity and MRI ultrasensitivity. These nanoparticles sustained drug release for three weeks, with magnetic cores not affecting formulation or release kinetics (Yang et al. 2007). Furthermore, polymer vesicles were widely used for imaging and drug delivery. For instance, A stable, pH-responsive polymer vesicle system was designed for targeted cancer therapy and imaging, delivering both doxorubicin (DOX) and superparamagnetic iron oxide (SPIO) nanoparticles. The vesicles featured folate for tumor targeting, pH-sensitive DOX release, and encapsulated SPIO for enhanced MRI contrast. This multifunctional system demonstrated high cellular uptake, increased cytotoxicity, and superior MRI performance compared to commercial SPIO agents (Yang et al. 2010a). Multifunctional wormlike polymer vesicles, loaded with superparamagnetic iron oxide (SPIO) nanoparticles for MRI contrast and doxorubicin (DOX) for targeted therapy, were developed for cancer theranostics. The folate-conjugated vesicles showed enhanced cellular uptake and cytotoxicity against tumor cells, while crosslinked inner PEG layers provided stability and improved MRI sensitivity over commercial agents. This system enables targeted cancer therapy and imaging, supporting advancements in medicine (Yang et al. 2010b). Another study presented a novel tumor-targeting polymersome system for MRI and chemotherapy. The doxorubicin-loaded, SPION-enhanced polymersomes, with chitosan and folate-functionalized nanogel layers, provided in vivo stability, pH-responsive drug release, and active targeting. Upon internalization by HeLa cells and exposure to magnetic fields, they achieved enhanced cytotoxicity via pH and hyperthermia-triggered DOX release. The SPION clustering effect resulted in superior MRI contrast, demonstrating their potential as an advanced cancer theranostic platform (Chiang et al. 2013). Hydrophobically modified maghemite (γ-Fe_2O_3) nanoparticles were encapsulated within poly (trimethylene carbonate)-b-poly (L-glutamic acid) (PTMC-b-PGA) vesicles using a nanoprecipitation method, achieving up to 70 wt% loading while controlling vesicle size (100 – 400 nm). Doxorubicin was also loaded simultaneously, and small-angle neutron scattering confirmed membrane deformation under a magnetic field. These superparamagnetic hybrid self-assemblies showed enhanced MRI contrast and enabled controlled drug release through radiofrequency magnetic hyperthermia, demonstrating their potential as multifunctional nanocarriers for combined therapy and imaging (Sanson et al. 2011).

13.3 Mechanisms of Localization and Targeting

Targeting with nanoparticles involves unique properties of nanoscale materials to enhance therapeutic efficacy and reduce side effects in drug delivery. This strategy employs various mechanisms to direct nanoparticles to specific sites in the body, allowing for more precise and controlled drug delivery. Targeting approaches, including active and passive targeting, have been employed based on the distinct characteristics of nanoparticles, such as their size, surface properties, and functionalization.

13.3.1 Passive Targeting

Passive targeting in nanoparticle-based drug delivery systems is a strategy that exploits the unique physiological characteristics of tumors to enhance therapeutic efficacy without relying on specific molecular interactions (Bazak et al. 2014). This approach depends on the abnormal vasculature and impaired lymphatic drainage commonly found in solid tumors, allowing nanoparticles to accumulate preferentially in these diseased tissues. The mechanism of passive targeting lies in the Enhanced Permeability and Retention (EPR) effect. (Mitchell et al. 2020)Tumor blood vessels, formed through rapid and disorganized angiogenesis, exhibit several structural abnormalities such as larger endothelial fenestrations, irregular basement membrane, lack of smooth muscle layer, and absence of functional pericytes. These features result in increased vascular permeability, allowing nanoparticles to extravasate more readily into the tumor interstitium (Alavi and Hamidi 2019a). Additionally, the poor lymphatic drainage in tumors contributes to the prolonged retention of nanoparticles within the tumor microenvironment. The effectiveness of passive targeting depends heavily on the physicochemical properties of the nanoparticles. The optimal size range is typically 20–200 nm, as particles smaller than 10 nm are rapidly cleared by renal filtration, while those larger than 200 nm are more likely to be recognized and cleared by the mononuclear phagocyte system (MPS). Neutral or slightly negative surface charges are preferred to minimize nonspecific interactions and prolong circulation time. Nonspherical particles, such as rod-shaped or disk-shaped nanoparticles, have shown improved tumor accumulation in some studies due to their unique flow dynamics and cellular internalization patterns. Surface modification, such as hydrophilic coatings like polyethylene glycol (PEG), reduces opsonization and clearance by the MPS (Alavi and Hamidi 2019b). While passive targeting has shown promise, several challenges remain. These include heterogeneity in the EPR effect across different tumor types and even within the same tumor, high interstitial fluid pressure in tumors impeding nanoparticle penetration, dense extracellular matrix acting as a physical barrier, and limited efficacy in metastatic lesions or poorly vascularized tumors. To address these limitations, researchers are exploring various approaches to enhance passive targeting. Strategies to improve passive targeting include vascular normalization using agents like bevacizumab to temporarily normalize tumor vasculature, potentially improving nanoparticle delivery. Modulation of the tumor microenvironment using enzymes like collagenase or hyaluronidase to degrade the extracellular matrix can enhance nanoparticle penetration (Rosenblum et al. 2018). The development of stimuli-responsive nanoparticles that respond to tumor-specific stimuli (e.g., pH, enzymes) for improved drug release and penetration is another promising avenue. Additionally, combining passive targeting with active targeting by incorporating tumor-specific ligands can enhance cellular uptake after passive accumulation (Fig. 13.2). In conclusion, passive targeting is a key strategy in nanomedicine for cancer therapy, with ongoing research in nanoparticle design and tumor biology promising more effective, targeted treatments and improved patient outcomes.

13.3.2 Active Targeting

Active targeting is another mechanism of nanoparticle delivery, and it has revolutionized cancer treatment strategies by enhancing the precision of drug delivery systems. This approach involves attaching specific molecules to drug carriers or therapeutic agents, enabling them to bind selectively to biomarkers found on tumor cells or within the tumor microenvironment, with targeted and effective treatment that minimizes damage to healthy tissues. Antibodies have been explored for targeting due to their specificity and binding affinity. Antibody-drug conjugates represent a significant advancement in this field, combining the targeting capabilities of antibodies with potent cytotoxic agents. These conjugates work by binding to specific cell surface receptors, triggering internalization, and releasing their payload within cancer cells. Nanoparticle-based tumor targeting with ultrasmall silica nanoparticles functionalized with anti-HER2 single-chain variable fragments was studied. These nanoparticles provide high tumor-targeting efficiency with favorable biodistribution and efficient renal clearance, offering potential for both tumor imaging and targeted drug delivery (Chen et al. 2018). Another approach is peptide targeting, which has emerged as a promising option (Accardo et al. 2013). Peptides offer several advantages over antibodies, including their smaller size, lower production costs, and reduced immunogenicity. Peptide-drug conjugates can be designed to target specific receptors overexpressed on cancer cells (Sharma et al. 2022). It has been reported that a simple synthetic strategy for preparing high-purity, stable water-soluble peptide-stabilized gold nanoparticles (AuNPs) using reduced glutathione (GSH) as a synthon for peptide growth on the AuNP surface was used. Ultracentrifugation purification and control of peptide length, charge, and hydrophobicity enabled the development of tailored, biocompatible AuNPs with minimal aggregation, making them suitable for biomedical applications (Wu et al. 2014). Another study introduced a polydopamine (PDA)-based surface modification method to develop targeted doxorubicin-loaded mesoporous silica nanoparticles (DOX-loaded MSNs@PDA-PEP) for bladder cancer therapy. The nanoparticles showed controlled drug release, enhanced cellular uptake via ligand-receptor recognition, and superior antitumor effects in vitro and in vivo compared to free DOX, making them a promising targeted drug delivery system for bladder cancer (Wei

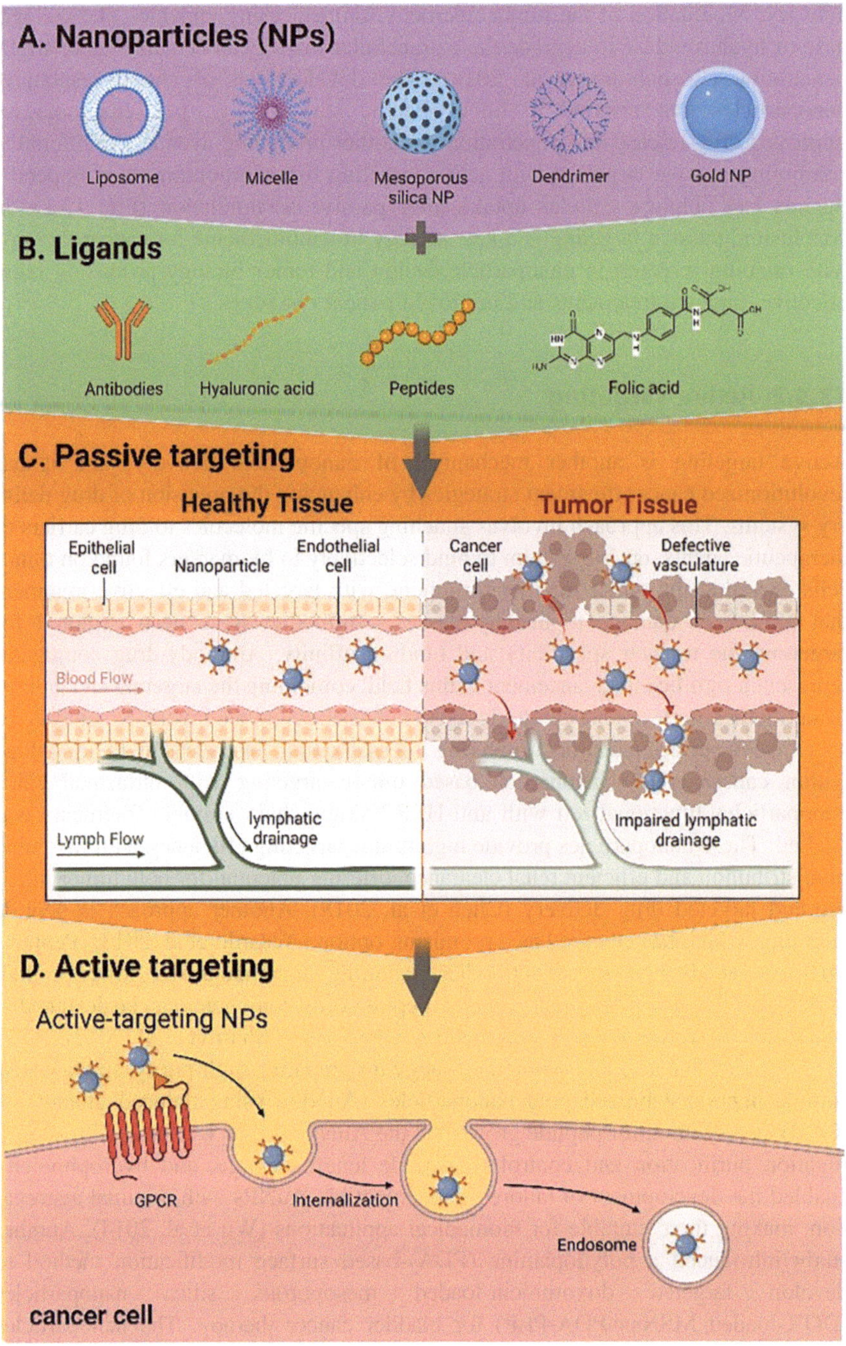

Fig. 13.2 Nanoparticle drug delivery system for targeted cancer therapy. Created with biorender.com

et al. 2017). Small molecules, aptamers, and various proteins have also been explored for their targeting potential. Folic acid has been extensively studied due to its affinity for folate receptors, which are overexpressed in many cancer types. The study explored cellulose nanocrystals (CNCs) as nanocarriers for targeted intracellular delivery, conjugating them with folic acid for specific binding to folate receptors on cancer cells. The CNCs exhibited significant folate-receptor-mediated uptake in KB and MDA-MB-468 cells, with clathrin-mediated endocytosis as the main internalization mechanism, demonstrating their potential for targeted cancer imaging and therapy (Bittleman et al. 2018). Aptamers, short DNA or RNA sequences, offer another interesting approach, as they can be selected to bind specific targets with high affinity and specificity (Sun and Zu 2015; Mahmoudian et al. 2024). The study introduced a smart imaging nanoprobe combining a magnetic nanocrystal with an aptamer targeting VEGFR2, overexpressed on angiogenic vasculature in glioblastoma. The aptamer-conjugated nanoprobe showed high MR imaging sensitivity and targeted VEGFR2 effectively in vivo, demonstrating potential for nontoxic, specific detection of glioblastoma angiogenesis (Kim et al. 2013). Factors including binding affinity, specificity, stability, and the ability to trigger internalization are considered while designing these systems (Mahmoudian et al. 2024). The choice of linker between the ligand and the drug is also crucial, as it affects the stability of the conjugate in circulation and controls drug release at the target site. The field of active targeting via ligand conjugation is composed for further advancements, promising more effective and less toxic cancer therapies in the future.

13.4 Preclinical and Clinical Translation of Nanoimaging Agents

Nanoparticle-based imaging agents have shown tremendous potential for imaging and treatment. However, their clinical translation has been slower compared to nanoparticle therapeutics. The preclinical development of nanoimaging agents focuses on creating biocompatible formulations with minimal organ accumulation and favorable excretion profiles. These agents are designed with tunable properties to enable multimodal imaging across different modalities and resolutions, and often incorporate multifunctionality for theranostic applications that combine imaging and therapy. Preclinical studies have demonstrated improved signal sensitivity and specificity in animal models, with nanoimaging agents showing the ability to report on molecular events, cell migration, and functional/anatomic consequences of cancer. However, the transition from preclinical success to clinical application has been challenging. To date, only a few nanoimaging agents have received regulatory approval. The first was 99mTc sulfur nanocolloid, approved in 1974 for lymphoscintigraphy, followed by several modifications of 99mTc nanocolloids for applications like bone marrow scintigraphy and lymph node mapping. Sulfur colloids are the most common carrier platform in the USA, used extensively for sentinel lymph node mapping in cancers like breast cancer and melanoma. After subcutaneous injection, these nanoparticles navigate the lymphatic system to identify sentinel lymph nodes, guiding surgeons in targeted removal, thereby reducing the risk of complications like lymphedema. In addition to lymph node mapping, 99mTc-sulfur colloids are administered intravenously to evaluate organs within the mononuclear phagocyte system, such as the liver, spleen, and bone marrow, where they accumulate due to their short circulation half-life. These nanoparticles are also used intraperitoneally for assessing diaphragmatic injury and shunt patency. Beyond sulfur colloids, stannous fluoride (99mTc-SnF2) colloids are used to label white blood cells, enabling the detection of infections and inflammatory conditions via SPECT imaging (Thakor et al. 2016).Currently, several nanoimaging agents are undergoing clinical trials, including gold shell-silica nanoparticles for photothermal ablation of prostate tumors, silica nanoparticles encapsulating fluorescent dyes for improved tumor detection and surgical guidance, and ferumoxytol (iron oxide nanoparticles) used off-label as an MRI contrast agent (Yao et al. 2023). The clinical translation of nanoimaging agents faces numerous challenges. Their complex design can lead to safety concerns, such as potential toxicity and immune reactions, making it difficult to ensure they are safe for patients. Producing these nanoparticles consistently on a large scale is another challenge, as small changes in their size or coating can affect how they work, leading to unreliable results. Additionally, nanoparticles often behave differently in humans than in animal studies, making it hard to predict their effectiveness and safety (Xu et al. 2023). Regulatory and commercial factors also play a role, with rigorous FDA safety and efficacy requirements for approval and less commercial incentive compared to nanotherapeutics. Despite these challenges, the outlook for nanoimaging agents remains promising. Key areas for advancement include improving nanoparticle design for optimal biodistribution and clearance, developing robust manufacturing processes, conducting thorough long-term safety studies, exploring novel multimodal and theranostic applications, and focusing on clinical needs to drive its translation.

13.5 Conclusion

In summary, nanoimaging agents offer significant potential in theranostics by enabling simultaneous disease diagnosis, treatment, and monitoring. These nanoparticles enhance imaging sensitivity and targeted drug delivery, improving outcomes with fewer side effects. While many agents show promise in preclinical studies, translating them into clinical use remains challenging due to safety and targeting limitations. Ongoing research focuses on optimizing nanoparticle designs to extend circulation, improve targeting, and enable multimodal imaging. Continued advancements in this field could transform diagnostic imaging and targeted therapies, paving the way for more precise and personalized healthcare.

References

Accardo A, Tesauro D, Morelli G (2013) Peptide-based targeting strategies for simultaneous imaging and therapy with nanovectors. Polym J 45(5):481–493. https://doi.org/10.1038/pj.2012.215

Ahmed N, Fessi H, Elaissari A (2012) Theranostic applications of nanoparticles in cancer. Drug Discov Today 17(17–18):928–934. https://doi.org/10.1016/J.DRUDIS.2012.03.010

Alavi M, Hamidi M (2019a) Passive and active targeting in cancer therapy by liposomes and lipid nanoparticles. Drug Metab Pers Ther 34(1). https://doi.org/10.1515/DMPT-2018-0032

Alavi M, Hamidi M (2019b) Passive and active targeting in cancer therapy by liposomes and lipid nanoparticles. Drug Metab Pers Ther 34(1). https://doi.org/10.1515/DMPT-2018-0032/ASSET/GRAPHIC/J_DMPT-2018-0032_FIG_005.JPG

Bazak R, Houri M, El Achy S, Hussein W, Refaat T (2014) Passive targeting of nanoparticles to cancer: a comprehensive review of the literature. Mol Clin Oncol 2(6):904. https://doi.org/10.3892/MCO.2014.356

Bittleman KR, Dong S, Roman M, Lee YW (2018) Folic acid-conjugated cellulose nanocrystals show high folate-receptor binding affinity and uptake by KB and breast cancer cells. ACS Omega 3(10):13952–13959. https://doi.org/10.1021/ACSOMEGA.8B01619/ASSET/IMAGES/LARGE/AO-2018-01619Y_0005.JPEG

Chen F, Ma K, Madajewski B, Zhuang L, Zhang L, Rickert K, Marelli M, Yoo B, Turker MZ, Overholtzer M, Quinn TP, Gonen M, Zanzonico P, Tuesca A, Bowen MA, Norton L, Subramony JA, Wiesner U, Bradbury MS (2018) Ultrasmall targeted nanoparticles with engineered antibody fragments for imaging detection of HER2-overexpressing breast cancer. Nat Commun 9(1):1–11. https://doi.org/10.1038/s41467-018-06271-5

Cheng Y, Dai Q, Morshed RA, Fan X, Wegscheid ML, Wainwright DA, Han Y, Zhang L, Auffinger B, Tobias AL, Rincón E, Thaci B, Ahmed AU, Warnke PC, He C, Lesniak MS (2014) Blood-brain barrier permeable gold nanoparticles: an efficient delivery platform for enhanced malignant glioma therapy and imaging. Small 10(24):5137. https://doi.org/10.1002/SMLL.201400654

Chiang WH, Huang WC, Chang CW, Shen MY, Shih ZF, Huang YF, Lin SC, Chiu HC (2013) Functionalized polymersomes with outlayered polyelectrolyte gels for potential tumor-targeted delivery of multimodal therapies and MR imaging. J Control Release 168(3):280–288. https://doi.org/10.1016/J.JCONREL.2013.03.029

Chiu SH, Gedda G, Girma WM, Chen JK, Ling YC, Ghule AV, Ou KL, Chang JY (2016) Rapid fabrication of carbon quantum dots as multifunctional nanovehicles for dual-modal targeted imaging and chemotherapy. Acta Biomater 46:151–164. https://doi.org/10.1016/J.ACTBIO.2016.09.027

Choi J, Yang J, Bang D, Park J, Suh JS, Huh YM, Haam S (2012) Targetable gold nanorods for epithelial cancer therapy guided by near-IR absorption imaging. Small 8(5):746–753. https://doi.org/10.1002/SMLL.201101789

Coughlin AJ, Ananta JS, Deng N, Larina IV, Decuzzi P, West JL (2013) Gadolinium-conjugated gold nanoshells for multimodal diagnostic imaging and photothermal cancer therapy. Small 10(3):556. https://doi.org/10.1002/SMLL.201302217

de Smet M, Langereis S, van den Bosch S, Grüll H (2010) Temperature-sensitive liposomes for doxorubicin delivery under MRI guidance. J Control Rel 143(1):120–127. https://doi.org/10.1016/J.JCONREL.2009.12.002

Du F, Zhang M, Gong A, Tan Y, Miao J, Gong Y, Zou S, Zhang L, Zhang L, Wu C, Sun M, Ju H (2017) Engineered gadolinium-doped carbon dots for magnetic resonance imaging-guided radiotherapy of tumors. Biomaterials 121:109–120. https://doi.org/10.1016/J.BIOMATERIALS.2016.07.008

Gao Y, Li Y, Chen J, Zhu S, Liu X, Zhou L, Shi P, Niu D, Gu J, Shi J (2015) Multifunctional gold nanostar-based nanocomposite: synthesis and application for noninvasive MR-SERS imaging-guided photothermal ablation. Biomaterials 60:31–41. https://doi.org/10.1016/J.BIOMATERIALS.2015.05.004

Guo W, Qiu Z, Guo C, Ding D, Li T, Wang F, Sun J, Zheng N, Liu S (2017) Multifunctional theranostic agent of Cu2(OH)PO4 quantum dots for photoacoustic image-guided photothermal/ photodynamic combination cancer therapy. ACS Appl Mater Interfaces 9(11):9348–9358. https://doi.org/10.1021/ACSAMI.6B15703

Hauck TS, Chan WCW (2007) Gold nanoshells in cancer imaging and therapy: towards clinical application. Nanomedicine (Lond) 2(5):735–738. https://doi.org/10.2217/17435889.2.5.735

Ji X, Peng F, Zhong Y, Su Y, He Y (2014) Fluorescent quantum dots: synthesis, biomedical optical imaging, and biosafety assessment. Colloids Surf B Biointerfaces 124:132–139. https://doi.org/10.1016/J.COLSURFB.2014.08.036

Jing L, Liang X, Deng Z, Feng S, Li X, Huang M, Li C, Dai Z (2014) Prussian blue coated gold nanoparticles for simultaneous photoacoustic/CT bimodal imaging and photothermal ablation of cancer. Biomaterials 35(22):5814–5821. https://doi.org/10.1016/J.BIOMATERIALS.2014.04.005

Karakoti AS, Shukla R, Shanker R, Singh S (2015) Surface functionalization of quantum dots for biological applications. Adv Colloid Interf Sci 215:28–45. https://doi.org/10.1016/J.CIS.2014.11.004

Karpuz M, Silindir-Gunay M, Ozer AY, Ozturk SC, Yanik H, Tuncel M, Aydin C, Esendagli G (2021) Diagnostic and therapeutic evaluation of folate-targeted paclitaxel and vinorelbine encapsulating theranostic liposomes for non-small cell lung cancer. Eur J Pharm Sci 156: 105576. https://doi.org/10.1016/J.EJPS.2020.105576

Karpuz M, Ozgenc E, Oner E, Atlihan-Gundogdu E, Burak Z (2024) 68Ga-Labeled, imatinib encapsulated, theranostic liposomes: formulation, characterization, and in vitro evaluation of anticancer activity. Drug Dev Res 85(1):e22136. https://doi.org/10.1002/DDR.22136

Kaur G, Kaur V, Kaur N, Kaur C, Sood K, Shanavas A, Sen T (2023) Design of Silica@Au hybrid nanostars for enhanced SERS and photothermal effect. ChemPhysChem 24(22):e202200809. https://doi.org/10.1002/CPHC.202200809

Khoury CG, Vo-Dinh T (2008) Gold nanostars for surface-enhanced Raman scattering: synthesis, characterization and optimization. J Phys Chem C Nanomater Interfaces 2008(112):18849. https://doi.org/10.1021/JP8054747

Kim B, Yang J, Hwang M, Choi J, Kim HO, Jang E, Lee JH, Ryu SH, Suh JS, Huh YM, Haam S (2013) Aptamer-modified magnetic nanoprobe for molecular MR Imaging of VEGFR2 on angiogenic vasculature. Nanoscale Res Lett 8(1):399. https://doi.org/10.1186/1556-276X-8-399

Kresse M, Wagner S, Pfefferer D, Lawaczeck R, Elste V, Semmler W (1998) Targeting of ultrasmall superparamagnetic iron oxide (USPIO) particles to tumor cells in vivo by using transferrin receptor pathways. Magn Reson Med 40(2):236–242. https://doi.org/10.1002/MRM.1910400209

Kuang Y, Zhang K, Cao Y, Chen X, Wang K, Liu M, Pei R (2017a) Hydrophobic IR-780 dye encapsulated in CRGD-conjugated solid lipid nanoparticles for NIR imaging-guided photothermal therapy. ACS Appl Mater Interfaces 9(14):12217–12226. https://doi.org/10.1021/ACSAMI.6B16705/ASSET/IMAGES/LARGE/AM-2016-16705G_0009.JPEG

Kuang Y, Zhang K, Cao Y, Chen X, Wang K, Liu M, Pei R (2017b) Hydrophobic IR-780 dye encapsulated in CRGD-conjugated solid lipid nanoparticles for NIR imaging-guided photothermal therapy. ACS Appl Mater Interfaces 9(14):12217–12226. https://doi.org/10.1021/ACSAMI.6B16705

Lee DE, Koo H, Sun IC, Ryu JH, Kim K, Kwon IC (2012) Multifunctional nanoparticles for multimodal imaging and theragnosis. Chem Soc Rev 41(7):2656–2672. https://doi.org/10.1039/C2CS15261D

Lim EK, Kim T, Paik S, Haam S, Huh YM, Lee K (2015) Nanomaterials for theranostics: recent advances and future challenges. Chem Rev 115(1):327–394. https://doi.org/10.1021/CR300213B

Liu Y, Ashton JR, Moding EJ, Yuan H, Register JK, Fales AM, Choi J, Whitley MJ, Zhao X, Qi Y, Ma Y, Vaidyanathan G, Zalutsky MR, Kirsch DG, Badea CT, Vo-Dinh T (2015) A plasmonic gold nanostar theranostic probe for in vivo tumor imaging and photothermal therapy. Theranostics 5(9):946. https://doi.org/10.7150/THNO.11974

Lutz JF, Stiller S, Hoth A, Kaufner L, Pison U, Cartier R (2006) One-pot synthesis of pegylated ultrasmall iron-oxide nanoparticles and their in vivo evaluation as magnetic resonance imaging contrast agents. Biomacromolecules 7(11):3132–3138. https://doi.org/10.1021/BM0607527

Mahmoudian F, Ahmari A, Shabani S, Sadeghi B, Fahimirad S, Fattahi F (2024) Aptamers as an approach to targeted cancer therapy. Cancer Cell Int 24(1):1–22. https://doi.org/10.1186/S12935-024-03295-4

Martina MS, Fortin JP, Ménager C, Clément O, Barratt G, Grabielle-Madelmont C, Gazeau F, Cabuil V, Lesieur S (2005) Generation of superparamagnetic liposomes revealed as highly efficient MRI contrast agents for in vivo imaging. J Am Chem Soc 127(30):10676–10685. https://doi.org/10.1021/JA0516460

Mitchell MJ, Billingsley MM, Haley RM, Wechsler ME, Peppas NA, Langer R (2020) Engineering precision nanoparticles for drug delivery. Nat Rev Drug Discov 20(2):101–124. https://doi.org/10.1038/s41573-020-0090-8

Olerile LD (2020) Further development of near-infrared mediated quantum dots and paclitaxel co-loaded nanostructured lipid carrier system for cancer theragnostic. Technol Cancer Res Treat:19. https://doi.org/10.1177/1533033820914308/ASSET/IMAGES/LARGE/10.1177_1533033820914308-FIG3.JPEG

Olerile LD, Liu Y, Zhang B, Wang T, Mu S, Zhang J, Selotlegeng L, Zhang N (2017) Near-infrared mediated quantum dots and paclitaxel co-loaded nanostructured lipid carriers for cancer theragnostic. Colloids Surf B Biointerfaces 150:121–130. https://doi.org/10.1016/J.COLSURFB.2016.11.032

Ranjan A, Jacobs GC, Woods DL, Negussie AH, Partanen A, Yarmolenko PS, Gacchina CE, Sharma KV, Frenkel V, Wood BJ, Dreher MR (2011) Image-guided drug delivery with magnetic resonance guided high intensity focused ultrasound and temperature sensitive liposomes in a rabbit Vx2 tumor model. J Control Rel 158(3):487. https://doi.org/10.1016/J.JCONREL.2011.12.011

Rosenblum D, Joshi N, Tao W, Karp JM, Peer D (2018) Progress and challenges towards targeted delivery of cancer therapeutics. Nat Commun 9:1–12. https://doi.org/10.1038/s41467-018-03705-y

Ryu JH, Koo H, Sun IC, Yuk SH, Choi K, Kim K, Kwon IC (2012) Tumor-targeting multi-functional nanoparticles for theragnosis: new paradigm for cancer therapy. Adv Drug Deliv Rev 64(13):1447–1458. https://doi.org/10.1016/J.ADDR.2012.06.012

Sanson C, Diou O, Thévenot J, Ibarboure E, Soum A, Brûlet A, Miraux S, Thiaudière E, Tan S, Brisson A, Dupuis V, Sandre O, Lecommandoux S (2011) Doxorubicin loaded magnetic polymersomes: theranostic nanocarriers for MR imaging and magneto-chemotherapy. ACS Nano 5(2):1122–1140. https://doi.org/10.1021/NN102762F/SUPPL_FILE/NN102762F_SI_004.MPG

Schleich N, Po C, Jacobs D, Ucakar B, Gallez B, Danhier F, Préat V (2014) Comparison of active, passive and magnetic targeting to tumors of multifunctional paclitaxel/SPIO-loaded nanoparticles for tumor imaging and therapy. J Control Rel 194:82–91. https://doi.org/10.1016/J.JCONREL.2014.07.059

Shanavas A, Rengan AK, Chauhan D, George L, Vats M, Kaur N, Yadav P, Mathur P, Chakraborty S, Tejaswini A, De A, Srivastava R (2018) Glycol chitosan assisted in situ reduction of gold on polymeric template for anti-cancer theranostics. Int J Biol Macromol 110:392–398. https://doi.org/10.1016/J.IJBIOMAC.2017.11.127

Shanavas A, Jain NK, Kaur N, Thummuri D, Prasanna M, Prasad R, Naidu VGM, Bahadur D, Srivastava R (2019) Polymeric core-shell combinatorial nanomedicine for synergistic anticancer therapy. ACS Omega. https://doi.org/10.1021/ACSOMEGA.9B02167/ASSET/IMAGES/MEDIUM/AO9B02167_M004.GIF

Sharma R, Borah SJ, Bhawna, Kumar S, Gupta A, Singh P, Goel VK, Kumar R, Kumar V (2022) Functionalized peptide-based nanoparticles for targeted cancer nanotherapeutics: a state-of-the-art review. ACS Omega 7(41):36092–36107. https://doi.org/10.1021/ACSOMEGA.2C03974/ASSET/IMAGES/LARGE/AO2C03974_0013.JPEG

Stark DD, Weissleder R, Elizondo G, Hahn PF, Saini S, Todd LE, Wittenberg J, Ferrucci JT (1988) Superparamagnetic iron oxide: clinical application as a contrast agent for mr imaging of the liver. Radiology 168(2):297–301. https://doi.org/10.1148/RADIOLOGY.168.2.3393649

Sun H, Zu Y (2015) Aptamers and their applications in nanomedicine. Small 11(20):2352. https://doi.org/10.1002/SMLL.201403073

Sun C, Sze R, Zhang M (2006) Folic acid-PEG conjugated superparamagnetic nanoparticles for targeted cellular uptake and detection by MRI. J Biomed Mater Res A 78(3):550–557. https://doi.org/10.1002/JBM.A.30781

Thakor AS, Jokerst JV, Ghanouni P, Campbell JL, Mittra E, Gambhir SS (2016) Clinically approved nanoparticle imaging agents. J Nucl Med 57(12):1833–1837. https://doi.org/10.2967/JNUMED.116.181362

Tong L, Wei Q, Wei A, Cheng JX (2009) Gold nanorods as contrast agents for biological imaging: optical properties, surface conjugation, and photothermal effects. Photochem Photobiol 85(1):21. https://doi.org/10.1111/J.1751-1097.2008.00507.X

Vigderman L, Zubarev ER (2013) Therapeutic platforms based on gold nanoparticles and their covalent conjugates with drug molecules. Adv Drug Deliv Rev 65(5):663–676. https://doi.org/10.1016/J.ADDR.2012.05.004

Volkov Y (2015) Quantum dots in nanomedicine: recent trends, advances and unresolved issues. Biochem Biophys Res Commun 468(3):419–427. https://doi.org/10.1016/J.BBRC.2015.07.039

Wang L, Bao J, Wang L, Zhang F, Li Y (2006) One-pot synthesis and bioapplication of amine-functionalized magnetite nanoparticles and hollow nanospheres. Chemistry 12(24):6341–6347. https://doi.org/10.1002/CHEM.200501334

Wei Y, Gao L, Wang L, Shi L, Wei E, Zhou B, Zhou L, Ge B (2017) Polydopamine and peptide decorated doxorubicin-loaded mesoporous silica nanoparticles as a targeted drug delivery system for bladder cancer therapy. Drug Deliv 24(1):681–691. https://doi.org/10.1080/10717544.2017.1309475

Wu RH, Nguyen TP, Marquart GW, Miesen TJ, Mau T, Mackiewicz MR (2014) A facile route to tailoring peptide-stabilized gold nanoparticles using glutathione as a synthon. Molecules 19(5):6754–6775. https://doi.org/10.3390/molecules19056754

Xu B, Li S, Shi R, Liu H (2023) Multifunctional mesoporous silica nanoparticles for biomedical applications. Signal Transduct Target Ther 8:1–28. https://doi.org/10.1038/s41392-023-01654-7

Yang J, Lee CH, Park J, Seo S, Lim EK, Song YJ, Suh JS, Yoon HG, Huh YM, Haam S (2007) Antibody conjugated magnetic PLGA nanoparticles for diagnosis and treatment of breast cancer. J Mater Chem 17(26):2695–2699. https://doi.org/10.1039/B702538F

Yang X, Grailer JJ, Rowland IJ, Javadi A, Hurley SA, Matson VZ, Steeber DA, Gong S (2010a) Multifunctional stable and PH-responsive polymer vesicles formed by heterofunctional triblock copolymer for targeted anticancer drug delivery and ultrasensitive MR imaging. ACS Nano 4(11):6805–6817. https://doi.org/10.1021/NN101670K/SUPPL_FILE/NN101670K_SI_001.PDF

Yang X, Grailer JJ, Rowland IJ, Javadi A, Hurley SA, Steeber DA, Gong S (2010b) Multifunctional SPIO/DOX-loaded wormlike polymer vesicles for cancer therapy and MR imaging. Biomaterials 31(34):9065–9073. https://doi.org/10.1016/J.BIOMATERIALS.2010.08.039

Yao L, Bojic D, Liu M (2023) Applications and safety of gold nanoparticles as therapeutic devices in clinical trials. J Pharm Anal 13(9):960. https://doi.org/10.1016/J.JPHA.2023.06.001

Biosafety and Reality of Nanoimaging and Therapeutics

14

Zain Siddiqui, Alev K. Erdi, and Berivan Cecen

Abstract

The rapid advancement of nanotechnology has greatly expanded the range of biomaterials in biomedicine, including nanoparticles (NPs) made from conventional organic materials and inorganic biomaterials or organic/inorganic hybrid NPs. This expansion demonstrates the potential of engineered inorganic material-based NPs in therapeutics. Recent advances in nanomaterial-based therapeutics have allowed for imaging-guided biosafety and nanoimaging, opening up new possibilities in drug delivery and developing various therapeutic systems. With their distinctive properties, inorganic nanomaterials, such as carbon, metal, ceramic, and semiconductor-based nanomaterials, offer a promising future for therapeutic applications. However, ensuring the safe utilization of nanomaterials in clinical applications remains a significant concern. In this review, we have gathered recent research on the biosafety of nanomaterials in various therapeutic applications, both in vitro and in vivo. We have also discussed how size, surface functionalization, test species, and test conditions influence the formulation of particles and their biocompatibility. Furthermore, we have discussed the progress and challenges in developing nanoimaging nanomaterials for therapeutic

Z. Siddiqui · A. K. Erdi
Department of Biomedical Engineering, Newark College of Engineering, New Jersey Institute of Technology, Newark, NJ, USA

B. Cecen (✉)
Department of Biomedical Engineering, Newark College of Engineering, New Jersey Institute of Technology, Newark, NJ, USA

Stem Cell and Tissue Engineering Application and Research Center (ISUKOK), İstinye University, Istanbul, Turkey
e-mail: cecen@rowan.edu

R. Prasad, M. Mohan Yallapu (eds.), *Advancements in Cancer Theranostics*,
https://doi.org/10.1007/978-981-95-0938-6_14

purposes. In conclusion, with proper precautions and safe conditions, nanomaterials have excellent potential for future therapeutic applications, instilling optimism and hope for the future of medicine.

Keywords

Nanomedicine · Nanotheranostics · Nanoimaging · Biosafety · Translational · Nanoparticles · Preclinical and clinical

14.1 Introduction

Nanoimaging and nanotherapeutics are rapidly advancing interdisciplinary fields that combine nanoscience, nanoengineering, and nanotechnology with life sciences (Tiwari et al. 2024; Tripathi et al. 2023). This convergence has produced promising outcomes for the medical sector and society and instilled a sense of confidence in the potential of nanomedicine (Monteiro-Riviere and Tran 2014; Nordmann 2004). Numerous nanomedicine platforms exist, but the most noteworthy ones include nanotechnology-based drug delivery systems and nanoagents (Liu et al. 2022a; Nguyen et al. 2021). Nanotherapeutics involves therapeutics with nanosized and multiple abilities, including targeted delivery, regulated release, enhanced transport efficiency, and the integration of various therapeutic methods, including multimodal diagnosis and treatment (Al-Thani et al. 2024; Conte and Cauda 2022). Nanobiomaterials have been extensively studied in various fields related to nanomedicine. These include drug delivery systems for transporting drugs and genes, diagnostic probes, tissue engineering materials, and ultrasound therapy or radiotherapy agents (Tiwari and Tiwari 2013; Veg et al. 2025). Alternatives in treatment, including photothermal, photodynamic, and therapies based on magnetic hyperthermia, have also been developed (Gavilán et al. 2021; Sharma et al. 2019). The goal is to ensure that nanotherapeutics can circulate in the body for a long time, show precise release behavior, target specific tissues with high penetration, enable imaging, and achieve a high target-to-background ratio (Hegde et al. 2023; Saroj and Rajput 2018). Various nanotherapeutics can localize diagnostic and therapeutic agents at particular disease sites, reducing undesired side effects. Developing new tools with enhanced imaging capabilities is crucial for early disease detection. Imaging agents that can detect cancer early in its early stages also benefit the medical community. Furthermore, nanotherapeutics are critical in treating severe diseases (Ali et al. 2021; Wong et al. 2020; Yetisgin et al. 2020).

Notably, most of the reported NPs evaluations are based on animal model results, with few clinical studies being conducted. This research gap must be addressed as it could significantly advance our understanding of nanomaterials and their potential in therapeutic applications (Tutty et al. 2022; Vashishat et al. 2024). The urgency of addressing this research gap is paramount to the future of nanomedicine. Various nanoimaging agents have been specifically engineered as diagnostic and therapeutic tools. However, specific nanomaterials have restricted cell penetration, which calls for improvements in their pharmacokinetics (Bhaskar et al. 2010; Tchuendem 2024). The applications of nanotherapeutics have advanced significantly in recent years. Summarizing the current developments in nanotherapeutics could help medical professionals make further progress. Evaluating the present use of imaging techniques and nanotherapeutics could drive the progress of nanomedicine. This paper reviews nanomaterials' distinct physical and chemical characteristics, possible uses in healthcare sectors, and ensuring human safety in clinical studies. It also discusses approaches for upcoming endeavors in nanomedicine research to optimize nanomedicine development.

14.2 BioSafety Evaluations

There is increasing concern regarding the human biosafety of nanomaterials intended for clinical applications (Su et al. 2018; Xu et al. 2020). Nanomaterials trigger distinct chemical and mechanical reactions within the human body, potentially leading to unforeseen side effects. Inorganic nanomaterials commonly show acute and chronic cytotoxicity influenced by dosage and exposure duration (Ahmad et al. 2016; Saifi et al. 2018). Thus, their administration conditions must be meticulously optimized. The decrease in particle size influences the surface chemistry of NPs, affecting the zeta potential, dielectric constant, and surface charges (Harish et al. 2023; Mu et al. 2014). Typically, the function of surface chemistry becomes more significant in single particles and smaller aggregates than in larger ones. Nanomaterials of small size, located within or near human cells, frequently remain in the body for long durations because macrophages cannot effectively perform phagocytosis (Gustafson et al. 2015; Revell 2006). These phenomena produce unwanted by-products and immune reactions, resulting in significant health risks like damage to cell membranes and organelle integrity, nuclear genotoxicity, and elevated intracellular reactive oxygen species (ROS) levels. The creation of ROS influences the toxicity of nanomaterials (Attarilar et al. 2020; Singh 2017; Tee et al. 2016). To grasp the underlying mechanisms of this toxicity, it is crucial to examine different toxicological factors and evaluation methods. In vitro assays provide limited data on material biosafety, yet they are essential for validation before conducting in vivo studies across various experimental conditions and species (Bernard et al. 2018; De Jong et al. 2020; Tirumala et al. 2021). Following this, in vivo assays should clarify biodistribution, clearance mechanisms, and possible adverse effects. Various in vitro and in vivo assays have been developed to deepen our understanding of the toxicity mechanisms linked to the biosafety of inorganic nanomaterials. These assessments pinpoint different types of cellular damage triggered by chemicals, physical injury, radiation, microorganisms, and other factors (Truong et al. 2024; Zambonino et al. 2023).

14.3 Biosafety of Nanotherapeutic Applications of Drug Carriers

14.3.1 Inorganic-Metal and Carbon-Based NPs

Inorganic NPs have remarkably succeeded as drug carriers and diagnostic tools (Paul and Sharma 2020; Unnikrishnan et al. 2023). Although it was previously thought that Inorganic NPs exhibited greater stability than organic ones, new evidence indicates that they may break down separately in living organisms, potentially causing toxic effects (Feliu et al. 2016; Soenen et al. 2015). As a result, many inorganic NPs are typically coated with biocompatible molecules. They provide a strong framework with particle diameters between 80 and 250 nm. It is distinguished by its internal porosity, which includes pore sizes ranging from 3 to 5 nm, a significant pore volume, an extensive surface area, and organized pore networks (Farasati Far et al. 2024; Radhakrishnan et al. 2022). These materials are relatively nontoxic, possess a high loading capacity, and demonstrate excellent cytocompatibility (Liu et al. 2020; Saber-Samandari and Saber-Samandari 2017). Bioreductive chemotherapy is predicated upon enzymes' activation of pharmaceutical agents, including quinone oxidoreductase and P450 reductase. Furthermore, it involves the identification of tumors characterized by a high abundance of these enzymes, as well as the differentiation between oxygen and pH levels in normative versus tumor tissues. As a result, the proposed system introduces an innovative approach to nanotheranostics (Janczy-Cempa et al. 2022; Robertson 1998; Seddon et al. 2004).

Carbon-based nanomaterials, including fullerenes, Carbon NPs, nanotubes, graphene, and nanodiamonds, have greatly enhanced biomedical applications, particularly in nanotheranostics (Lai et al. 2023; Sargazi et al. 2022). Carbon nanotubes are highly structured, pseudo-one-dimensional carbon allotropes and can be categorized into single-walled or multiwalled types. Single-walled are formed from a single rolled-up layer of graphite, whereas multiwalled consist of multiple concentric cylindrical shells made from graphite sheets (Prasanth 2015; Thakur and Thakur 2018). Carbon nanotubes can penetrate diverse cells to deliver drugs or other agents to targeted tissues (Fig. 14.1a). Given their potential in cancer therapies like PDT, photothermal therapy (PTT), radiotherapy, and chemotherapy are utilized to treat various conditions investigated as tumor theranostics. They also serve as innovative contrast agents for magnetic resonance imaging (MRI) (Li et al. 2017; Wang et al. 2020). This new nanoplatform, owing to its bioimaging capabilities and controlled drug delivery, holds promise as an effective nanotheranostic (Chowdhury et al. 2016; Ma et al. 2016). Nonetheless, additional in vivo studies are necessary to validate these findings. Nanodiamonds, characterized by a truncated octahedral shape and a 2–8 nm diameter, exhibit remarkable chemical stability, high hardness, stiffness, strength, small size, extensive surface area, and significant adsorption capacity (Chen et al. 2007).

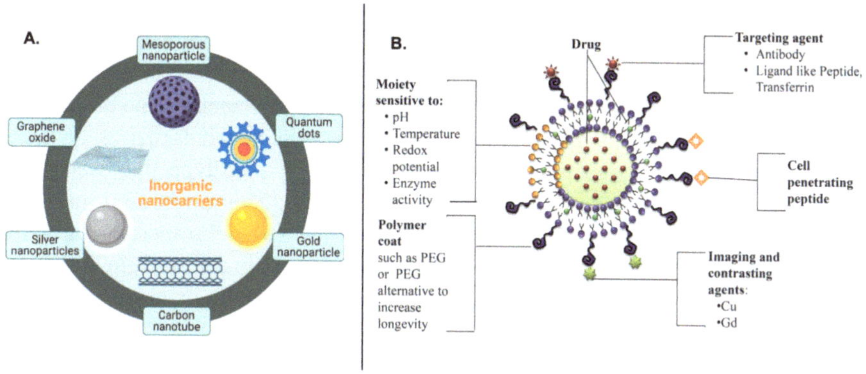

Fig. 14.1 (**a**) Inorganic nanocarrier examples include mesoporous and silica NPs with a size under 1 μm, and pores ranging from 2 to 50 nm. Quantum dots are colloidal fluorescent semiconductor nanocrystals measuring 2–10 nm. Silver and gold NPs, with sizes 1–100 nm, are nontoxic and have a high surface area and adjustable optical properties. Carbon nanotubes are formed from rolled coaxial graphite sheets measuring less than 100 nm, while graphene oxide consists of a single atomic layer of carbon with a thickness of 1 nm (Kaushik et al. 2022). (**b**) Feasible alteration possibilities of liposomes for targeting, diagnosis, or theranostics (Saraf et al. 2020)

14.3.2 Polymer, Liposomes, and Dendrimers as Nanotherapeutics

Polymers represent some of the most cost-effective and manageable carriers, prized for their essential properties like biocompatibility, biodegradability, and resistance to degradation. Synthetic and natural macromolecules can serve as nanotheranostics, although modifications are required to improve their imaging capabilities and therapeutic functions (Chen et al. 2017; Silva et al. 2019). Chitosan, an essential linear polysaccharide, presents various benefits such as low toxicity, affordable pricing, antimicrobial properties, and antioxidant traits. In nanotheranostics, chitosan can act as a coating agent or matrix for polymers (Balan et al. 2019; Bandara et al. 2020; Siafaka et al. 2021). Its carboxyl and amino groups facilitate easy modifications, allowing different molecules to produce significant effects. Palmitoylated grafted chitosan has been utilized as polymeric shells for magnetic NPs (Jain et al. 2013; Khalili et al. 2022).

Liposomes are characterized as organized, compatible with living systems, and capable of breaking down naturally occurring lipid carriers (Liu et al. 2022b; Teixeira et al. 2017). Liposomes consist of one or several layers of natural or synthetic lipids, along with an aqueous core (Fig. 14.1b). These entities have functioned as carriers for a multitude of active hydrophilic or lipophilic molecules. Radio-labeled liposomes containing radionuclides provide diagnostic, monitoring, and therapeutic capabilities. In addition to their imaging capability from radiolabeling, these liposomes are an effective tool for selecting the most suitable therapeutic approach for individual patients. Liposomes are typically enhanced with active agents for improved biocompatibility. For instance, liposomes coated with D-α-tocopheryl polyethylene glycol succinate containing vitamin E are commonly formulated. Studies have focused on long-circulating liposomes sensitive to pH and loaded with GD and poly-lysine (Agarwal et al. 2024) for cancer nanotheranostics (Arora and Jaglan 2017). These formulations have been proven to increase animal survival rates and enhance tumor uptake. Gadolinium diethylenetriamine pentaacetic acid salt was used to modify liposomes and their PEGylated forms as an MRI contrast agent, while zinc phthalocyanine served as a prototype photosensitizer (Düzgüneş et al. 2018; Skupin-Mrugalska et al. 2018). The results indicated that these liposomal formulations could function as imaging agents. Furthermore, a novel system utilizing liposomes filled with the anticancer drug cinobufagin has been created and evaluated, incorporating carbon dots as a potentially practical anticancer nanotheranostic approach.

Dendrimers are essential nanostructured materials because they can easily modify external groups with antibodies, peptides, or proteins. These dendritic polymers encapsulated paclitaxel, showcasing a controlled drug release mechanism (Fig. 14.2). In vivo, MRI scans performed on rats showed a favorable distribution of unimolecular micelles in the liver and kidneys, characterized by significant positive contrast enhancement and an extended blood circulation time. Thus, dendrimers are promising candidates for alternative nanotheranostics. Multifunctional Poly(amidoamine) dendrimers, which incorporate gold NPs and α-tocopheryl succinate, have been created for CT imaging and treatment aimed at cancer agents. Poly(amidoamine) dendrimers were modified with fluorescein isothiocyanate and PEGylated folic acid, serving as templates for creating gold nanoplatforms. The resulting nanoplatforms showed a strong affinity for cancer cells expressing high levels of receptors, effectively targeting CT imaging in both in vitro experiments and an in vivo model of a xenografted tumor (Kesharwani et al. 2019; Li and Kono 2018).

14.3.3 Inorganic-Mesoporous Silica NPs with Magnetic Response

Magnetic nanomaterials represent a vibrant field in biomedical research, which is evident from the numerous nanosystems used in clinical trials. Typically, these materials consist of crystalline iron oxides with engineered surfaces that minimize aggregation and enhance resistance to dissolution (Araújo et al. 2024; Reddy et al. 2012). The most developed type of these nanometric magnetic materials is superparamagnetic iron oxide NPs. They gain magnetization in the presence of external magnetic fields but return to a nonmagnetic state once the fields are removed. Iron oxide NPs behave differently depending on the external magnetic field (Koksharov 2009; Kumar et al. 2021). Magnetized NPs can be controlled and directed toward the magnet when exposed to steady magnetic fields. Under alternating magnetic fields, NP experience changes in magnetization. If the oscillation intensity is sufficient, this continual reorientation leads to magnetic resistance, which is then released as heat. This results in heat-induced apoptosis in living organisms, known as magnetic hyperthermia. Iron oxide NPs are commonly employed as contrast agents in MRI because they can be easily tailored for specific

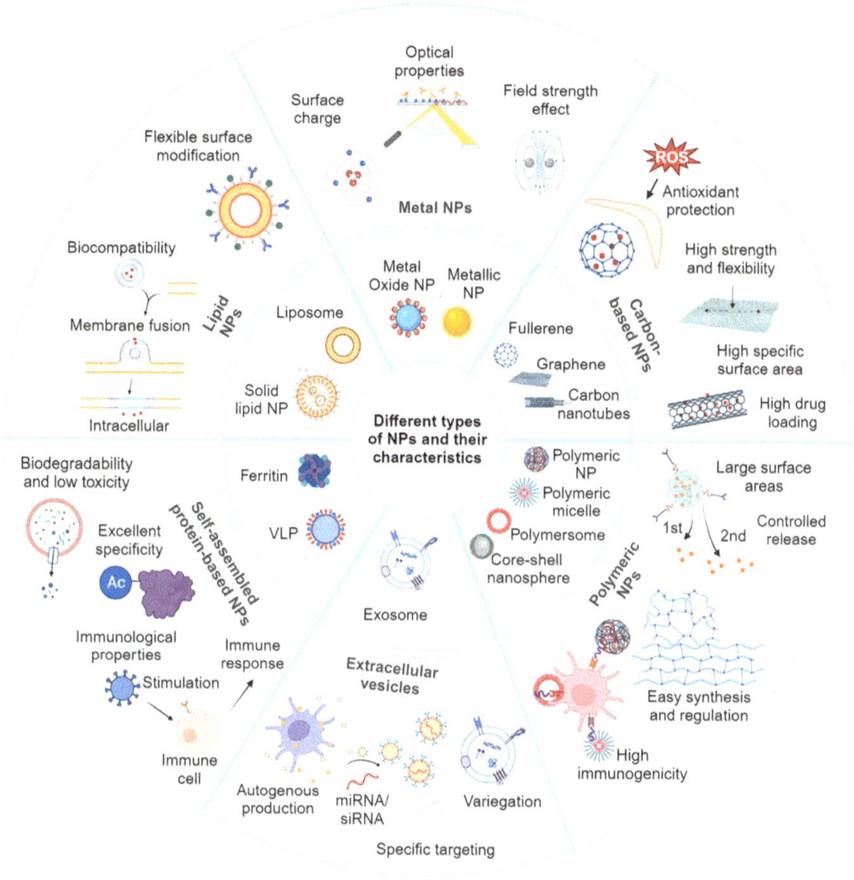

Fig. 14.2 Lipid NPs exhibit remarkable biocompatibility and a range of versatile surface modification options. Metallic NPs, including metals such as gold and their oxides, exhibit remarkable optical, electronic, and magnetic properties, making them suitable for biological imaging, PTT, and sensing applications. Carbon-based nanomaterials are characterized by their large surface area, significant drug-loading capacity, as well as their exceptional strength and chemical stability, allowing them to withstand oxidative environments. Polymer NPs, derived from various polymeric materials, display diverse structures and characteristics (Huang et al. 2024)

tissue or cell targeting. Regarding the biosafety of clinically significant NPs, it is noteworthy that iron oxide crystals smaller than 20 nm are rapidly eliminated from the body while maintaining excellent biocompatibility. They require a surface coating to prolong circulation times and evade the immune system (Chen et al. 2022; Nowak-Jary and Machnicka 2024).

14.4 Diagnostic Imaging Using Nanomedicine

14.4.1 Nanoimaging Applications for Cancer

NPs have been used for molecular imaging, especially MRI, optical, and luminescence imaging, and positron emission tomography (PET). Due to their size, they can evaluate target-specific accumulation and biodistribution. Various NPs, including gold, polymer iodine, and recently F-Macroflor-modified poly glucose NPs, have been used for nanoimaging applications in cancer (Abedin et al. 2023; Parvez et al. 2022; Siddique and Chow 2020).

Nanotechnology has revolutionized cancer imaging by enabling earlier detection, improved diagnostic accuracy, and enhanced monitoring of treatment response. Imaging agents based on NPs present significant benefits compared to traditional methods. Their enhanced sensitivity and specificity come from NPs designed to target cancer cells and biomarkers with great affinity, facilitating earlier tumor detection. These NPs' distinct optical and magnetic characteristics also allow for more sensitive imaging than conventional contrast agents. Iron oxide NPs improve MRI contrast, aiding in identifying smaller metastases (Fig. 14.3).

Meanwhile, gold NPs boost CT imaging signals, enhancing tumor visualization. The engineered accumulation of NPs can selectively target cancer cells and biomarkers, facilitating more precise imaging techniques (Kumar et al. 2019; Luo et al. 2021). Specifically, iron oxide NPs directed toward folate receptor alpha significantly enhance MRI detection of breast tumors. Human epidermal growth

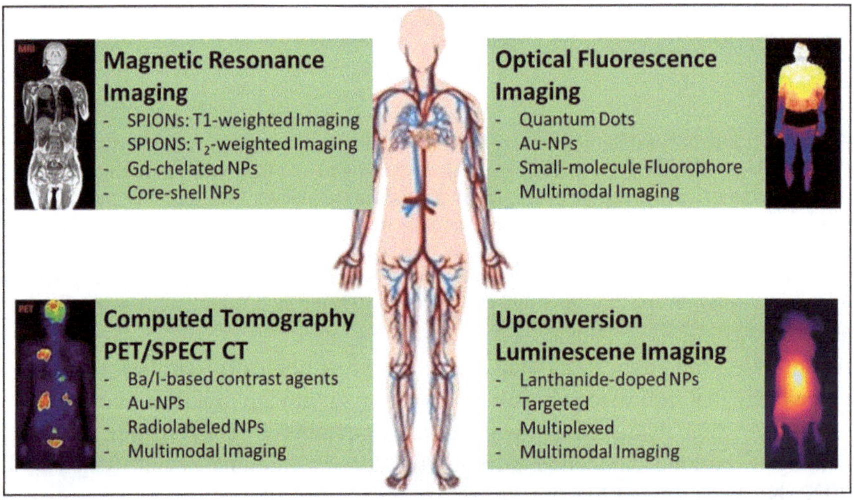

Fig. 14.3 Notable nanotechnology products in clinical practice or demonstrating potential applications include iron oxide NPs, which are approved for enhancing MRI contrast. Additionally, several colloidal derivatives of 99mTc serve as contrast agents for PET imaging. Conversely, optical fluorescence and up-conversion luminescence imaging exhibit encouraging preclinical findings and are presently being explored (Singh and Amiji 2022)

factor receptor 2-targeted silica NPs enhance PET imaging of breast cancer. This targeted approach allows for detecting tumors earlier than nontargeted imaging agents. Multimodal imaging refers to integrating various imaging modalities into a single platform, such as MRI, PET, and optical imaging (Martí-Bonmatí et al. 2010; Pichler et al. 2008). This approach offers a thorough combination of anatomical and functional data, significantly improving the sensitivity and accuracy of cancer detection. Molecular and functional imaging of NPs allows for visualizing molecular activities and cellular processes linked to cancer. Gold NPs that target cathepsin enhance CT imaging by highlighting increased cathepsin activity in tumors. Glucose-functionalized gold NPs provide CT imaging opportunities to observe heightened glucose uptake in tumors, which may help differentiate them from inflammatory responses (Pontico et al. 2023; Wang et al. 2023). The enhanced signal-to-noise ratio results from NPs' high payload capacity, which enables the delivery of a more significant number of contrast-generating molecules with each targeting event. This improvement significantly amplifies the signal-to-noise ratio, facilitating the detection of smaller tumors or sparse biomarkers that conventional agents may overlook (Giger et al. 2013; Zhu et al. 2024).

The improved retention of NPs with more extended blood circulation enables their tumor accumulation through the enhanced permeability and retention effect (Ejigah et al. 2022; Shinde et al. 2022). This leads to more prolonged and more precise imaging signals. Utilizing these mechanisms, NPs-based imaging agents offer possibilities for earlier cancer detection, increased diagnostic accuracy, and better monitoring of treatment response than conventional imaging techniques. Sentinel lymph node mapping utilizes technetium-99 m nanocolloids in NPs-enhanced lymphoscintigraphy, assisting in surgical planning by outlining lymph node drainage patterns. Fluorescent NPs assist surgeons in visualizing tumor boundaries during resection, thanks to image-guided interventions. Magnetic NPs facilitate MRI-guided focused ultrasound ablation of tumors. While promising, translating nanoimaging agents to the clinic remains challenging. These include optimizing biodistribution, minimizing toxicity, and standardizing manufacturing. Nonetheless, NPs-based imaging holds immense potential to transform cancer detection and management (Abdellatif et al. 2021; Das 2023).

14.4.2 Nanoimaging Applications for Neurological Disorders

NPs have also been used in imaging modalities within neurological disorders. This is primarily due to the function of NPs and their ability to reach the central nervous system due to their size (Feng et al. 2015; Zhang et al. 2021). Further, gold NPs allow for sufficient contrast within primary brain tumors like glioblastoma and meningiomas, as the vascular nature of these tumors provides a highway-like environment for the NPs to penetrate efficiently. Iron oxide NPs have also been used for nanoimaging within these indications due to the increased spatial resolution of tumors when imaging with MRI. These iron oxide NPs can be further functionalized with albumin or folate to increase glioma cells' targeting (Felenji et al. 2022; Wang et al. 2016).

Nanotechnology has become vital for diagnosing, imaging, and treating neurological disorders. Iron oxide NPs are widely utilized to enhance brain imaging, while manganese oxide NPs improve the contrast and spatial resolution in MRI scans for brain tumors and various neurological conditions. Multimodal imaging combines different methods, including MRI, PET, and optical imaging, to provide detailed anatomical and functional information. Additionally, NPs designed with specific targeting ligands improve the visualization of biomarkers associated with neurodegenerative diseases, brain tumors, and other disorders (Padmanabhan et al. 2016; Thangam et al. 2021). These versatile NPs combine diagnostic imaging with therapeutic drug administration, enabling image-guided treatment and monitoring. While nanoimaging has substantial potential for neurological applications, challenges persist in optimizing biodistribution, minimizing toxicity, and translating these technologies into clinical use. Ongoing research focuses on developing biocompatible NPs that enhance blood-brain barrier penetration and targeted delivery effectiveness (Fig. 14.4).

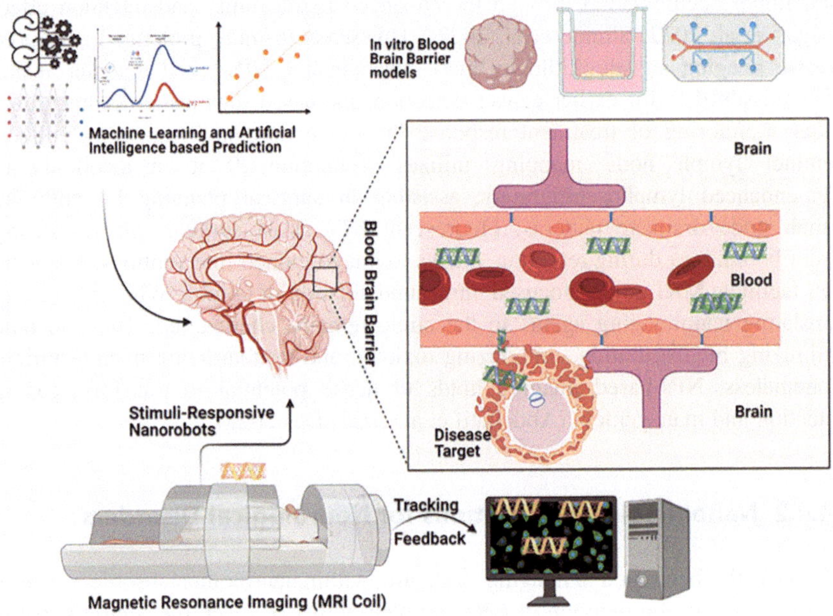

Fig. 14.4 The blood-brain barrier neurovascular unit's biochemical signaling and structural hierarchy, in place, have been the primary targets for recapitulation into the in vitro modules (Singh et al. 2021)

14.5 Challenges

Ensuring the safety of inorganic nanomaterials is essential for their broader application in therapy. This review classifies inorganic nanotherapeutics into five categories: carbon, metal, two-dimensional, ceramic, and semiconductor nanomaterials (Hasnain et al. 2023; Song et al. 2024). It aims to evaluate biosafety studies conducted both in vitro and in vivo. While there is research regarding the toxicity of these materials, several unresolved issues remain (Ahmad et al. 2022). Although we understand the toxicity of inorganic nanomaterials, several concerns persist. Research on toxicity primarily focuses on spherical NPs, leaving other forms, like nanorods, nanocages, and nanorings, that need more examination. Human trials are necessary for clinical applications. Variations in experimental conditions and standards for toxicity assessments lead to inconsistent results. More clinical trials are needed to address these issues and clarify long-term toxicity and potential side effects. Real-time monitoring of therapeutic responses in nanotheranostics can provide patients with up-to-date information on their treatment progress, allowing adjustments to the nanotheranostic properties as needed. We are moving toward personalized treatments by generating and analyzing large datasets using artificial intelligence. Combining various approaches in nanomedicine is essential for personalization, facilitating the interaction of multiple therapeutics to tackle the challenges in the field. Innovative models for assessing systemic toxicity are crucial to revealing the inherent toxicity mechanisms of inorganic nanomaterials across different cell types and animal models. We expect that researchers with diverse expertise will advance the area of inorganic nanomaterial-based theranostics by addressing current challenges and pursuing key recommendations.

14.6 Conclusion

Extensive research has encouraged the transition of nanotechnology and nanobiomedical investigations of NPs from in vitro assays to in vivo evaluations utilizing small animal models. These studies primarily examine the biosafety assessments and therapeutic and diagnostic uses of chemically engineered NPs, especially in oncology (Chen et al. 2013). Biosafety assessments are critical prerequisites for imaging diagnostics and treatment procedures concerning diseases. Regarding materials, specially engineered and manufactured NPs are vital in tackling clinical challenges (Chen et al. 2013; Aljabali and Pal 2022). Given the intricate physiological settings in vivo, the structural, compositional, and morphological features of NPs need careful tuning and optimization to comply with increasingly stringent standards for in vivo applications. This encompasses factors such as particle size, pore structure, aggregation state, surface characteristics, degradation rate, circulation time, and penetration within biological tissues (Smith 2017). The degradation behaviors of NPs require further clarification and adjustment for therapeutic uses. While MSNs can be converted into clearable silicate species, their biodegradation rate may often be extended (Croissant et al. 2017; Zhang et al.

2024). The therapeutic process requires regular dosing intervals to attain the desired pharmacological effect accumulation. Due to the carriers' slow degradation rate, consistent injections will likely result in substantial in vivo NPs accumulation, which could lead to unexpected toxicity, especially over time (Chenthamara et al. 2019; Zhang et al. 2020). Creating NPs with hollow interiors can improve drug-loading capacity and minimize administration frequency, significantly decreasing carrier accumulation. To address the accumulation issues of NPs in organs, it is essential to adjust the framework compositions to better manage their degradation behaviors, which should be thoroughly explored and implemented (Vuong 2022; Yang et al. 2020; Zou et al. 2021). It is essential to advance surface engineering techniques to address the nonspecific absorption of mononuclear phagocytic and nontarget cells (Kim et al. 2017; Nam et al. 2013). While PEGylated options may suffice for specific applications, the rapid and significant aggregation of NPs in organs, including the liver, remains an unavoidable concern. This phenomenon has emerged as one of the principal challenges researchers face in this field (Ahmad et al. 2016; Santhanakrishnan et al. 2024). The inquiry into novel targeting ligands designed to modify the surfaces of NPs significantly augments drug accretion within exact cells or subcellular organelles while simultaneously resulting in a marked reduction of accumulation in other organs (Behzadi et al. 2017; Jhaveri and Torchilin 2016; Liu et al. 2024). Identifying innovative and more efficient targeting strategies that enhance the endocytosis of active tumor cells while decreasing nonspecific uptake by healthy cells and tissues is essential. This approach addresses the diverse needs of various patient diseases within the framework of personalized medicine (Akhter et al. 2021; Vyas et al. 2022).

References

Abdellatif AA et al (2021) Nano-scale delivery: A comprehensive review of nano-structured devices, preparative techniques, site-specificity designs, biomedical applications, commercial products, and references to safety, cellular uptake, and organ toxicity. Nanotechnol Rev 10(1): 1493–1559

Abedin A et al (2023) Recent advancement of nanotechnology in bio applications. In: Advanced materials for multidisciplinary applications. Springer, pp 59–104

Agarwal H et al (2024) Theranostic nanoparticles for detection and treatment of pancreatic cancer. Wiley Interdiscip Rev Nanomed Nanobiotechnol 16(4):e1983

Ahmad MZ et al (2016) Toxicity of inorganic nanoparticles used in targeted drug delivery and other biomedical application: An updated account on concern of biomedical nanotoxicology. J Nanosci Nanotechnol 16(8):7873–7897

Ahmad A, Imran M, Sharma N (2022) Precision nanotoxicology in drug development: current trends and challenges in safety and toxicity implications of customized multifunctional nanocarriers for drug-delivery applications. Pharmaceutics 14(11):2463

Akhter MH et al (2021) Receptor-based targeting of engineered nanocarrier against solid tumors: Recent progress and challenges ahead. Biochim Biophys Acta-Gen Subjects 1865(2):129777

Ali ES et al (2021) Targeting cancer cells with nanotherapeutics and nanodiagnostics: Current status and future perspectives. In: Seminars in cancer biology. Elsevier

Aljabali AA, Pal K (2022) Bionanotechnology: next-generation therapeutic tool. Bentham Science Publishers

Al-Thani AN et al (2024) Nanoparticles in cancer theragnostic and drug delivery: a comprehensive review. Life Sci 122899

Araújo E et al (2024) Advances in surface design and biomedical applications of magnetic nanoparticles. Adv Colloid Interf Sci 103166

Arora D, Jaglan S (2017) Nanocarriers for resveratrol delivery. Nanosci Food Agric 5:123–138

Attarilar S et al (2020) The toxicity phenomenon and the related occurrence in metal and metal oxide nanoparticles: a brief review from the biomedical perspective. Front Bioeng Biotechnol 8: 822

Balan V, Malihin S, Verestiuc L (2019) Chitosan-based systems for theranostic applications. In: Functional chitosan: drug delivery and biomedical applications, pp 343–384

Bandara S et al (2020) Agricultural and biomedical applications of chitosan-based nanomaterials. Nano 10(10):1903

Behzadi S et al (2017) Cellular uptake of nanoparticles: journey inside the cell. Chem Soc Rev 46(14):4218–4244

Bernard M et al (2018) Biocompatibility of polymer-based biomaterials and medical devices–regulations, in vitro screening and risk-management. Biomater Sci 6(8):2025–2053

Bhaskar S et al (2010) Multifunctional Nanocarriers for diagnostics, drug delivery and targeted treatment across blood-brain barrier: perspectives on tracking and neuroimaging. Part Fibre Toxicol 7:1–25

Chen M et al (2007) Engineering multifunctional biologically-amenable nanomaterials for interfacial therapeutic delivery and substrate-based cellular interrogation. In: 2007 IEEE 7th International Symposium on BioInformatics and BioEngineering. IEEE

Chen Y, Chen H, Shi J (2013) In vivo bio-safety evaluations and diagnostic/therapeutic applications of chemically designed mesoporous silica nanoparticles. Adv Mater 25(23):3144–3176

Chen H et al (2017) Rethinking cancer nanotheranostics. Nat Rev Mater 2(7):1–18

Chen C et al (2022) Ultrasmall superparamagnetic iron oxide nanoparticles: a next generation contrast agent for magnetic resonance imaging. Wiley Interdiscip Rev Nanomed Nanobiotechnol 14(1):e1740

Chenthamara D et al (2019) Therapeutic efficacy of nanoparticles and routes of administration. Biomater Res 23(1):20

Chowdhury MR et al (2016) Cancer nanotheranostics: strategies, promises and impediments. Biomed Pharmacother 84:291–304

Conte M, Cauda V (2022) Multimodal therapies against pancreatic ductal adenocarcinoma: a review on synergistic approaches toward ultimate nanomedicine treatments. Adv Therapeut 5(11):2200079

Croissant JG, Fatieiev Y, Khashab NM (2017) Degradability and clearance of silicon, organosilica, silsesquioxane, silica mixed oxide, and mesoporous silica nanoparticles. Adv Mater 29(9): 1604634

Das KP (2023) Nanoparticles and convergence of artificial intelligence for targeted drug delivery for cancer therapy: current progress and challenges. Front Med Technol 4:1067144

De Jong W, Carraway J, Geertsma R (2020) In vivo and in vitro testing for the biological safety evaluation of biomaterials and medical devices. In: Biocompatibility and performance of medical devices. Elsevier, pp 123–166

Düzgüneş N et al (2018) Photodynamic therapy of cancer with liposomal photosensitizers. Ther Deliv 9(11):823–832

Ejigah V et al (2022) Approaches to improve macromolecule and nanoparticle accumulation in the tumor microenvironment by the enhanced permeability and retention effect. Polymers 14(13): 2601

Farasati Far B, AdibAmini S, Pourmolaei A (2024) Cytotoxicity and biocompatibility of metal–organic frameworks. In: Logic for metal−organic framework selection: MOFs for biomedical applications. ACS Publications, pp 69–105

Felenji H et al (2022) Folic acid-conjugated iron oxide magnetic nanoparticles based on bovine serum albumin (BSA) for targeted delivery of curcumin to suppress liver cancer cells. Chem Afr 5(5):1627–1639

Feliu N et al (2016) In vivo degeneration and the fate of inorganic nanoparticles. Chem Soc Rev 45(9):2440–2457

Feng X et al (2015) Central nervous system toxicity of metallic nanoparticles. Int J Nanomedicine:4321–4340

Gavilán H et al (2021) Magnetic nanoparticles and clusters for magnetic hyperthermia: optimizing their heat performance and developing combinatorial therapies to tackle cancer. Chem Soc Rev 50(20):11614–11667

Giger ML, Karssemeijer N, Schnabel JA (2013) Breast image analysis for risk assessment, detection, diagnosis, and treatment of cancer. Annu Rev Biomed Eng 15(1):327–357

Gustafson HH et al (2015) Nanoparticle uptake: the phagocyte problem. Nano Today 10(4): 487–510

Harish V et al (2023) Cutting-edge advances in tailoring size, shape, and functionality of nanoparticles and nanostructures: a review. J Taiwan Inst Chem Eng 149:105010

Hasnain MS, Nayak AK, Aminabhavi TM (2023) Inorganic nanosystems: theranostic nanosystems, vol 2. Academic Press

Hegde M et al (2023) Nanoparticles in the diagnosis and treatment of cancer metastases: current and future perspectives. Cancer Lett 556:216066

Huang Y et al (2024) Nanotechnology's frontier in combatting infectious and inflammatory diseases: prevention and treatment. Signal Transduct Target Ther 9(1):34

Jain A et al (2013) A new horizon in modifications of chitosan: syntheses and applications. Crit Rev Ther Drug Carrier Syst 30(2)

Janczy-Cempa E et al (2022) Significance of specific oxidoreductases in the design of hypoxia-activated prodrugs and fluorescent turn off–on probes for hypoxia imaging. Cancer 14(11):2686

Jhaveri A, Torchilin V (2016) Intracellular delivery of nanocarriers and targeting to subcellular organelles. Expert Opin Drug Deliv 13(1):49–70

Kaushik N et al (2022) Nanocarrier cancer therapeutics with functional stimuli-responsive mechanisms. J Nanobiotechnol 20(1):152

Kesharwani P et al (2019) Dendrimer-entrapped gold nanoparticles as promising nanocarriers for anticancer therapeutics and imaging. Prog Mater Sci 103:484–508

Khalili L et al (2022) Smart active-targeting of lipid-polymer hybrid nanoparticles for therapeutic applications: recent advances and challenges. Int J Biol Macromol 213:166–194

Kim J et al (2017) Co-coating of receptor-targeted drug nanocarriers with anti-phagocytic moieties enhances specific tissue uptake versus non-specific phagocytic clearance. Biomaterials 147:14–25

Koksharov YA (2009) Magnetism of nanoparticles: effects of size, shape, and interactions. Magn Nanoparticles:197–254

Kumar V et al (2019) Advanced functional structure-based sensing and imaging strategies for cancer detection: possibilities, opportunities, challenges, and prospects. Adv Funct Mater 29(16):1807859

Kumar S, Kumar M, Singh A (2021) Synthesis and characterization of iron oxide nanoparticles (Fe2O3, Fe3O4): a brief review. Contemp Phys 62(3):144–164

Lai C et al (2023) Current advances and prospects in carbon nanomaterials-based drug deliver systems for cancer therapy. Curr Med Chem 30(24):2710–2733

Li X, Kono K (2018) Functional dendrimer–gold nanoparticle hybrids for biomedical applications. Polym Int 67(7):840–852

Li X et al (2017) Cancer-associated, stimuli-driven, turn on theranostics for multimodality imaging and therapy. Adv Mater 29(23):1606857

Liu S et al (2020) Current applications of poly (lactic acid) composites in tissue engineering and drug delivery. Compos Part B 199:108238

Liu M et al (2022a) Nanotechnology-based drug delivery system for anticancer active ingredients from traditional Chinese medicines: a review. Am J Chin Med 50(08):2011–2032

Liu P, Chen G, Zhang J (2022b) A review of liposomes as a drug delivery system: current status of approved products, regulatory environments, and future perspectives. Molecules 27(4):1372

Liu X, Li M, Woo S (2024) Subcellular drug distribution: exploring organelle-specific characteristics for enhanced therapeutic efficacy. Pharmaceutics 16(9):1167

Luo D et al (2021) Recent development of gold nanoparticles as contrast agents for cancer diagnosis. Cancer 13(8):1825

Ma Y et al (2016) Cancer-targeted nanotheranostics: recent advances and perspectives. Small 12(36):4936–4954

Martí-Bonmatí L et al (2010) Multimodality imaging techniques. Contrast Media Mol Imaging 5(4):180–189

Monteiro-Riviere, N.A., Tran C.L. (2014) Progress toward Nanomedicine.

Mu Q et al (2014) Chemical basis of interactions between engineered nanoparticles and biological systems. Chem Rev 114(15):7740–7781

Nam J et al (2013) Surface engineering of inorganic nanoparticles for imaging and therapy. Adv Drug Deliv Rev 65(5):622–648

Nguyen TT et al (2021) Nanotechnology-based drug delivery for central nervous system disorders. Biomed Pharmacother 143:112117

Nordmann A (2004) Converging technologies: shaping the future of European societies. Office for Official Publications of the European Communities Luxembourg

Nowak-Jary J, Machnicka B (2024) Comprehensive analysis of the potential toxicity of magnetic iron oxide nanoparticles for medical applications: cellular mechanisms and systemic effects. Int J Mol Sci 25(22):12013

Padmanabhan P et al (2016) Nanoparticles in practice for molecular-imaging applications: an overview. Acta Biomater 41:1–16

Parvez S et al (2022) Dodging blood brain barrier with "nano" warriors: novel strategy against ischemic stroke. Theranostics 12(2):689

Paul W, Sharma CP (2020) Inorganic nanoparticles for targeted drug delivery. Biointegr Med Implant Mater:333–373

Pichler BJ, Judenhofer MS, Pfannenberg C (2008) Multimodal imaging approaches: pet/ct and pet/mri. Mol Imaging I:109–132

Pontico M et al (2023) 18F-fluorodeoxyglucose (18F-FDG) functionalized gold nanoparticles (GNPs) for plasmonic photothermal ablation of cancer: a review. Pharmaceutics 15(2):319

Prasanth, R., et al (2015) The high-impact engineering materials of the millennium

Radhakrishnan D et al (2022) The emergence of nanoporous materials in lung cancer therapy. Sci Technol Adv Mater 23(1):225–274

Reddy LH et al (2012) Magnetic nanoparticles: design and characterization, toxicity and biocompatibility, pharmaceutical and biomedical applications. Chem Rev 112(11):5818–5878

Revell P (2006) The biological effects of nanoparticles. Nanotechnol Percept 2(3):283–298

Robertson N (1998) The sensitivity of human tumour cells to quinone-based bioreductive anticancer agents. The University of Manchester (United Kingdom)

Saber-Samandari S, Saber-Samandari S (2017) Biocompatible nanocomposite scaffolds based on copolymer-grafted chitosan for bone tissue engineering with drug delivery capability. Mater Sci Eng C 75:721–732

Saifi MA, Khan W, Godugu C (2018) Cytotoxicity of nanomaterials: using nanotoxicology to address the safety concerns of nanoparticles. Pharmaceut Nanotechnol 6(1):3–16

Santhanakrishnan KR, Koilpillai J, Narayanasamy D (2024) PEGylation in pharmaceutical development: current status and emerging trends in macromolecular and immunotherapeutic drugs. Cureus 16(8)

Saraf S et al (2020) Advances in liposomal drug delivery to cancer: an overview. J Drug Deliv Sci Technol 56:101549

Sargazi S et al (2022) Aptamer-conjugated carbon-based nanomaterials for cancer and bacteria theranostics: a review. Chem Biol Interact 361:109964

Saroj S, Rajput SJ (2018) Composite smart mesoporous silica nanoparticles as promising therapeutic and diagnostic candidates: recent trends and applications. J Drug Deliv Sci Technol 44:349–365

Seddon B, Kelland LR, Workman P (2004) Bioreductive prodrugs for cancer therapy. Suicide Gene Ther: Methods Rev:515–542

Sharma S et al (2019) Nanoparticles-based magnetic and photo-induced hyperthermia for cancer treatment. Nano Today 29:100795

Shinde VR et al (2022) Enhanced permeability and retention effect: A key facilitator for solid tumor targeting by nanoparticles. Photodiagn Photodyn Ther 39:102915

Siafaka PI et al (2021) Current update on nanoplatforms as therapeutic and diagnostic tools: a review for the materials used as nanotheranostics and imaging modalities. Asian J Pharmaceut Sci 16(1):24–46

Siddique S, Chow JC (2020) Application of nanomaterials in biomedical imaging and cancer therapy. Nano 10(9):1700

Silva CO et al (2019) Current trends in cancer nanotheranostics: metallic, polymeric, and lipid-based systems. Pharmaceutics 11(1):22

Singh, B. (2017) Major concerns on antibiotic resistance and antibiotic-induced complications–paradigm shift to nanaoscale and secondary metabolites from edible plants for rapid killing of bacteria as an imperative

Singh A, Amiji MM (2022) Application of nanotechnology in medical diagnosis and imaging. Curr Opin Biotechnol 74:241–246

Singh AV et al (2021) Emerging application of nanorobotics and artificial intelligence to cross the BBB: advances in design, controlled maneuvering, and targeting of the barriers. ACS Chem Neurosci 12(11):1835–1853

Skupin-Mrugalska P et al (2018) Theranostic liposomes as a bimodal carrier for magnetic resonance imaging contrast agent and photosensitizer. J Inorg Biochem 180:1–14

Smith KJ (2017) Understanding sieving and clogging in ultrathin membrane filtration. University of Rochester

Soenen SJ et al (2015) (Intra) cellular stability of inorganic nanoparticles: effects on cytotoxicity, particle functionality, and biomedical applications. Chem Rev 115(5):2109–2135

Song K et al (2024) Functional inorganic nanomaterials for optical cancer theranostics. Chem Eng J 485:150067

Su H et al (2018) Potential applications and human biosafety of nanomaterials used in nanomedicine. J Appl Toxicol 38(1):3–24

Tchuendem LM (2024) Investigating the immune cell response to albumin-derived perfluorocarbon-based artificial oxygen carriers (A– AOCs). Dissertation. Universität Duisburg-Essen, Duisburg, Essen

Tee JK et al (2016) Oxidative stress by inorganic nanoparticles. Wiley Interdiscip Rev Nanomed Nanobiotechnol 8(3):414–438

Teixeira M, Carbone C, Souto E (2017) Beyond liposomes: Recent advances on lipid based nanostructures for poorly soluble/poorly permeable drug delivery. Prog Lipid Res 68:1–11

Thakur VK, Thakur MK (2018) Chemical functionalization of carbon nanomaterials. CRC Press, Warentown, NJ

Thangam R, Paulmurugan R, Kang H (2021) Functionalized nanomaterials as tailored theranostic agents in brain imaging. Nano 12(1):18

Tirumala MG et al (2021) Novel methods and approaches for safety evaluation of nanoparticle formulations: a focus towards in vitro models and adverse outcome pathways. Front Pharmacol 12:612659

Tiwari A, Tiwari A (2013) Nanomaterials in drug delivery, imaging, and tissue engineering. John Wiley & Sons

Tiwari H et al (2024) Advancing era and rising concerns in nanotechnology-based cancer treatment. ACS Chem Health Safe 31(2):153–161
Tripathi D, Hajra K, Maity D (2023) Recent advancement of bio-inspired nanoparticles in cancer theragnostic. J Nanotheranost 4(3):299–322
Truong TT et al (2024) Precision-engineered metal and metal-oxide nanoparticles for biomedical imaging and healthcare applications. Adv Colloid Interf Sci:103263
Tutty MA, Movia D, Prina-Mello A (2022) Three-dimensional (3D) liver cell models-a tool for bridging the gap between animal studies and clinical trials when screening liver accumulation and toxicity of nanobiomaterials. Drug Deliv Transl Res 12(9):2048–2074
Unnikrishnan G et al (2023) Exploration of inorganic nanoparticles for revolutionary drug delivery applications: a critical review. Discover Nano 18(1):157
Vashishat A et al (2024) Alternatives of animal models for biomedical research: a comprehensive review of modern approaches. Stem Cell Rev Rep 20(4):881–899
Veg E et al (2025) The role of nanomaterials in diagnosis and targeted drug delivery. Chem Biodivers 22(1):e202401581
Vuong DLM (2022) Degradation mechanisms of nanoparticle-based drug delivery systems: the case of polymer conjugate and metal-organic framework. Université Paris-Saclay
Vyas D, Patel M, Wairkar S (2022) Strategies for active tumor targeting-an update. Eur J Pharmacol 915:174512
Wang X et al (2016) Synthesis of tumor-targeted folate conjugated fluorescent magnetic albumin nanoparticles for enhanced intracellular dual-modal imaging into human brain tumor cells. Anal Biochem 512:8–17
Wang X et al (2020) Recent progress of chemodynamic therapy-induced combination cancer therapy. Nano Today 35:100946
Wang Q et al (2023) Surface hydrolysis-designed AuNPs-zwitterionic-glucose as a novel tool for targeting macrophage visualization and delivery into infarcted hearts. J Control Release 356: 678–690
Wong XY et al (2020) Nanomaterials for nanotheranostics: tuning their properties according to disease needs. ACS Nano 14(3):2585–2627
Xu L et al (2020) Silver nanoparticles: synthesis, medical applications and biosafety. Theranostics 10(20):8996
Yang W et al (2020) Soft and condensed nanoparticles and nanoformulations for cancer drug delivery and repurpose. Adv Therapeut 3(1):1900102
Yetisgin AA et al (2020) Therapeutic nanoparticles and their targeted delivery applications. Molecules 25(9):2193
Zambonino MC et al (2023) Biogenic selenium nanoparticles in biomedical sciences: properties, current trends, novel opportunities and emerging challenges in theranostic nanomedicine. Nano 13(3):424
Zhang A et al (2020) Absorption, distribution, metabolism, and excretion of nanocarriers in vivo and their influences. Adv Colloid Interf Sci 284:102261
Zhang X et al (2021) Advances in nanomedicines for diagnosis of central nervous system disorders. Biomaterials 269:120492
Zhang Y et al (2024) Strategies to regulate the degradation and clearance of mesoporous silica nanoparticles: a review. Int J Nanomed 19:5859
Zhu X et al (2024) Stimuli-responsive 19F MRI probes: From materials design to in vitro detection and in vivo diagnosis. TrAC Trends Anal Chem 117607
Zou Y et al (2021) Tailored mesoporous inorganic biomaterials: assembly, functionalization, and drug delivery engineering. Adv Mater 33(2):2005215

Cancer Imaging and Theranostics Intervention: Current Status and Future Perspective

15

Bela Jain and Ankita Singh

Abstract

Theranostics have been a breakthrough in oncology, offering a dual approach that integrates diagnosis and targeted therapy within a single platform. This advancement has refined personalized medicine, enabling clinicians to visualize tumors, assess their molecular characteristics, and deliver precise treatments tailored to individual patients. Current cancer imaging modalities, including PET, MRI, and CT, have been enhanced by the development of radiopharmaceuticals and nanoparticle-based agents that not only identify malignant tissues with high specificity but also serve as vehicles for therapeutic delivery. So far, various classes of theranostics agents have been reported for solid tumor imaging and therapeutics in combination. Among them, radiolabeled antibodies, peptides, nanoparticles, etc. have gained tremendous attention in treating various cancers, notably in neuroendocrine tumors and prostate cancer. On the other hand, the integration of molecular imaging with theranostics approaches has improved the accuracy of tumor staging, patient stratification, and treatment planning, leading to better outcomes. Such agents augment the real-time monitoring of therapeutic efficacy and early detection of resistance, providing a methodology that can guide treatment adjustments, while minimizing organ toxicity. Despite these advancements, challenges remain, including the need for more robust biomarkers, improved targeting specificity, and the development of agents that can overcome

B. Jain (✉)
Department of Nuclear Medicine, All India Institute of Medical Sciences, New Delhi, India

A. Singh
Department of Pathology, Sanjay Gandhi Post Graduate Institute of Medical Sciences, Lucknow, India

R. Prasad, M. Mohan Yallapu (eds.), *Advancements in Cancer Theranostics*,
https://doi.org/10.1007/978-981-95-0938-6_15

tumor heterogeneity and resistance mechanisms. Herein, various advancements in theranostics enabling precise and effective cancer imaging and therapy have been discussed.

Keywords

Nanotheranostics · Theranostics intervention · Solid tumor · Bioimaging · Translational · Nanoparticles medicine

15.1 Introduction

15.1.1 Cancer: A Global Health Crisis

Cancer is a complex and heterogeneous group of diseases that are characterized by uncontrolled cell growth and spread to other parts of the body. It is one of the leading causes of morbidity and mortality worldwide. The global burden of cancer has been steadily increasing, with an estimated 19.98 million new cancer cases and almost 9.74 million cancer-related deaths in 2022 alone, according to the World Health Organization (WHO) data (Bray et al. 2024). Cancer can develop in almost any organ or tissue, and it is classified into over 100 different types, including carcinomas (cancers of epithelial cells), sarcomas (cancers of connective tissues), leukemias (blood cancers), lymphomas, and brain tumors. The heterogeneity of cancer at the genetic, molecular, and histological levels makes it a challenging disease to diagnose and treat. Cancer arises due to the accumulation of genetic mutations that drive uncontrolled cell division, resistance to cell death, and the ability to invade surrounding tissues. These mutations can occur in the genes regulating cell growth, differentiation, and apoptosis (programmed cell death), or in genes responsible for maintaining the integrity of the genome, leading to genomic instability. In addition, the ability of tumor cells to metastasize, or spread to other organs, makes cancer particularly lethal.

Early detection of cancer is crucial for improving survival rates. The earlier the cancer is diagnosed, the more effective the treatment can be. However, detecting cancer at an early stage is challenging, as many tumors may not exhibit symptoms until they are well advanced.

15.1.2 Challenges in Early Diagnosis and Concept of Theranostics

While screening methods like mammography for breast cancer and colonoscopy for colorectal cancer exist, not all types of cancer have reliable screening tests. Many cancers, such as pancreatic and ovarian cancer, often present with few or no symptoms until they have progressed to an advanced stage, making diagnosis at an early stage difficult. Some tumors can be difficult to visualize, especially in organs with high tissue density or deep anatomical locations (e.g., liver, pancreas, or brain). Therefore, achieving early diagnosis requires advanced imaging technologies that can detect cancer at the molecular and cellular level, even before structural changes to tissues are visible. The term "theranostics" was coined by Dr. Rafael Yuste, a neuroscientist and researcher at Columbia University, in the early 2000s. The word is a combination of "therapy" and "diagnostics," reflecting the merging of diagnostic and therapeutic approaches, particularly in the context of personalized medicine. The concept of theranostics has its roots in the growing field of molecular imaging and nanomedicine, where scientists began to explore how diagnostic tools could be used to guide and enhance targeted therapies. Over time, theranostics has evolved to refer to the integration of diagnostic imaging, molecular biology, and targeted therapy, particularly in the treatment of cancer. One of the primary advantages of theranostics is the ability to guide therapy based on precise information from imaging. By using molecular imaging, theranostics can detect changes at the cellular or molecular level, allowing for earlier intervention and treatment. This is crucial for aggressive cancers that metastasize quickly or for cancers where symptoms appear late in the disease process.

The application of theranostics in clinical practice has already demonstrated substantial benefits, particularly in terms of early diagnosis, delivering targeted therapy, treatment monitoring, and testing treatment effectiveness with minimized side effects. **Personalized medicine**, also known as precision medicine, focuses on tailoring medical treatment to the individual characteristics of each patient. This approach includes factors like genetic makeup, lifestyle, and environment, allowing clinicians to choose the most appropriate treatment and diagnostic tools for each patient. Theranostic techniques are central to personalized medicine, as they provide a real-time view of a patient's cancer, enabling clinicians to adjust therapy as needed and monitor its effectiveness.

15.1.3 Principles of Cancer Imaging

Medical imaging plays a crucial role in diagnosing, staging, and monitoring the progression of cancer. The goal of cancer imaging is to detect abnormalities in tissues, assess the extent of the disease, monitor treatment responses, and plan surgical interventions. Cancer imaging techniques use different physical principles to visualize internal structures in the body and provide insight into the cellular and molecular changes associated with cancer. These modalities are broadly categorized into anatomical imaging, functional imaging, and molecular imaging (Table 15.1).

The choice of imaging modality often depends on the location of the tumor, the type of cancer, the stage of the disease, and the clinical goal. Each modality has its unique advantages and is used in different clinical scenarios and shall be discussed in the sections below.

15.1.3.1 X-Ray and Computed Tomography (CT)

X-rays are used to create images of the inside of the body. In oncology, chest X-rays are commonly used to detect lung cancer or metastatic lesions in the lungs.

Table 15.1 Types of imaging techniques

Imaging type	Purpose	Key techniques
Anatomical imaging	Provides structural information about tissues and organs	X-ray, CT, MRI, ultrasound
Functional imaging	Assesses physiological or metabolic activity within tissues	PET, fMRI, SPECT
Molecular imaging	Visualizes molecular and cellular processes, tumor-specific biomarkers	PET/CT with molecular radiotracers, MRS, fluorescence imaging, nanoparticle imaging, bioluminescence

Mammography is a specialized X-ray imaging technique used to detect breast cancer. Computed Tomography (CT) uses X-rays to create cross-sectional images of the body. Opposite the X-ray source are **detectors** that measure the intensity of X-rays after they pass through the body. The X-ray tube and detectors rotate around the patient, capturing multiple projections (views) from different angles. As the X-rays pass through the body, they are **attenuated** (weakened) depending on the density and composition of the tissues. Different tissues attenuate X-rays to varying degrees, e.g., bones have a high attenuation coefficient and appear white on imaging. The raw data from the detectors (attenuation profiles) are processed using advanced algorithms. Unlike traditional X-rays, CT provides more detailed, three-dimensional images and allows clinicians to assess tumor size, shape, location, and metastasis (Seeram 2015).

It helps to determine whether cancer has spread to lymph nodes or other organs and is essential for planning treatment strategies, including surgery and radiation therapy. CT is commonly used to evaluate cancers in the lungs, abdomen, liver, and colorectal regions.

15.1.3.2 Magnetic Resonance Imaging (MRI)

Magnetic Resonance Imaging (MRI) is a noninvasive imaging technique, without using ionizing radiation (as in X-rays or CT scans). It uses strong magnetic fields and radiofrequency waves to create cross-sectional images of organs, tissues, and blood vessels. MRI is especially useful in imaging the brain, spinal cord, soft tissue tumors (e.g., breast cancer), and cancers in organs like the liver, prostate, and pancreas. The principle of MRI is based on the interaction of hydrogen nuclei (protons) in the body with a strong magnetic field and radiofrequency (RF) pulses. The body is primarily composed of water, which contains hydrogen atoms (protons). When the body is placed inside a strong magnetic field (usually 1.5–3 Tesla), the hydrogen protons align either parallel or antiparallel to the magnetic field. A short RF pulse is applied, which provides energy to the hydrogen protons, causing them to "flip" out of alignment with the magnetic field. After the RF pulse is turned off, the hydrogen protons relax back to their original alignment with the magnetic field. During this relaxation, they release energy in the form of radiofrequency signals, which are converted into electrical signals. A computer then processes these signals to form an image. The different relaxation times (T1 and T2) of various tissues (fat, muscle, water, etc.) give MRI its excellent contrast and ability to differentiate between tissue types (Minhas and Oliver 2022). **Magnetic Resonance Spectroscopy** (MRS) is also an MRI-based technique used to detect biochemical changes in tissues. It can identify specific metabolites in tumors, providing insight into the metabolic activity of cancer cells. MRS detects the levels of metabolites in the tissue, such as choline, creatine, and N-acetylaspartate (NAA), by measuring their magnetic resonance signals. MRS is particularly useful in evaluating brain tumors, including gliomas, and assessing treatment responses (Gharzeddine et al. 2019). **Functional MRI (fMRI)** measures changes in blood flow associated with neural activity, and it can be used to map important brain functions (such as speech or movement) in relation to tumor location, while planning surgeries or radiation therapy. There are other several specialized types of MRI, each designed for specific purposes or to visualize particular aspects of the body (Yousaf et al. 2018).

15.1.3.3 Positron Emission Tomography (PET)

Positron Emission Tomography (PET) is a nuclear medicine imaging technique that uses radioactive tracers to detect metabolic activity in the body. Cancer cells typically have higher metabolic rates than normal cells, and PET imaging can identify these areas of abnormal activity, often before tumors become visible on CT or MRI scans. PET uses a small amount of radioactive material (called a radiotracer) that is injected into the body. This radiotracer is typically a form of glucose (e.g., fluorodeoxyglucose or FDG) tagged with a positron-emitting isotope. The radiotracer accumulates in areas with high metabolic activity (such as tumors). The radioactive isotopes decay and emit positrons (the antimatter counterpart of electrons). When a positron encounters an electron, they annihilate each other, producing two gamma rays that move in opposite directions. The PET scanner detects these gamma rays, and a computer processes the signals to create images that show the distribution of the radiotracer, reflecting areas of high metabolic activity (e.g., cancer, inflammation) (Fonti et al. 2019). These tracers typically consist of a biologically active molecule attached to a positron-emitting isotope (such as fluorine-18, carbon-11, oxygen-15, or nitrogen-13). The type of tracer used depends on the biological process being studied (Table 15.2). Fluorodeoxyglucose (FDG)—[^{18}F] FDG is the most commonly used PET tracer and is a glucose analog. It mimics glucose but is not metabolized completely, accumulating in cells that have high glucose consumption, such as cancer cells, which tend to have higher metabolic activity.

PET and CT scans are performed sequentially or simultaneously. The computer fuses the functional information from the PET scan (showing metabolic activity) with the detailed anatomical images from the CT scan. This fusion allows for accurate localization of metabolic abnormalities in relation to anatomical structures. PET can be combined with CT (PET/CT) to provide both functional and anatomical information in one scan. This fusion allows for precise localization of cancerous tissues and is invaluable for treatment planning, staging, and monitoring responses to therapy. An emerging technique, **PET/MRI** is particularly useful in brain tumors

Table 15.2 Summary of key PET tracers and their applications

Tracer	Target	Applications
^{18}F-FDG	Glucose metabolism	Cancer detection, neurological disorders, cardiology
^{18}F-Fluoride	Bone metabolism	Bone metastasis, fractures, osteoporosis
^{11}C-Methionine	Protein synthesis	Brain tumors, neurological diseases
^{18}F-DOPA	Dopamine synthesis	Parkinson's disease, brain tumor imaging
^{15}O-Water	Blood flow	Brain and cardiac perfusion studies
^{11}C-Acetate	Fatty acid metabolism	Heart studies, tumors
^{68}Ga-DOTATATE	Somatostatin receptors	Neuroendocrine tumors
^{68}Ga-PSMA	Prostate-specific membrane antigen	Prostate cancer imaging

and pediatric cancers. MRI provides high-resolution anatomical images, while PET assesses metabolic activity, enhancing the diagnostic capabilities in complex cancers.

15.1.3.4 Gamma Imaging or Single Photon Emission Computed Tomography (SPECT)

Single Photon Emission Computed Tomography (SPECT) is another nuclear imaging technique that uses radioactive tracers to assess tissue function. After the radiotracer is injected, it decays by emitting gamma rays. Unlike positron emission in PET, which produces two gamma rays in opposite directions, SPECT detects single gamma photons emitted from the radiotracer inside the body. Unlike PET, which detects positrons, SPECT detects gamma rays emitted from these radiotracers. Combining SPECT with CT imaging enhances the ability to pinpoint the exact location of abnormal metabolic activity seen in SPECT imaging. This fusion allows for improved cancer diagnosis and monitoring of treatment efficacy (Bybel et al. 2008). SPECT is less commonly used for direct cancer detection, but it can help in detecting tumor recurrence, metastasis, and assessing the effectiveness of treatments such as chemotherapy or radiotherapy.

Common SPECT radiotracers include Technetium-99 m (^{99m}Tc), Iodine-123 (^{123}I), and Thallium-201 (^{201}Tl), which are often used in various diagnostic applications. The radiotracer is typically designed to target a specific biological process, such as blood flow, cellular metabolism, or receptor activity, and will accumulate in areas with high metabolic or functional activity (Table 15.3).

15.1.3.5 Ultrasound

Ultrasound imaging uses sound waves to create real-time images of internal organs. It is often used in oncology for imaging soft tissues such as the liver, kidneys, thyroid, and breasts. Ultrasound is a noninvasive, cost-effective, and portable modality, although its resolution is lower compared to MRI or CT. Ultrasound can be used for tumor detection, guiding biopsies, and monitoring treatment response.

Table 15.3 Summary of SPECT radiopharmaceuticals

Tracer	Target	Applications
^{99m}Tc-MIBI	High metabolic activity	Breast cancer, lung cancer, head and neck cancers
^{99m}Tc-Sestamibi	Mitochondrial activity	Breast cancer, solid tumors, monitoring multidrug resistance
^{99m}Tc-Colloids	Reticuloendothelial system (RES)	Imaging lymph node metastases, assessing lymphatic spread
^{201}Tl	High metabolic activity	Lung cancer, head and neck cancer, melanoma
^{123}I-FIAU	DNA synthesis	Brain tumors, solid tumors with increased DNA synthesis
^{123}I-MIBG	Norepinephrine transporter	Neuroendocrine tumors, pheochromocytomas, neuroblastomas
^{111}In-Octreotide	Somatostatin receptors	Neuroendocrine tumors, staging, and treatment planning
^{111}In-Labeled Antibodies	Tumor antigens	Lymphomas, colorectal cancer, ovarian cancer

15.1.3.6 Optical Imaging

Optical imaging uses light (infrared or visible) to visualize tissues and cellular structures. While this technique is primarily used in research and preclinical studies, it is becoming more relevant in clinical oncology. Optical imaging modalities include fluorescence imaging, bioluminescence imaging, and near-infrared fluorescence (NIRF) imaging. **Fluorescence and Bioluminescence** are often used in animal models to track tumor growth, evaluate metastasis, and monitor responses to treatments. Bioluminescence relies on the emission of light from genetically engineered cells that express luciferase, while fluorescence imaging uses fluorescent dyes that bind to specific cancer markers (like organic dyes, quantum dots, and fluorescent proteins). **Near-Infrared Fluorescence (NIRF) Imaging** allows for deeper tissue penetration than visible light and reduced background autofluorescence, enhancing imaging of deeper tissues. It is particularly useful for visualizing cancerous tissues in the brain, breast, and other organs. Fluorescence imaging agents that selectively bind to tumor-specific biomarkers are increasingly being used for real-time visualization of small tumors during surgery or diagnostic procedures. For example, 5-aminolevulinic acid (5-ALA), which fluoresces when metabolized by tumor cells, is used for visualizing gliomas during brain surgery. Such techniques have the potential to improve the accuracy of tumor detection and enhance surgical resection (Martelli et al. 2016). The clinical application of these procedures faces two primary challenges. First, the use of fluorescence imaging in humans is limited by physical constraints, which could be addressed by developing new fluorophores or employing advanced imaging techniques such as endoscopic or intraoperative methods. Second, many of the probes have not yet received approval for human studies, despite significant progress in evaluating their sensitivity, specificity, and safety profiles, including the absence of toxic or adverse effects. Once these agents are approved for clinical use, optical imaging has the potential to revolutionize tumor treatment by enabling real-time, image-guided surgeries for more complete tumor resections and providing noninvasive tumor characterization without the risks associated with ionizing radiation.

15.1.4 Liquid Biopsy

The use of liquid biopsy, involving the detection of **circulating tumor DNA (ctDNA)**, **circulating tumor cells (CTCs)**, or **exosomes** in blood or other bodily fluids, is a rapidly growing field in cancer diagnostics. Liquid biopsy, when combined with molecular imaging, offers an opportunity for noninvasive, early detection of tumors. Research is underway to develop imaging agents that can specifically target ctDNA or RNA markers present in the bloodstream, providing a real-time, noninvasive method for monitoring tumor presence and response to therapy. This approach holds significant potential in detecting cancers at earlier stages, especially those that are difficult to biopsy (Nikanjam et al. 2022). Imaging technologies form the backbone of cancer diagnosis, staging, and treatment monitoring. As imaging modalities continue to evolve, they are becoming increasingly sophisticated, offering not only detailed anatomical images but also functional and molecular insights into cancer biology. Molecular imaging, in particular, has paved the way for more personalized and precise approaches to cancer care, allowing clinicians to detect tumors earlier, choose more effective treatments, and monitor their success in real time.

15.1.5 The Role of Molecular Targeting in Theranostics

At the core of theranostic approaches is molecular targeting. Cancer cells often exhibit specific markers or alterations in cellular processes that are not present or are less prominent in normal cells. By identifying and targeting these unique molecular features, theranostic agents can be designed to selectively bind to cancer cells, allowing for more efficient diagnosis and treatment. These molecular targets can include overexpressed proteins, mutated genes, or altered signaling pathways that are specific to the cancer type. Molecular imaging techniques, such as PET, MRI, and SPECT, can identify these targets in the early stages of cancer, even before physical changes to the tissue structure are apparent. Theranostic agents, which are often conjugates of imaging tracers and therapeutic agents, can bind to these markers, providing both visualization of the tumor and delivery of localized treatment. Theranostic agents can be broadly classified into several categories based on their mechanisms of action. These include:

15.1.5.1 Radiopharmaceuticals

Nuclear medicine imaging is mainly based on the principle of using radioactive isotopes linked to specific molecules (i.e., radiopharmaceutical agents or radiotracers) to assess key biologic pathways, especially the pathophysiologic features of diseases. Therapy in nuclear medicine acts in the same way: Radiopharmaceutical agents are used to target diseased tissue, with radiation applied at the cellular level by way of a specific chemical and/or biologic affinity. The imaging or therapeutic capability of each radioisotope is determined according to the type of radiation emitted because electromagnetic radiation (i.e., γ rays) can be detected by imaging systems (scintigraphy, SPECT, and PET), and particulate irradiations have cytotoxic (i.e., therapeutic) properties. There are two main forms of particulate radiation used for therapeutic applications: radiation with α particles and radiation with β particles (which are electrons). These two forms of radiation share a common feature: a high transfer of energy to tissues that leads to severe cellular injury due to DNA damage, as shown in Table 15.4. The α particles have higher energy and much greater mass. Some isotopes (e.g., Lutetium 177 [^{177}Lu]) can emit both electromagnetic and particulate radiation, making simultaneous treatment and imaging possible. Therefore, combining the emitting characteristics of different radioisotopes with specific molecules that target key cellular pathways is the mechanism of nuclear medicine theranostics (Marin et al. 2020). Classically, isotopes of a single element are used in the theranostic concept, such as Iodine [^{123}I/ ^{131}I] or Copper [^{64}Cu/^{67}Cu]. These are called true theranostic pairs. Hybrid theranostic pairs use different radionuclides for either therapeutic or diagnostic applications, using a combination of two binding sites in the same molecule, e.g., Gallium-Lutetium[^{68}Ga/^{177}Lu] or Gallium-Actinium [^{68}Ga/^{225}Ac]. Using this concept, a consistent pharmacological behavior of the resulting radioconjugate is achieved. The rationale behind this approach is to address the shortcomings of existing therapeutic radionuclides, which are limited by their inherent constraints. This approach aims to bridge this gap by providing a means to overcome the lack of suitable nuclides.

15.1.5.2 Commonly Used Radiopharmaceuticals in Clinical Practice

Prostate-Specific Membrane Antigen (PSMA)-Targeting Agents

Example: *^{68}Ga-PSMA-11* (diagnostic) and *^{177}Lu-PSMA-617* (therapeutic)

- **Diagnostic Use:** Gallium-68 PSMA is a PET imaging agent that binds to PSMA, a receptor highly expressed in prostate cancer cells. It enables precise localization of metastases as shown in Fig. 15.1.
- **Therapeutic Use:** Lutetium-177 PSMA delivers beta radiation to PSMA-positive cancer cells, selectively killing them while sparing healthy tissues.
- **Significance:** Revolutionized the management of advanced prostate cancer by providing a clear path from imaging to treatment.

Peptide Receptor Radionuclide Therapy (PRRT)

Example: *68Ga-DOTATATE* (diagnostic) and *177Lu-DOTATATE* (therapeutic)

- **Diagnostic Use:** Gallium-68 DOTATATE binds to somatostatin receptors, which are overexpressed in neuroendocrine tumors (NETs). PET imaging helps identify tumor burden and receptor status.

Table 15.4 Summary of different theranostic radiopharmaceuticals

Vector	Drug name	Target	Imaging label	Therapy label (particle)	Cancer type (approved year)
Small molecule	FAPI	Fibroblast activation protein	^{68}Ga, ^{18}F, ^{99m}Tc, ^{64}Cu	^{90}Y (beta), ^{231}Bi ^{212}Pb (alpha)	Multiple cancer entities
	OncoFAP, BiOncoFAP	Fibroblast activation protein	^{68}Ga	^{177}Lu	Multiple cancer entities
	MIBG	Norepinephrine transporter	^{123}I, ^{124}I, ^{131}I	^{131}I (beta), ^{125}I (Auger)	Neuroendocrine, including neuroblastoma, pheochromocytoma, ganglioneuroblastomas, ganglioneuromas, paragangliomas, carcinoid tumors
	Radioiodine	Sodium iodide symporter	^{123}I, ^{124}I, ^{131}I (salt solution)	^{131}I (beta)	Thyroid cancer (first used in 1941 for hyperthyroidism)
	Radium-223	Newly formed bone matrix (osteoblastic metastases)	^{18}F, ^{99m}Tc	^{223}Ra (alpha)	Castrate-resistant prostate cancer, symptomatic bone metastases, and no known metastatic disease (2013)
	Strontium-89	Newly formed bone matrix (osteoblastic metastases	^{18}F,^{99m}Tc	^{89}Sr (beta)	Relief of bone pain in patients with confirmed osteoblastic metastatic bone lesions (1991)
Peptide	Lu-177 oxodotreotide	SSTR, SSTR2—agonist mode	^{68}Ga, ^{18}F	^{177}Lu (beta)	Unresectable or metastatic, progressive, well-differentiated neuroendocrine tumors in adults (2018)
	^{90}Y/^{177}Lu-DOTATATE	SSTR, SSTR2—agonist mode	^{68}Ga, ^{18}F	^{90}Y (50%; beta), ^{177}Lu (50%; beta)	Metastatic and inoperable neuroendocrine tumor
	^{213}Bi-DOTATOC	SSTR, SSTR2—agonist mode	^{68}Ga	^{213}Bi (alpha)	Liver metastases refractory to treatment with ^{90}Y/^{177}Lu-DOTATOC
	PSMA-617/ PNT2002/^{131}I-	PSMA	^{68}Ga, ^{18}F		PSMA-positive, metastatic, castration-resistant prostate cancer

	1095/^{177}Lu– PSMA-R2			^{177}Lu, ^{131}I (beta) ^{225}Ac, ^{213}Bi, ^{212}Pb (alpha)	
	Pentixather	C-X-C chemokine receptor type 4	^{68}Ga	^{177}Lu (beta)	CXCR4-positive multiple myeloma and lymphoma
	^{177}Lu-GRPR peptide	GRPR/bombesin receptor	^{68}Ga, ^{64}Cu	^{177}Lu (beta)	Gastrin-releasing peptide receptor (GRPR)-positive, castrate-resistant prostate cancer
Antibody	^{90}Y-ibritumomab	CD20	^{86}Y ^{111}In-ibritumomab	^{90}Y (beta)	Adult patients with relapsed or refractory, low-grade or follicular B-cell non-Hodgkin lymphoma (2002)
	^{131}I-iobenguane	Norepinephrine transporter	^{123}I, ^{124}I, ^{131}I	^{131}I (beta)	Pheochromocytoma and paraganglioma (2018)
	^{227}Th-BAY 2315497	PSMA	^{68}Ga	^{227}Th (alpha)	PSMA-positive, metastatic, castration-resistant prostate cancer alone or in combination with darolutamide
Nanoparticle medical device for local injection	Microspheres—glass, resin, PLLA	Tumor vessels	^{99m}Tc	^{166}Ho, ^{90}Y (beta)	Unresectable hepatocellular carcinoma and liver metastases

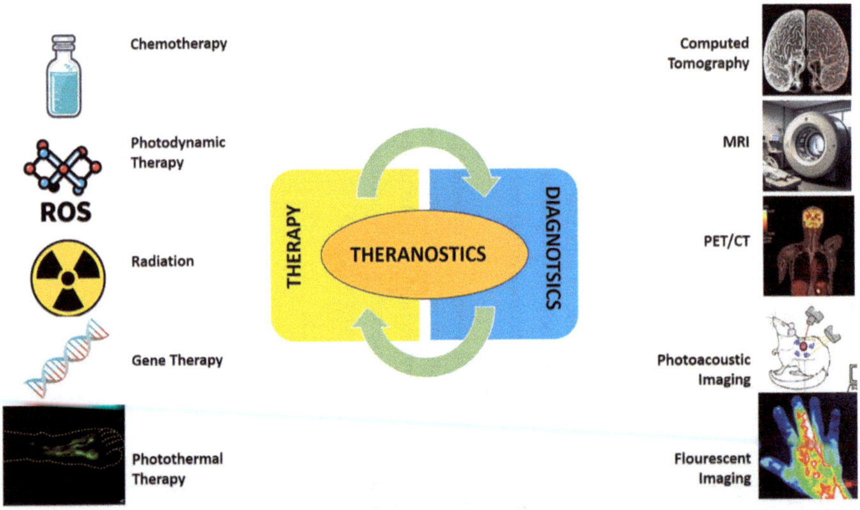

Fig. 15.1 Various therapeutic and diagnostic modalities for cancer imaging and treatment

- **Therapeutic Use:** Lutetium-177 DOTATATE delivers therapeutic radiation, targeting somatostatin receptor-positive cells to reduce tumor size and progression.
- **Significance:** PRRT has transformed outcomes for patients with NETs, improving survival and quality of life.

Radiolabeled Antibodies for Hematologic Malignancies

Example: *Ibritumomab tiuxetan* (Zevalin)

- **Dual Use:** Zevalin is a monoclonal antibody labeled with Yttrium-90 (therapeutic) or Indium-111 (diagnostic). It targets CD20, a marker on B-cells, for imaging and treatment of non-Hodgkin lymphoma.
- **Mechanism:** Antibody-mediated targeting ensures selective radiation delivery to cancer cells.
- **Significance:** First FDA-approved radiopharmaceutical for treating hematologic cancers.

Alpha-Particle Theranostics

Example: *225Ac-PSMA* (therapeutic)

- **Mechanism:** Actinium-225 emits alpha particles, which have a short path but high energy, causing irreparable DNA damage in cancer cells.
- Used for prostate cancer patients resistant to Lutetium-177 therapy.
- **Emerging Role:** Alpha therapies are gaining attention due to their ability to overcome resistance in certain cancer types.

Iodine-Based Theranostics

Example: *I-123/I-124/I-131 for Thyroid Cancer*

- **Diagnostic Use:** Iodine-123 purely γ-ray emitter, is also well suited for use in SPECT and hybrid imaging in conjunction with CT (SPECT/CT). Iodine-124, a promising diagnostic agent, is a positron emitter and is therefore used in PET.
- **Therapeutic Use:** ^{131}I is a dual (β- and γ-ray) emitting isotope with a low cost and long half-life (~8 days) that is used for therapy and whole-body diagnostic scanning to identify remnant thyroid tissue or metastatic thyroid cancer.
- **Significance:** Pioneered the concept of theranostics, providing the foundation for modern approaches. The applications for ^{131}I therapy, which may be superimposed in the same patient, consist of ablation of residual normal tissue, adjuvant treatment of occult disease, and treatment of known local-regional or, especially, metastatic disease.

^{131}I-MIBG for Neural Crest-Derived Tumors

- The molecular analog of norepinephrine, MIBG labeled with ^{131}I (^{131}I-MIBG), is another classic theranostic agent that is suitable for gamma camera and SPECT imaging, and it has been used in clinical settings since 1981.
- ^{131}I-MIBG enters the neuroendocrine cells of the sympathetic nervous system by means of endocytosis or diffusion and remains stored in neurosecretory vesicles.
- It has high sensitivity (>90%) and specificity (>95%) for the detection of neural crest-derived (neuroectodermic) tumors.

Breast Cancer

- Breast cancer, particularly HER2-positive subtypes, has also seen significant advancements in theranostics.
- HER2 Imaging and Therapy: HER2-targeted imaging using agents like trastuzumab labeled with a radioactive isotope allows for the assessment of tumor response during treatment.
- Moreover, HER2-targeted therapy has shown remarkable success in patients with HER2-positive breast cancer, and monitoring the therapy with imaging techniques ensures that tumors do not develop resistance.

Radiopharmaceuticals are the most widely used aspect of theranostics today, offering both diagnostic and therapeutic capabilities. The advent of alpha emitters and the integration of radiopharmaceuticals with nanoparticles represent significant advancements in this field. However, their full potential is constrained by challenges such as limited isotope availability, imprecise dosimetry, radiation-induced toxicity (particularly in organs with high uptake like the kidneys and salivary glands), and high costs (Nelson et al. 2024).

15.1.6 Nanoparticles for Drug Delivery

Nanoparticles are engineered particles with dimensions in the nanometer range (typically between 1 and 100 nm). Due to their small size, nanoparticles can penetrate biological barriers and accumulate in tumor tissues more effectively than conventional drugs. Their unique properties, such as high surface area and the ability to be easily functionalized, make them ideal candidates for theranostic applications (Table 15.5). Nanoparticles can be loaded with chemotherapeutic agents, and their surface can be modified with specific ligands, antibodies, or peptides that allow for targeting cancer cells. Once administered, these nanoparticles preferentially accumulate in the tumor due to the **enhanced permeability and retention (EPR)** effect,

Table 15.5 Various types of nanoparticles and their uses

Type of nanoparticle	Composition	Uses in theranostics	Example applications
Metallic nanoparticles	Gold, silver, iron oxide	– Imaging (MRI, CT, photoacoustic) – Photothermal therapy – Drug delivery	– Gold nanoparticles for tumor ablation – Iron oxide nanoparticles for MRI contrast enhancement
Lipid-based nanoparticles	Liposomes, solid lipid	– Drug and gene delivery – Encapsulation of hydrophobic agents – mRNA vaccines	– Lipid nanoparticles in COVID-19 mRNA vaccines – Liposomes for chemotherapy (e.g., Doxil)
Polymeric nanoparticles	PLGA, PEGylated polymers	– Controlled drug release – Encapsulation of imaging agents – Biodegradable drug carriers	– PLGA nanoparticles for sustained drug release – PEGylated particles to improve blood circulation
Quantum dots	Semiconductor materials	– High-resolution optical imaging – Fluorescent labeling	– Imaging in cancer diagnostics – Fluorescence-guided surgery
Magnetic nanoparticles	Iron oxides (Fe_3O_4)	– MRI contrast agents – Magnetically guided drug delivery – Hyperthermia therapy	– Iron oxide nanoparticles for brain imaging – Magnetite for thermal ablation of tumors
Carbon-based nanoparticles	Graphene, carbon tubes	– Imaging – Drug delivery – Photothermal therapy	– Graphene for molecular imaging – Carbon nanotubes for targeted chemotherapy
Silica nanoparticles	Mesoporous silica	– Controlled release of drugs – Imaging agent delivery	– Mesoporous silica nanoparticles for co-delivery of drugs and fluorescent dyes
Hybrid nanoparticles	Combination of materials	– Multimodal imaging – Dual or synergistic therapy (e.g., chemo-photothermal)	– Gold-silica nanoshells for imaging and therapy – Magnetic-lipid hybrids for targeted cancer therapy

a phenomenon where leaky tumor vasculature allows for greater accumulation of particles. This targeted drug delivery system improves the concentration of the therapeutic agent at the tumor site while reducing systemic toxicity. Nanocarriers can encapsulate chemotherapy drugs, protecting them from premature degradation and ensuring their delivery directly to cancer cells. This approach can help reduce the side effects of chemotherapy, improve the therapeutic index, and allow for real-time monitoring of drug distribution using molecular imaging. Gold Nanoparticles absorb light in the near-infrared spectrum, converting it into heat to ablate tumor cells (**photothermal therapy**). Nanoparticles loaded with photosensitizers can generate reactive oxygen species under light exposure, inducing cancer cell death (**photodynamic therapy**) (Nelson et al. 2024; Hou et al. 2020). In addition to delivering therapy, nanoparticles can also serve as imaging agents. Magnetic nanoparticles, for instance, can be used for MRI imaging, while gold nanoparticles can enhance the resolution of CT imaging. Magnetic nanoparticles, used in MRI and magnetic particle imaging (MPI), offer high sensitivity and resolution for tumor detection. These particles can also be used for targeted drug delivery, where therapeutic agents are carried directly to tumor cells using the magnetic field, reducing off-target effects. Gold nanoparticles can be conjugated with specific ligands, such as antibodies or peptides, to target tumor cells. Additionally, gold nanoparticles can be used for photothermal therapy, where localized heating kills tumor cells after being activated by near-infrared light (Thalluri et al. 2023). The ability to use the same nanoparticle for both diagnosis and therapy makes them excellent candidates for theranostic strategies. Nanoparticles can be functionalized with both imaging agents (e.g., fluorescent dyes, MRI contrast agents) and therapeutic molecules (e.g., chemotherapeutic drugs, gene therapy agents). These dual-functional nanoparticles enable the simultaneous diagnosis and treatment of cancer, allowing for monitoring of treatment progress and adjusting therapeutic strategies in real time. These nanocarriers can be designed to release their payload in response to specific stimuli, such as pH changes, temperature, or enzyme activity, which are characteristic of the tumor microenvironment (Liu et al. 2017). These, often referred to as "smart nanoparticles," enable highly selective delivery of both diagnostic and therapeutic agents, minimizing toxicity to healthy tissues. Nanoparticles are also being used to deliver gene editing tools like CRISPR/Cas9 or RNA-based therapies to cancer cells (Naeem et al. 2021). The ability to modify the genetic makeup of tumors to enhance their susceptibility to treatment is an emerging strategy in cancer theranostics, with nanoparticle-based delivery systems playing a central role in their development. Lipid nanoparticles are used for the delivery of mRNA (e.g., COVID-19 vaccines) or siRNA for gene silencing. They ensure the stability and targeted release of genetic material (Kon et al. 2023). Some nanoparticles may accumulate in organs like the liver, spleen, or kidneys, causing long-term toxicity. Complex designs and multifunctionality require extensive safety and efficacy testing, delaying regulatory approval. Consistent production at clinical-grade quality and scale remains a hurdle.

15.1.7 Antibody-Based Therapies and Conjugates

Antibodies can be engineered to specifically bind to tumor-associated antigens (TAAs) or receptors that are overexpressed on the surface of cancer cells. The specificity of antibodies allows them to selectively target and deliver therapy directly to cancer cells, sparing normal tissues.

Monoclonal Antibodies (mAbs) are designed to target specific tumor antigens and can be used either as standalone therapies or as conjugates with chemotherapy agents or radionuclides. For example, the antibody **trastuzumab (Herceptin)** is used for HER2-positive breast cancer, and it targets the HER2 receptor on the surface of tumor cells.

These antibodies can also be combined with other therapeutic agents as follows:

- **Antibody-Drug Conjugates (ADCs)**: ADCs are a class of biopharmaceutical drugs that combine the specificity of antibodies with the cytotoxicity of chemotherapy agents. In ADCs, the monoclonal antibody is conjugated to a cytotoxic drug, which is delivered specifically to cancer cells when the antibody binds to its target antigen. Once internalized, the cytotoxic drug is released, inducing cell death. The ADC **Kadcyla** (trastuzumab emtansine), which targets HER2-positive breast cancer cells, is an example of this approach (Chau et al. 2019).
- **Theranostic Antibody Conjugates**: By linking diagnostic imaging agents (such as radionuclides or fluorescence markers) to monoclonal antibodies, theranostic antibody conjugates allow for both the visualization of tumors and the targeted delivery of therapy. For example, an antibody can be conjugated to a radiolabeled compound that is used for imaging via PET/CT while also carrying a cytotoxic drug for therapy (Sharma et al. 2022).

15.1.8 Clinical Implementation of Cancer Theranostics

The clinical implementation of theranostics is an evolving process that involves not only the development of new theranostic agents but also the integration of these agents into standard clinical practice. The clinical use of theranostic agents represents a significant shift from traditional, one-size-fits-all approaches to a more personalized, precision-based methodology. The clinical implementation of theranostics requires a well-coordinated multidisciplinary approach, integrating diagnostic and therapeutic elements seamlessly into clinical practice. It involves close collaboration between oncologists, radiologists, nuclear medicine specialists, and pathologists to ensure that diagnostic imaging, molecular profiling, and therapy are appropriately matched.

- **Molecular Imaging**: Theranostic strategies start with the use of advanced diagnostic imaging to determine the molecular and genetic makeup of the tumor. This helps clinicians identify biomarkers specific to cancer cells and understand their heterogeneity, which is essential for selecting the most appropriate therapeutic interventions. As discussed previously, molecular imaging techniques, such as PET, MRI, and SPECT, help identify specific tumor markers. For example, in prostate cancer, **PSMA-PET imaging** enables visualization of prostate-specific membrane antigen (PSMA) on the tumor surface, guiding both diagnosis and therapeutic intervention.

- **Genomic Profiling**: Complementing molecular imaging, genomic profiling allows for the identification of mutations, gene expression changes, and specific tumor markers that may guide the use of targeted therapies. For example, genomic tests for the **HER2** receptor in breast cancer determine whether patients would benefit from **HER2-targeted therapies** such as trastuzumab (Herceptin).
- **Precision Medicine**: Once imaging and molecular data are collected, theranostic agents can be used to target the tumor more precisely. Personalized treatment regimens are then chosen based on the patient's specific tumor biology, tumor location, genetic mutations, and response to previous therapies. Precision medicine approaches, enabled by theranostics, allow oncologists to choose the most appropriate therapies based on an individual's unique molecular profile. For example, patients with **nonsmall cell lung cancer (NSCLC)** harboring mutations in the **EGFR** gene may benefit from targeted therapies like **Erlotinib** or **Gefitinib**, and their treatment can be monitored using imaging techniques such as ^{18}F-FDG PET/CT to assess tumor response.
- **Therapeutic Modality Integration**: In practice, this often means combining molecular imaging with specific treatments such as chemotherapy, radiation, or immunotherapy. For example, patients with breast cancer overexpressing **HER2** can receive **trastuzumab-based therapy**, and the treatment's effectiveness can be monitored using **HER2-targeted imaging**.
- **Tumor Regression Monitoring**: One of the key advantages of theranostics is the ability to monitor treatment responses in real time. Traditional imaging techniques such as CT and MRI often provide delayed insights into treatment efficacy, while molecular imaging provides a much earlier understanding of how the tumor is responding. PET scans, for example, can provide insight into tumor metabolism and help oncologists track changes in glucose uptake by tumors over time, reflecting the effect of therapy on tumor viability. ^{18}F-FDG PET/CT is frequently used to monitor treatment response, especially in **lymphoma** and **colorectal cancer**, where high metabolic activity correlates with aggressive tumor growth.
- **Adaptive Treatment Strategy**: Real-time imaging also allows for more adaptive treatment strategies. If a tumor does not respond to the selected therapy, the clinician can quickly switch to an alternative treatment plan based on the new imaging findings. In neuroendocrine tumors, for example, **DOTATATE PET imaging** is used to track the success of **Peptide Receptor Radionuclide Therapy (PRRT)**.

15.1.9 Future Directions and Emerging Trends in Cancer Theranostics

15.1.9.1 Advancements in Molecular Imaging for Theranostics

Hybrid imaging, which combines two or more imaging modalities into one, is an exciting direction for theranostic applications. The combination of imaging modalities, such as PET/CT or PET/MRI, can provide more comprehensive and accurate information about the tumor's metabolism, morphology, and functional characteristics. The development of new molecular imaging probes that can specifically target unique cancer biomarkers is a key area of focus for future theranostic applications. Novel tracers with greater specificity, sensitivity, and versatility could further enhance early detection and personalized treatment planning. Researchers are developing more advanced molecular probes that target specific tumor markers such as EGFR, VEGF, and PD-L1. These tracers will not only improve diagnostic accuracy but will also facilitate the monitoring of therapy targeting these markers. Fluorescent imaging agents are being developed to guide surgeons during tumor resections. These probes bind specifically to tumor cells, making the tumor visible during surgery under a special light source. This technology could lead to more complete removal of tumors, reducing recurrence rates. Photoacoustic imaging, an emerging molecular imaging technique, uses laser-induced ultrasound signals to produce high-resolution images of tissue. This technique combines the spatial resolution of ultrasound with the molecular specificity of optical imaging by using targeted probes that bind specifically to cancer cells or their associated biomarkers. This enables the noninvasive visualization of tumors in real time, with high sensitivity, and can be combined with therapeutic approaches, such as photodynamic therapy (PDT).

15.1.9.2 Newer Nanotechnology and Nanomedicine in Cancer Theranostics

Nanotechnology plays a pivotal role in the future of cancer theranostics by enabling the design of nanoparticles that can simultaneously deliver drugs and serve as diagnostic agents. These nanoparticles can be engineered to target tumors with high specificity, enhancing the delivery of therapeutic agents while minimizing side effects. To ensure that nanoparticles deliver their therapeutic payloads specifically to cancer cells, they can be coated with ligands, antibodies, or peptides that bind to tumor-associated antigens.

Smart Nanoparticles: Nanoparticles can be designed to release their therapeutic payload in response to specific triggers such as changes in pH, temperature, or the presence of enzymes in the tumor microenvironment. For example, pH-sensitive nanoparticles may release chemotherapy agents preferentially in the acidic tumor microenvironment, sparing normal tissues.

Multifunctional Nanoparticles: The development of multifunctional nanoparticles is a major trend in theranostics. These nanoparticles combine diagnostic and therapeutic properties within a single particle, allowing for real-time monitoring of treatment while delivering drugs directly to cancer cells. For example, **magnetic nanoparticles** may be used for both MRI imaging and drug delivery, while **gold nanoparticles** could serve as both CT contrast agents and carriers for chemotherapy drugs. Additionally, a single nanoparticle may carry both a chemotherapy drug and a gene therapy agent, allowing for synergistic treatment.

Real-Time Imaging with Nanoparticles: Nanoparticle-based theranostic agents, which can target tumor cells and provide both diagnostic imaging and therapeutic treatment, offer real-time monitoring capabilities as shown in Fig. 15.1. These nanoparticles can be designed to fluoresce or emit signals during imaging, allowing clinicians to track tumor growth or shrinkage over time and adjust treatment regimens accordingly.

The future of cancer theranostics will involve more precise and controlled drug delivery systems. Nano- and microcarriers such as liposomes, dendrimers, and polymeric nanoparticles are being developed to target tumor cells more efficiently while reducing off-target effects.

15.1.9.3 Genetic Approaches in Theranostics

The development of gene editing technologies, such as CRISPR-Cas9, is revolutionizing the future of cancer theranostics. Gene editing has the potential to directly alter the genetic makeup of cancer cells, rendering them more sensitive to treatment or correcting underlying genetic mutations. For example, cancer cells could be engineered to express certain receptors that allow for targeted therapy with theranostic agents. For example, **CRISPR-Cas9** could be used to knock out specific oncogenes in tumors, rendering them more sensitive to therapies such as targeted radiation or immunotherapy. Gene editing may also enable the development of personalized cancer vaccines, which train the immune system to recognize and attack specific cancer cells. These vaccines can be designed based on the unique genetic mutations found in an individual's tumor, ensuring a tailored therapeutic response, to stimulate the patient's immune system to recognize and destroy tumor cells expressing specific mutations or antigens. Molecular imaging could help track the progress of the vaccine and evaluate the immune response.

- **Neoantigen-Based Vaccines**: One of the most exciting developments in cancer immunotherapy is the creation of **neoantigen-based vaccines**, which are personalized to the unique mutations present in an individual's cancer. These vaccines can stimulate the patient's immune system to recognize and destroy cancer cells expressing the specific mutated proteins.
- **Theranostic Vaccines**: The combination of personalized cancer vaccines with imaging agents can lead to theranostic vaccines that both stimulate an immune response and enable real-time monitoring of the immune system's activity. Such vaccines could improve patient outcomes by not only treating the cancer but also providing clinicians with valuable data on immune response and tumor dynamics.

15.1.9.4 Immunotherapy and Theranostics

Immunotherapy has shown tremendous promise in treating certain types of cancer by harnessing the body's immune system to attack tumor cells. The future of cancer theranostics will likely involve the integration of immunotherapy with molecular imaging to monitor immune response and optimize treatment regimens.

Immunotherapy has revolutionized cancer treatment, but its effects are often difficult to predict and monitor. By combining immunotherapy with theranostic imaging agents, clinicians can assess immune response and predict the efficacy of treatments such as **checkpoint inhibitors** or **CAR-T cell therapy**.

Immuno-PET: Immuno-PET, using radiolabeled antibodies that bind to immune checkpoints like **PD-L1** or **CTLA-4**, is a promising tool for monitoring immunotherapy response. This approach could enable real-time tracking of immune infiltration into tumors, allowing clinicians to adjust therapies based on tumor response.

15.1.9.5 Theranostics in Rare and Difficult-to-Treat Cancers

Theranostic approaches are particularly promising for cancers that are rare or difficult to treat due to limited treatment options and complex biology. Pancreatic cancer, known for its poor prognosis, has benefited from theranostic strategies aimed at improving early diagnosis and delivering targeted therapies. **PSMA-targeted imaging** and therapy, for example, are being explored for its application in pancreatic cancer, allowing for earlier detection and more effective treatment as shown in Fig. 15.2.

15.1.9.6 Artificial Intelligence and Machine Learning in Cancer Theranostics

Artificial Intelligence (AI) and machine learning (ML) algorithms are increasingly becoming integral tools in cancer theranostics, enabling more precise image analysis, enhanced diagnostic capabilities, and optimized therapeutic decision-making. AI and ML can analyze large datasets from various imaging modalities (e.g., PET, MRI, CT) and provide faster, more accurate interpretations of images, leading to improved diagnosis and treatment outcomes. Artificial intelligence (AI) and machine learning (ML) are poised to transform cancer imaging and theranostics. AI algorithms can

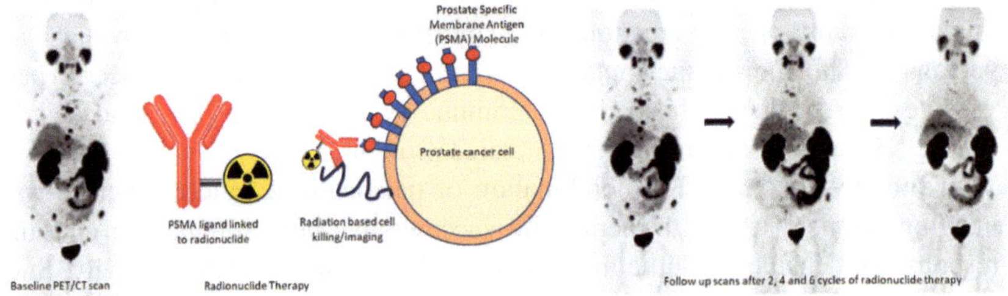

Fig. 15.2 Representation of radiopharmaceutical therapy, using Ga-68 PSMA for imaging and Lu-177 PSMA for therapy. The images on the right represent treatment response after subsequent Lu-177PSMA therapy

analyze vast amounts of imaging data, identify patterns in tumors, and predict treatment outcomes with greater precision than ever before. AI can help automate the process of tumor segmentation, which involves distinguishing tumor tissue from healthy tissue in imaging scans. This is crucial for assessing the size, shape, and location of the tumor before and after treatment. AI-enhanced segmentation improves accuracy, reduces human error, and speeds up the decision-making process in clinical practice. Machine learning models can also predict how tumors will respond to certain therapies, enabling clinicians to adjust treatment plans proactively.

15.1.10 Challenges and Limitations of Cancer Theranostics

Despite its potential, several hurdles need to be overcome for cancer theranostics to reach its full clinical potential. These challenges are multifaceted, ranging from technical issues related to imaging and nanoparticle design to clinical obstacles posed by tumor heterogeneity, resistance mechanisms, and financial constraints. The lack of large-scale validation and regulatory frameworks also limits the wider adoption of this transformative approach in routine clinical oncology.

15.1.10.1 Technical Challenges

- **Integration of Multimodal Imaging Techniques:** Multimodal imaging is a cornerstone of cancer theranostics, combining various imaging techniques to improve both diagnostic accuracy and therapeutic efficacy. For example, integrating positron emission tomography (PET) with computed tomography (CT) or magnetic resonance imaging (MRI) allows for better localization of tumors, functional information, and assessment of therapy response. However, there are significant technical challenges in achieving seamless integration of these imaging modalities:
- **Synchronous Imaging**: Each imaging modality relies on different technologies and principles, which can lead to discrepancies when trying to fuse them together. PET, for example, uses radioactive tracers, while MRI uses magnetic fields, and CT relies on X-rays. The physical and technological differences between these modalities can cause challenges in aligning the images accurately, potentially leading to misinterpretation of tumor location and size. Misalignment during the fusion process may impact diagnosis and treatment planning.
- **Image Fusion and Processing**: Combining images from multiple modalities requires sophisticated software and algorithms. While advances in software have made this fusion more reliable, inconsistencies in resolution, image acquisition times, and spatial orientation can still cause issues. These variations can lead to difficulties in interpreting the data, especially when subtle differences in the tumor microenvironment need to be assessed.
- **Increased Complexity and Cost**: The requirement for multiple imaging systems, combined with the need for specialized expertise to interpret multimodal images, increases the complexity and cost of theranostic applications. These financial and operational barriers can limit the widespread use of multimodal imaging in clinical settings.

- **Limited Sensitivity for Small Tumors**: Current imaging technologies, such as MRI, PET, and CT, are often limited in their ability to detect small tumors, especially those less than a few millimeters in size. This limitation is particularly problematic in the early detection of cancer when the chances for successful treatment are higher. The sensitivity of molecular imaging agents needs to be significantly improved to allow the detection of early-stage tumors and micrometastases.
- **Resolution and Sensitivity**: One of the biggest challenges in imaging is achieving the necessary spatial and temporal resolution to accurately detect small or early-stage tumors. For example, PET imaging provides excellent functional data but has relatively low resolution compared to other modalities like MRI. On the other hand, MRI offers high-resolution anatomical images but may lack the ability to directly visualize metabolic or molecular changes within a tumor.
- **Contrast Agents and Tracers**: Many of the imaging agents used in theranostics, such as radiopharmaceuticals for PET scans, still have limitations in terms of specificity, sensitivity, and safety. Developing new tracers that can specifically target cancer cells with minimal off-target binding is a significant challenge. Additionally, the short half-lives of some radiopharmaceuticals and their potential side effects can limit their practical use in clinical settings.
- **Hybrid Imaging**: Hybrid imaging techniques, such as PET/MRI and SPECT/CT, which combine functional and anatomical imaging, hold great promise but face challenges in terms of integration, complexity, and cost. These technologies require highly specialized equipment and trained personnel, which may not be readily available in all healthcare settings.

15.1.10.2 Challenges in Nanoparticle Design and Functionality

- **Toxicity and Biocompatibility**: One of the main concerns with the use of nanoparticles is their potential toxicity. While nanoparticles have shown promise in preclinical studies, some materials may induce inflammatory responses or cause damage to healthy tissues. Additionally, nanoparticles can accumulate in organs such as the liver and spleen, leading to toxicity if they are not adequately cleared from the body. Ensuring the biocompatibility of nanoparticles while maximizing their therapeutic potential is an ongoing challenge.
- **Optimization of Targeting and Specificity**: For theranostic nanoparticles to be effective, they must selectively target cancer cells without affecting surrounding healthy tissue. Achieving this specificity is difficult because tumors exhibit heterogeneity in terms of receptor expression, genetic mutations, and other factors. Additionally, the tumor microenvironment often has a complex and dynamic nature, making it harder for nanoparticles to deliver drugs or imaging agents directly to tumor cells without being cleared by the immune system or accumulating in nontarget tissues.
- **Controlled Release Mechanisms**: Many nanoparticles are designed to release therapeutic agents at the tumor site, but achieving precise control over the release is a challenge. For instance, nanoparticles can be engineered to respond to specific stimuli such as changes in pH or the presence of certain enzymes, but ensuring

that the release is both timely and localized to the tumor site remains an area of active research. Overcoming these challenges is crucial to achieving optimal therapeutic outcomes.

- **Scalability and Manufacturing**: Producing nanoparticles for clinical use on a large scale is costly and time-consuming. The synthesis of nanoparticles that are uniform in size and shape, and that can be produced at a consistent quality level, is a major obstacle. Furthermore, nanoparticle production must meet stringent regulatory standards to ensure safety and efficacy.
- **Standardization of Imaging Protocols:** For theranostic approaches to be effective across different institutions and clinical settings, there needs to be standardization in imaging protocols. Variability in imaging techniques, patient preparation, and data analysis can lead to inconsistent results and misinterpretations. Standardized guidelines for imaging protocols and the use of theranostic agents are necessary to ensure consistent results across diverse clinical environments.
- **Data Interpretation**: The integration of imaging results into clinical decision-making requires sophisticated data analysis and interpretation. However, the complexity of imaging data and the lack of standardized interpretation tools can pose challenges. For instance, AI-based models for imaging interpretation are still in the early stages and require validation and standardization for widespread use.

15.1.10.3 Challenges in Radiopharmaceutical Availability

- **Limited Isotope Production Facilities and Complex Synthesis**: The production of key isotopes, such as Technetium-99 m (Tc-99 m), relies on a small number of nuclear reactors worldwide. Any disruption in these reactors (e.g., maintenance or shutdowns) leads to global shortages. The establishment and maintenance of radiopharmaceutical production facilities, including cyclotrons and reactors, require significant investment.
- **Short Half-Life**: Many radiopharmaceuticals have short half-lives, necessitating continuous production and rapid distribution. For example, Tc-99 m has a half-life of only six hours.
- **Logistical Challenges:** The need for rapid transportation of radiopharmaceuticals from production facilities to medical centers is hindered by geographical constraints, customs delays, and transportation regulations.
- **Licensing and Compliance:** Handling radioactive materials requires specialized licenses and compliance with radiation safety standards, limiting the number of facilities eligible for production and use.

15.1.10.4 Clinical Challenges

- **Tumor Heterogeneity and Evolution:** Cancer is a highly heterogeneous disease, both at the tumor level and among metastatic sites. Tumor heterogeneity presents one of the greatest clinical challenges in cancer theranostics as shown in Fig. 15.3. Tumors are composed of different subpopulations of cancer cells, each with distinct genetic and phenotypic features. This diversity complicates both diagnosis and treatment, as different regions of the same tumor or metastatic lesions in distant organs may respond differently to therapy.

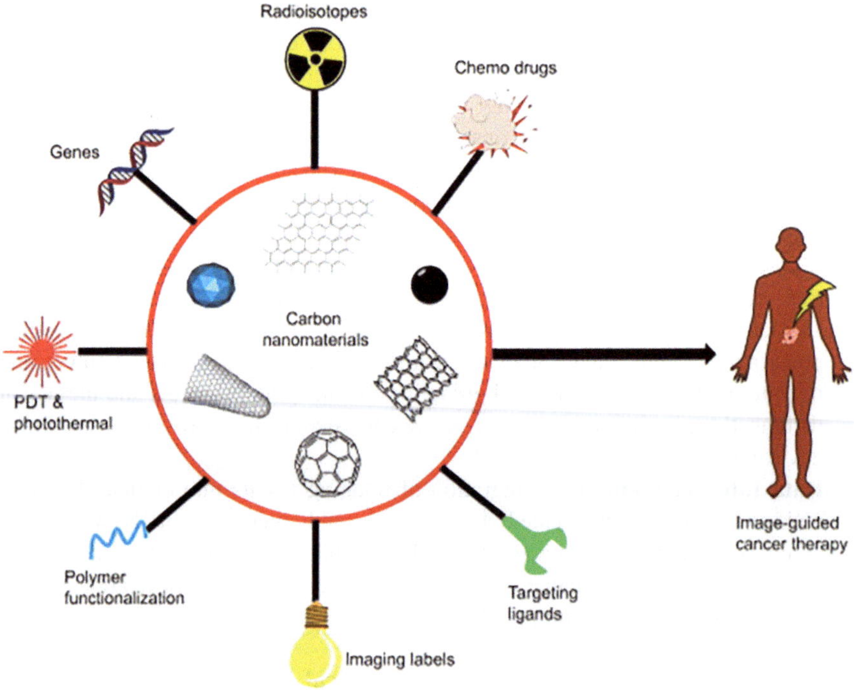

Fig. 15.3 Nanoparticle use in theranostics (Image courtesy: Dutta, S.D., Ganguly, K., Bandi, R., Alle, M. (2021). A New Era of Cancer Treatment: Carbon Nanotubes as Drug Delivery Tools. In: Kim, J.C., Alle, M., Husen, A. (eds) Smart Nanomaterials in Biomedical Applications. Nanotechnology in the Life Sciences)

- **Variability in Biomarker Expression**: One of the primary goals of cancer theranostics is to identify and target specific tumor biomarkers, such as overexpressed receptors or mutated proteins. However, the expression of these biomarkers can vary significantly across different regions of the same tumor or between primary and metastatic tumors. This variability can result in some tumor cells evading detection or treatment due to insufficient expression of the target biomarker.
- **Tumor Evolution and Resistance**: Tumors can evolve over time, developing resistance to targeted therapies. This is particularly problematic when using theranostic approaches, as resistance mechanisms may evolve during treatment, rendering initial therapeutic strategies ineffective. Moreover, some tumors may develop new mutations or alter their metabolic pathways to evade imaging detection, further complicating treatment monitoring.

15.1.10.5 Clinical Validation and Translation to Practice

- **Lack of Large-Scale Clinical Trials**: Despite the numerous theranostic agents in development, there is a lack of large-scale, multicenter clinical trials demonstrating the efficacy and safety of these agents in diverse patient populations. Many trials are small, single-center studies with limited sample sizes, making it difficult to generalize the findings to broader populations.

- **Longer Development Timelines**: The approval process for novel theranostic agents is time-consuming, as it requires demonstrating both diagnostic and therapeutic efficacy. This dual requirement often leads to longer development timelines compared to conventional therapies. Additionally, the lack of standardized guidelines for clinical trials involving theranostics further complicates their validation.
- **Regulatory Approvals and Combination Treatments**: The regulatory approval process for theranostic agents that combine imaging and therapy is complex. In many cases, these agents require approval from multiple regulatory bodies (e.g., FDA for drugs, EMA for imaging agents), which can delay their availability for clinical use.
- **Need for Specialized Expertise:** Successful clinical implementation of theranostics requires specialized training for healthcare professionals. Oncologists, radiologists, nuclear medicine specialists, and radiopharmacists must collaborate closely to ensure the proper use of theranostic agents and accurate interpretation of imaging results.
- **Multidisciplinary Collaboration**: Effective theranostics requires the collaboration of various specialists, including oncologists, radiologists, nuclear medicine physicians, and pharmacists. However, the integration of these specialists into a coordinated care team may not be feasible in all healthcare settings. Ensuring that clinicians from different disciplines work together seamlessly to develop personalized treatment plans is a key barrier to clinical implementation.
- **Training and Expertise**: The complexity of theranostic procedures requires specialized training for healthcare professionals, including the interpretation of molecular imaging results, the selection of appropriate therapies, and the administration of radiopharmaceuticals. In many clinical settings, the necessary expertise may not be readily available, which can hinder the adoption of these technologies.
- **Infrastructure and Resources**: Theranostics often relies on advanced imaging equipment, radiopharmaceutical production facilities, and specialized staff. Many healthcare facilities, particularly in low-resource settings, may lack the infrastructure and resources necessary to implement these technologies on a wide scale.

15.1.10.6 Financial and Economic Challenges

One of the significant barriers to the widespread adoption of cancer theranostics is the high cost associated with the development and implementation of these advanced technologies. Several factors contribute to the high cost:

- **Cost of Nanoparticles and Imaging Agents**: The synthesis and production of nanomaterials and radiopharmaceuticals are complex and expensive. Developing high-quality, biocompatible nanoparticles that can both deliver therapeutic agents and serve as imaging probes requires substantial investment in research, development, and manufacturing.

- **Expensive Imaging Equipment**: Multimodal imaging technologies such as PET/CT, PET/MRI, and hybrid imaging systems are costly to acquire and maintain. These imaging systems require specialized training, skilled operators, and regular maintenance, adding to the overall cost of cancer theranostics.
- **Cost of Personalized Therapies**: Personalized treatments, including those using theranostics, often involve bespoke drug regimens tailored to the genetic and molecular profile of the patient's tumor. These treatments are often more expensive than standard chemotherapy, which may limit their accessibility, particularly in lower-resource settings.

15.1.10.7 Regulatory and Ethical Challenges

- **Dual Regulatory Pathways**: Theranostic agents often need to undergo dual regulatory processes: one for the diagnostic imaging agent and another for the therapeutic component. This dual pathway can result in longer approval times and increased development costs.
- **Need for Standardized Guidelines**: The lack of standardized guidelines for the development and clinical trial design of theranostic agents complicates the regulatory process. The field would benefit from unified standards that streamline the approval process and clarify regulatory expectations.

15.1.10.8 Ethical Considerations

Theranostics raises several ethical and legal concerns, particularly in the context of gene editing, AI, and personalized medicine.

- **Gene Editing**: The use of technologies like **CRISPR-Cas9** for cancer therapy involves altering the genetic makeup of patients or tumors, which raises significant ethical questions. Concerns about the long-term effects of gene editing, unintended genetic consequences, and the potential for off-target mutations must be addressed.
- **AI and Data Privacy**: The use of AI in theranostics necessitates the collection and analysis of large amounts of patient data, raising concerns about data privacy and security. There must be clear legal frameworks in place to protect patient confidentiality while ensuring that AI algorithms are used responsibly and ethically.
- **Informed Consent and Patient Autonomy**: Given the complexity of theranostic procedures, it is crucial that patients fully understand the potential risks and benefits before undergoing treatment. Ethical guidelines need to ensure that patients can make informed decisions without undue pressure or coercion.

15.1.11 Conclusion and Future Perspectives in Cancer Theranostics

Cancer theranostics is a rapidly evolving field that merges diagnostic imaging and therapeutic interventions into a single, integrated approach to cancer management. This innovative concept enables the personalized treatment of cancer, allowing clinicians to precisely target tumors while simultaneously monitoring the effectiveness of treatment in real time. The shift toward personalized medicine will continue to accelerate, with theranostics playing a pivotal role in identifying the best treatments for each patient based on their genetic, molecular, and tumor-specific profiles. The development of personalized cancer vaccines targeting neoantigens (mutated proteins found in tumors) is gaining momentum. These vaccines can be combined with theranostic agents to both stimulate the immune system and monitor immune response in real time. While the future of cancer theranostics is promising, several challenges must be addressed to ensure its widespread adoption and successful clinical implementation. Addressing technical and clinical barriers with improving imaging resolution and sensitivity and handling tumor heterogeneity, combining genomic, proteomic, and imaging data, will be key to understanding the full complexity of tumors and their responses to treatment. The development of cost-effective methods for producing theranostic nanoparticles on a large scale will be crucial for making these therapies widely available. Cancer theranostics represents one of the most exciting frontiers in modern oncology, offering the potential to revolutionize the diagnosis and treatment of cancer. By combining molecular imaging with targeted therapies, theranostics offers personalized, precise, and highly effective cancer care. As advancements in technology, AI, nanomedicine, and personalized medicine continue to evolve, theranostics will play an increasingly vital role in transforming the landscape of cancer treatment. Overcoming technical, clinical, financial, and regulatory barriers will require continued collaboration across disciplines, along with the development of new technologies and strategies to address these issues. Through these efforts, cancer theranostics has the potential to revolutionize cancer care, providing patients with more personalized, effective, and less toxic treatment options.

References

Bray F, Laversanne M, Sung H, Ferlay J, Siegel RL, Soerjomataram I et al (2024) Global cancer statistics 2022: GLOBOCAN estimates of incidence and mortality worldwide for 36 cancers in 185 countries. CA Cancer J Clin 74(3):229–263. Available from: https://pubmed.ncbi.nlm.nih.gov/38572751/

Bybel B, Brunken RC, DiFilippo FP, Neumann DR, Wu G, Cerqueira MD (2008) SPECT/CT imaging: clinical utility of an emerging technology. Radiographics 28(4):1097–1113. Available from: https://pubmed.ncbi.nlm.nih.gov/18635631/

Chau CH, Steeg PS, Figg WD (2019) Antibody-drug conjugates for cancer. Lancet 394(10200): 793–804. Available from: https://pubmed.ncbi.nlm.nih.gov/31478503/

Fonti R, Conson M, Del Vecchio S (2019) PET/CT in radiation oncology. Semin Oncol 46(3): 202–209. Available from: https://pubmed.ncbi.nlm.nih.gov/31378377/

Gharzeddine K, Hatzoglou V, Holodny AI, Young RJ (2019) MR perfusion and MR spectroscopy of brain neoplasms. Radiol Clin North Am 57(6):1177–1188. Available from: https://pubmed.ncbi.nlm.nih.gov/31582043/

Hou YJ, Yang XX, Liu RQ, Zhao D, Guo CX, Zhu AC et al (2020) Pathological mechanism of photodynamic therapy and photothermal therapy based on nanoparticles. Int J Nanomedicine 15:6827–6838. Available from: https://pubmed.ncbi.nlm.nih.gov/32982235/

Kon E, Ad-El N, Hazan-Halevy I, Stotsky-Oterin L, Peer D (2023) Targeting cancer with mRNA-lipid nanoparticles: key considerations and future prospects. Nat Rev Clin Oncol 20(11): 739–754. Available from: https://pubmed.ncbi.nlm.nih.gov/37587254/

Liu JP, Wang TT, Wang DG, Dong AJ, Li YP, Yu HJ (2017) Smart nanoparticles improve therapy for drug-resistant tumors by overcoming pathophysiological barriers. Acta Pharmacol Sin 38(1): 1–8. Available from: https://pubmed.ncbi.nlm.nih.gov/27569390/

Marin JFG, Nunes RF, Coutinho AM, Zaniboni EC, Costa LB, Barbosa FG et al (2020) Theranostics in nuclear medicine: emerging and re-emerging integrated imaging and therapies in the era of precision oncology. Radiographics 40(6):1715–1740. https://doi.org/10.1148/rg.2020200021

Martelli C, Lo Dico A, Diceglie C, Lucignani G, Ottobrini L (2016) Optical imaging probes in oncology. Oncotarget 7(30):48753–48787. Available from: https://pubmed.ncbi.nlm.nih.gov/27145373/

Minhas AS, Oliver R (2022) Magnetic resonance imaging basics. Adv Exp Med Biol 1380:47–82. Available from: https://pubmed.ncbi.nlm.nih.gov/36306094/

Naeem M, Hoque MZ, Ovais M, Basheer C, Ahmad I (2021) Stimulus-responsive smart nanoparticles-based CRISPR-Cas delivery for therapeutic genome editing. Int J Mol Sci 22-(20) Available from: https://pubmed.ncbi.nlm.nih.gov/34681959/

Nelson BJB, Krol V, Bansal A, Andersson JD, Wuest F, Pandey MK (2024) Aspects and prospects of preclinical theranostic radiopharmaceutical development. Theranostics 14(17):6446–6470. Available from: https://pubmed.ncbi.nlm.nih.gov/39479448/

Nikanjam M, Kato S, Kurzrock R (2022) Liquid biopsy: current technology and clinical applications. J Hematol Oncol 15(1) Available from: https://pubmed.ncbi.nlm.nih.gov/36096847/

Seeram E (2015) Computed tomography-E-Book: physical principles. In: Clinical applications, and quality control. Computed tomography physical principles, clinical applications, and quality control physical principles, clinical applications, and quality control, pp 47–94. [cited 2024 Dec 25] Available from: https://www.elsevier.com/books/computed-tomography/seeram/978-0-323-31288-2

Sharma R, Suman SK, Mukherjee A (2022) Antibody-based Radiopharmaceuticals as theranostic agents: an overview. Curr Med Chem 29(38):5979–6005. Available from: https://pubmed.ncbi.nlm.nih.gov/35674298/

Thalluri C, Swain K, Pattnaik S (2023) Rise of gold nanoparticles as carriers of therapeutic agents. Acta Chim Slov 70(4):467–478. Available from: https://pubmed.ncbi.nlm.nih.gov/38124649/

Yousaf T, Dervenoulas G, Politis M (2018) Advances in MRI methodology. Int Rev Neurobiol 141: 31–76. Available from: https://pubmed.ncbi.nlm.nih.gov/30314602/

MIX
Papier aus verantwortungsvollen Quellen
Paper from responsible sources
FSC® C105338

If you have any concerns about our products,
you can contact us on
ProductSafety@springernature.com

In case Publisher is established outside the EU,
the EU authorized representative is:
Springer Nature Customer Service Center GmbH
Europaplatz 3, 69115 Heidelberg, Germany

Printed by Libri Plureos GmbH
in Hamburg, Germany